SCHAUM'S OUTLINE OF

THEORY AND PROBLEMS

OF

REINFORCED CONCRETE DESIGN

•

BY

NOEL J. EVERARD, MSCE, Ph.D.
Professor of Engineering Mechanics and Structures
The University of Texas at Arlington

AND

JOHN L. TANNER III, MSCE
Technical Consultant
Texas Industries, Inc.

•

SCHAUM'S OUTLINE SERIES
McGRAW-HILL BOOK COMPANY
New York, St. Louis, San Francisco, Toronto, Sydney

ISBN 07-019770-9

7 8 9 10 11 12 13 14 15 SH SH 7 5 4

Preface

This book was written to serve as a supplement to current texts in structural design using reinforced concrete and as a handy reference book for professional architects and structural engineers. Moreover, the statements of theory and principle are sufficiently complete that the book could be used as a text by itself.

The contents are divided into chapters covering duly-recognized areas of theory and study. Each chapter begins with clear statements of pertinent definitions, principles and theorems together with illustrative and descriptive material. This is followed by graded sets of solved and supplementary problems. The solved problems illustrate and amplify the theory, present methods of analysis, provide practical examples, and bring into sharp focus those fine points which enable the student to apply the basic principles correctly and confidently. When more than one accepted procedure for handling a problem exists, the authors have in some cases adopted what they feel to be the best; in other cases, alternative procedures are shown; and in a number of situations there is some innovation in treatment. As a result, while this book will not mesh precisely with any one text, the authors feel that it can be a very valuable adjunct to all. Numerous proofs of theorems and derivations of formulas are included among the solved problems.

In Chapter 1 the component materials related to reinforced concrete are discussed, considering those factors which affect its strength and acceptability. Then, in Chapters 2 and 3 a comprehensive review of structural analysis is presented, employing simplified procedures that will be found useful in later chapters. Chapter 4 actually begins the study of reinforced concrete design principles, and Chapters 4 through 12 present the subject matter normally covered in a first course in reinforced concrete design.

Since we are now in a period of transition from the traditional working stress design (WSD) procedure to the ultimate strength design (USD) method, both approaches are given equal attention in this book. Some instructors prefer to cover the USD method first, while others present the WSD method as an introduction to reinforced concrete design. Further, in many countries the WSD method is seldom used. Hence the authors considered it desirable to separate the two procedures wherever feasible. In some cases both methods are discussed simultaneously; for example, in shear and bond stresses the working stress method represents nothing more than the ultimate strength procedure with a selected safety factor applied thereto.

Chapter 4 presents a detailed treatment of the general WSD method including the mechanics, analysis, and proportioning of flexural members, and a discussion of safety provisions. In Chapter 5 the same material is discussed with respect to the USD method.

Shear and diagonal tension are covered in Chapter 6, and bond stresses in Chapter 7. In both chapters the WSD and USD procedures are given in parallel discussions for the reason stated above. Most of the material presented in these chapters is relatively new inasmuch as these facets of reinforced concrete theory have experienced extensive revision due to recent research.

Columns are covered in four separate chapters to provide further flexibility in using the book. Chapter 8 presents the basic ideas and principles affecting columns that are not altered by changing from the WSD method to the USD approach. The theory and problems related to the WSD method are then explained in Chapter 9, while the USD approach is covered in Chapter 10. Hence the study of either method may be omitted without losing continuity.

The study of long columns requires some knowledge of one of the methods of short column design, but is independent of which of the two basic approaches is used, except in isolated cases. Thus long columns are treated separately in Chapter 11. This chapter provides material long known but little used in the past in reinforced concrete design, and also keeps pace with the most recent research and methods.

Chapter 12 treats of reinforced concrete footings, including symmetrically loaded foundations on soil and piles. Eccentrically loaded footings and multiple column footings are also considered.

Two-way slabs are covered in detail in Chapter 13, and flat slabs and flat plates are discussed in Chapter 14. Often this material is not considered in detail in a first course. However, the reader should find little difficulty in mastering the comprehensive treatment presented here.

The study of retaining walls, Chapter 15, requires some knowledge of soil mechanics, in addition to the basic principles discussed in Chapters 3 through 7. Accordingly, a review of factors affecting soil pressure is included in Chapter 15.

Chapter 16 contains new material on torsion of reinforced concrete that has not appeared elsewhere. The method used for torsion design is related to the plastic theory and is consistent with the present state of the art in reinforced concrete design. However, it is recognized that the simplified approach is not necessarily representative of the true complex state of combined stresses that exists in the presence of flexure, shear, thrust and torsion. At the time of writing this text, the true conditions of such combined stresses have not been established by research.

Throughout the text the specifications of the Building Code Requirements for Reinforced Concrete, ACI 318-63 (ACI Code), have been used as a basis of design. Since the ACI Code is used as a standard in the United States and many other countries, the authors considered it advisable to follow that manual in this text.

Many of the charts, tables and other design aids derive from original work of the authors. Some were prepared especially for this text, and others were prepared for ACI Committees and reproduced here. The authors wish to thank the American Concrete Institute and the Portland Cement Association and their members for permission to reproduce copyrighted material, including many charts and tables, and for very helpful suggestions.

The authors are deeply indebted to many associates, especially to Mr. Nabil Hadawi, for invaluable assistance and critical review of parts of the manuscript. To their wives, Courtney Everard and Marge Tanner, the authors wish to express their gratitude for unfailing assistance and understanding in this endeavor.

NOEL J. EVERARD
JOHN L. TANNER

Arlington, Texas
April, 1966

CONTENTS

CONTENTS

Chapter 1

Materials and Components for Reinforced Concrete Construction

DEFINITIONS

Concrete is a non-homogeneous manufactured stone composed of graded, granular inert materials which are held together by the action of cement and water. The inert materials usually consist of gravel or large particles of crushed stone, and sand or pulverized stone. Manufactured lightweight materials are also used. The inert materials are called *aggregates.* The large particles are called *coarse aggregates* and the small particles are called *fine aggregates.*

Concrete behaves very well when subjected to compressive forces, but ruptures suddenly when small tension forces are applied. Therefore in order to utilize this material effectively, steel reinforcement is placed in the areas subjected to tension.

Reinforced concrete is a composite material which utilizes the concrete in resisting *compression forces,* and some other material, usually steel bars or wires, to resist the *tension forces.* Steel is also often used to assist the concrete in resisting compression forces. Concrete is always assumed to be incapable of resisting tension, even though it can resist a small amount of tension.

A number of definitions are presented in the following list taken from the *Building Code Requirements for Reinforced Concrete,* ACI 318-63, of the American Concrete Institute. (This code is usually referred to as the ACI Code.)

Admixture – A material other than portland cement, aggregate, or water added to concrete to modify its properties.

Aggregate – Inert material which is mixed with portland cement and water to produce concrete.

Aggregate, lightweight – Aggregate having a dry, loose weight of 70 lb/ft^3 or less.

Building official – City Engineer, Plan Examiner, etc.

Column – An upright compression member the length of which exceeds three times its least lateral dimension.

Combination column – A column in which a structural steel member, designed to carry the principal part of the load, is encased in concrete of such quality and in such a manner that an additional load may be placed thereon.

Composite column – A column in which a steel or cast-iron structural member is completely encased in concrete containing spiral and longitudinal reinforcement.

Composite concrete flexural construction – A precast concrete member and cast-in-place reinforced concrete so interconnected that the component elements act together as a flexural unit.

Compressive strength of concrete (f'_c) – Specified compressive strength of concrete in pounds per square inch (psi). Compressive strength is determined by tests of standard (6″ × 12″) cylinders made and tested in accordance with ASTM (American Society for Testing and Materials) specifications at 28 days or such earlier age as the concrete is to receive its full service load or maximum stress.

Concrete – A mixture of portland cement, fine aggregate, coarse aggregate and water.

Concrete, structural lightweight – A concrete containing lightweight aggregate (weighing 90 to 115 pcf when hardened).

Deformed bar – A reinforcing bar conforming to "Specifications for Minimum Requirements for the Deformations of Deformed Steel Bars for Concrete Reinforcement" (ASTM A-305) or "Specifications for Special Large Size Deformed Billet-Steel Bars for Concrete Reinforcement" (ASTM A-408). Welded wire fabric with welded intersections not farther apart than 12″ in the direction of the principal reinforcement and with cross wires not more than six gage numbers smaller in size than the principal reinforcement may be considered equivalent to a deformed bar when used in slabs.

Effective area of concrete – The area of a section which lies between the centroid of the tension reinforcement and the compression face of the flexural member.

Effective area of reinforcement – The area obtained by multiplying the right cross-sectional area of the reinforcement by the cosine of the angle between the axis of the reinforcement and the direction for which the effectiveness is to be determined.

Pedestal – An upright compression member whose height does not exceed three times its average least lateral dimension.

Plain bar – Reinforcement that does not conform to the definition for a deformed bar.

Plain concrete – Concrete that does not conform to the definition for reinforced concrete.

Precast concrete – A plain or reinforced concrete element cast in other than its final position in the structure.

Prestressed concrete – Reinforced concrete in which there have been introduced internal compressive stresses of such magnitude and distribution that the tension stresses resulting from service loads are counteracted to a desired degree.

Reinforced concrete – Concrete containing reinforcement, designed on the assumption that the two materials act together in resisting forces.

Reinforcement – Steel bars used in concrete to resist tension forces.

Service dead load – The calculated dead weight supported by a member.

Service live load – The live load (specified by the general building code) for which a member must be designed.

Splitting tensile strength – The results of splitting tests of cylinders.

Stress – Force per unit area.

Surface water – Water carried by an aggregate, excluding that water held by absorption within the aggregate particles themselves.

Yield strength or yield point (f_y) – Specified minimum yield strength or yield point of reinforcement in pounds per square inch. Yield strength or yield point shall be determined in tension according to applicable ASTM specifications.

MATERIALS FOR CONCRETE

Cement as used in plain or reinforced concrete has the ability to form a paste when mixed with water. The paste hardens with passage of time, holding all of the larger inert particles together in a common bond. Cement may be obtained from nature (natural cement) or it may be manufactured. When manufactured, cement usually conforms to certain specifications of the ASTM. When the material is so manufactured, it is classified as *Portland Cement,* and concrete made using this material is called *Portland Cement Concrete,* or simply, concrete.

The particular ASTM specifications which apply to cement are C-150, C-175, C-205 and C-340.

In general, Portland Cement is manufactured using definite proportions of various calcareous materials, which are burned to form clinkers. The clinkers are pulverized to a powder-like form, which then becomes cement.

Portland Cement is generally available in a number of different types:

(1) *Normal Portland Cement* – used for general purposes when specific properties are not required.

(2) *Modified Portland Cement* – for use when low heat of hydration is desired, such as in mass concrete, huge piers, heavy abutments and heavy retaining walls, particularly when the weather is hot. (Type 1 may be more desirable in cold weather.)

(3) *High early strength Portland Cement* – for use when very high strength is desired at an early age.

(4) *Low heat of hydration Portland Cement* – for use in large masses such as dams. Low heat of hydration is desirable to reduce cracking and shrinkage.

(5) *Sulfate resistant Portland Cement* – for use when the structure will be exposed to soil or water having a high alkali content.

(6) *Air-entrained Portland Cement* – for use when severe frost action is present, or when salt application is used to remove snow or ice from the structure.

Table 1.1 shows the variations in strength of identical mixtures using different types of cement, with the exception of air-entrained cement.

TABLE 1.1

APPROXIMATE RELATIVE STRENGTHS OF CONCRETE AS AFFECTED BY TYPE OF CEMENT

Type of portland cement	Compressive strength – percent of strength of normal portland cement concrete		
	3 days	28 days	3 months
1 – Normal	100	100	100
2 – Modified	80	85	100
3 – High-early-strength	190	130	115
4 – Low-Heat	50	65	90
5 – Sulfate-resistant	65	65	85

It should be noted here that other types of cement are also available. Some cements are made using blast furnace slag, while others consist of a mixture known as *Portland-Pozzolan Cement*. Pozzolan cement is generally a natural cement. These types of cement have rather special areas of application which will not be discussed in detail in this text.

AGGREGATES

Aggregates form the bulk of the concrete components. Fine aggregates consist of sand or other fine grained inert material usually less than $\frac{1}{4}''$ maximum size. Coarse aggregates consist of gravel or crushed rock usually larger than $\frac{1}{4}''$ size and usually less than 3″ size.

WATER

Water is an important ingredient in the concrete mixture. The water must be clean and free from salts, alkalis or other minerals which react in an undesirable manner with the cement. Thus sea water is not recommended for use in mixing concrete.

WATER-CEMENT RATIO

The *water-cement ratio* is the most important single factor involved in mixing concrete. Using average materials, the strength of the concrete and all of the other desirable properties of concrete are directly related to the water-cement ratio.

ULTIMATE STRENGTH AND DESIGN OF CONCRETE MIXES

The *ultimate strength* of concrete is generally defined as the compressive strength of a molded concrete cylinder or prism after proper curing for 28 days and is designated as f'_c, measured in psi. Using average materials, the ultimate strength in compression for various water-cement ratios can be predicted with reasonable accuracy using Table 1.2.

TABLE 1.2

COMPRESSIVE STRENGTH OF CONCRETE FOR VARIOUS WATER-CEMENT RATIOS

Water-Cement Ratio		Probable Compressive Strength at 28 days, psi	
By Wt.	Gal/sack	Non-air-entrained	Air-entrained
0.35	4.0	6000	4800
0.44	5.0	5000	4000
0.53	6.0	4000	3200
0.62	7.0	3200	2600
0.71	8.0	2500	2000
0.80	9.0	2000	1800

Strengths are based on 6″ x 12″ cylinders moist cured under standard conditions for 28 days.

Concrete mixtures can be designed to provide a given strength and other desirable properties by properly proportioning all of the materials. Two general methods are currently in use: (1) The arbitrary proportions method, and (2) The trial batch method.

The *arbitrary proportions method* is based on the assumption that average materials will be used and that ordinary weight aggregates (i.e. sand and gravel or crushed stone) will be used. When lightweight structural aggregates are used, or if the design strength exceeds 4000 psi, this method may not be used. If the aggregates are properly graded and proportioned the 28 day compressive strength of concrete may be based on the water-cement ratios shown in Table 1.3.

TABLE 1.3

ARBITRARY PROPORTIONS METHOD FOR CONCRETE STRENGTH

Maximum permissible water-cement ratios

Specified compressive strength at 28 days, psi f'_c	Maximum permissible water-cement ratio*			
	Non-air-entrained concrete		Air-entrained concrete	
	U.S. gal. per 94-lb bag of cement	Absolute ratio by weight	U.S. gal. per 94-lb bag of cement	Absolute ratio by weight
2500	$7\frac{1}{4}$	0.642	$6\frac{1}{4}$	0.554
3000	$6\frac{1}{2}$	0.576	$5\frac{1}{4}$	0.465
3500	$5\frac{3}{4}$	0.510	$4\frac{1}{2}$	0.399
4000	5	0.443	4	0.354

*Including free surface moisture on aggregates.

The *trial batch method* is a scientific method of proportioning concrete. Various proportions of fine and coarse aggregate are used with different predetermined water-cement ratios in order to provide a mixture having the desired *consistency* and a probable strength about 15 percent higher than the proposed design strength. Test specimens are molded for each mix and are tested at 28 days to establish the value of f'_c for each. The most desirable mix is adopted for use. A number of factors which enter into the trial batch method are discussed in the following paragraphs.

FINENESS MODULUS

The *fineness modulus* indicates the relative fineness of the aggregates. The aggregates are sieved using standard screens and the weight of all of the particles larger than a given size are tabulated. The percent retained on sieves No. 4, 8, 16, 30, 50 and 100 are tabulated, as well as that passing the No. 100 sieve. The sum of the weights retained on all of the sieves larger than a given sieve are accumulated and then added together and divided by 100. The result, always larger than 1.0, is the fineness modulus.

Specifications for sieve analysis are usually provided as part of a general specification for concrete on major projects. One example of the sieve analysis specifications is that of the U. S. Dept. of Interior, which is shown in Table 1.4.

TABLE 1.4

RANGES OF PERCENTAGES OF AGGREGATE SIZES FOR CONCRETE CONSTRUCTION

Sieve Size	Percent Retained (Cumulative)
No. 4	0 to 5
No. 8	10 to 20
No. 16	20 to 40
No. 30	40 to 70
No. 50	70 to 88
No. 100	92 to 98

AGGREGATE SIZE AND EFFECTS ON CONCRETE

Aggregate gradation enters into the consideration of the strength and workability of concrete. Table 1.5 illustrates the effects of aggregate gradation on the cement requirement for quality concrete.

TABLE 1.5

EFFECTS OF AGGREGATE GRADATION ON CEMENT REQUIREMENT

Grading of coarse aggregate (percent by weight)			Optimum* amount of sand	Cement required at percent of sand indicated—sacks per cu. yd.	
No. 4–$\frac{3}{8}$ in.	$\frac{3}{8}$–$\frac{3}{4}$ in.	$\frac{3}{4}$–$1\frac{1}{2}$ in.	Percent	Optimum	35 percent
35.0	00.0	65.0	40	5.4	5.7
30.0	17.5	52.5	41	5.4	5.8
25.0	30.0	45.0	41	5.4	6.2
20.0	48.0	32.0	41	5.4	6.0
00.0	40.0	60.0	46	5.4	7.0

*Amount giving best workability with aggregates used. Water content 6.3 gal. per sack of cement.

The average quantities of fine and coarse aggregate required for a unit volume of concrete are functions of the fineness modulus and the maximum size of the coarse aggregate. Values obtained experimentally are shown in Tables 1.6(a) and 1.6(b) for the coarse and fine aggregates respectively.

The type of construction often dictates the desirable maximum size of aggregate. Recommended values are listed in Table 1.7.

TABLE 1.6(a)
VOLUME OF COARSE AGGREGATE PER UNIT VOLUME OF CONCRETE*

(From report of A.C.I. Committee 613: Recommended Practice for Selecting Proportions for Concrete.)

Maximum Size of Aggregate, in.	Volume of dry-rodded coarse aggregate per unit volume of concrete for different fineness modulus of sand					
	2.40	2.60	2.80	3.00	3.20	3.40
$\frac{3}{8}$	0.46	0.44	0.42	0.40	0.38	0.36
$\frac{3}{4}$	0.65	0.63	0.61	0.59	0.57	0.55
1	0.70	0.68	0.66	0.64	0.62	0.60
$1\frac{1}{2}$	0.76	0.74	0.72	0.70	0.68	0.66
2	0.79	0.77	0.75	0.73	0.71	0.69
3	0.84	0.82	0.80	0.78	0.76	0.74

*These volumes are selected from empirical relationships to produce concrete with a degree of workability suitable for usual reinforced construction. For less workable concrete such as required for concrete pavement construction, they may be increased approximately 10%.

TABLE 1.6(b)
APPROXIMATE PERCENTAGES OF SAND FOR DIFFERENT GRADINGS AND MAXIMUM SIZES OF COARSE AGGREGATE

Maximum Size of Coarse Aggregate	Cement Factor, sacks per cu. yd. for Rounded Aggregate				Cement Factor, sacks per cu. yd. for Angular Aggregate			
	4	5	6	7	4	5	6	7
	Fine Sand – F. M. 2.3 to 2.4							
$\frac{3}{4}$	39	37	34	32	46	44	42	39
1	38	36	33	31	45	43	41	38
$1\frac{1}{2}$	36	34	31	29	42	40	38	36
2	35	33	30	28	41	39	37	35
3	33	31	29	27	39	37	35	34
	Medium Sand – F. M. 2.6 to 2.7							
$\frac{3}{4}$	42	40	37	34	49	47	45	42
1	40	38	36	33	47	45	43	41
$1\frac{1}{2}$	38	36	34	31	45	43	41	39
2	36	34	33	30	43	41	39	37
3	34	32	31	29	41	39	37	35
	Coarse Sand – F. M. 3.0 to 3.1							
$\frac{3}{4}$	47	44	41	38	55	52	49	47
1	45	42	39	37	53	50	47	45
$1\frac{1}{2}$	42	40	37	35	50	47	45	43
2	40	38	35	33	47	45	43	41
3	37	35	33	31	44	42	40	38

(Table from N.R.M.C.A. Publication No. 52, Calculating the Proportions for Concrete.)

TABLE 1.7
MAXIMUM SIZES OF AGGREGATES RECOMMENDED FOR VARIOUS TYPES OF CONSTRUCTION

	Maximum Size Aggregates, in.			
Minimum Dimension of Section, in.	Reinforced Walls, Beams and Columns	Unreinforced Walls	Heavily Reinforced Slabs	Lightly Reinforced or Unreinforced Slab
$2\frac{1}{2}$–5	$\frac{1}{2}$–$\frac{3}{4}$	$\frac{3}{4}$	$\frac{3}{4}$–1	$\frac{3}{4}$–$1\frac{1}{2}$
6–11	$\frac{3}{4}$–$1\frac{1}{2}$	$1\frac{1}{2}$	$1\frac{1}{2}$	$1\frac{1}{2}$–3
12–29	$1\frac{1}{2}$–3	3	$1\frac{1}{2}$–3	3

CONSISTENCY

The consistency of concrete is important since it is necessary to have the concrete flow freely around corners and between the reinforcing bars. The standard method of determining the relative consistency of concrete is the *slump test.* In this test, a standard slump cone is filled in three layers, rodding each layer 25 times. The concrete is smoothed off at the top of the cone. The cone is then lifted vertically, permitting the concrete to slump downward. The distance between the original and final surfaces of the concrete is called the slump and is measured in inches. Recommended slumps for various types of construction are listed in Table 1.8.

TABLE 1.8
RECOMMENDED SLUMPS FOR VARIOUS TYPES OF CONSTRUCTION*

Types of Construction	Slump, in.**	
	Max.	Min.
Reinforced Foundation Walls and Footings	5	2
Plain Footings and Caissons	4	1
Slabs, Beams, and Reinforced Walls	6	3
Building Columns	6	3
Pavements	3	2
Heavy Mass Construction	3	1

*Adapted from Table 4 of the 1940 Joint Committee Report on Recommended Practice and Standard Specifications for Concrete and Reinforced Concrete.

**When high frequency vibration is used, the values given should be reduced by about one-third.

Experience has shown that the slump of concrete is directly related to the water-cement ratio and the sizes, gradation and quantities of the aggregates. Table 1.9 indicates the approximate mixing water requirements for the variables which affect the slump of concrete.

TABLE 1.9
APPROXIMATE MIXING WATER REQUIREMENTS FOR DIFFERENT SLUMPS AND MAXIMUM SIZES OF AGGREGATES

Slump, in.	Water, gals., per cu. yd. of concrete						
	$\frac{3}{8}''$	$\frac{3}{4}''$	$1''$	$1\frac{1}{2}''$	$2''$	$3''$	$6''$
	Unit water requirement without admixture						
1–2	41	37	35	33	31	29	27
3–4	45	40	38	36	34	32	30
5–6	48	42	40	38	36	34	32
Approximate amt. of entrapped air, %	3	2	1.5	1	0.5	0.3	0.2
	Unit water requirements for air-entrained concrete						
1–2	39	35	33	31	29	27	25
3–4	43	38	36	34	32	30	28
5–6	45	40	38	36	34	32	30
Recommended avg. total air, %	8	6	5	4.5	4	3.5	3

ENTRAINED AIR AND DISPERSING AGENTS

Air is entrapped naturally in concrete during the mixing process. Experience has shown that if air is entrained artificially using *air-entraining agents* the quantity of mixing water required to produce a given consistency will be reduced. The entrained air reduces friction between the particles and lessens the need for water as a lubricant. *Dispersing agents* break the surface tension in water bubbles and accomplish some of the same effects as air-entraining agents.

Among the many products commercially available are PDA-Protex dispersing agent and Protex air-entraining agent. The Protex Company has provided Tables 1.10(a), (b), (c), (d) in order to show the results of experiments with air-entrained concrete and non-air-entrained concrete. The tables provide data pertaining to the water-cement ratio (W/C), gallons of water per sack of cement (G/S), gallons per cubic yard (G/Y) and the cement factor (CF) in sacks of cement per cubic yard of concrete.

The tables provide the probable *over-design* strength required to insure against *under-design.* The indicated water-cement ratios and compressive strengths are from "ACI Recommended Practice for Selecting Proportions for Concrete" (ACI 613-54). The unit water contents are derived from local job experience.

TABLE 1.10(a)

Required compressive strength—2000 psi
Coefficient of variation – 15%
Maximum size aggregate – $1\frac{1}{2}''$

Percent of Tests Above Req. Str.	Required Design Strength	W/C G/S	3″–4″ Slump				5″–6″ Slump			
			Plain		PDA		Plain		PDA	
			Water G/Y	CF S/Y	Water G/Y	CF S/Y	Water G/Y	CF S/Y	Water G/Y	CF S/Y
70	2180	8.75	36	4.1	32	3.7	38	4.35	34	3.9
80	2300	8.5	36	4.25	32	3.8	38	4.5	34	4.0
90	2480	8.0	36	4.5	32	4.0	38	4.75	34	4.25
99	3100	7.25	36	5.0	32	4.4	38	5.25	34	4.7
Maximum size aggregate – $\frac{3}{4}''$										
70	2180	8.75	40	4.6	36	4.1	42	4.8	38	4.35
80	2300	8.5	40	4.75	36	4.25	42	4.9	38	4.5
90	2480	8.0	40	5.0	36	4.5	42	5.25	38	4.75
99	3100	7.25	40	5.5	36	5.0	42	5.8	38	5.25

TABLE 1.10(b)

Required compressive strength—2500 psi
Coefficient of variation – 15%
Maximum size aggregate – $1\frac{1}{2}''$

Percent of Tests Above Req. Str.	Required Design Strength	W/C G/S	3″–4″ Slump				5″–6″ Slump			
			Plain		PDA		Plain		PDA	
			Water G/Y	CF S/Y	Water G/Y	CF S/Y	Water G/Y	CF S/Y	Water G/Y	CF S/Y
70	2700	7.75	36	4.6	32	4.1	38	4.9	34	4.4
80	2850	7.5	36	4.8	32	4.25	38	5.1	34	4.5
90	3100	7.25	36	5.0	32	4.4	38	5.3	34	4.7
99	3850	6.25	36	5.8	32	5.1	38	6.1	34	5.45
Maximum size aggregate – $\frac{3}{4}''$										
70	2700	7.75	40	5.15	36	4.6	42	5.4	38	4.9
80	2850	7.5	40	5.35	36	4.8	42	5.6	38	5.1
90	3100	7.25	40	5.5	36	5.0	42	5.8	38	5.3
99	3850	6.25	40	6.4	36	5.8	42	6.7	38	6.1

TABLE 1.10(c)

Required compressive strength—3000 psi
Coefficient of variation—15%
Maximum size aggregate—$1\frac{1}{2}''$

Percent of Tests Above Req. Str.	Required Design Strength	W/C G/S	3″–4″ Slump				5″–6″ Slump			
			Plain		PDA		Plain		PDA	
			Water G/Y	CF S/Y	Water G/Y	CF S/Y	Water G/Y	CF S/Y	Water G/Y	CF S/Y
70	3250	7.0	36	5.1	32	4.5	38	5.4	34	4.8
80	3450	6.75	36	5.3	32	4.7	38	5.6	34	5.0
90	3700	6.25	36	5.8	32	5.1	38	6.1	34	5.4
99	4600	5.5	36	6.5	32	5.8	38	6.9	34	6.2
Maximum size aggregate—$\frac{3}{4}''$										
70	3250	7.0	40	5.7	36	5.1	42	6.0	38	5.4
80	3450	6.75	40	5.9	36	5.3	42	6.25	38	5.6
90	3700	6.25	40	6.4	36	5.8	42	6.7	38	6.1
99	4600	5.5	40	7.3	36	6.5	42	7.6	38	6.9

TABLE 1.10(d)

Required compressive strength—3500 psi
Coefficient of variation—15%
Maximum size aggregate—$1\frac{1}{2}''$

Percent of Tests Above Req. Str.	Required Design Strength	W/C G/S	3″–4″ Slump				5″–6″ Slump			
			Plain		PDA		Plain		PDA	
			Water G/Y	CF S/Y	Water G/Y	CF S/Y	Water G/Y	CF S/Y	Water G/Y	CF S/Y
70	3800	6.25	36	5.75	32	5.1	38	6.1	34	5.45
80	4000	6.0	36	6.0	32	5.35	38	6.35	34	5.65
90	4300	5.75	36	6.25	32	5.6	38	6.6	34	5.9
99	5400	4.75	36	7.6	32	6.75	38	8.0	34	7.0
Maximum size aggregate—$\frac{3}{4}''$										
70	3800	6.25	40	6.4	36	5.75	42	6.7	38	6.1
80	4000	6.0	40	6.7	36	6.0	42	7.0	38	6.35
90	4300	5.75	40	7.0	36	6.25	42	7.3	38	6.6
99	5400	4.75	40	8.4	36	7.6	42	8.85	38	7.9

APPROXIMATE METHODS FOR MIX DESIGN FOR CONCRETE

The scientific methods presented for design of concrete using the trial batch method cannot always be followed because of the lack of trained personnel and equipment. When such is the case, a conservative method is available. The proportions shown in Table 1.11, although not precise, will prove to be satisfactory for small work and for work at locations where the more scientific process cannot be used.

Fig. 1-1 shows the usual range in proportions of materials used in concrete. These ranges can be used as a guide for concrete mix design.

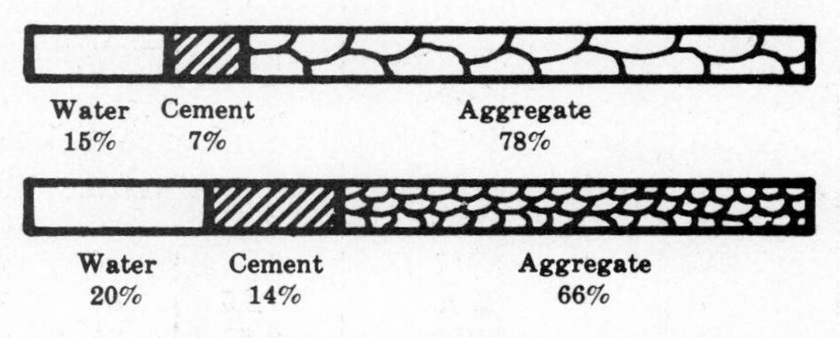

Range in proportions of materials usually used in concrete. Upper bar represents lean mix of stiff consistency with large aggregate. Lower bar represents rich mix of wet consistency with small aggregate.

Fig. 1-1

TABLE 1.11

CONCRETE MIXES FOR SMALL JOBS

Maximum size of aggregate, in.	Mix designation	Approximate bags of cement per cu yd of concrete	Aggregate, lb per 1-bag batch			
			Sand		Gravel or crushed stone	Iron blast furnace slag
			Air-entrained concrete	Concrete without air		
$\frac{1}{2}$	A	7.0	235	245	170	145
	B	6.9	225	235	190	165
	C	6.8	225	235	205	180
$\frac{3}{4}$	A	6.6	225	235	225	195
	B	6.4	225	235	245	215
	C	6.3	215	225	265	235
1	A	6.4	225	235	245	210
	B	6.2	215	225	275	240
	C	6.1	205	215	290	255
$1\frac{1}{2}$	A	6.0	225	235	290	245
	B	5.8	215	225	320	275
	C	5.7	205	215	345	300
2	A	5.7	225	235	330	270
	B	5.6	215	225	360	300
	C	5.4	205	215	380	320

Mix B is the starting point. If the mix appears to be *under-sanded,* change to mix A. If the mix is *over-sanded,* change to mix C. In all cases the mixes refer to dry materials and there must be adjustments for water in the aggregates.

The water-cement ratio may be obtained from Table 1.3 for a given strength of concrete.

WATER IN THE AGGREGATES

The water-cement ratio must be carefully preserved if the results are to be reproducible from one batch to another. For this reason, *free water* in the aggregates must be considered as part of the mixing water. Since the free water varies from time to time in a given stock pile of aggregates, it is necessary to determine the free water content several times each day and the added water adjusted accordingly. A definite procedure is available for determining the free water in the aggregates.

A representative sample of aggregates is weighed. The surfaces of the particles are then dried to a *saturated-surface dry state* in an oven or pan, or by pouring alcohol on the aggregates and setting afire. The dried aggregates are then weighed. The percentage moisture (by weight of the aggregate) is obtained using the equation

$$p = 100(W_w - W_D)/W_D \tag{1.1}$$

in which p = percentage of moisture by dry weight, W_w = wet weight of the material, and W_D = dry weight of the material. The percentage of *surface water* is deducted from the total water required in order to obtain the desired water-cement ratio.

QUANTITY OF CONCRETE OBTAINED

The quantity of concrete or *yield* of a trial batch is used to predict the quantity of concrete to be obtained from the job mix. The yield is predicted using the *absolute volume method* to obtain the absolute volume of each of the component materials (i.e. gravel, sand, cement and water) using the equation

$$V_a = \text{(weight of loose material)}/(\text{SG})(W_u) \qquad (1.2)$$

in which V_a = absolute volume of material (ft^3), SG = specific gravity of the material, and W_u = unit weight of water (62.4 lb/ft^3).

MAKING, CURING AND TESTING SPECIMENS

Making Cylinders

Test cylinders are made using the trial batch method in order to obtain the true strength of the manufactured material. The cylinders are also used during construction to insure that the strength of the concrete is maintained at the desired level.

The fresh concrete is placed in a mold 6″ in diameter and 12″ high in 3 layers. Each layer is rodded 25 times. The concrete is trowled smooth at the top surface and the cylinder allowed to air cure for 24 hours, after which it is placed in a *damp room* under controlled humidity and temperature to age for 28 days. After the curing period has elapsed, the cylinders are tested in a testing machine to determine the compressive strength f'_c, the modulus of elasticity E_c, and often the complete stress-strain diagram.

Curing Cylinders

Proper curing of specimens is vitally important to the strength of the concrete. *Moist curing* is the most desirable method, as shown in Figure 1-2.

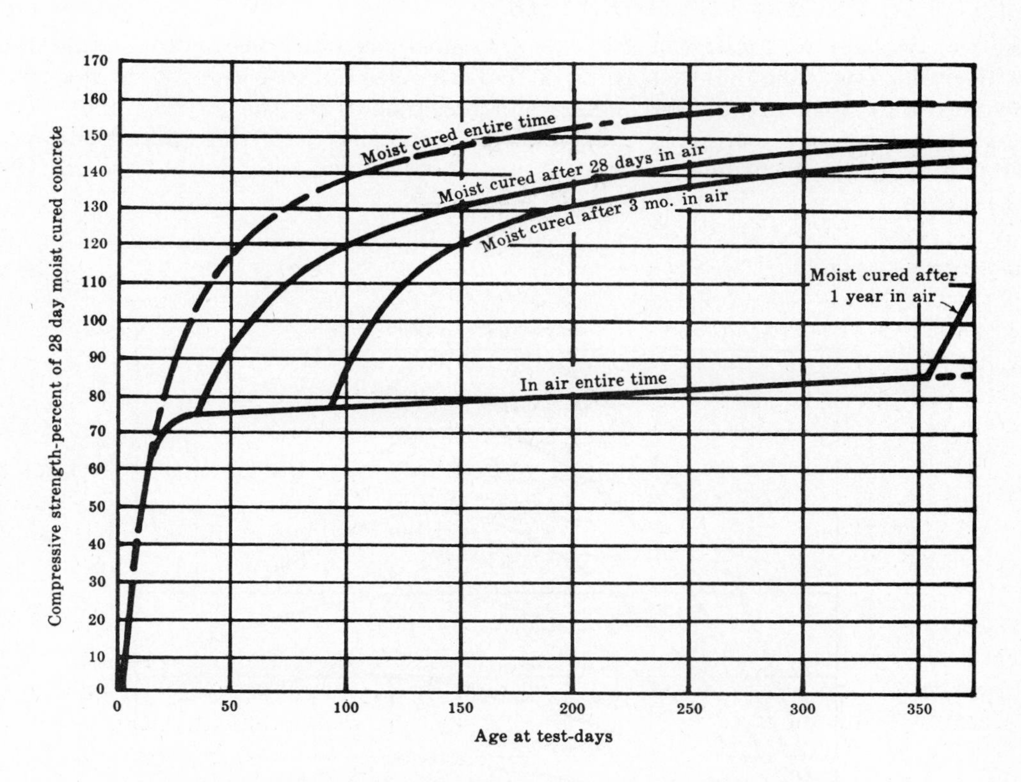

Fig. 1-2. Effects of Moist Curing on the Strength of Concrete

Temperature of curing also affects concrete strength, as seen from Fig. 1-3 below, since strength and other desirable properties of concrete improve more rapidly at normal temperatures than at low temperatures.

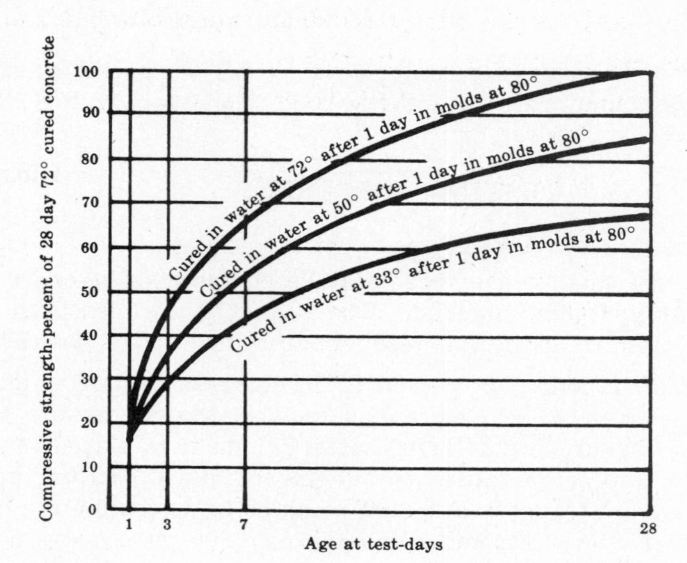

Fig. 1-3. Effects of Temperature on the Strength of Concrete

STRUCTURAL DESIGN CONSIDERATIONS

The *compression stress-strain diagram* provides the most important single factor for use in deriving equations for designing structural elements of reinforced concrete. The stress-strain diagrams are plotted using data obtained from the 28 day tests of concrete cylinders. During the loading process, loads in pounds and the corresponding strains (inches per inch) are recorded. The loads are transformed into direct stresses (P/A) and the stress-strain diagram is then plotted.

Fig. 1-4 shows a series of typical stress-strain diagrams obtained using different strengths of concrete.

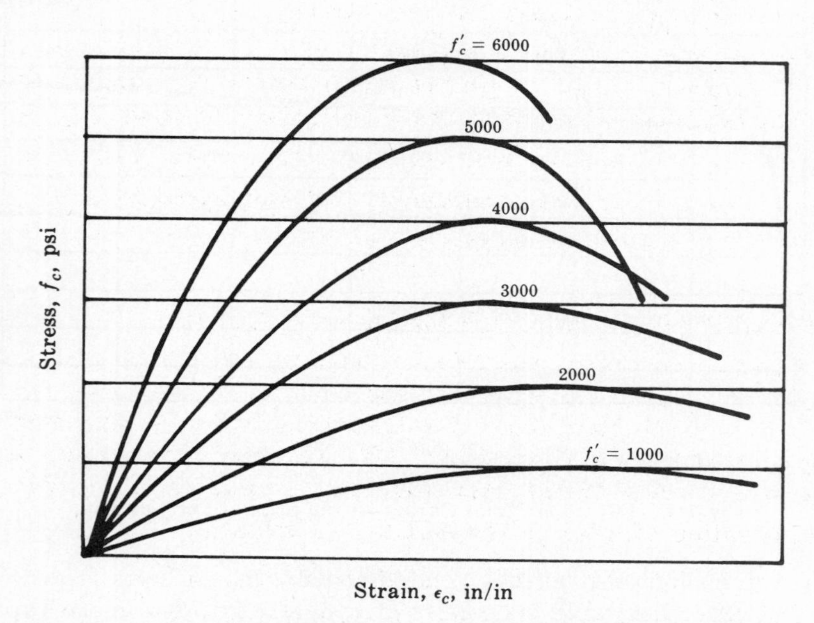

Fig. 1-4. Typical Stress-Strain Diagrams for Concrete

The stress-strain diagrams for concrete indicate three distinct ranges:

(1) The initial range, which is very nearly linear.

(2) The intermediate range, in which there is increasing curvature, ultimately reaching a point of maximum stress, f'_c.

(3) The final range, in which strain continues to increase while the load-carrying capacity decreases.

Fig. 1-5 presents the usually accepted plot of the stress-strain diagram for concrete subjected to axial load and flexure which was developed statistically. Certain properties of the curve must be described.

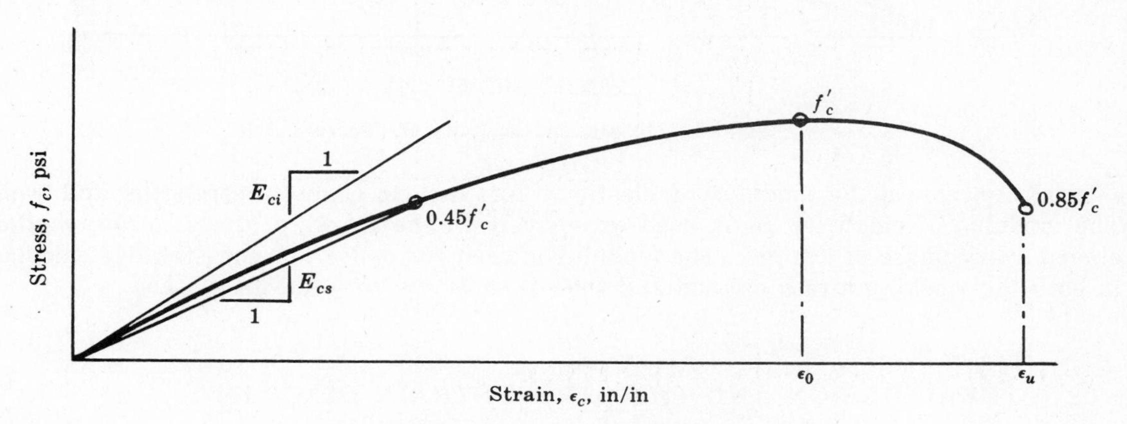

Fig. 1-5. Stress-Strain Diagram

(1) The tangent to the curve at its origin is called the *initial tangent modulus of elasticity*, E_{ci}, psi.

(2) A line drawn from the origin to a point on the curve at which $f_c = 0.45f'_c$ is called the *secant modulus of elasticity*, psi.

(3) For low strength concrete E_{ci} and E_{cs} differ widely. For high strength concrete there is practically no difference between the two values.

(4) For lightweight aggregate concrete the initial slope is somewhat less than that for normal weight concrete. The maximum stress occurs for larger strain values for lightweight concrete when f'_c is the same for both types of concrete.

(5) Definitions:
ϵ_c = unit strain in the concrete for any stress, f_c
ϵ_o = strain corresponding to maximum stress, f'_c
ϵ_u = ultimate strain at rupture

MODULUS OF ELASTICITY OF CONCRETE

In general practice the *secant modulus of elasticity* is usually used and is simply referred to as E_c. Numerous experimental equations have been proposed for obtaining the value of E_c. One of the most recent empirical equations was presented by Pauw and was adopted by the ACI Code Committee. This equation represents a modification of Pauw's equation which was obtained from statistical correlation of test data.

The modified equation adopted by the ACI Code Committee is

$$E_c = 33\,W^{1.5}\sqrt{f'_c} \qquad (1.3)$$

in which E_c = modulus of elasticity (lb/in^2), W = unit weight of the concrete (lb/ft^3), and f'_c = ultimate strength of the concrete (lb/in^2).

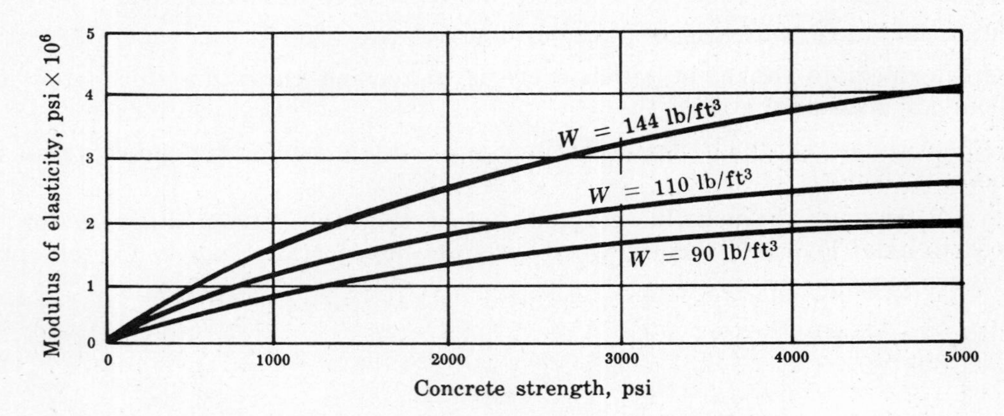

Fig. 1-6. Modulus of Elasticity of Concrete

Fig. 1-6 shows the modulus of elasticity for various concrete strengths and weights. The modulus of elasticity E_c is used extensively in the *working stress design method* in almost every phase of design. The modulus is used for deflection and stability calculations in both the working stress method and the *ultimate strength design method.*

STRUCTURAL DESIGN AND THE STRESS-STRAIN DIAGRAM

In order to derive equations for the design of reinforced concrete structural elements it is necessary to describe the stress-strain function mathematically. After establishing an equation for the stresses in the concrete, it is only necessary to integrate the equation to obtain the magnitude and point of application of the total compression force.

Fig. 1-7 illustrates the various stages of development of the stress-strain function as the loads are increased. Selection of a design method depends on which stage is to be used as a basis of proportioning the structural elements.

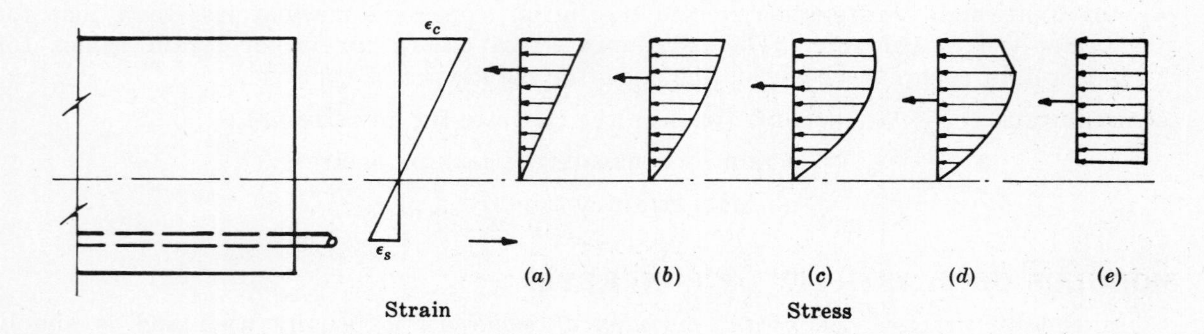

Fig. 1-7

Fig. 1-7(*a*) is used in the working stress design method, limiting f_c to a maximum value $0.45f'_c$. The diagram is considered to be linear, and the relationship between the elastic modulii of steel and concrete is used to transform the steel into equivalent concrete. The *modular ratio* is $n = E_s/E_c$.

Fig. 1-7(*d*) is used in the ultimate strength design method, using $0.85f'_c$ as the maximum stress. This is based on experimental data, using statistical methods for correlation.

In general practice an *equivalent rectangular stress-block* (*e*) is used to approximate the effects of the true ultimate stress-strain diagram. The limits of the block are defined so as to provide a total force in compression identical to that developed by the true diagram and to locate the force at its true point of application.

The ACI Code permits the use of any mathematical expression for describing the stress function, providing the resulting equations will be in general agreement with comprehensive test results. The rectangular stress block fulfills these requirements.

TENSILE STRENGTH OF CONCRETE

The tensile strength of plain concrete is rather small compared to the compressive strength, and tension in the concrete is always neglected in design practice. It is important, however, to give some consideration to the tensile strength with regard to combined stresses which cause *diagonal tension failure.*

Beam tests have been utilized for obtaining the tensile strength of concrete in past years. A method developed recently for obtaining a measure of the tensile strength is called the *split-cylinder test.* Standard concrete cylinders are loaded along the sides until the cylinder splits. The stress at which splitting occurs is designated as the *split cylinder strength,* f_{sp}. This value is used to determine a design factor

$$F_{sp} = f_{sp}/\sqrt{f'_c} \qquad (1.4)$$

This constant is used in connection with *shear stresses* in concrete design to guard against diagonal tension failure.

REINFORCING STEEL

Steel reinforcing for concrete consists of bars, wires and welded wire fabric, all of which are manufactured in accord with ASTM specifications. The most important properties of reinforcing steel are:

(1) modulus of elasticity, E_s, psi
(2) tensile strength, psi
(3) yield point stress, f_y, psi
(4) steel grade designation
(5) size or diameter of the bar or wire.

Fig. 1-8 illustrates properties (1), (2) and (3). Tables 1.12 and 1.13 provide information pertaining to properties (4) and (5).

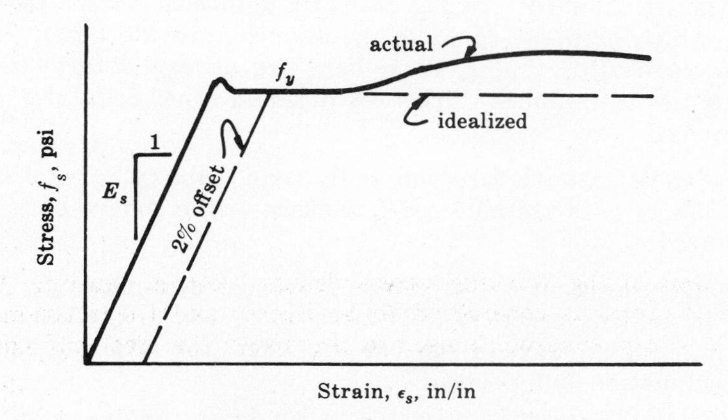

Fig. 1-8. Stress-Strain Diagram for Reinforcing Steel

The stress-strain diagram is *idealized* by assuming that stress is constant in the *plastic region* and equal to f_y.

The yield point is established by drawing a line parallel to the elastic region at some offset therefrom, usually about 0.002 in/in.

The idealized stress-strain diagram consists of an elastic portion and a purely plastic portion, i.e. a sloping line and a horizontal line.

The ASTM designation, tensile strength and yield point stress for reinforcing steel bars are listed in Table 1.12. For all of the types of steel listed the modulus of elasticity E_s is generally considered to be 29×10^6 psi.

TABLE 1.12

PROPERTIES OF REINFORCING STEEL

Type steel	ASTM desig.	Size nos. available	Grade	f_y, psi	Tensile strength, psi
Billet	A-15	2 to 11	Structural	33,000	55,000 to 75,000
			Intermediate	40,000	70,000 to 90,000
			Hard	50,000	80,000 minimum
Billet	A-408	14-S, 18-S	Structural	33,000	55,000 to 75,000
			Intermediate	40,000	70,000 to 90,000
			Hard	50,000	80,000 minimum
Billet	A-432	3 to 11 and 14-S, 18-S	ASTM A-431	60,000	90,000 minimum
High Strength Billet	A-431	3 to 11 and 14-S, 18-S	ASTM A-432	75,000	100,000 minimum
Rail Steel	A-16	2 to 11 3 to 11	Regular Special	50,000 60,000	80,000 90,000

Reinforcing bars are manufactured as *plain or deformed bars.* Deformed bars have ribbed projections which grip the concrete in order to provide better *bond* between the two materials. In the United States, main bars are always deformed. Plain bars are used for spirals and ties in columns. Practices differ in other countries, depending on the availability of deformed bars.

Formerly, bars were manufactured in both round and square shapes. In modern practice, however, square bars are not used. In place of the square bars, *equivalent round bars* are manufactured.

Bars are no longer designated by diameter or side dimension as in the past. Bar numbers are used in modern practice. For bars up to No. 8, the number coincides with the number of eighths of an inch in the bar diameter; for larger bars the numbers are used merely for designation purposes.

Table 1.13 indicates the sizes, numbers and various properties of the types of reinforcing bars currently used in the United States.

TABLE 1.13

PROPERTIES OF REINFORCING BARS*

Standard A305 Reinforcing Bars					
Bar Sizes		**Weight** pounds per foot	**Nominal Dim.—Round Sec.**		
Old (inches)	New (numerals)		Diameter (inches)	Cross sec. area, sq. in.	Perimeter (inches)
(1/4)	#2	.167	.250	.05	.786
(3/8)	#3	.376	.375	.11	1.178
(1/2)	#4	.668	.500	.20	1.571
(5/8)	#5	1.043	.625	.31	1.963
(3/4)	#6	1.502	.750	.44	2.356
(7/8)	#7	2.044	.875	.60	2.749
(1)	#8	2.670	1.000	.79	3.142
[1]	#9	3.400	1.128	1.00	3.544
[1 1/8]	#10	4.303	1.270	1.27	3.990
[1 1/4]	#11	5.313	1.410	1.56	4.430

Special Deformed Round Steel Bars ASTM Designation (A 408-58T)					
Bar Sizes		**Weight** pounds per foot	**Nominal Dim.—Round Sec.**		
Old (inches)	New (numerals)		Diameter (inches)	Cross sec. area, sq. in.	Perimeter (inches)
[1 1/2]	(14S)	7.65	1.693	2.25	5.32
[2]	(18S)	13.60	2.257	4.00	7.09

*Reproduced with permission of the Laclede Steel Co., St. Louis, Mo.

The bars listed in Tables 1.12 and 1.13 are used primarily for main reinforcement in beams, slabs, columns, footings, walls and other structural elements. Bar sizes Nos. 2, 3, 4 and 5 are often used for *spirals and lateral ties* in columns and as *stirrups* in beams. Spiral bars are usually plain, without deformations.

Reinforcing steel rods having diameters less than $\frac{1}{4}''$ are referred to as *wires,* and the sizes are designated by the AW & S gage number. Table 1.14 lists the pertinent data for wires. (*Note.* The smaller the gage number, the larger the diameter.)

TABLE 1.14

GAGE AND DIAMETERS OF REINFORCING WIRES

Gage number	Equivalent diameter, in.
0000000	0.4900
000000	0.4615
00000	0.4305
0000	0.3938
000	0.3625
00	0.3310
0	0.3065

Gage number	Equivalent diameter, in.
1	0.2830
2	0.2625
3	0.2437
4	0.2253
5	0.2070
6	0.1920
7	0.1770

Gage number	Equivalent diameter, in.
8	0.1620
9	0.1483
10	0.1350
11	0.1205
12	0.1055
13	0.0915
14	0.0800

Table 1.15 provides data concerning the cross-sectional areas of wires used in grid patterns for slabs. Each intersection is welded, hence the grids are referred to as *welded wire mesh.*

TABLE 1.15

SECTIONAL AREAS OF WELDED WIRE FABRIC

(Area in square inches per foot of width for various spacings of wire)

Steel Wire Gage Numbers	Wire			Center to Center Spacing, in Inches							
	Diameter (inches)	Area (square inches)	Weight (pounds per foot)	2	3	4	6	8	10	12	16
0000000	.4900	.18857	.6404	1.131	.754	.566	.377	.283	.226	.189	.141
000000	.4615	.16728	.5681	1.004	.669	.502	.335	.251	.201	.167	.125
00000	.4305	.14556	.4943	.873	.582	.437	.291	.218	.175	.146	.109
0000	.3938	.12180	.4136	.731	.487	.365	.244	.183	.146	.122	.091
000	.3625	.10321	.3505	.619	.413	.310	.206	.155	.124	.103	.077
00	.3310	.086049	.2922	.516	.344	.258	.172	.129	.103	.086	.065
0	.3065	.073782	.2506	.443	.295	.221	.148	.111	.089	.074	.055
1	.2830	.062902	.2136	.377	.252	.189	.126	.094	.075	.063	.047
2	.2625	.054119	.1838	.325	.216	.162	.108	.081	.065	.054	.041
$\frac{1}{4}$"	.2500	.049087	.1667	.295	.196	.147	.098	.074	.059	.049	.037
3	.2437	.046645	.1584	.280	.187	.140	.093	.070	.056	.047	.035
4	.2253	.039867	.1354	.239	.159	.120	.080	.060	.048	.040	.030
5	.2070	.033654	.1143	.202	.135	.101	.067	.050	.040	.034	.025
6	.1920	.028953	.09832	.174	.116	.087	.058	.043	.035	.029	.022
7	.1770	.024606	.08356	.148	.098	.074	.049	.037	.030	.025	.018
8	.1620	.020612	.07000	.124	.082	.062	.041	.031	.025	.021	.015
9	.1483	.017273	.05866	.104	.069	.052	.035	.026	.021	.017	.013
10	.1350	.014314	.04861	.086	.057	.043	.029	.021	.017	.014	.011
11	.1205	.011404	.03873	.068	.046	.034	.023	.017	.014	.011	.009
12	.1055	.0087417	.02969	.052	.035	.026	.017	.013	.010	.009	.007
13	.0915	.0065755	.02233	.039	.026	.020	.013	.010	.008	.007	.005
14	.0800	.0050266	.01707	.030	.020	.015	.010	.008	.006	.005	.004
15	.0720	.0040715	.01383	.024	.016	.012	.008	.006	.005	.004	.003
16	.0625	.0030680	.01042	.018	.012	.009	.006	.005	.004	.003	.002

Two wire sizes used in the same mesh may not have gage differentials of more than six gages. For example, No. 6 gage could be used with No. 12 wire, but not with No. 13 wire.

Specifications for wire reinforcement are as follows:

Tensile strength, 80,000 psi minimum

f_y = yield point stress = 0.8 × tensile strength, psi.

In testing specimens of wire, the reduction in area may not exceed 30% of the original area. For wire having over 100,000 psi tensile strength, the reduction in area may not exceed 25% of the original area.

BENDING OF REINFORCING BARS FOR STRUCTURAL CONSIDERATIONS

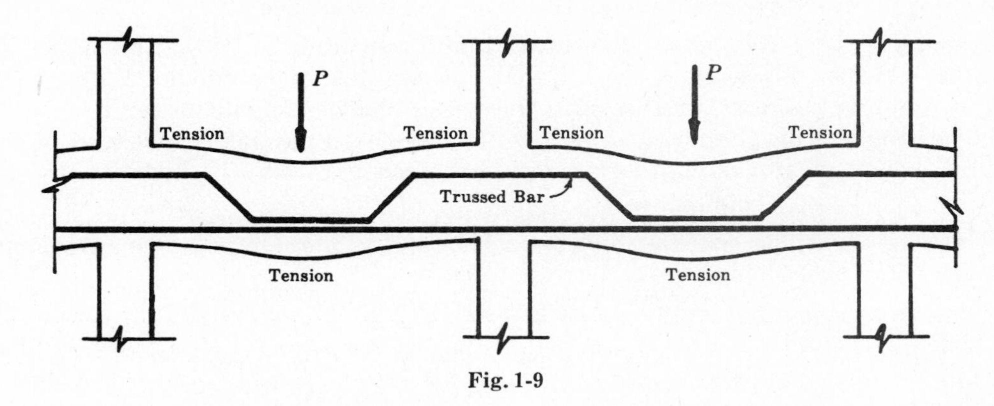

Fig. 1-9

Reinforcement is usually placed in the *tension zone* since the concrete is assumed to be incapable of resisting tension. Bars are, however, also often used in the compression zone. It is economical to use *bent-up bars* in such cases. Further, in order to anchor the bars properly to resist *bond stresses,* hooks are often provided at the ends of bars as shown in Fig. 1-10. Bars must be embedded properly in order to provide *anchorage.* Bending or *trussing* of bars is often used to accomplish this purpose.

Ties or spirals are always used in columns to hold the main vertical bars in place. *Stirrups* are often used in beams to hold the longitudinal steel in place, but the most important function of stirrups is to resist *diagonal tension* stresses. The concrete acting alone cannot always resist all of the shear stress which is present, thus the stirrups are used.

Fig. 1-10 shows typical examples of bar bending for various purposes.

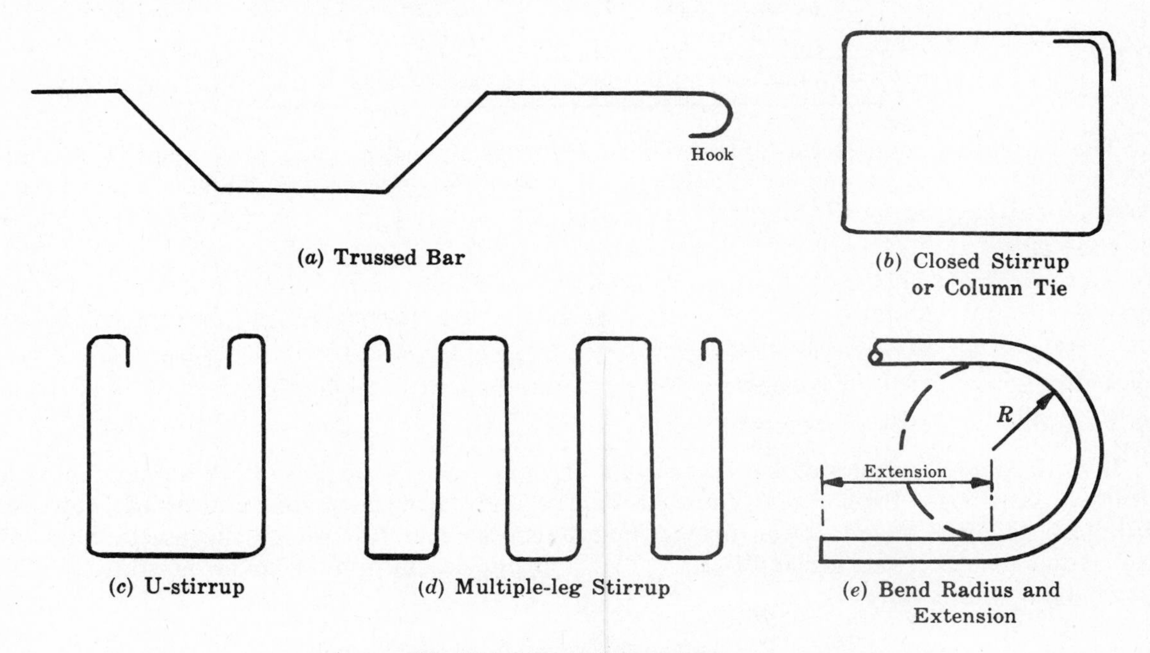

Fig. 1-10. Typical Bent Bars

Numerous shapes of bent bars are used. Some are used so frequently that pattern numbers have been standardized. Bent bars are therefore often called for by pattern number, indicating the various center to center dimensions in a tabular form.

Radii of bend have been standardized and are specified in the ACI Code. The bend radii and extensions are usually specified in terms of the diameter of the size bar being bent, with minimum extensions beyond the bend being specified.

Bars should be *bent cold* in a fabricating plant. Bending bars on the job is usually not permitted except in special cases. Under no circumstances should a bar be heated since this introduces residual stresses into the bar. Torches should never be used to cut bars to the desired length. Bending of bars which have been partially embedded in concrete is prohibited except when the engineer states that such bars may be bent.

Standard hooks consist of one of the following:

(*a*) A 180 degree turn plus an extension of at least *4 bar diameters.* The extension must be at least $2\frac{1}{2}$ inches long.

(*b*) A 90 degree turn plus an extension of at least *12 bar diameters.*

(*c*) For *stirrups and ties,* a 90 degree or 135 degree turn plus an extension of at least 6 bar diameters. The extension must be at least $2\frac{1}{2}$ inches long. The inside radius of bend shall be at least 1 bar diameter.

For *structural and intermediate grade steel,* bars in the size range No. 6 through No. 11 must have a *radius of bend* which is not less than $2\frac{1}{2}$ bar diameters. For steel other than structural and intermediate grades, and for bars other than stirrups or ties, the minimum bend radius must conform to Table 1.16.

TABLE 1.16

MINIMUM RADII OF BEND

Bar Sizes (numbers)	Minimum Radii of Bend
3, 4 or 5	$2\frac{1}{2}$ bar diameters
6, 7 or 8	3 bar diameters
9, 10 or 11	4 bar diameters
14 S or 18 S	5 bar diameters

The radius of bend is closely controlled in order to insure that high *residual stresses* will not develop in the bar as a result of the bending operations. This is the primary reason for discouraging bending of bars at the job site and for requiring that bars should be bent *cold.*

Reinforcement should be bent and placed as accurately as possible. For flexural members, walls and columns, the following variations are permitted in placement of bars (compared to specified dimensions):

(*a*) For effective depth of concrete $\leq 24''$, an error of $\pm\frac{1}{4}''$ is allowed.

(*b*) For effective depth of concrete $> 24''$, an error of $\pm\frac{1}{2}''$ is allowed.

Longitudinal locations of bends or ends of bars may be displaced ±2 inches from the specified locations. Proper minimum clearances and cover must be maintained, however. Minimum cover is specified for fireproofing purposes and to protect the steel from corrosion due to atmospheric conditions. The appropriate minimum cover is stated in the paragraphs which follow.

BAR SPACING AND CONCRETE COVER FOR STEEL

Figs. 1-11 through 1-15 illustrate the key dimensions related to cover and spacing for reinforcement for the various types of members illustrated and listed. The dimensions X and Z must be at least equal to those stated.

For Beams or Girders

$X \geq$ diameter of main bars
$X \geq 1\frac{1}{3} \times$ maximum aggregate size
$X \geq 1''$
$Z \geq$ diameter of main bars, always.

When not exposed to weather or earth,
$Z \geq 1\frac{1}{2}''$

When exposed to weather or earth,
$Z \geq 1\frac{1}{2}''$, bars No. 5 or smaller
$Z \geq 2''$, bars larger than No. 5.

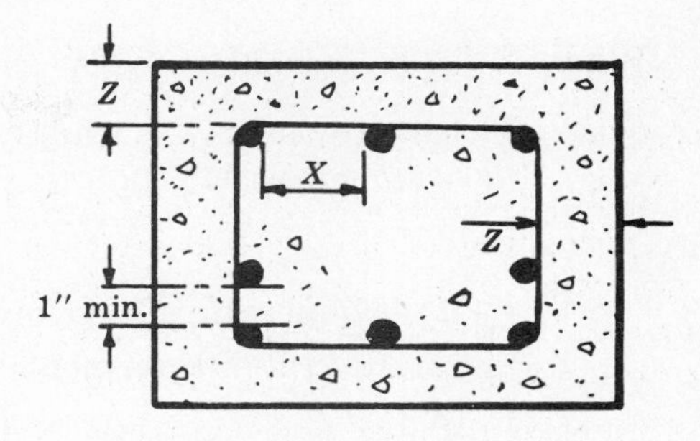

Fig. 1-11. **Bar Spacing and Cover for Beams or Girders**

Columns with Spirals or Ties

$X \geq 1\frac{1}{2}$ diameters of vertical bars
$X \geq 1\frac{1}{2} \times$ maximum aggregate size
$X \geq 1\frac{1}{2}''$
$Z \geq$ diameter of vertical bars, always.

When not exposed to weather or earth,
$Z \geq 1\frac{1}{2}''$
$Z \geq 1\frac{1}{2} \times$ maximum aggregate size.

When exposed to weather or earth,
$Z \geq 2''$, unless the local building code requires greater clearance.

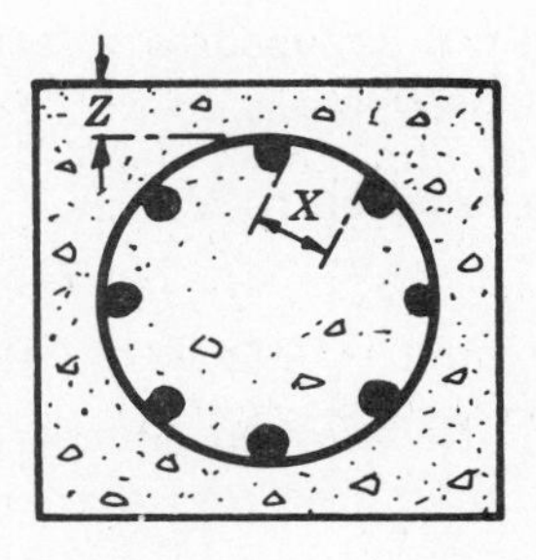

Fig. 1-12. **Bar Spacing and Cover for Columns**

Footings

$X \geq 2''$ when side forms are used
$X \geq 3''$ without side forms.

When exposed to weather or earth,
$Z \geq 2''$ (formed surfaces)
Z or X is always greater than the bar diameter.

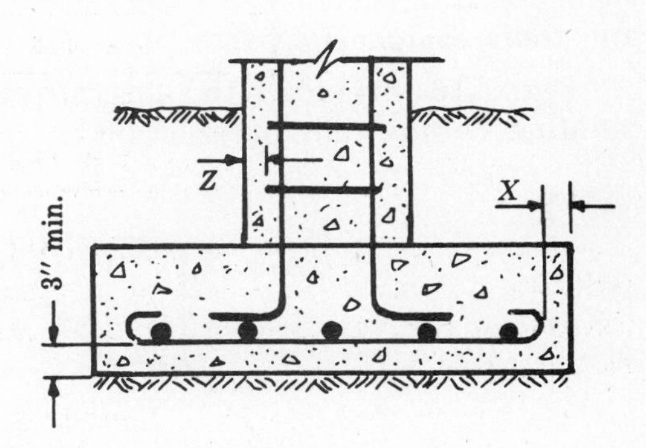

Fig. 1-13. **Bar Spacing and Cover for Footings**

Slabs and Joists

$Y \leq 3t$
$Y \leq 18''$
$Y \geq$ bar diameter
$Y \geq 1\frac{1}{3} \times$ max. aggregate size
$Y \geq 1''$

When exposed to weather,
$Z \geq 1\frac{1}{2}''$, bars No. 5 or smaller
$Z \geq 2''$, for bars larger than No. 5

When not exposed to weather,
$Z \geq \frac{3}{4}''$

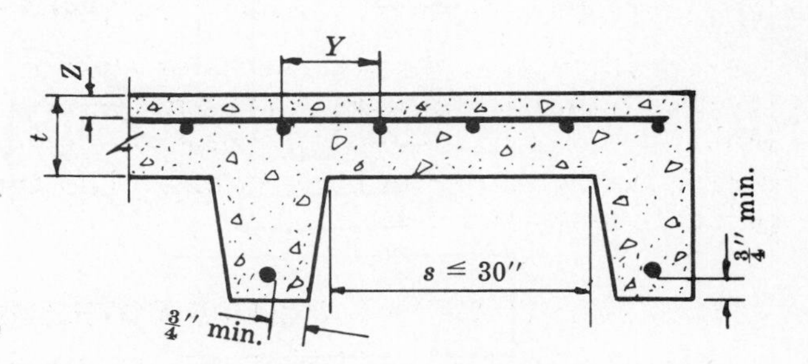

Fig. 1-14. **Bar Spacing and Cover for Slabs and Joists**

When $S > 30''$, consider a joist as a beam.

Note. These provisions also apply to solid slabs without joists, but do not apply to *flat slab* construction.

Walls

$Y \leq 5t$ (Temperature Reinforcement)
$Y \leq 3t$ (Main Reinforcement)
$Y \leq 18''$
$Y \geq$ bar diameter
$Y \geq 1\frac{1}{3} \times$ maximum aggregate size
$Y \geq 1''$

When not exposed to weather or earth,
$Z \geq \frac{3}{4}''$

When exposed to weather or earth,
$Z \geq 1\frac{1}{2}''$, bars No. 5 or smaller
$Z \geq 2''$, bars larger than No. 5.

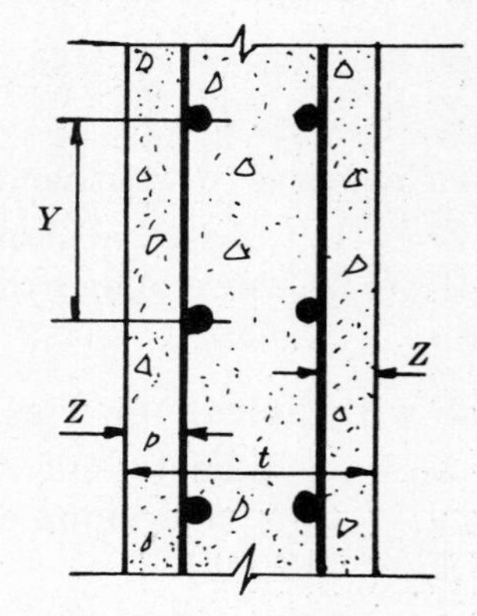

Fig. 1-15. Walls (Including Retaining Walls)

Note. Reinforced concrete walls carrying reasonable concentric loads are designed as columns and are subject to the same bar spacing requirements as for columns.

IMPORTANT COMPONENTS OF STRUCTURES

In order to fully understand the Building Code provisions and to design structural elements, it is necessary to have a thorough understanding of the types of structures used and their component parts.

Fig. 1-16 through 1-18 illustrate the various types of elements usually encountered in building design and construction.

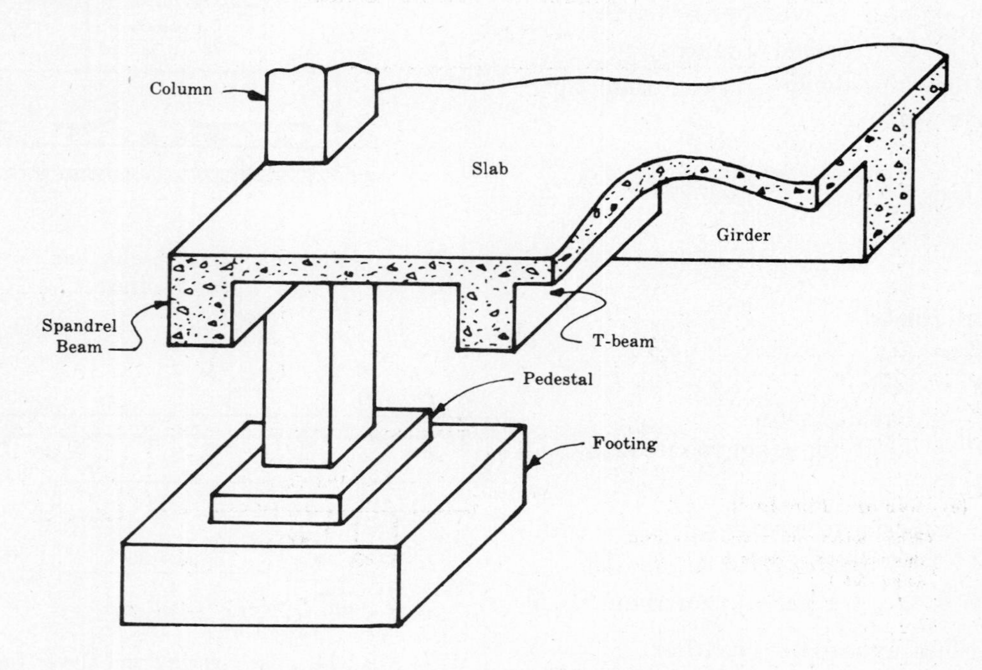

Fig. 1-16. Beam and Slab Construction

Notes for Fig. 1-16:

(1) The slab shown delivers load primarily to the T-beams and is called a *one way slab*.

(2) The girder receives loads primarily from the T-beams as concentrated loads, then delivers the loads to the columns.

(3) Pedestals are used to spread the load over a large area of the footing. The pedestal receives the column loads.

(4) Spread footings bear directly on the earth.

(5) The girder is also a T-beam (in shape, even if not considered so in structural action).

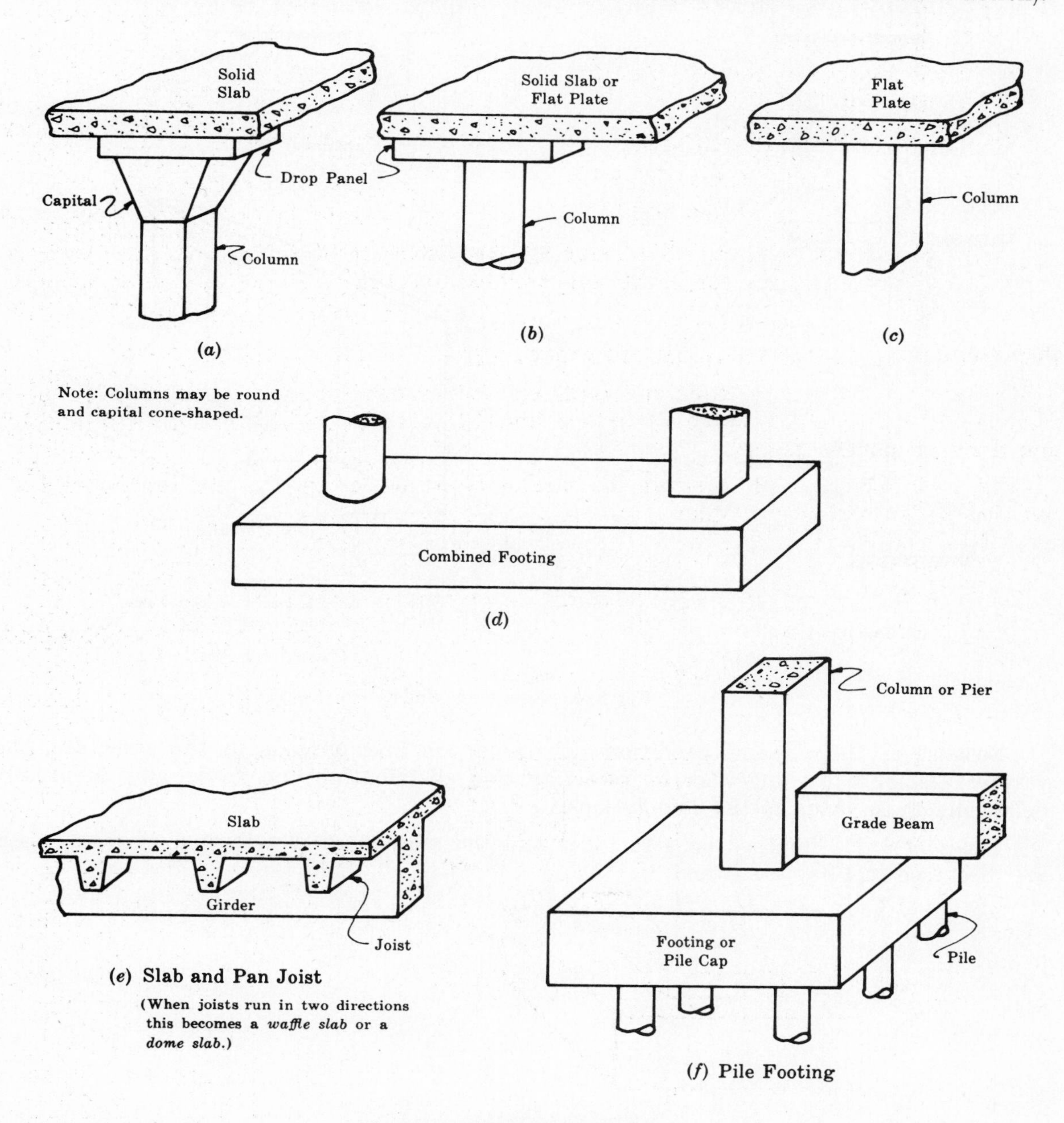

(*e*) **Slab and Pan Joist**

(When joists run in two directions this becomes a *waffle slab* or a *dome slab*.)

(*f*) **Pile Footing**

Fig. 1-17. Structural Systems

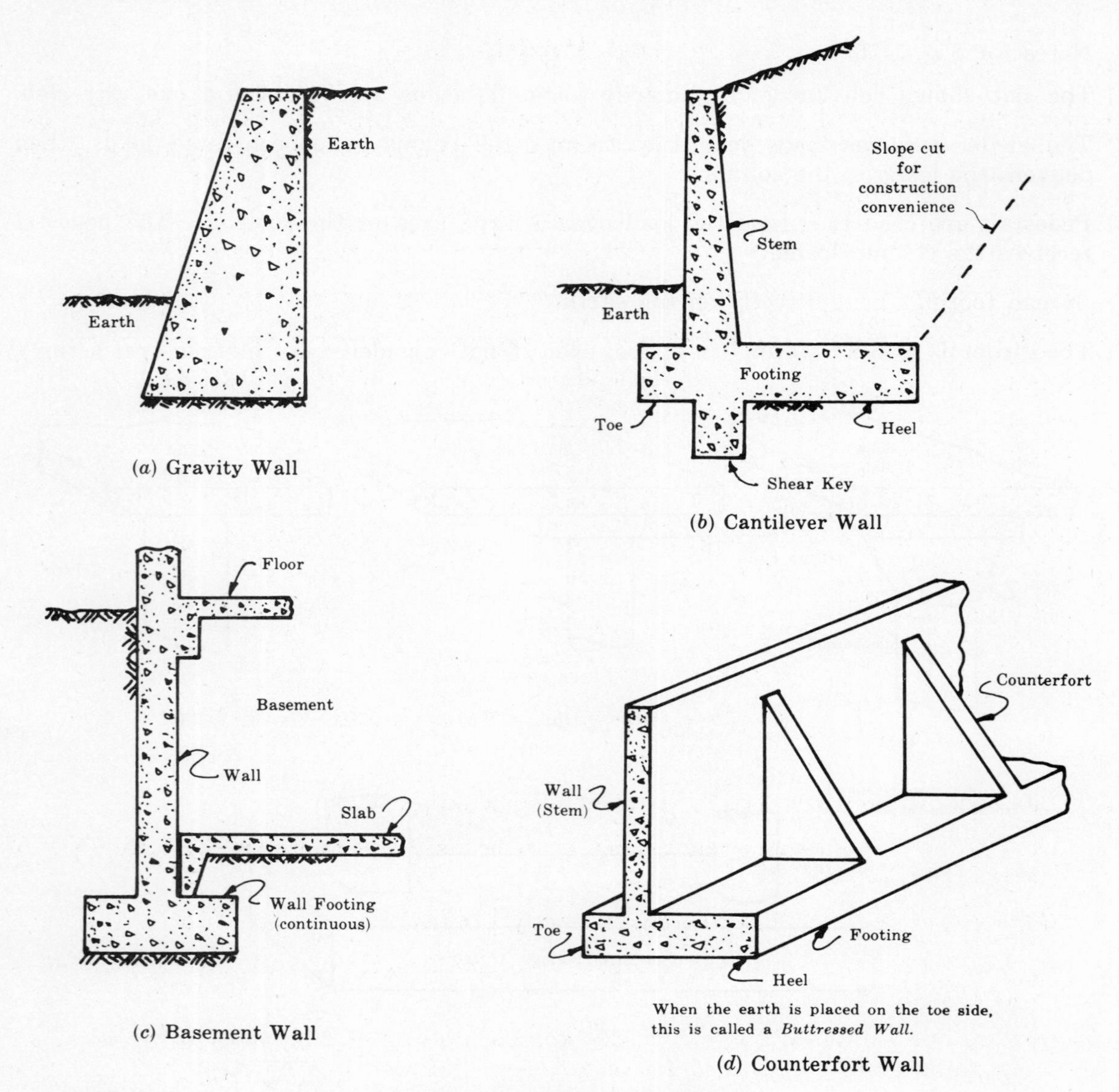

(*a*) Gravity Wall

(*b*) Cantilever Wall

(*c*) Basement Wall

(*d*) Counterfort Wall

Fig. 1-18. Retaining Walls

Because of their shape, conditions of connection and purpose in the structure, the various elements act together in receiving and distributing the loads and eventually delivering those loads to the foundations.

Types of loads imposed on structures and the method of distribution of those loads are discussed in Chapters 2 and 3.

Solved Problems

1.1. A sieve analysis was performed using 100 pounds of aggregate. The *weights retained on each sieve* are shown in Table 1.17. Determine the fineness modulus of the aggregate.

TABLE 1.17

Sieve No.	4	8	16	30	50	100	Dust
Weight retained, lb	3	12	20	20	25	18	2

Sieve size	Percent retained on Sieve No.	Percent retained (cumulative)
No. 4	3	0 + 3 = 3
No. 8	12	12 + 3 = 15
No. 16	20	20 + 15 = 35
No. 30	20	20 + 35 = 55
No. 50	25	25 + 55 = 80
No. 100	18	18 + 80 = 98
Passing No. 100	2	Sum = 286
	100%	

Fineness modulus = 286/100 = 2.86

1.2. Design a concrete mix using the following assumptions:

(*a*) Type 1, non-air-entrained cement is to be used. Assume specific gravity 3.15 for the cement.

(*b*) Coarse and fine aggregates are properly graded.

(*c*) The coarse aggregate has a bulk-dry specific gravity of 2.68 and an *absorption* coefficient of 0.5 percent.

(*d*) The fine aggregate has a bulk-dry specific gravity of 2.64, an absorption of 0.7 percent, and fineness modulus of 2.8.

(*e*) The concrete will be used below ground and will not be exposed to severe weathering or sulfate attack. Structural specifications require a 28 day compressive strength of 3500 psi.

The following computations have been taken from *ACI Standard-Recommended Practice for Selecting Proportions for Concrete*, ACI 613-54.

On the basis of Table 1.8 as well as previous experience it is determined that under the conditions of placement to be employed, a slump of 3 to 4″ should be used. A locally available No. 4 to $1\frac{1}{2}''$ coarse aggregate will be used. The dry rodded weight of the coarse aggregate is 100 pcf.

The proportions may be computed as follows:

(1) Since the structure will not be exposed to severe conditions, non-air-entrained cement is used and the water-cement ratio will be established solely on the basis of strength requirements.

(2) From Table 1.2, the water-cement ratio needed for $f'_c = 3500$ psi is about 6.6 gal/sack.

(3) Approximate amount of mixing water for 3 to 4″ slump with $1\frac{1}{2}''$ aggregate is 36 gal/yd³ from Table 1.9.

(4) Cement content is 36/6.6 = 5.5 sacks/yd³.

(5) Estimate quantity of coarse aggregate from Table 1.6(a) for fineness modulus 2.8 and maximum size aggregate $1\frac{1}{2}''$. Use 0.72 ft^3 per ft^3 of concrete. For a cubic yard of concrete use $0.72(27) = 19.4\ ft^3$. Weight of coarse aggregate $= 19.4(100) = 1940$ lb.

(6) Approximate entrapped air from Table 1.9 is 0.01 times the volume of concrete. ($1\ yd^3 = 27\ ft^3$)

(7) Tabulate the quantities as follows:

Solid volume of cement $= 5.5(94)/(3.15)(62.4)$	$=$	$2.63\ ft^3$
Volume of water $= 36/7.5$	$=$	4.80
Solid volume coarse aggregate $= 1940/(2.68)(62.4)$	$=$	11.60
Volume of entrapped air $= 0.01(27)$	$=$	0.27
Total volume of solid ingredients except sand	$=$	$19.30\ ft^3$
Solid volume of sand required $= 27 - 19.30$	$=$	$7.70\ ft^3$
Required weight dry sand $= 7.70(2.64)(62.4)$	$=$	1270 lb

(8) Estimated batch quantities per cubic yard are:

Cement 5.5 sacks $= 5.5(94)$	$=$	517 lb
Water 36 gallons	$=$	300 lb
Sand (dry basis)	$=$	1270 lb
Coarse aggregate (dry basis)	$=$	1940 lb

The batch weights must be adjusted in the field to account for the free water in the sand and gravel. It is also possible that some minor adjustments in proportions might be made in the field as a result of experience with the concrete so produced.

Chapter 2

Gravity Loads

FORCES, SHEAR, MOMENTS AND REACTIONS

NOTATION

b = width of a beam or column, inches
C = moment coefficient of $w(L')^2$
I = moment of inertia, in^4
K = Stiffness factor, in^3
L = any span length, usually center-to-center of supports, feet
L = long span for two-way slabs, feet
L' = clear span length, ft. (average of 2 adjacent spans for negative M)
m = ratio of short span to long span, two-way slabs
M = bending moment, ft-kips or ft-pounds
P = any concentrated load, kips or pounds
P_{DL} = dead load concentrated load, kips or pounds
P_{LL} = live load concentrated load, kips or pounds
P_{TL} = total load concentrated load, kips or pounds
q = any uniformly distributed load, kips/ft
R = any reaction, kips or pounds
S = short span length, two-way slabs, feet
t = total depth of a member, inches
V = shear force, kips or pounds
w = any uniformly distributed load, kips/ft or $kips/ft^2$ (ksf)
w_{DL} = uniformly distributed dead load
w_{LL} = uniformly distributed live load
w_{TL} = uniformly distributed total load

STRUCTURAL ANALYSIS AND DESIGN

The analysis of structures deals with the determination of loads, reactions, shear and bending moments. Structural design deals with the proportioning of members to resist the applied forces. The sequence involved in creating a structure, then, involves analysis first and then design.

The ACI Code currently requires that analysis be made using the *elastic theory*, whereas structural design may be accomplished using either the *working stress method* or the *ultimate strength method.*

EXACT AND APPROXIMATE METHODS OF ANALYSIS

There exist methods which provide for an exact mathematical analysis of structures. Such methods as slope-deflection and moment distribution may *always* be used to analyze concrete structures. In some cases it is absolutely necessary to use the exact methods. In the most common cases, however, it is sufficiently accurate to use approximate methods.

The ACI Code contains approximate coefficients for calculating shears and moments, which can be used when (and only when) specified conditions have been satisfied.

Since exact methods are studied in *statically indeterminate structures,* a prerequisite to the study of reinforced concrete, only the approximate methods will be discussed in this text. An exception exists in the case of the cantilever moment distribution method, which is an exact mathematical method when certain conditions are satisfied.

CONTINUOUS BEAMS AND SLABS

Approximate coefficients of shear and bending moment may be utilized when the following conditions are satisfied:

(1) Adjacent *clear* spans may not differ in length by more than 20% of the shorter span.

(2) The ratio of live load to dead load may not exceed 3.

(3) The loads *must be* uniformly distributed.

Beam and Slab Coefficients

When conditions (1), (2) and (3) are satisfied, the following listed approximate formulas which are stated in the ACI Code may be used to determine shear forces and bending moments in continuous beams and *one-way slabs.* (A one-way slab is one which distributes its load to two end supports only. A two-way slab distributes its load to four supports, one support existing along each of the ends and sides.)

For Positive Moment

End spans:

If discontinuous end is unrestrained .... $w(L')^2/11$ (2.1)

If discontinuous end is integral with the support .... $w(L')^2/14$ (2.2)

Interior spans .... $w(L')^2/16$ (2.3)

For Negative Moment

Negative moment at exterior face of first interior support:

Two spans .... $w(L')^2/9$ (2.4)

More than two spans .... $w(L')^2/10$ (2.5)

Negative moment at other faces of interior supports .... $w(L')^2/11$ (2.6)

Negative moment at face of all supports for (*a*) slabs with spans not exceeding 10 ft and (*b*) beams and girders where the ratio of sum of column stiffnesses to beam stiffness exceeds 8 at each end of the span .... $w(L')^2/12$ (2.7)

Negative moment at interior faces of exterior supports, for members built integrally with their supports:

Where the support is a spandrel beam or girder .... $w(L')^2/24$ (2.8)

Where the support is a column .... $w(L')^2/16$ (2.9)

Shear Forces

Shear in end members at first interior support .... $1.15w(L')/2$ (2.10)

Shear at all other supports .... $w(L')/2$ (2.11)

End Reactions

Reactions to a supporting beam, column or wall are obtained as the sum of the shear forces acting on both sides of the support.

Integral and unrestrained supports are illustrated in Fig. 2-1. Exterior and interior supports are shown in Fig. 2-2.

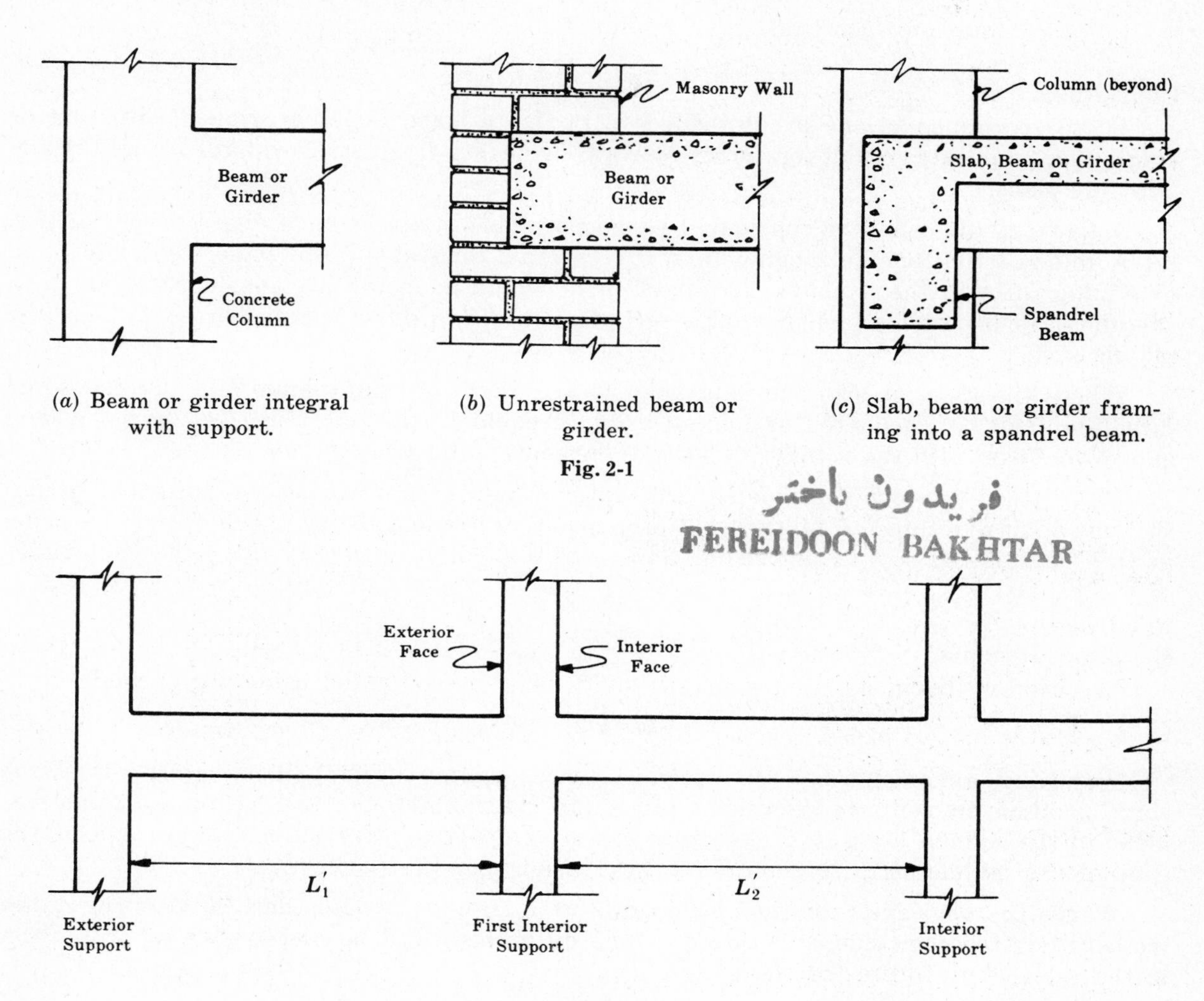

(*a*) Beam or girder integral with support.

(*b*) Unrestrained beam or girder.

(*c*) Slab, beam or girder framing into a spandrel beam.

Fig. 2-1

Fig. 2-2

TWO-WAY SLABS

The ACI Code provides three separate approximate methods for use in determining shear and moments in slabs which distribute their loads to four supports.

Traditionally, the procedure known as *Method 2* has been used almost exclusively in engineering practice. This method has been devised considering the *theory of elasticity* and the results of experiments. The method applies *only* when (*a*) the loads are uniformly distributed and (*b*) the ratio of live load to dead load does not exceed 3.

The following notes relative to the analysis of two-way slabs are reproduced from the ACI Code:

C = moment coefficient for two-way slabs as given in Table 2.1

m = ratio of short span to long span for two-way slabs

S = length of short span for two-way slabs. The span shall be considered as the center-to-center distance between supports or the clear span plus twice the thickness of slab, whichever value is the smaller. (Since the beam width is not known for certain at the outset, it is sufficient and safe to use the center-to-center span.)

w = total uniform load, lb/ft^2 or kips/ft^2

Limitations

These recommendations are intended to apply to slabs (solid or ribbed), isolated or continuous, supported on all four sides by walls or beams, in either case built monolithically with the slabs.

A two-way slab shall be considered as consisting of strips in each direction as follows: (*a*) A middle strip one-half panel in width, symmetrical about the panel centerline and extending through the panel in the direction in which moments are considered and (*b*) a column strip one-half panel in width, occupying the two quarter-panel areas outside the middle strip.

Where the ratio of short to long span is less than 0.5, the middle strip in the short direction shall be considered as having a width equal to the difference between the long and short span with the remaining area representing the two column strips.

The critical sections for moment calculations are referred to as the principal design sections and are located as follows: (*a*) For negative moment, along the edges of the panel at the faces of the supporting beams and (*b*) for positive moment, along the centerlines of the panels.

Bending Moments

The bending moments for the middle strips shall be computed using the formula

$$M = CwS^2 \tag{2.12}$$

The average moments per foot of width in the column strip shall be two-thirds of the corresponding moments in the middle strip. In determining the spacing of the reinforcement in the column strip, the moment may be assumed to vary from a maximum at the edge of the middle strip to a minimum at the edge of the panel.

Where the negative moment on one side of a support is less than 80 percent of the moment on the other side, two-thirds of the difference shall be distributed in proportion to the relative stiffnesses of the slabs.

Shear

The shear stresses in the slab may be computed on the assumption that the load is distributed to the supports in accordance with equations (*2.13*) or (*2.14*) given below.

Supporting Beams

The loads on the supporting beams for a two-way rectangular panel may be assumed as that load contained within the tributary areas of the panel bounded by the intersection of 45-degree lines from the corners with the median line of the panel parallel to the long side. (See Fig. 2-3.)

Equivalent Uniform Loads

The bending moments may be determined approximately by using an *equivalent uniform load* per lineal foot of beam for each panel supported as follows:

For the short span: $$\frac{wS}{3} \tag{2.13}$$

For the long span: $$\frac{wS}{3}\,\frac{(3-m^2)}{2} \tag{2.14}$$

TABLE 2.1
MOMENT COEFFICIENTS FOR TWO-WAY SLABS, METHOD 2

Moments	Short span						Long span, all values of m
	Values of m						
	1.0	0.9	0.8	0.7	0.6	0.5 and less	
Case 1 – Interior panels							
Negative moment at –							
Continuous edge	0.033	0.040	0.048	0.055	0.063	0.083	0.033
Discontinuous edge	–	–	–	–	–	–	–
Positive moment at midspan	0.025	0.030	0.036	0.041	0.047	0.062	0.025
Case 2 – One edge discontinuous							
Negative moment at –							
Continuous edge	0.041	0.048	0.055	0.062	0.069	0.085	0.041
Discontinuous edge	0.021	0.024	0.027	0.031	0.035	0.042	0.021
Positive moment at midspan	0.031	0.036	0.041	0.047	0.052	0.064	0.031
Case 3 – Two edges discontinuous							
Negative moment at –							
Continuous edge	0.049	0.057	0.064	0.071	0.078	0.090	0.049
Discontinuous edge	0.025	0.028	0.032	0.036	0.039	0.045	0.025
Positive moment at midspan	0.037	0.043	0.048	0.054	0.059	0.068	0.037
Case 4 – Three edges discontinuous							
Negative moment at –							
Continuous edge	0.058	0.066	0.074	0.082	0.090	0.098	0.058
Discontinuous edge	0.029	0.033	0.037	0.041	0.045	0.049	0.029
Positive moment at midspan	0.044	0.050	0.056	0.062	0.068	0.074	0.044
Case 5 – Four edges discontinuous							
Negative moment at –							
Continuous edge	–	–	–	–	–	–	–
Discontinuous edge	0.033	0.038	0.043	0.047	0.053	0.055	0.033
Positive moment at midspan	0.050	0.057	0.064	0.072	0.080	0.083	0.050

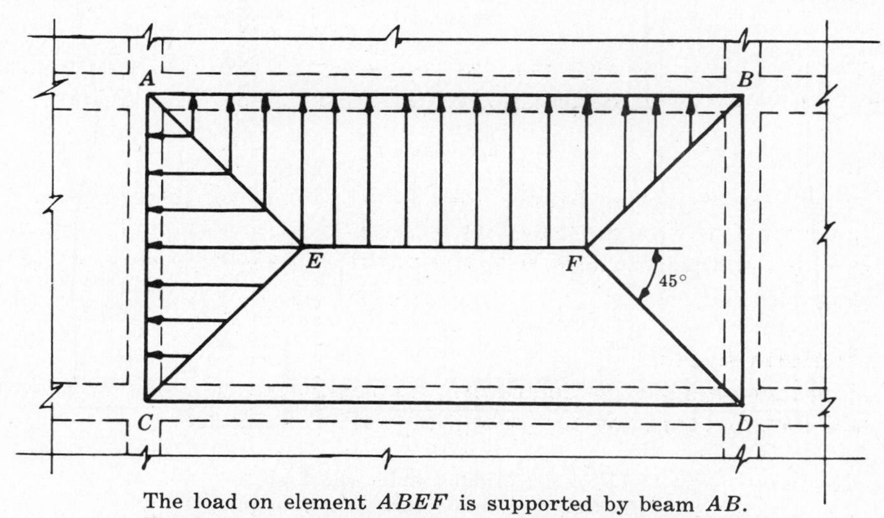

The load on element *ABEF* is supported by beam *AB*.
The load on element *AEC* is supported by beam *AC*.

Fig. 2-3

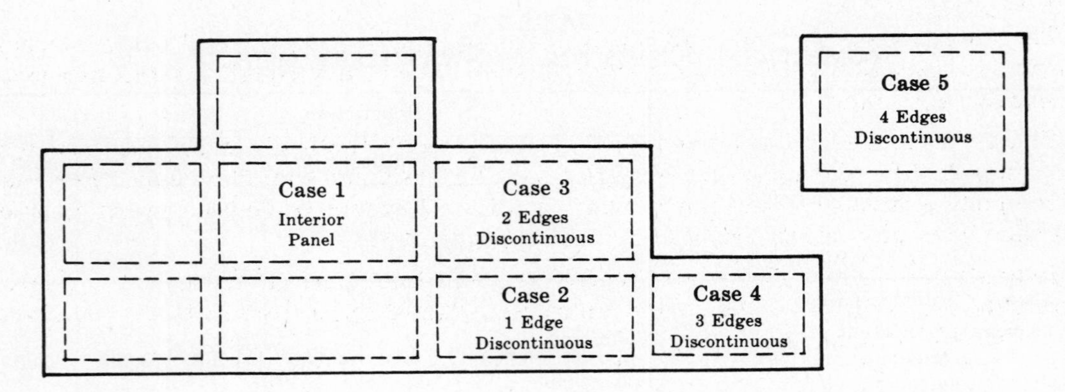

Fig. 2-4

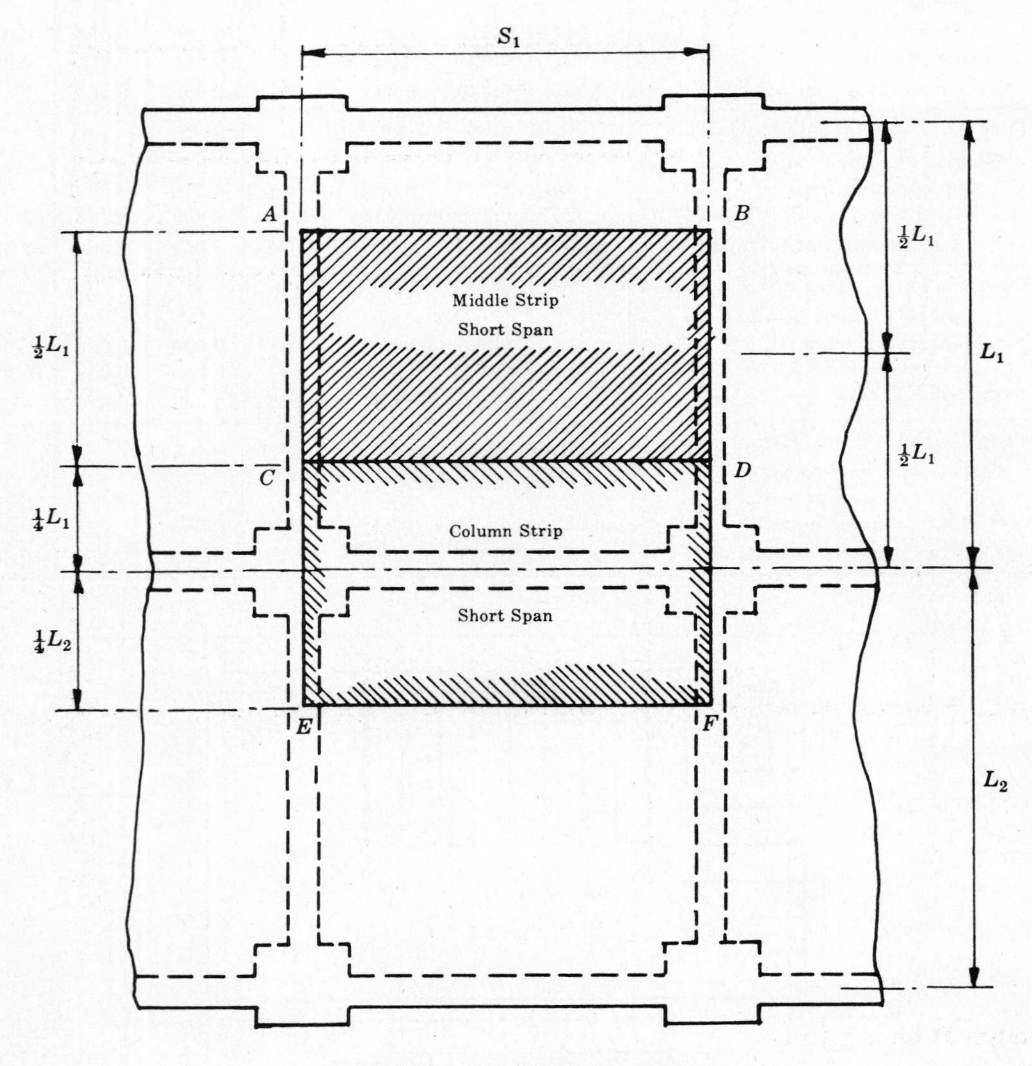

$ABCD$ = Middle strip, short span
$CDEF$ = Column strip, short span

Fig. 2-5

FRAME ANALYSIS

The approximate methods previously discussed apply to *usual conditions* of construction, where loads are uniformly distributed, spans are nearly equal and the live loads are not excessively high.

When the approximate methods do not apply, mathematically exact methods must be used. The ACI Code does, however, permit the utilization of simplifications. For example, the complete analysis of the frame shown in Fig. 2-6 would be time-consuming, unless an electronic computer would be used for the analysis.

In analyzing the beams for level *EFGH*, the ACI Code permits the use of the substitute frame shown in Fig. 2-7. The columns may be assumed to be *fixed* at the floors above and below the level in question.

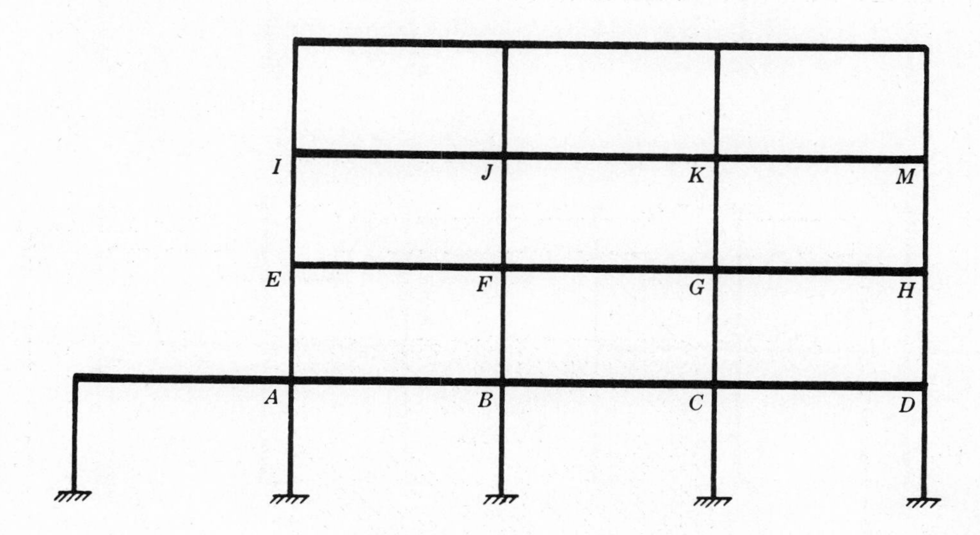

Fig. 2-6

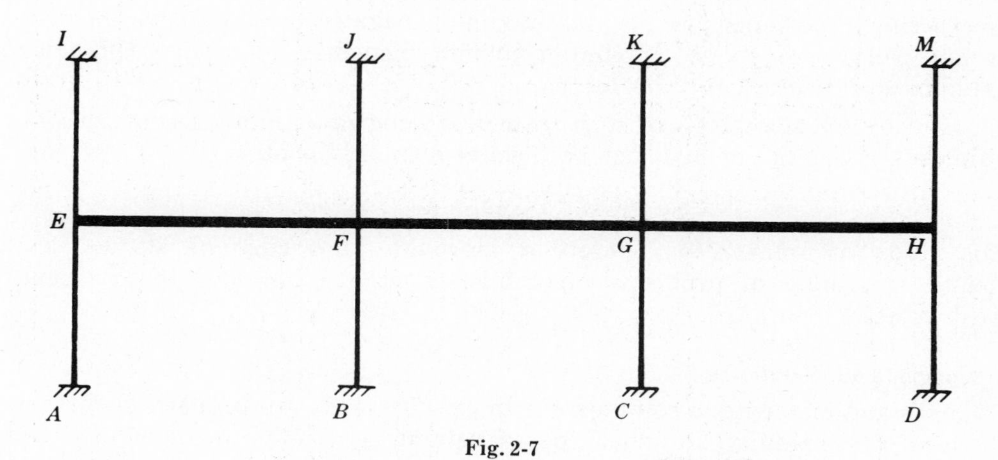

Fig. 2-7

Maximum Design Values

The maximum shear force, reaction and bending moment *do not occur* when the entire frame is loaded. This becomes apparent by observing the deflections of a loaded frame.

Fig. 2-8 shows the deflection diagrams due to four different loading conditions, each of which must be considered in determining the maximum shear, moment and reactions. In each case, *all spans are loaded with the dead load* which consists of the weight of the materials.

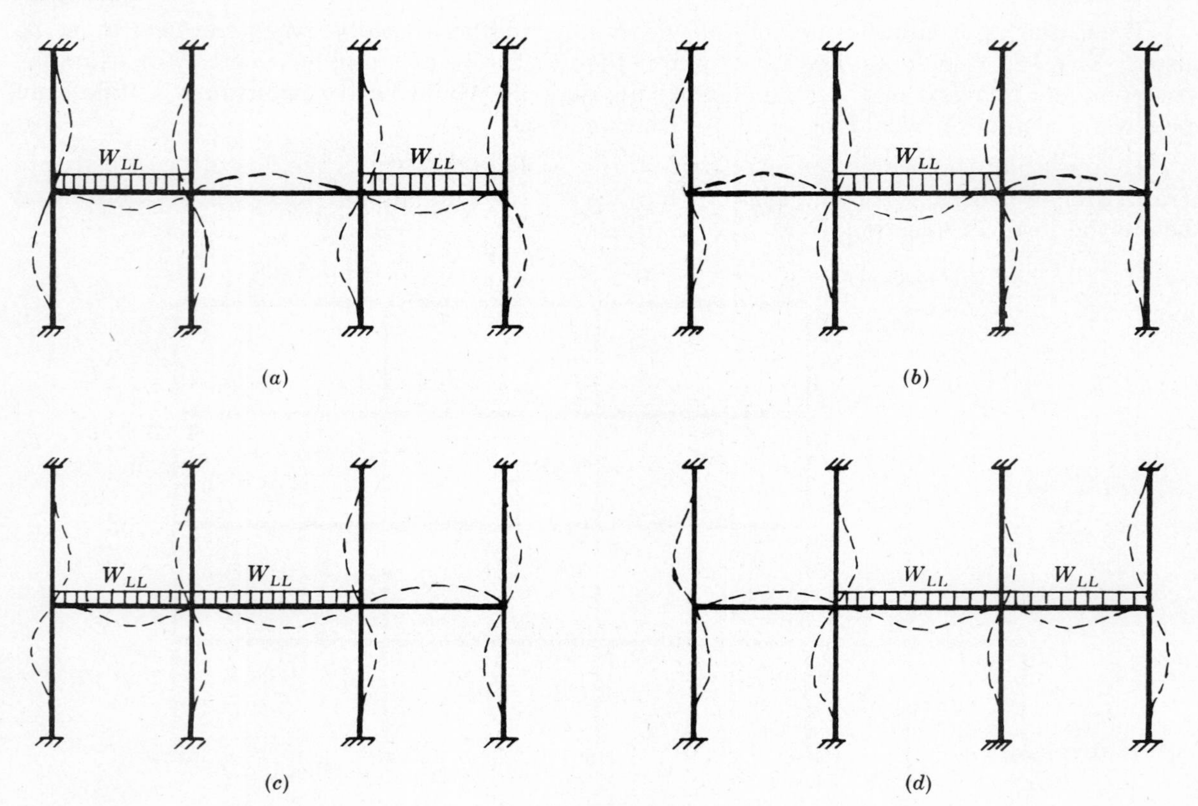

Fig. 2-8

Those loading conditions provide the maximum negative and positive moments at the supports and centers of spans, maximum column moments, maximum column reactions and maximum shear forces for the beams.

Methods of complete analysis of such frames are discussed in texts concerning statically indeterminate structures and will not be discussed in this book.

Statics

In the study of reinforced concrete a thorough knowledge of statics is assumed. Nevertheless, a number of problems involving statics are provided in this text for the purpose of review.

Column Forces and Moments

Axial loads and moments in columns due to gravity loads are properly determined using the principles of continuity and methods of solving statically indeterminate structures. Under normal circumstances, column moments on interior columns are relatively small when span lengths are nearly equal. However, the exterior columns may have reasonably large moments accompanied by relatively small axial loads.

When the approximate methods are used for obtaining beam moments, the column moments are also determined by approximate methods. The differences in the beam moments are distributed to the columns above and below the joint in the ratio of the *stiffnesses* of the columns.

It must be emphasized here, however, that approximate methods usually provide for excessively large structural members. Approximate methods should not be used when an exact mathematical analysis is available.

Solved Problems

2.1. A two span beam (Fig. 2-9) is supported by spandrel beams at the outer edges and by a column in the center. Dead load (including beam weight) is 1.0 kip/ft and live load is 2.0 kips/ft on both beams. Calculate all critical shear forces and bending moments for the beams. The torsional resistance of the spandrel beam is not sufficient to cause restraint of beam ABC at the masonry walls.

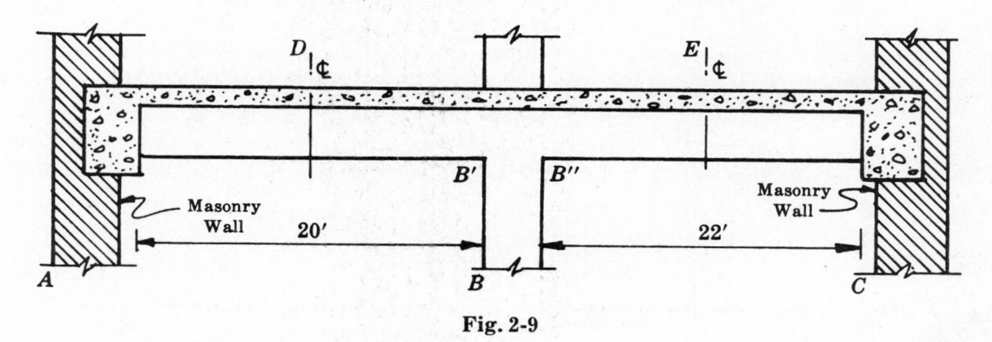

Fig. 2-9

(*a*) Loads are uniformly distributed, (*b*) $LL/DL = 2 < 3$, (*c*) $(L_2' - L_1')/L_1' = (22-20)/20 = 0.1 < 0.2$. ACI coefficients apply.

Bending Moments

$M_{AB} = -3(20)^2/24 = -50$ ft-kips, $M_{BA} = -(3/9)[(20+22)/2]^2 = -147$ ft-kips $= M_{BC}$.

$M_{CB} = -3(22)^2/24 = -60.5$ ft-kips, $M_D = 3(20)^2/11 = 109$ ft-kips, $M_E = 3(22)^2/11 = 132$ ft-kips.

Shear Forces

$V_A = 3(20)/2 = 30$ kips $\qquad V_{B'} = 1.15(3)(20)/2 = 34.5$ kips

$V_C = 3(22)/2 = 33$ kips $\qquad V_{B''} = 1.15(3)(22)/2 = 37.95$ kips

Reactions

$R_A = V_A = 30$ kips, $\quad R_B = V_{B'} + V_{B''} = 72.45$ kips, $\quad R_C = V_C = 33$ kips.

2.2. Given, conditions identical to those of Problem 2.1, except that A and B are built integrally with the columns. Determine the critical moments, shears and column reactions.

Refer to equations (*2.2*), (*2.3*), (*2.4*) and (*2.9*).

$M_{AB'} = -3(20)^2/16 = -75$ ft-kips $\qquad M_E = 3(22)^2/14 = 103.7$ ft-kips

$M_{BA} = -3(21)^2/9 = -147$ ft-kips $\qquad M_D = 3(20)^2/14 = 85.7$ ft-kips

All shear forces and reactions are identical to those for Problem 2.1 for the corresponding sections.

2.3. Given, a five span beam as shown in Fig. 2-10. Total dead load is 1.5 kips/ft and total live load is 2.5 kips/ft. Calculate the critical moments, shear forces and column reactions. The beam is built into a girder at A and a column at E. The masonry wall at A does not offer restraint to beam AB.

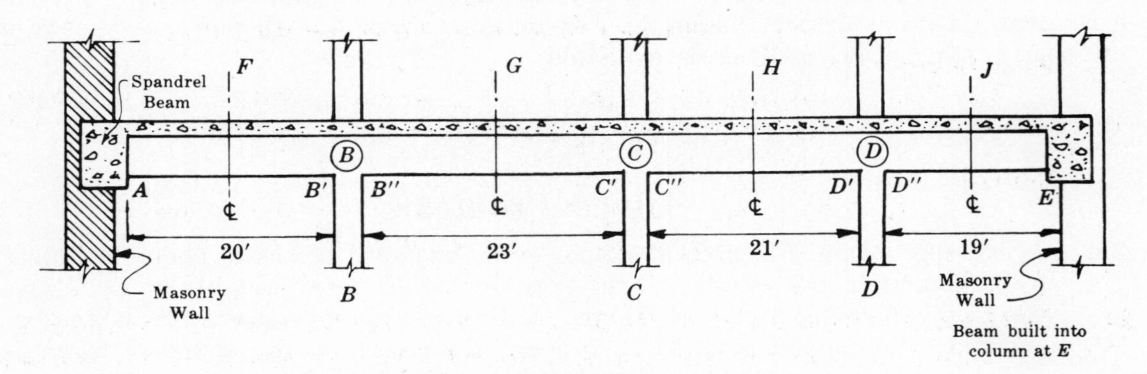

Fig. 2-10

Investigation indicates that the criteria are satisfied for the ACI Code coefficients. Note that the average spans are used for negative moments at supports. Total load $w_{TL} = 1.5 + 2.5 = 4.0$ kips/ft. Moments in foot kips are:

$$M_{BA} = -(4/10)(21.5)^2 = -184.9 \qquad M_{CB} = -(4/11)(22)^2 = -176.0$$
$$M_{DC} = -(4/11)(20)^2 = -145.4 \qquad M_{DE} = -(4/10)(20)^2 = -160.0$$
$$M_{ED} = -(4/16)(19)^2 = -90.25 \qquad M_{AB} = -(4/24)(20)^2 = -66.6$$
$$M_F = (4/11)(20)^2 = 145.4 \qquad M_G = (4/16)(23)^2 = 132.25$$
$$M_H = (4/16)(21)^2 = 110.0 \qquad M_J = (4/14)(19)^2 = 103.0$$
$$M_{CD} = -(4/11)(22)^2 = -176.0$$

Short cuts may be introduced by making use of common quantities in the moment equations. For example,

$$M_H = (21/23)^2(M_G) = 110.0 \quad \text{and} \quad M_{DE} = (11/10)(M_{DC}) = -160.0$$

Shear forces in kips:

$$V_A = 4(20/2) = 40.0 \qquad V_{B'} = 1.15\,V_A = 46.0 \qquad V_{B''} = 4(23/2) = 46.0 = V_{C'}$$
$$V_E = 4(19/2) = 38.0 \qquad V_{D''} = 1.15\,V_E = 43.7 \qquad V_{C''} = 4(21/2) = 42.0 = V_{D'}$$

Reactions in kips:

$$R_A = V_A = 40.0 \qquad R_C = V_{C'} + V_{C''} = 88.0 \qquad R_E = V_E = 38.0$$
$$R_B = V_{B'} + V_{B''} = 92.0 \qquad R_D = V_{D'} + V_{D''} = 85.7$$

2.4. Fig. 2-11 shows the cross-section of a 4.5 inch thick one-way slab. Live load is 100 psf, and the floor covering weighs 1.5 psf. Determine the shear forces and the reactions delivered to the supporting beams. The concrete weighs 150 lb/ft³, or 12.5 psf/inch thickness.

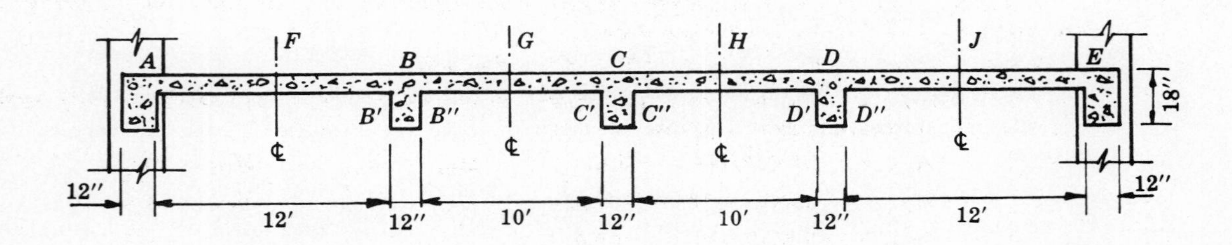

Fig. 2-11

A check will show that the ACI coefficients may be used.

$$w_{TL} = 100 + 1.5 + (4.5)(12.5) = 157.75 \text{ (use 158 lb/ft}^2\text{)}$$

Shear Forces:

$$V_A = V_E = 158(12/2) = 948 \text{ lb/ft} \qquad V_{D''} = V_{B'} = 158(12/2)(1.15) = 1090 \text{ lb/ft}$$

$$V_{B''} = V_{D'} = 158(10/2) = 790 \text{ lb/ft} \qquad V_{C'} = V_{C''} = 158(10/2) = 790 \text{ lb/ft}$$

Reactions:

$$R_D = R_B = V_{B'} + V_{B''} = 1880 \text{ lb/ft} \qquad R_C = V_{C'} + V_{C''} = 1580 \text{ lb/ft}$$

$$R_A = R_E = V_A = V_E = 948 \text{ lb/ft}$$

2.5. Determine the moments at A, B', B'', C', F and G for the slabs of Problem 2.4.

Beams A and E are spandrel beams, so $M = -w(L')^2/24$ in the slab at A and E. For negative moments at supports use average of adjacent clear spans. For positive moment use clear span. Moments in ft-lb are:

$$M_A = -158(12)^2/24 = -948 \qquad M_{B'} = -158(11)^2/10 = -1912$$

$$M_{B''} = (10/11)(M_{B'}) = -1738 \qquad M_{C'} = -158(10)^2/11 = -1437$$

$$M_F = 158(12)^2/14 = 1625 \qquad M_G = 158(10)^2/16 = 988$$

2.6. If beam B of Fig. 2-11 is 20 ft long and is simply supported on brick walls, determine the end reactions and the moment at the center of the beam. (Dead load + live load of beam B = 225 lb/ft.)

End reactions are $w_{TL}(L/2) = 2205(20/2) = 22{,}050$ lb to the wall.

M at center $= wL^2/8 = 2205(20)^2/8 = 110{,}250$ ft-lb.

2.7. Fig. 2-12 shows a continuous slab supported on intermediate beams. Live load is 100 psf and dead load is 50 psf. Determine all slab shear forces and moments and the reactions to the beams.

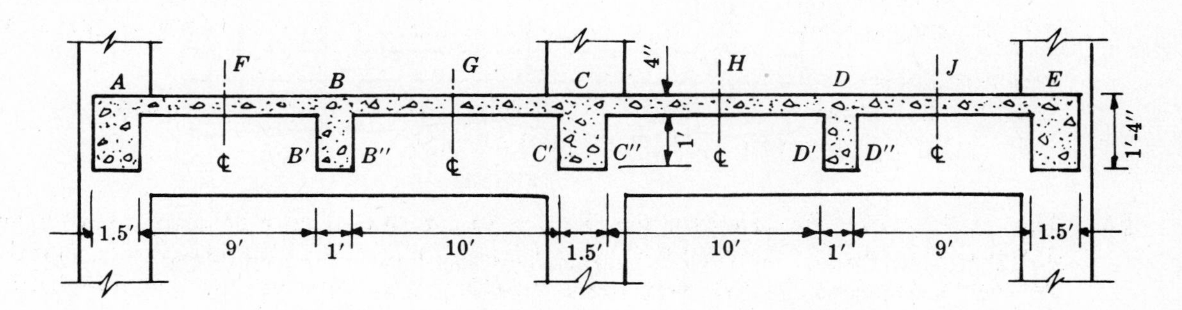

Fig. 2-12

A check shows that ACI coefficients apply. Note symmetry and calculate half of the values. Use moments in ft-lb/ft of supporting beam.

Total load $w_{TL} = w_{LL} + w_{DL} = 100 + 50 = 150$ psf.

For negative moments, note that slab spans do not exceed 10 ft, and equation (*2.7*) applies. Use average of adjacent span lengths.

$$M_{AB} = -150(9)^2/12 = -1013 \qquad M_{ED} = M_{AB} \qquad M_{DE} = M_{BA}$$

$$M_{BA} = -150(9.5)^2/12 = -1128 \qquad M_{CD} = M_{CB} \qquad M_{DC} = M_{BC}$$

$$M_{CB} = -150(10)^2/12 = -1250$$

For positive moments, ends supported on girders, or integral with supports when considering slabs:

$$M_F = M_J = 150(9)^2/14 = 868 \qquad M_G = M_H = 150(10)^2/16 = 938$$

Shear forces in lb/ft of supporting beams:

$$V_A = V_E = 150(9/2) = 675 \qquad V_{B''} = V_{C'} = 150(10/2) = 750$$

$$V_{B'} = V_{D''} = 1.15\,V_A = 776 \qquad V_{C''} = V_{D'} = 150(10/2) = 750$$

Reactions in lb/ft of supporting beam:

$$R_A = R_E = V_A = 675 \qquad R_B = R_D = V_{B'} + V_{B''} = 1526 \qquad R_C = V_{C'} + V_{C''} = 1500$$

2.8. Determine the total design loads for beams A, C and E of Problem 2.7.

Loads are delivered from the slab in lb/ft of beam. Clear spans were used for slab analysis, so added loads over the beams must be determined.

$$\text{Added } DL = (4+12)(1)(150)/12 = 200 \text{ lb/ft}^2 \qquad \text{Added } LL = (100)(1)(1) = 100 \text{ lb/ft}^2$$

$$\text{Added total load} = 200 + 100 = 300 \text{ lb/ft}^2$$

Beams A, C and E: added $w_{TL} = 1.5(300) = 450$ lb/ft

$$w_A = w_E = 450 + 675 = 1125 \text{ lb/ft} \qquad w_C = 450 + 1500 = 1950 \text{ lb/ft}$$

2.9. The two-way slab shown in Fig. 2-13 is an interior slab and is 4″ thick. The live load is 50 psf. Determine the slab moments for designing the middle strip in the short direction at D, C and E. Assume all adjacent slabs are identical to the one shown.

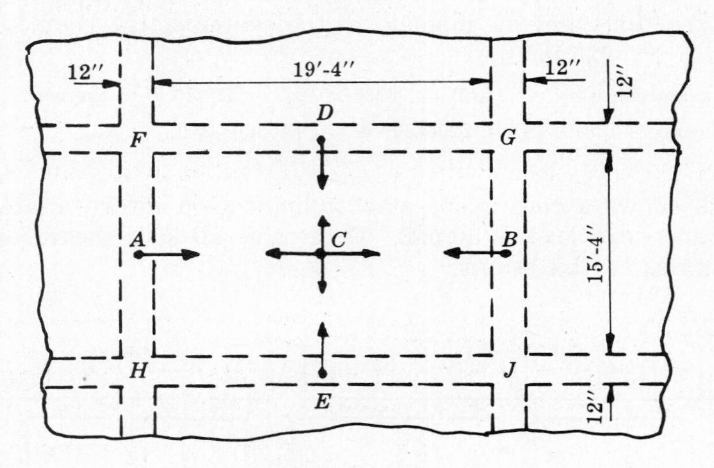

Fig. 2-13

L is lesser of: $(19'-4'')+(2)(4'') = (20'-0'')$ and $(19'-4'')+(2)(12'')/2 = (20'-4'')$. S is lesser of: $(15'-4'')+(2)(4'') = (16'-0'')$ and $(15'-4'')+(2)(12'')/2 = 16'-4''$. Then $L = 20'$ and $S = 16'$, so $m = S/L = 0.8$.

From Page 31, Table 2.1, Case 1, $m = 0.8$ obtain: for negative M, continuous edge, $C = 0.048$; for positive M, $C = 0.036$.

Slab weight $= w_{DL} = 4(12.5) = 50$ psf. $w_{TL} = w_{DL} + w_{LL} = 100$ psf.

At D and E: $M = -Cw_{TL}S^2 = -0.048(100)(16)^2 = -1229$ ft-lb/ft.

At C: $M = +Cw_{TL}S^2 = 0.036(100)(16)^2 = 922$ ft-lb/ft.

2.10. Determine the moments at A, B and C for the middle strip, long direction, for the slab of Fig. 2-13.

For the long direction, for all values of m, $C = 0.033$ for negative moment and $C = 0.025$ for positive moment (Table 2.1, Page 31).

At A and B: $M = -Cw_{TL}S^2 = -0.033(100)(16)^2 = -845$ ft-lb/ft.

At C: $M = +Cw_{TL}S^2 = 0.025(100)(16)^2 = 640$ ft-lb/ft.

Note. The *design moments* may be reduced to those values which exist at the *face of the support,* in accord with the ACI Code. The reductions in moments are usually small and the calculations are lengthy. Thus the reduction is usually unwarranted considering the fact that the design coefficients are somewhat approximate.

2.11. Use the solutions obtained in Problem 2.9 to determine the moments in the column strips, short direction.

The column strip width is (2)(20/4) = 10 ft. The Code requires that the *average moments* in the column strip shall be two-thirds of the moments in the adjacent middle strip. The Code permits varying the moment from a maximum at the edge of the middle strip to a minimum at the support. Two solutions are possible:

(*a*) Using average moment of 2/3 of middle strip moment:

At support (both sides of G or J), $M = -0.667(1229) = -819$ ft-lb/ft.

At center (both sides of B), $M = -0.667(922) = -615$ ft-lb/ft.

(*b*) If the average column strip moments are 2/3 of the middle strip moments, the column strip moments may be varied from $(3/3)M$ to $(1/3)M$.

At G, $M = -0.333(1229) = -409$ ft-lb/ft. At B, $M = 0.333(922) = 307$ ft-lb/ft.

If (*a*) is used, the reinforcement is spaced uniformly over the column strip. If (*b*) is used, the reinforcement spacing is *gradually increased* from the middle strip to the support. The latter provides better load distribution, but the former is more practical from the viewpoint of simplicity in construction.

2.12. Determine the design moment for point D, middle strip, long span for the slab shown in Fig. 2-14. Dead load is 75 psf and live load is 100 psf. The slab has 6″ constant thickness. Beams are 12″ wide.

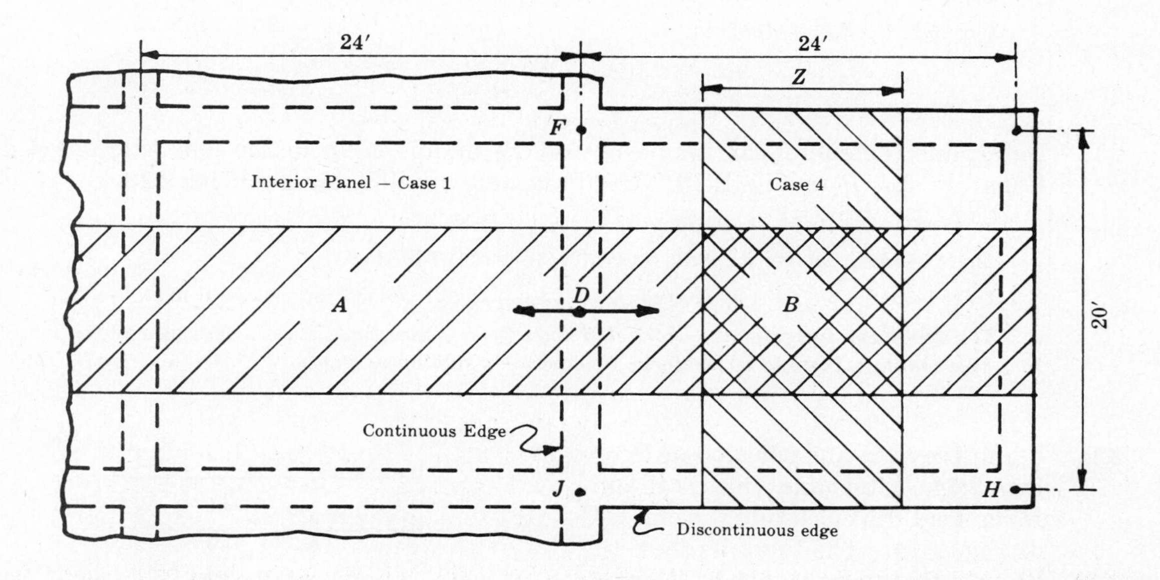

Fig. 2-14

Center to center spans are identical to clear spans plus twice the slab thickness. The values of C (for $M = Cw_{TL}S^2$) are independent of $m = S/L$ for the long span. Moments are in ft-lb/ft width of slab.

From Table 2.1: Slab A, $C_D = 0.033$; slab B, $C_D = 0.058$.

$$M'_D = -0.033(175)(20)^2 = -2310 \qquad M''_D = -0.058(175)(20)^2 = -4060$$

$$\text{Ratio } M'_D/M''_D = 2310/4060 = 0.569 < 0.80; \text{ distribution necessary.}$$

Stiffness I/L is the same for the two slabs. Thus distribution factors are 0.5 for each side. Difference in moments $= |M''_D| - |M'_D| = 1750$. Distribute 2/3 of the difference to each side.

$$M''_D = -4060 + (2/3)(1750)(0.5) = -3477 \qquad M'_D = -2310 - (2/3)(1750)(0.5) = -2893$$

2.13. Determine the width of the middle strips (X_A and X_B) for slabs A and B in the short direction and the width of the column strip (Y) between the two middle strips for the slabs shown in Fig. 2-15.

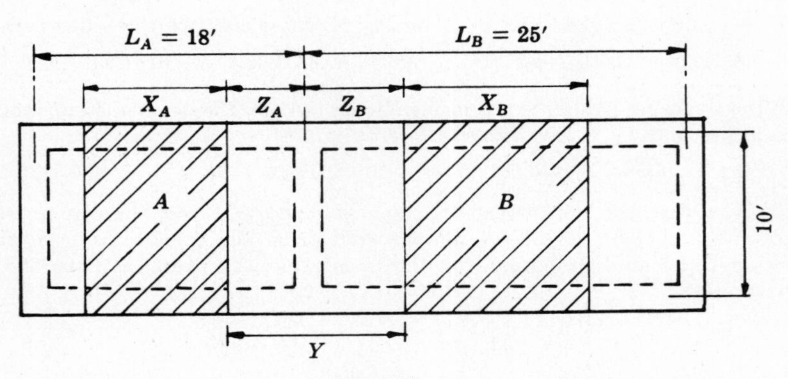

Fig. 2-15

For slab A, $m = 10/18 = 0.555 > 0.5$. For slab B, $m = 10/25 = 0.4 < 0.5$.

For slab A, $X_A = 18/2 = 9'$. For slab B, $X_B = 25 - 10 = 15'$.

For slab A, $Z_A = 18/4 = 4.5'$. For slab B, $Z_B = (25 - 15)/2 = 5'$.

$$Y = 4.5 + 5.0 = 9.5'$$

2.14. Determine the end shear for design of the middle strip in the long and short directions for slab B in Fig. 2-14. Use data and solution for Problem 2.12.

$m = S/L = 20/24 = 0.833$

Short beam: $w'_S = w_{TL}S/3 = 175(20/3) = 1167$ lb/ft

Long beam: $w'_L = (w_{TL}S/3)(3 - m^2)/2 = 1167[3 - (0.833)^2]/2 = 1346$ lb/ft

The end shear force for the *short slab span* $= w'_L$ since that force is delivered to the long beam. In a like manner, the end shear force for the *long slab span* $= w'_S$.

2.15. Using the data and solution of Problem 2.14, determine the total uniform load for designing beam JH of Fig. 2-14. The total depth of the beam is 24″ and the slab is 6″ thick.

The load to beam $JH = w'_L +$ weight of the shaded portion of the beam shown in Fig. 2-16. The area is $2(1) - (0.5)(0.5) = 1.75$ ft^2.

Weight of shaded portion $= 1.75(150)$
$= 263$ lb/ft

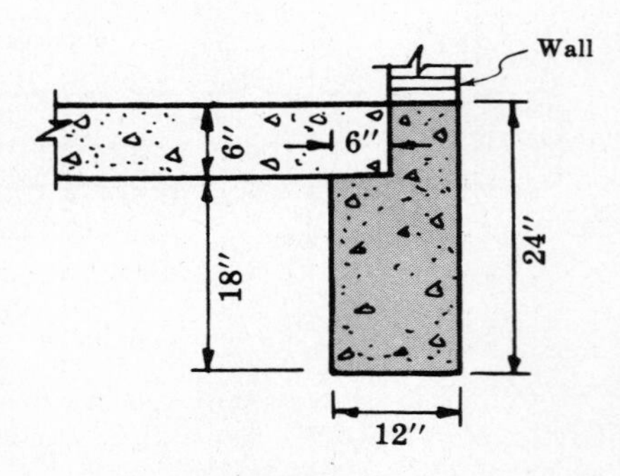

Fig. 2-16

$$w_{TL} = 1346 + 263 = 1609 \text{ lb/ft}$$

Note that additional live load was not added since the wall rests on the slab and live load cannot be placed on this 6″ width. The weight of the wall would also be added to the load on the beam.

Note. In this problem and other problems in this chapter the dead load and live load have been combined. This procedure is normal for use with *working stress design.* When using the ultimate strength design method, the dead load and live load are treated separately since *load factors* differ for dead load and live load.

2.16. Determine the uniform load for designing beam FJ of Fig. 2-14. Use data from Problem 2.14. The beam below the slab is 12″ wide and 18″ deep.

The beam receives load from two sides, as shown in Fig. 2-17.

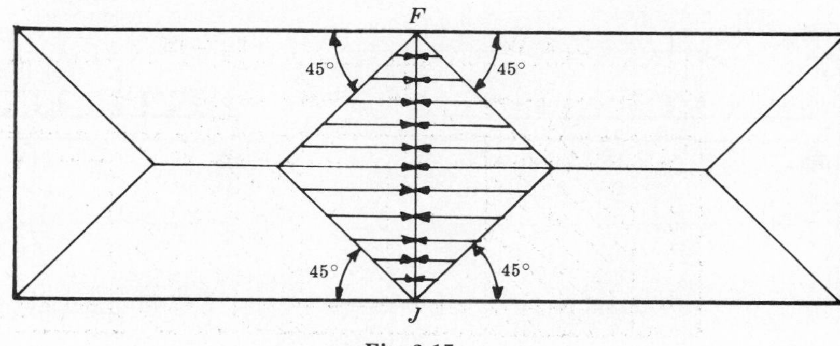

Fig. 2-17

$$w_{TL} = 2w_S' + \text{weight of beam below slab} = 2(1167) + (1.5)(1)(150) = 2559 \text{ lb/ft.}$$

2.17. Derive equation (*2.13*) for the equivalent uniform load w_S' delivered to the short beam (AC) due to a uniform load w psf on the slab. Do not include the weight of the beam. See Fig. 2-18.

The load is delivered to beam AC from area AEC. The maximum load is delivered along line EG and equals $wS/2$. Fig. 2-19(*a*) shows the actual loading condition and Fig. 2-19(*b*) shows the *equivalent* loading condition. (Load from one side only considered.)

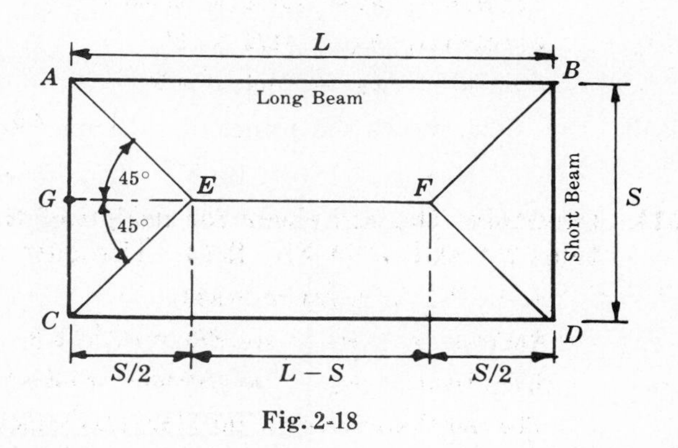

Fig. 2-18

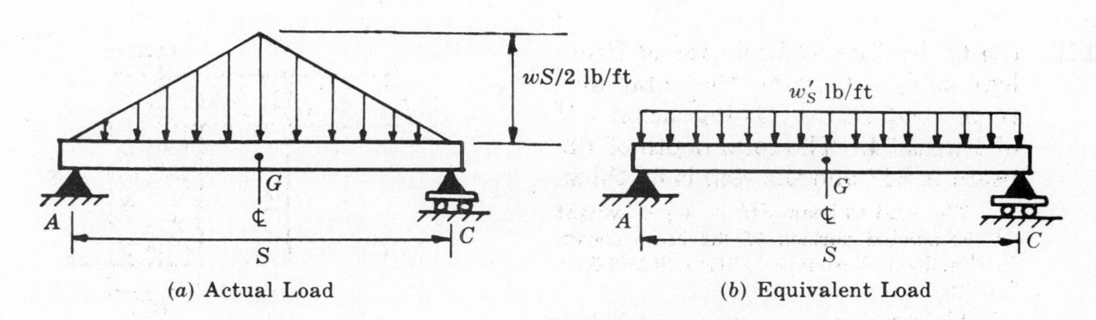

Fig. 2-19

Using the actual load condition, $R_A = (wS/2)(S/2)(1/2) = wS^2/8$.

The moment $M_G = (wS^2/8)(S/2) - (wS^2/8)(S/6) = wS^3/24$.

Using the equivalent load condition, $M_G = w_S'(S^2/8)$.

Considering M_G identical for both conditions, $wS^3/24 = w_S'(S^2/8)$ or $w_S' = wS/3$.

2.18. Derive equation (*2.14*) for the equivalent uniform load w_L' for the *long beam* (*CD*) shown in Fig. 2-18, due to a uniform load w psf. Do not include the weight of the beam.

The actual load condition is shown in Fig. 2-20(*a*) and the equivalent load condition is shown in Fig. 2-20(*b*). (Load from one side only considered.)

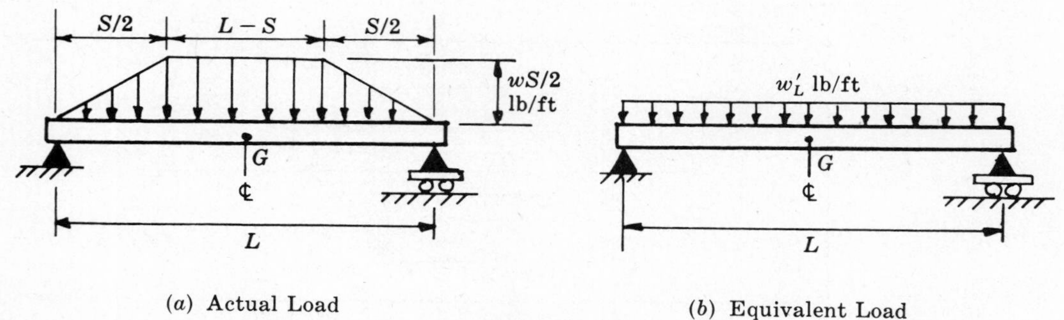

(*a*) Actual Load (*b*) Equivalent Load

Fig. 2-20

Using statics to obtain reactions, the moment M_G can be obtained.

Due to actual load conditions, $M_G = (wS^2/8)(S/3) + (wS/2)[(L-S)/2][(L+S)/4]$. Substituting $m = S/L$, obtain $M_G = (wSL^2)(3-m^2)/48$.

Due to the equivalent load conditions, $M_G = w_L'(L^2/8)$.

Equating the actual moment and the equivalent moment, obtain $w_L' = (wS/3)(3-m^2)/2$.

2.19. Fig. 2-21 shows the joints of a frame and the beam end moments.

$$M_{AB} = -110 \text{ ft-kips} \qquad M_{BA} = -150 \text{ ft-kips} \qquad M_{BC} = -100 \text{ ft-kips}$$

Stiffness factors are shown on the figure. Determine the column moments.

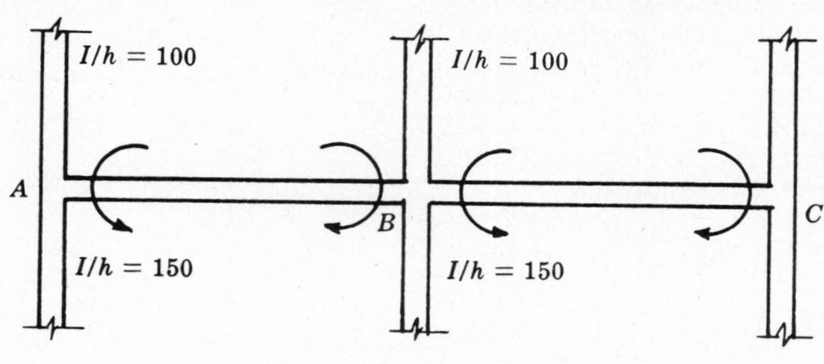

Fig. 2-21

At joint A: For column above, $M = 110[100/(100+150)] = 44$ ft-kips
For column below, $M = 110[150/(100+150)] = 66$ ft-kips

At joint B: For column above, $M = (150-100)[100/(100+150)] = 20$ ft-kips
For column below, $M = (150-100)[150/(100+150)] = 30$ ft-kips

The axial loads in the columns are obtained by summing the shear forces on both sides of the column at all floor levels above the joint in question. The weight of the column above the joint must also be added thereto.

Supplementary Problems

2.20. Determine slab shear force at A, B', B'' and C' in Fig. 2-22. Express answers in kips/ft of supporting beam. *Ans.* $V_A = 2.25$ $V_{B'} = 2.59$ $V_{B''} = 2.38$ $V_{C'} = 2.38$

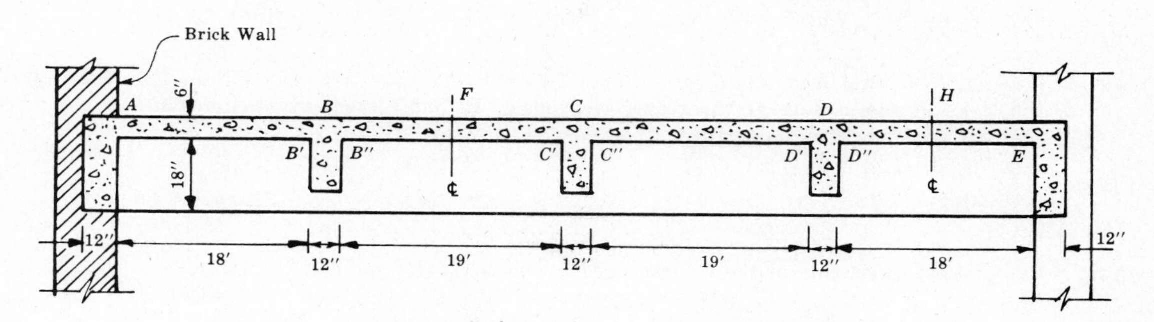

$w_{DL} = 0.075$ kips/ft^2 $w_{LL} = 0.175$ kips/ft^2

Fig. 2-22

2.21. Determine the negative moments in the slab of Fig. 2-22 at A and B'. Express answers in ft-kips/ft of supporting beam. *Ans.* $M_A = -3.38$ $M_{B'} = -8.56$

2.22. Determine the positive moments in the slab of Fig. 2-22 at F and H. Express answers in ft-kips/ft of supporting beam. *Ans.* $M_F = +5.64$ $M_H = +5.79$

2.23. Determine the shear forces and bending moments in the slab of Fig. 2-22 at D'' and E. *Ans.* $V_{D''} = 2.59$ $M_{D''} = -8.56$ $V_E = 2.25$ $M_E = -3.38$

2.24. Determine the width (X) of the middle strip, short direction for each slab shown in Fig. 2-23. Express answers in feet. *Ans.* $X_{\text{I}} = 12.0$ $X_{\text{II}} = 10.0$ $X_{\text{III}} = 10.0$ $X_{\text{IV}} = 10.0$ $X_{\text{V}} = 10.0$

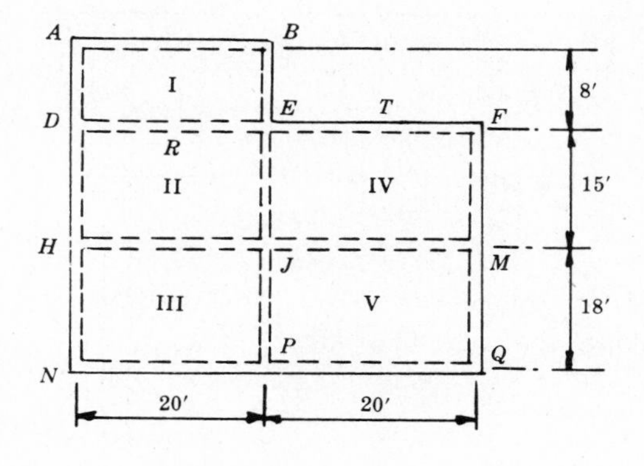

Data for Fig. 2-23:

(1) Slab thickness = 6″

(2) All beams are 12″ wide and project 16″ below the slab

(3) w_{LL} = 150 psf

(4) All beams are built into columns

Fig. 2-23

2.25. Determine the width (Y) of the column strips along DE and HJM for the slabs in Fig. 2-23. Express answers in feet. *Ans.* $Y_{DE} = 5.75$ $Y_{HJM} = 8.25$

2.26. Determine the negative moments for the middle strip, long span at the intersection of slabs II and IV in Fig. 2-23. Express answers in ft-kips/ft. *Ans.* $M_{\text{II}} = -2.07$ $M_{\text{IV}} = -2.48$

2.27. Determine the negative moments for the middle strip, short span at the intersection of slabs I and II in Fig. 2-23. (*Hint.* Distribution is necessary.) Express answers in ft-kips/ft. *Ans.* $M_{\text{I}} = -2.09$ $M_{\text{II}} = -2.60$

2.28. Determine the positive moments for the middle strips (long and short spans) for slab V in Fig. 2-23. Express answers in ft-kips/ft. *Ans.* Long span $M = +2.70$ Short span $M = +3.13$

2.29. Calculate the design dead and live equivalent uniform loads delivered from the slab to beams AD and DH in Fig. 2-23. Express answers in kips/ft.
Ans. Beam AD: $w'_{DL} = 0.20$ $w'_{LL} = 0.40$ Beam DH: $w'_{DL} = 0.375$ $w'_{LL} = 0.75$

2.30. Calculate the total load shear forces and moments for beams DE and EF in Fig. 2-23. Points R and T lie at the centers of the respective spans. Include necessary portions of the beam weight. Express shear in kips, moments in ft-kips.
Ans. $V_D = 24.26$ $V_{ED} = 27.90$ $V_{EF} = 18.5$ $V_F = 16.08$ $M_{DE} = -54.5$
$M_{ED} = -97.5$ $M_{EF} = -64.33$ $M_{FE} = -36.19$ $M_R = +62.5$ $M_T = +41.35$

2.31. Calculate the moments at the column ends for columns GD and DA in Fig. 2-24.
Ans. $M_{GD} = 90$ ft-kips $M_{DG} = 36$ ft-kips $M_{DA} = 60$ ft-kips

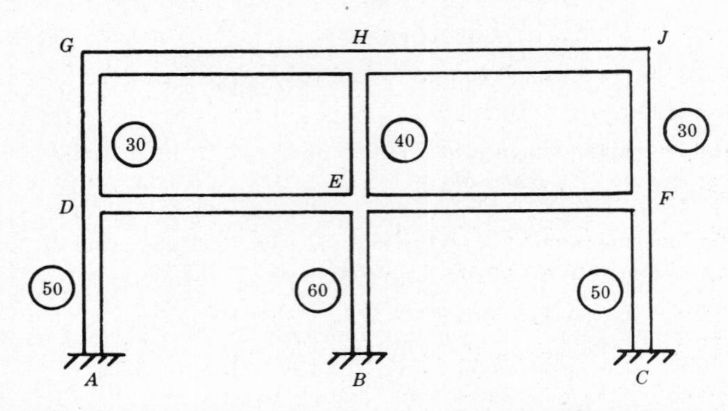

Beam	DE	ED	EF	FE	GH	HG	HJ	JH
Moment ft-kips	−96	−100	−130	−88	−90	−100	−120	−70

Stiffness factors (I/L) are shown in circles on the figure.

Fig. 2-24

2.32. Calculate the moments at the column ends for columns HE and EB in Fig. 2-24.
Ans. $M_{HE} = 20$ ft-kips $M_{EH} = 12$ ft-kips $M_{EB} = 18$ ft-kips

2.33. Calculate the moments at the column ends for columns JF and FC in Fig. 2-24.
Ans. $M_{JF} = 70$ ft-kips $M_{FJ} = 33$ ft-kips $M_{FC} = 55$ ft-kips

Chapter 3

Lateral Loads

FORCES, SHEAR, MOMENTS AND REACTIONS

NOTATION

CK = modified stiffness factor, in^3
h = column height, feet
H_i = lateral load at any joint (i)
I = moment of inertia, in^4
J = a joint factor relating column moments and girder moments
k = a column or beam factor related to stiffness, I/L
K = stiffness factor, (I/L) in^3
L = length of beam, feet
M = bending moment, ft-kips
M^F = fixed-end moment, ft-kips
N = relative moments in girders, ft-kips
Q = relative moments in columns, ft-kips
V = story shear due to lateral loads, kips

GENERAL NOTES

Forces due to wind, earthquakes or soil pressure must be considered in the analysis of structures. Methods used for determining the shears, moments and reactions due to loads caused by such conditions are studied in courses dealing with *statically indeterminate structures.* Such procedures include the slope-deflection method and the moment distribution method, a knowledge of which is prerequisite to the study of reinforced concrete structural analysis.

Courses pertaining to statically indeterminate structures provide only the basic principles one must master in order to analyze reinforced concrete structures for lateral loads. These principles must be expanded in the study of reinforced concrete analysis and design, particularly with reference to lateral loads. Therefore several methods for such analysis will be discussed and illustrated in this chapter.

PORTAL METHOD

In using the portal method for lateral load analysis one assumes that a *point of contraflexure* exists at the midpoint of every member. The structure becomes *statically determinate,* and only the principles of statics are involved in obtaining the shears, moments and axial forces.

This method provides excellent results for the intermediate stories of high-rise structures. The solutions obtained for the upper two stories and the lower two stories are often in error by as much as 50 percent. However, for the purpose of illustration of the method, simple structural frameworks will be utilized in this text.

Assumptions in the Portal Method

(1) A point of contraflexure (zero moment) occurs at the midpoint of all members.

(2) The total horizontal force resisted at any level is the sum of all horizontal forces applied above that level.

(3) The total horizontal force at any level is distributed so that the interior columns resist twice as much horizontal force as the exterior columns.

(4) Vertical forces are obtained by statics using assumptions 1, 2 and 3.

The *free-body diagram* for the analysis of the structure shown in Fig. 3-1(*a*) is illustrated in Fig. 3-1(*b*).

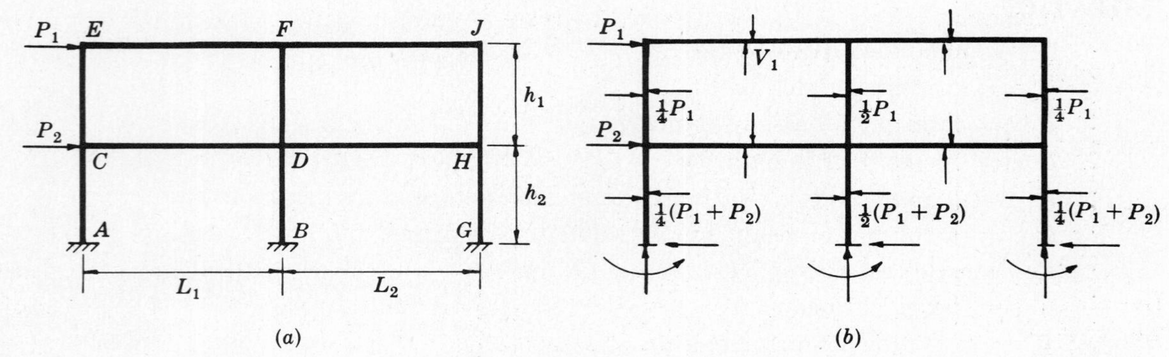

Fig. 3-1

The vertical forces are obtained using statics, as shown in Fig. 3-2. Summing moments about E, we obtain

$$V_1 = P_1 h_1/4L_1 \qquad (3.1)$$

If a section is cut horizontally at E, we obtain from statics

$$M_E = P_1 h_1/8L_1 \qquad (3.2)$$

Similar equations are used for the interior columns, with $H = P/2$ rather than $P/4$ as for the exterior columns.

Since the applied forces P_1 and P_2 are known and the lengths of all members are known, the horizontal and vertical forces and the joint moments may be obtained using statics.

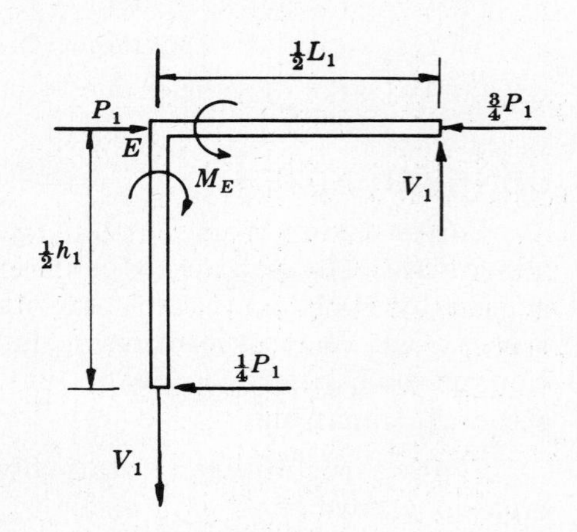

Fig. 3-2

Inaccuracies in the Portal Method

The portal method provides somewhat inaccurate results due to a number of factors, most of which are related to the disregard of the stiffness of the members. Regardless of the inaccuracies involved, the method is important for use in the preliminary stages of a design in order to obtain approximate sizes of members for use with more accurate methods.

SWAY MOMENT DISTRIBUTION

When certain requirements are fulfilled, the *sway moment distribution method* provides an *exact* mathematical solution. The requirements for an exact solution are:

(1) In any two-column bay and at any floor level, the column stiffness factors must be identical.

(2) Loads must be applied laterally at the joints.

(3) The stiffness factors ($K = I/L$) for the girders are modified by multiplying by $C = 6$, i.e. $CK = 6I/L$. The girder carryover factors then become *zero*.

(4) The stiffness factors for the columns are I/h.

(5) All column carryover factors are -1.0.

(6) When the structure consists of more than one bay, the structure must be separated into several single bay structures for which (1) is satisfied. The shear is divided in proportion to the relative stiffnesses of the columns of the single bay structures. The final solution consists of superimposing the separate solutions for the single bay structures, one upon another, in order to obtain the results for the complete original structure.

(7) The fixed-end moment, M^F, on any column is obtained from the equation

$$M^F = -Vh/4 \qquad (3.3)$$

where V = sum of the horizontal forces on the single bay structure, above the mid-height of the story in question, kips

h = height of the story in question, center to center of girders, feet.

In cases where the single bay structures do not have identical column stiffnesses at any given story, it is satisfactory to use the average stiffness of the two columns. Although the solution will not be exact, it will be far more correct than a solution obtained using the portal method. Further, use of the average stiffness factors in the sway moment distribution will provide a solution which is as accurate as any of the approximate methods of gravity load analysis which are usually accepted in practice.

THE FACTOR METHOD

The factor method provides another excellent procedure for determining moments in frames subjected to lateral loads. Although not an exact method, the solution will usually be close to that which would be obtained using a mathematically exact method. The good results are obtained because this method is an abbreviated moment distribution method. All stiffness factors are taken as I/L, without modification for special conditions. The process is outlined as follows:

(1) Sketch the structure, showing centerlines and dimensions.

(2) Record the stiffness factors in a circle on each member. If the analysis is a preliminary one, some assumption must be made for the stiffness of each member. In general, stiffness factors for columns gradually increase from the roof downward. Beam stiffness factors do not vary greatly from one story to another.

(3) Calculate the *sum of the stiffness factors* ($K = I/L$) for all members meeting at any joint. Record these factors in a box at each joint, as shown in Fig. 3-3.

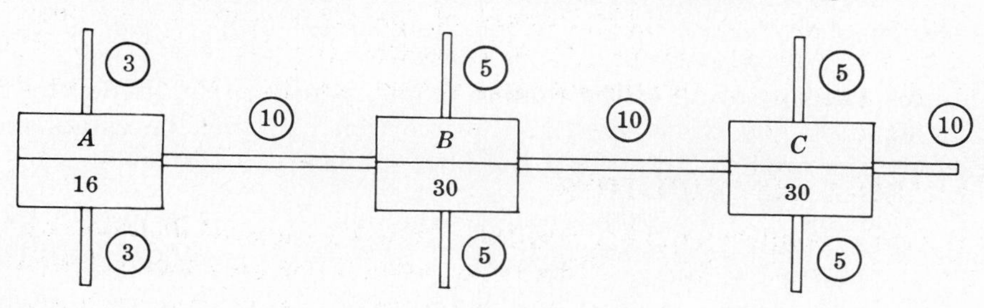

Fig. 3-3

(4) Calculate k *factors* for all beams or girders using

$$k_{AB} = \frac{\Sigma K_{\text{col}} \text{ at } A}{\Sigma K \text{ at } A} \tag{3.4}$$

and

$$k_{BA} = \frac{\Sigma K_{\text{col}} \text{ at } B}{\Sigma K \text{ at } B} \tag{3.5}$$

where ΣK_{col} = sum of I/h values for *columns* meeting at the joint

ΣK = sum of I/L values for *all members* meeting at the joint.

(5) On another line drawing of the centerlines of the members, record the appropriate *beam* k *factors* at the end of each beam. The beam factors will be identical on both sides of any joint.

(6) Calculate and record the column k factors using the equation

$$k_{\text{column}} = 1.0 - k_{\text{beam}} \tag{3.6}$$

The column factors will be identical for the columns above and below any particular joint. For the fixed ends of column bases, use $k = 1.0$. For pinned ends use $k = 0.0$.

(7) A k factor now appears at each end of every beam and column as illustrated in Fig. 3-4.

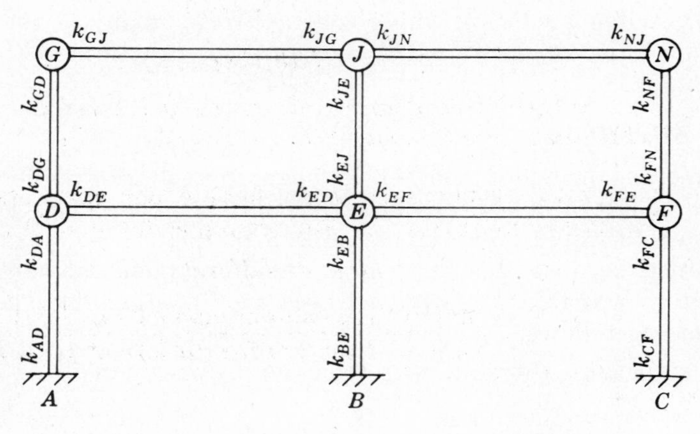

Fig. 3-4

(8) Add one-half of the k factor at the far end of each member to the k factor at the near end, i.e. at joint G. Calculate $k_{GD} + (0.5)k_{DG}$ for the column.

(9) Multiply the sums obtained in step 8 by the value of I/L or I/h for each member. *The resulting products are the Q factors which are proportional to the final moments.*

(10) By statics, the sum of all of the column moments in a story divided by the story height will be equal to the story shear. A proportion can thus be established between the Q factors, the applied forces and the final moments.

Considering Fig. 3-5 below, and noting that ΣH represents the sum of all of the applied horizontal forces between the roof level and the story in question, we can write

$$\Sigma H = [(M_{AD} + M_{DA})/h_{AD}] + [(M_{EB} + M_{BE})/h_{BE}] + [(M_{FC} + M_{CF})/h_{CF}] \tag{3.7}$$

or

$$\Sigma H = H_A + H_B + H_C \tag{3.8}$$

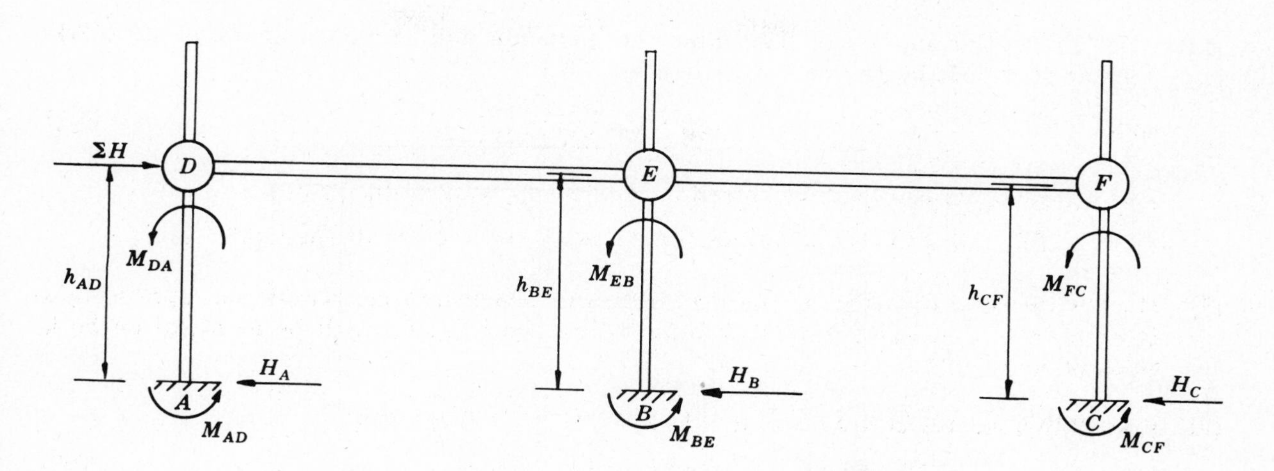

Fig. 3-5

(11) Since the Q factors are proportional to the final moments in the columns, the moments may be obtained by multiplying the Q factor at the end of any column by the *story factor* N, or, for end A of column AD,

$$M_{AD} = N(Q_{AD}) \tag{3.9}$$

in which

$$N = \Sigma H(h)/\Sigma Q_{\text{columns}} \tag{3.10}$$

The value $\Sigma Q_{\text{columns}}$ refers to the sum of all of the Q factors at the top and bottom of all columns *at the story being considered.*

(12) The girder Q factors are proportional to the final girder moments, so these values may be obtained using the joint factor J. For example, for end D of girder DE,

$$M_{DE} = J_D(Q_{DE}) \tag{3.11}$$

in which

$$J_D = \frac{\text{sum of final column moments at joint } D}{\text{sum of } Q \text{ factors for girders at joint } D} \tag{3.12}$$

PRELIMINARY AND FINAL ANALYSIS

It is always necessary to make a preliminary analysis using assumed dimensions and stiffness factors when employing either the cantilever moment distribution or the Q factor method.

The portal method provides a simple means of determining moments and axial loads for use in making a preliminary design. Thereafter, the more precise methods may be employed to obtain a reasonably accurate solution for the lateral load problem.

Solved Problems

3.1. Use the *portal method* to determine the moments and reactions at all joints in the upper story of the frame shown in Fig. 3-6.

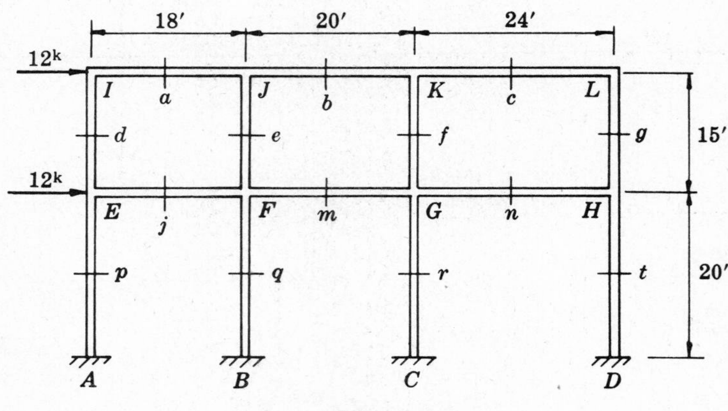

Fig. 3-6

The *free-body diagrams* for the elements of the upper story are shown in Fig. 3-7. Points of contraflexure are assumed to exist at the midpoints of all members. The upper story shear force (12 kips) is divided so that the exterior columns resist 2 kips each and the interior columns resist 4 kips each.

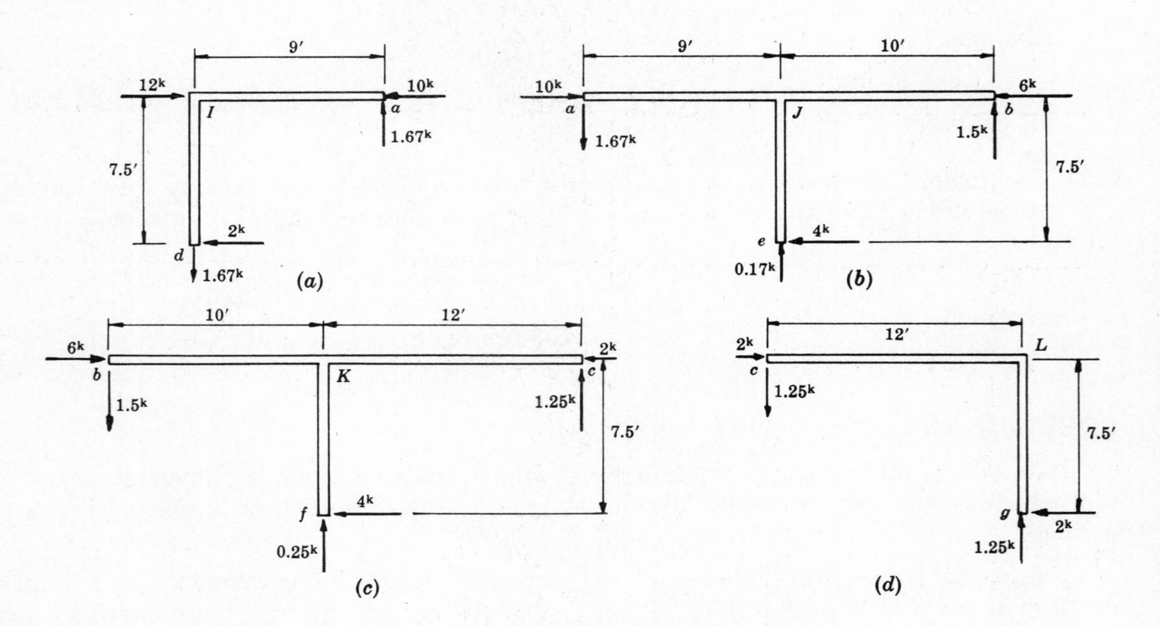

Fig. 3-7

From (*a*), ΣM about $I = 0$. $V_a = 2(7.5)/9 = 1.67$ kips.

From (*b*), ΣM about $J = 0$. $4(7.5) - 1.67(9) - 10V_b = 0$ and $V_b = 1.5$ kips.

From (*c*), ΣM about $K = 0$. $4(7.5) - 1.5(10) - 12V_c = 0$ and $V_c = 1.25$ kips.

From (*d*), ΣM about $c = 0$. $2(7.5) - 12V_g = 0$ and $V_g = 1.25$ kips.

The moments may now be computed using statics. Moments are in ft-kips.

$M_{Ia} = 9(1.67) = 15.0$ $M_{Lc} = 1.25(12) = 15.0$ $M_{Jb} = 1.5(10) = 15.0$ $M_{Je} = -4(7.5) = -30.0$

$M_{Ja} = 9(1.67) = 15.0$ $M_{Id} = -2(7.5) = -15.0$ $M_{Lg} = -2(7.5) = -15.0$

3.2. Given a 3-bay, 2-story structural frame as shown in Fig. 3-8, with lateral loads of 12 kips at the second floor and roof joints. Determine all moments using the *sway moment distribution* method.

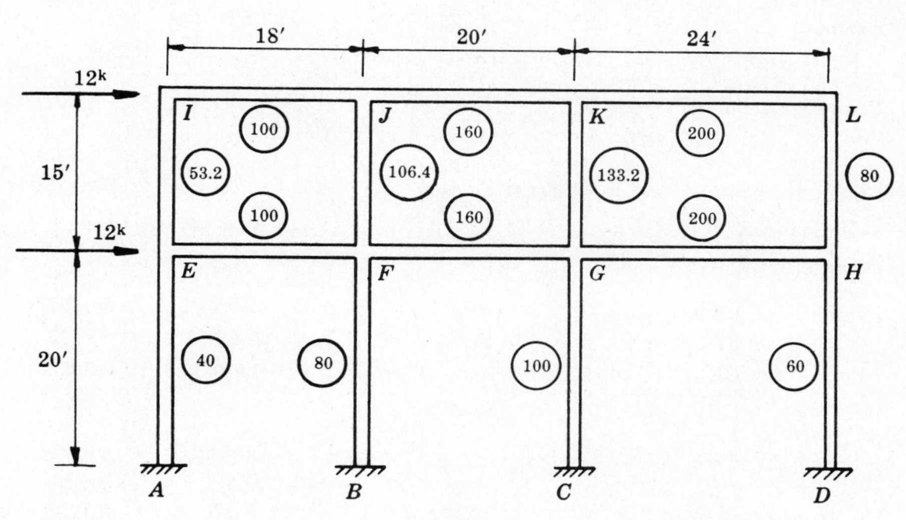

Fig. 3-8

(1) Separate the frame into three single bay frames having *symmetrical column stiffness factors* (or nearly so) at any floor level. This is shown in Fig. 3-9.

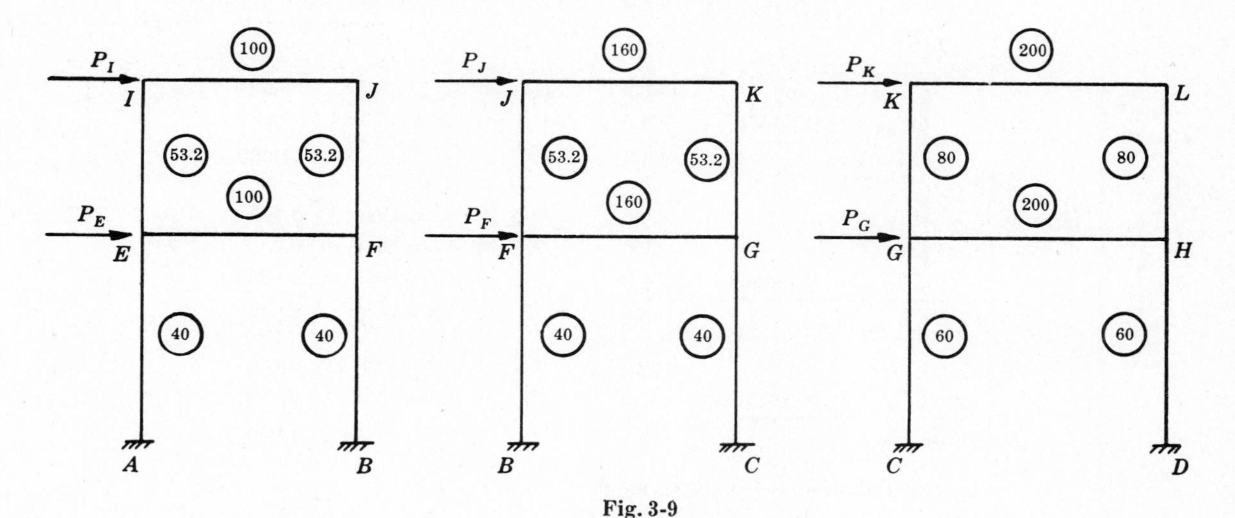

Fig. 3-9

(2) Calculate the load distribution constants at the roof level:

$$P_I = 53.2/(53.2 + 53.2 + 80) \times 12 = 3.425 \text{ kips}$$
$$P_J = 53.2/(53.2 + 53.2 + 80) \times 12 = 3.425 \text{ kips}$$
$$P_K = 80/(53.2 + 53.2 + 80) \times 12 = 5.150 \text{ kips}$$
$$\text{Total} = 12.000 \text{ kips}$$

(3) Calculate load distribution constants at second floor level:

$$P_E = 40/(40 + 40 + 60) \times 12 = 3.428 \text{ kips}$$
$$P_F = 40/(40 + 40 + 60) \times 12 = 3.428 \text{ kips}$$
$$P_G = 60/(40 + 40 + 60) \times 12 = 5.144 \text{ kips}$$
$$\text{Total} = 12.000 \text{ kips}$$

(4) Calculate fixed-end moments on columns, $M^F = -Vh/4$:

Frame 1 Col's. EI, FJ; $M^F = -(3.425)(15/4) = -12.84$ ft-kips
Col's. AE, BF; $M^F = -(3.425 + 3.428)(20/4) = -34.27$ ft-kips

Frame 2 Col's. FJ, GK; $M^F = -(3.425)(15/4) = -12.84$ ft-kips
Col's. BF, CG; $M^F = -(3.425 + 3.428)(20/4) = -34.27$ ft-kips

Frame 3 Col's. GK, HL; $M^F = -(5.15)(15/4) = -19.31$ ft-kips
Col's. CG, DH; $M^F = -(5.15 + 5.144)(20/4) = -51.47$ ft-kips

(5) Calculate modified (sway) stiffness factors (CK) and distribution factors (D):
Columns: $CK = K = I/h$
Girders: $CK = 6K = 6I/L$

Frame 1

$CK_{EA} = 40.0$ | $D_{EA} = 40/693.2 = 0.0577$
$CK_{EI} = 53.2$ | $D_{EI} = 53.2/693.2 = 0.0765$
$CK_{EF} = (6)(100) = 600.0$ | $D_{EF} = 600/693.2 = 0.8658$
$\Sigma CK_E = 693.2$ | $\Sigma = 1.0000$

$CK_{IE} = 53.2$ | $D_{IE} = 53.2/653.2 = 0.0814$
$CK_{IJ} = (6)(100) = 600.0$ | $D_{IJ} = 600/653.2 = 0.9186$
$\Sigma CK_I = 653.2$ | $\Sigma = 1.0000$

Frame 2

$CK_{FB} = 40.0$ | $D_{FB} = 40/1053.2 = 0.038$
$CK_{FJ} = 53.2$ | $D_{FJ} = 53.2/1053.2 = 0.051$
$CK_{FG} = (6)(160) = 960.0$ | $D_{FG} = 960/1053.2 = 0.911$
$\Sigma CK_F = 1053.2$ | $\Sigma = 1.000$

$CK_{JF} = 53.2$ | $D_{JF} = 53.2/1013.2 = 0.053$
$CK_{JK} = (6)(160) = 960.0$ | $D_{JK} = 960/1013.2 = 0.947$
$\Sigma CK_J = 1013.2$ | $\Sigma = 1.000$

Frame 3

$CK_{GC} = 60.0$ | $D_{GC} = 60/1340 = 0.045$
$CK_{GK} = 80.0$ | $D_{GK} = 80/1340 = 0.060$
$CK_{GH} = (6)(200) = 1200.0$ | $D_{GH} = 1200/1340 = 0.895$
$\Sigma CK_G = 1340.0$ | $\Sigma = 1.000$

$CK_{KG} = 80.0$ | $D_{KG} = 80/1280 = 0.062$
$CK_{KL} = (6)(200) = 1200.0$ | $D_{KL} = 1200/1280 = 0.938$
$\Sigma CK_K = 1280.0$ | $\Sigma = 1.000$

The moment distribution is tabulated as shown in Fig. 3-10.

Frame 1	I (0.0814)	E (0.0765)	E (0.0577)	A (0.0)
M^F	−12.84	−12.84	−34.27	−34.27
	−3.60	+3.60	+2.72	−2.72
	+1.34	−1.34	+0.08	−0.08
		+0.10		
	−15.10	−10.48	−31.47	−37.07

Frame 2	J′ (0.053)	F′ (0.051)	F′ (0.038)	B′ (0.0)
M^F	−12.84	−12.84	−34.27	−34.27
	−2.40	+2.40	+1.79	−1.79
	+0.81	−0.81	+0.03	−0.03
	−14.43	−11.21	−32.45	−36.09

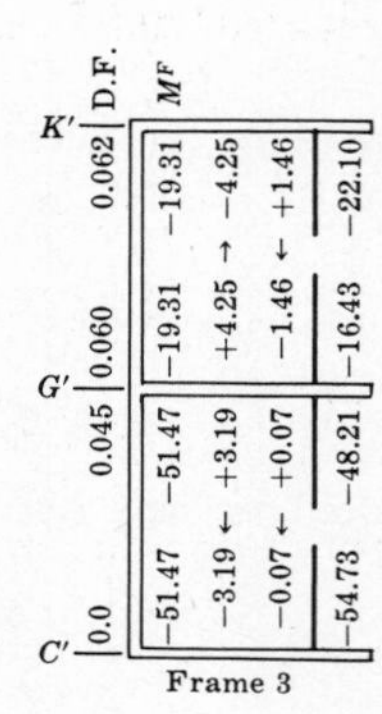

Fig. 3-10

The beam moments are obtained as the algebraic sum of the column moments at the joint in question, as shown in Fig. 3-11.

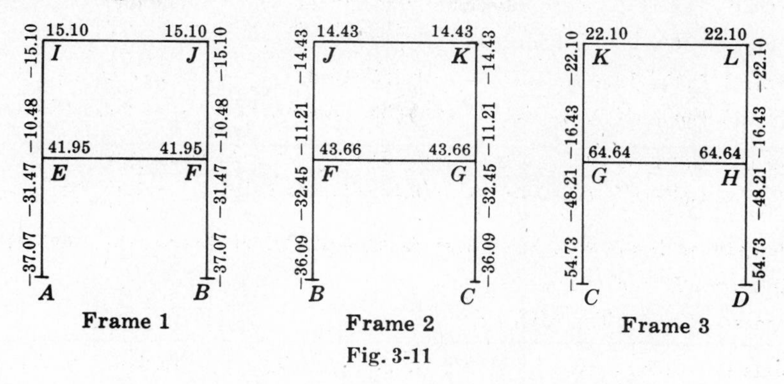

Fig. 3-11

The final moments are obtained by putting the separate bents together (i.e. superposition) and adding the corresponding moments algebraically.

The final solution is shown on Fig. 3-12.

Since the cantilever moment distribution method provides an *exact solution* in this case, it is of interest to compare the results of this problem with those of Problem 3.1. Note the excellent results obtained at joints I and J using the portal method. Also, note that the portal method yields an error of about 50 percent at joint L, because of the differences in stiffness factors for the various columns.

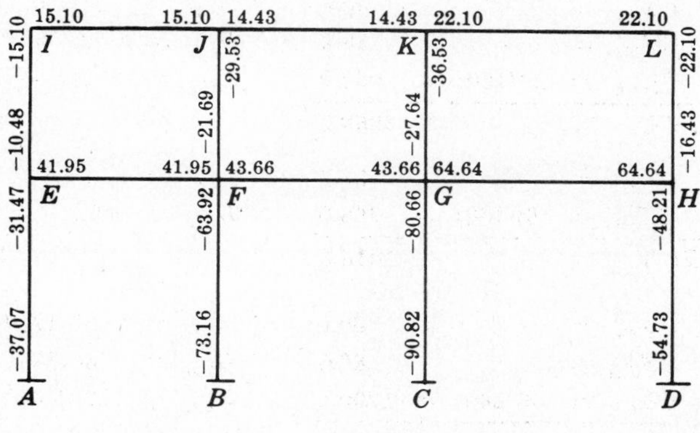

Fig. 3-12

3.3. A two-story, three-bay concrete frame has been designed on the basis of a preliminary structural analysis. The dimensions and preliminary stiffness factors (I/L) are recorded (encircled) in Fig. 3-13. Determine the beam and column end moments due to lateral loads of 12 kips applied at the second story and at the roof. Use the Q factor method.

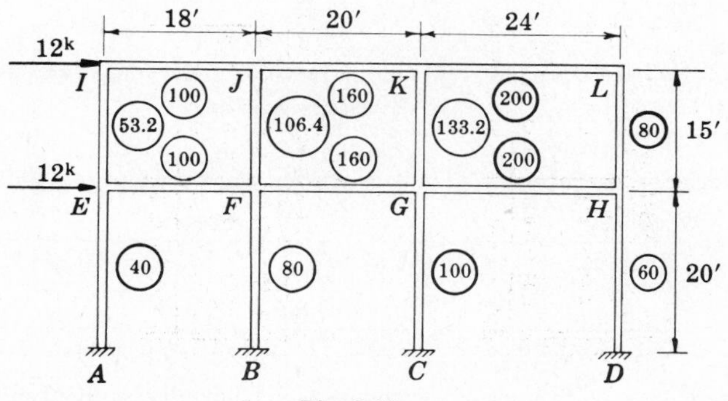

Fig. 3-13

The calculations are made in accord with the outline presented in the theory section, items (1) through (12). The results are shown on the figures which follow.

Calculation of relative moments:

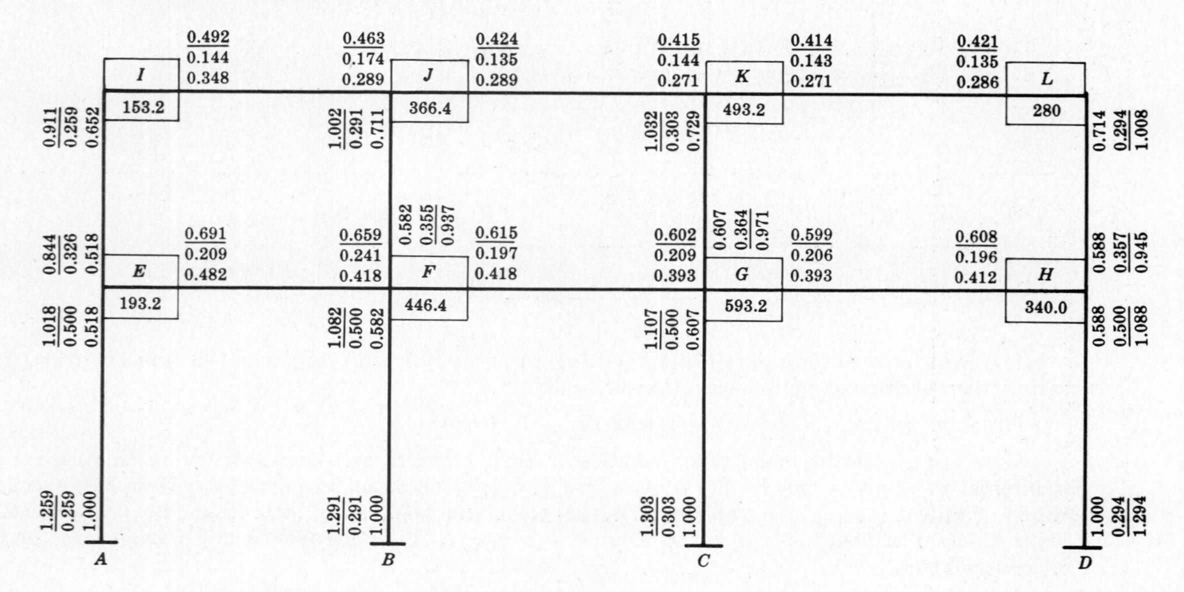

Fig. 3-14

Column factors $\times$ stiffness of element $= M'_C =$ Relative column M

Girder factors $\times$ stiffness of element $= M'_G =$ Relative girder M

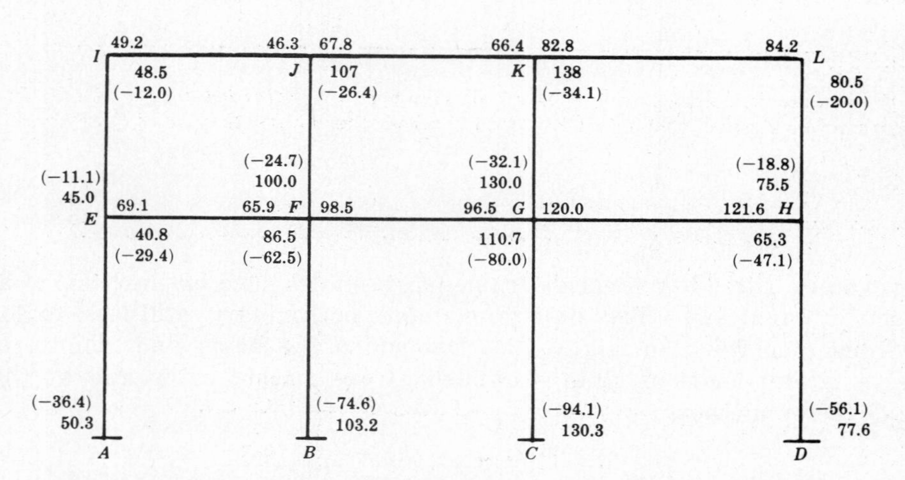

Fig. 3-15. Relative Moments or Q factors, and Final Moments in Parentheses. (All values are negative.)

The N factors for the upper and lower stories are obtained as follows:

$$N_U = \frac{H(h)}{\Sigma Q \text{ cols.}} = \frac{(12)(15)}{48.5 + 45 + 107 + 100 + 138 + 130 + 80.5 + 75.5} = 0.248$$

$$N_L = \frac{(24)(20)}{40.8 + 50.3 + 86.5 + 103.2 + 110.7 + 130.3 + 65.3 + 77.6} = 0.723$$

The final moments at the column ends are obtained by multiplying the appropriate Q factors and the N factors. The column moments in ft-kips are recorded in parentheses in Fig. 3-15.

Joint factors for obtaining final moments at the girder ends are calculated for each joint using the equation

$$J = \frac{\Sigma \text{ column moments at joint}}{\Sigma\ Q \text{ factors of girders at joint}}$$

The girder moments are calculated as the product of the J factor and the Q factor at the end of the girder. The results are listed in Table 3.1.

TABLE 3.1

Joint	ΣM_{col} ft-kips	ΣQ-factors (Girders)	J factor	Girder Moments ft-kips
E	40.5	69.1	0.590	$M_{EF} = (0.590)(69.1) = 40.8$
F	87.2	164.4	0.532	$M_{FE} = (0.532)(65.9) = 35.0$ $M_{FG} = (0.532)(98.5) = 52.4$
G	112.1	216.5	0.520	$M_{GF} = (0.520)(96.5) = 50.2$ $M_{GH} = (0.520)(120.0) = 62.4$
H	65.9	121.6	0.543	$M_{HG} = (0.543)(121.6) = 66.0$
I	12.0	49.2	0.246	$M_{IJ} = (0.246)(49.2) = 12.1$
J	26.4	114.1	0.235	$M_{JI} = (0.235)(46.3) = 10.9$ $M_{JK} = (0.235)(67.8) = 15.9$
K	34.1	149.2	0.232	$M_{KJ} = (0.232)(66.4) = 15.4$ $M_{KL} = (0.232)(82.8) = 19.2$
L	20.0	84.2	0.240	$M_{LK} = (0.24)(84.2) = 20.2$

The final moments at the ends of all members are recorded in Fig. 3-16. Moments obtained using the cantilever moment distribution method are shown thereon in parentheses.

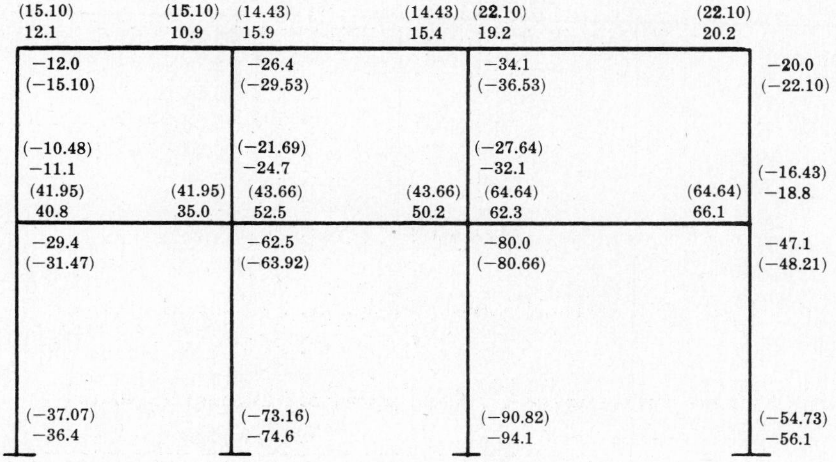

(15.10) 12.1 | (15.10) 10.9 | (14.43) 15.9 | (14.43) 15.4 | (22.10) 19.2 | (22.10) 20.2

−12.0 (−15.10) | −26.4 (−29.53) | −34.1 (−36.53) | −20.0 (−22.10)

(−10.48) −11.1 | (−21.69) −24.7 | (−27.64) −32.1 | (−16.43) −18.8

(41.95) 40.8 | (41.95) 35.0 | (43.66) 52.5 | (43.66) 50.2 | (64.64) 62.3 | (64.64) 66.1

−29.4 (−31.47) | −62.5 (−63.92) | −80.0 (−80.66) | −47.1 (−48.21)

(−37.07) −36.4 | (−73.16) −74.6 | (−90.82) −94.1 | (−54.73) −56.1

Fig. 3-16

One may note the results obtained using the Q factor method are generally closer to the true solution than those obtained using the Portal method.

Supplementary Problems

3.4. Use the *Portal method* to find the moments in all of the columns and girders of the frame shown in Fig. 3-17. The answers are listed in Table 3.2.

3.5. Use the *Cantilever moment distribution method* to find the moments in all of the columns and girders of the frame shown in Fig. 3-17. The answers are given in Table 3.2.

3.6. Use the *Q factor method* to find the moments in the columns and girders of the frame shown in Fig. 3-17. The answers are listed in Table 3.2.

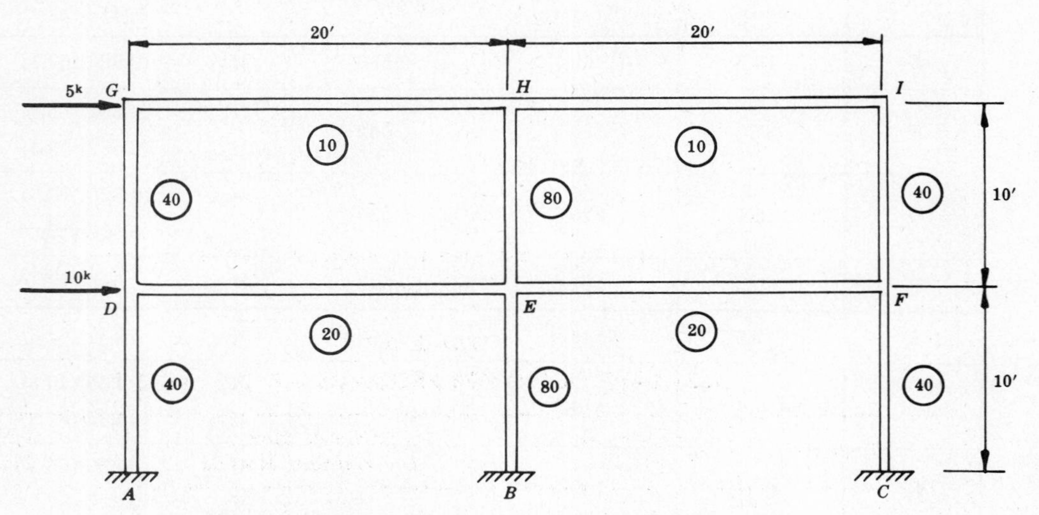

Stiffness factors are shown in circles on each member.

Fig. 3-17

TABLE 3.2

MOMENTS (ft-kips) FOR PROBLEMS 3.4, 3.5 and 3.6

Moment	Portal Method	Cantilever Moment Distribution Method	Q Factor Method
AD	−18.75	−24.72	−22.92
DA	−18.75	−12.78	−14.58
DG	−6.25	−5.14	−6.25
GD	−6.25	−7.33	−6.25
BE	−37.50	−49.44	−45.83
EB	−37.50	−25.56	−29.16
EH	−12.50	−10.28	−12.50
HE	−12.50	−14.64	−12.50
CF	−18.75	−24.72	−22.92
FC	−18.75	−12.78	−14.58
FI	−6.25	−5.14	−6.25
IF	−6.25	−7.33	−6.25
DE	+25.00	+17.92	+20.83
ED	+25.00	+17.92	+20.83
EF	+25.00	+17.92	+20.83
FE	+25.00	+17.92	+20.83
GH	+6.25	+7.33	+6.25
HG	+6.25	+7.33	+6.25
HI	+6.25	+7.33	+6.25
IH	+6.25	+7.33	+6.25

3.7. Find all end moments in the frame shown in Fig. 3-18 using the *Portal method.* Answers are listed in Table 3.3.

3.8. Solve Problem 3.7 using the *Cantilever moment distribution method.*

3.9. Solve Problem 3.7 using the *Q factor method.*

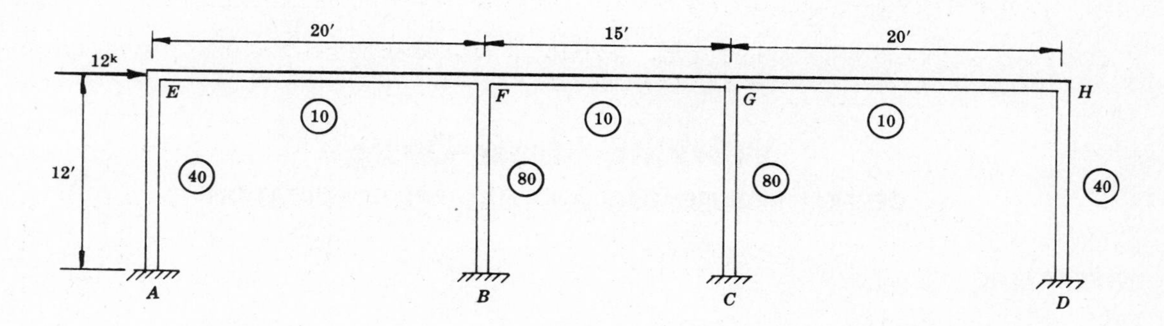

Stiffness factors are shown in circles on each member.

Fig. 3-18

TABLE 3.3

MOMENTS (ft-kips) FOR PROBLEMS 3.7, 3.8 and 3.9

Moment	Portal Method	Cantilever Moment Distribution Method	Q Factor Method
AE	−12.0	−16.8	−14.67
EA	−12.0	−7.4	−9.33
BF	−24.0	−33.6	−29.33
FB	−24.0	−14.8	−18.66
CG	−24.0	−33.6	−29.33
GC	−24.0	−14.8	−18.66
DH	−12.0	−16.8	−14.67
HD	−12.0	−7.4	−9.33
EF	+12.0	+7.4	+9.33
FE	+12.0	+7.4	+9.33
FG	+12.0	+7.4	+9.33
GF	+12.0	+7.4	+9.33
GH	+12.0	+7.4	+9.33
HG	+12.0	+7.4	+9.33

Chapter 4

Working Stress Design

GENERAL REQUIREMENTS AND FLEXURAL COMPUTATIONS

NOTATION

A_s = area of tension reinforcement

A'_s = area of compression reinforcement

b = width of compression face of flexural member

b' = width of web in I and T-sections

d = distance from extreme compression fiber to centroid of tension reinforcement

d' = distance from extreme compression fiber to centroid of compression reinforcement

E_c = modulus of elasticity of concrete

E_s = modulus of elasticity of steel = 29,000,000 psi

f_c = compressive stress in concrete

f'_c = compressive strength of concrete

f_y = yield strength of reinforcement

I = moment of inertia of beam or column

K = stiffness factor = EI/l

l = span length of slab or beam

n = ratio of modulus of elasticity of steel to modulus of elasticity of concrete

p = ratio of area of tension reinforcement to effective area of concrete in rectangular beam or in web of flanged member

p' = A'_s/bd

t = thickness of flexural member

v = shear stress

v_c = shear stress carried by concrete

w = weight of concrete, lb/ft^3

INTRODUCTION

The vast majority of reinforced concrete structures in America have been proportioned based on a straight-line theory which is called working stress design. And although ultimate strength design techniques are rapidly supplanting working stress design, the student or the designer should be proficient in both. Hence the fundamentals of the straight-line theory are presented here. Ultimate strength design will be discussed thoroughly in other chapters.

When using working stress design techniques, members are proportioned so that they may sustain the anticipated real loads induced (working or design loads) without the stresses in the concrete or reinforcing exceeding the proportional limits of the individual materials. Although the stress-strain diagram of concrete does not exhibit an initial straight-line portion, it is still assumed that Hooke's law does apply to concrete.

This leads to the basic assumptions in working stress design required for the following derivations and discussion:

(1) Plane sections before bending remain plane after bending.

(2) Both the concrete and reinforcing steel obey Hooke's law.

(3) Strain is proportional to the distance from the neutral axis. See (1) above.

(4) The tensile strength of the concrete is neglected.

(5) Perfect bond or adhesion is developed between the concrete and reinforcing steel so that there is no slippage between the two materials.

(6) The other basic assumptions concerning deformation and flexure of homogeneous members are valid.

(7) The modulus of elasticity of concrete shall be $w^{1.5}\,33\sqrt{f_c'}$ in psi.

(8) The modulus of elasticity of steel reinforcement shall be 29,000,000 psi.

Unfortunately the ordinary formulas for computing stresses and deformations in members of homogeneous, isotropic materials do not apply to the design of reinforced concrete. Many rather complex expressions can be developed for the members which are a composite of concrete and reinforcing steel.

It will become apparent that the developed expressions are rather lengthy and would prove to be quite cumbersome in the routine design of reinforced concrete. Tables, charts and other design aids will be discussed and derived which will greatly facilitate the solution of the derived formulas. However, so that there will be a thorough understanding of the principles involved, the fundamentals will be studied first and then we will consider the proportioning of members by using various design aids.

ALLOWABLE STRESSES

In working stress design all of the margin of safety is provided for by the fact that the calculated stresses in the members are such that they are considerably below the yield stress or ultimate stress of the various materials when the member is subjected to the anticipated service loads of the structure.

Table 4.1 and 4.1(*a*) give the allowable stress for concrete for working stress design as contained in the 1963 ACI Building Code. Unless otherwise noted, these will be the allowable stresses used throughout the straight-line theory portions of this book. These tables are reproduced with permission of the ACI.

TABLE 4.1

ALLOWABLE STRESSES IN CONCRETE

Description		Allowable stresses: For any strength of concrete in accordance with Section 502	For strength of concrete shown below: $f'_c =$ 2500 psi	$f'_c =$ 3000 psi	$f'_c =$ 4000 psi	$f'_c =$ 5000 psi
Modulus of elasticity ratio: n		$\frac{29{,}000{,}000}{w^{1.5}\,33\sqrt{f'_c}}$				
For concrete weighing 145 lb per cu ft (see Section 1102)	n		10	9	8	7
Flexure: f_c						
Extreme fiber stress in compression	f_c	$0.45f'_c$	1125	1350	1800	2250
Extreme fiber stress in tension in plain concrete footings and walls	f_c	$1.6\sqrt{f'_c}$	80	88	102	113
Shear: v (as a measure of diagonal tension at a distance d from the face of the support)						
Beams with no web reinforcement	v_c	$1.1\sqrt{f'_c}$	55	60	70	78
Joists with no web reinforcement	v_c	$1.2\sqrt{f'_c}$	61	66	77	86
Members with vertical or inclined web reinforcement or properly combined bent bars and vertical stirrups	v	$5\sqrt{f'_c}$	250	274	316	354
Slabs and footings (peripheral shear, Section 1207)	v_c	$2\sqrt{f'_c}$	100	110	126	141
Bearing: f_c						
On full area		$0.25f'_c$	625	750	1000	1250
On one-third area or less*		$0.375f'_c$	938	1125	1500	1875

*This increase shall be permitted only when the least distance between the edges of the loaded and unloaded areas is a minimum of one-fourth of the parallel side dimension of the loaded area. The allowable bearing stress on a reasonably concentric area greater than one-third but less than the full area shall be interpolated between the values given.

TABLE 4.1(a)

ALLOWABLE STRESSES IN CONCRETE (METRIC)

Description		Allowable stresses, kg per sq cm				
		For any strength of concrete in accordance with Section 502	For strength of concrete shown below			
			$f'_c =$ 176 kg per sq cm	$f'_c =$ 211 kg per sq cm	$f'_c =$ 281 kg per sq cm	$f'_c =$ 352 kg per sq cm
Modulus of elasticity ratio: n		$\frac{2{,}039{,}000}{w^{1.5}\,4.270\sqrt{f'_c}}$				
For concrete weighing 2.323 t per cu m (see Section 1102)	n	$\frac{478{,}000}{w^{1.5}\sqrt{f'_c}}$	10	9	8	7
Flexure: f_c						
Extreme fiber stress in compression	f_c	$0.45f'_c$	79	94.8	126.5	158.2
Extreme fiber stress in tension in plain concrete footings and walls	f_c	$0.424\sqrt{f'_c}$	5.63	6.16	7.11	7.95
Shear: v (as a measure of diagonal tension at a distance d from the face of the support)						
Beams with no web reinforcement	v_c	$0.292\sqrt{f'_c}$	3.87	4.24	4.89	5.49
Joists with no web reinforcement	v_c	$0.318\sqrt{f'_c}$	4.22	4.62	5.33	5.96
Members with vertical or inclined web reinforcement or properly combined bent bars and vertical stirrups	v	$1.325\sqrt{f'_c}$	17.6	19.3	22.2	24.9
Slabs and footings (peripheral shear, Section 1207)	v_c	$0.530\sqrt{f'_c}$	7.03	7.73	8.84	9.92
Bearing: f_c						
On full area		$0.25f'_c$	44	52.7	70.3	88.0
On one-third area or less*		$0.375f'_c$	66	79.1	105.5	132.0

*This increase shall be permitted only when the least distance between the edges of the loaded and unloaded areas is a minimum of one-fourth of the parallel side dimension of the loaded area. The allowable bearing stress on a reasonably concentric area greater than one-third but less than the full area shall be interpolated between the values given.

Special provisions for lightweight structural concrete concerning values of n and v_c will be discussed in later sections.

In addition to the above, the 1963 ACI Building Code gives the following for allowable stresses in the reinforcement:

(1) *In Tension*

Billet steel or axle-steel bars, structural grade .... 18,000 psi

Main reinforcement, $\frac{3}{8}$ in. or less in diameter, in one-way slabs of not more than 12 ft span, 50% of the minimum specified yield by the ASTM, but not to exceed .... 30,000 psi

For deformed bars, yield strengths of 60,000 psi or more, for bars #11 and smaller .... 24,000 psi

For all other .... 20,000 psi

(2) *In Compression, Vertical Column Reinforcement*

Spiral columns, 40% of the minimum yield, but not to exceed ..... 30,000 psi

Tied columns, 85% of value for spiral columns, but not to exceed.. 25,500 psi

Composite and combination columns:

Structural steel sections

ASTM A-36 .... 18,000 psi

ASTM A-7 .... 16,000 psi

Cast iron sections .... 10,000 psi

(3) *In Compression, Flexural Members*

The compression reinforcement is transformed by an effective modular ratio of $2n$ and the computed stress shall not exceed the allowable tensile stress.

(4) *Spirals*

Hot rolled rods, intermediate grade .... 40,000 psi

Hot rolled rods, hard grade .... 50,000 psi

Hot rolled rods, ASTM A-432 and cold-drawn wire .... 60,000 psi

Because effects due to wind loads and earthquake forces are intermittent and of short duration, most building codes permit a greater allowable stress when the structure must sustain these forces. Hence the 1963 ACI Building Code, like almost all modern codes, permits members to be proportioned for stresses one-third greater than those stated above when the applied stresses are due to wind or earthquake forces. However, the section so proportioned must be capable of resisting the combinations of dead and live loads if the allowable stresses are based on 100% of the values tabulated.

EFFECT OF REINFORCEMENT IN CONCRETE

As previously stated, the tensile strength of concrete is very low, being about 1/10 the compressive strength. Hence it is assumed in structural design that this tensile strength is nil. Therefore it is necessary to strengthen or reinforce concrete members where they are subjected to tensile stresses. This reinforcement is usually accomplished by the embedment of steel bars or rods which must then resist almost 100% of the tensile forces.

Fig. 4-1 is a schematic elevation of a reinforced concrete beam. As the magnitude of the vertical loads is increased, the elongation of the bottommost fibers of the beam exceeds the ultimate tensile strain of the concrete and it cracks. Since the strain in the section is proportional to the distance from the neutral axis, as the magnitude of the loads continues to increase, the tensile stresses increase and the cracks continue to increase in number and spread upward toward the neutral axis. These cracks should be perpendicular to the direction of the maximum principal tensile stress in the concrete. Hence the inclination of these cracks is a function of the flexural, shear and axial stresses to which the section is subjected.

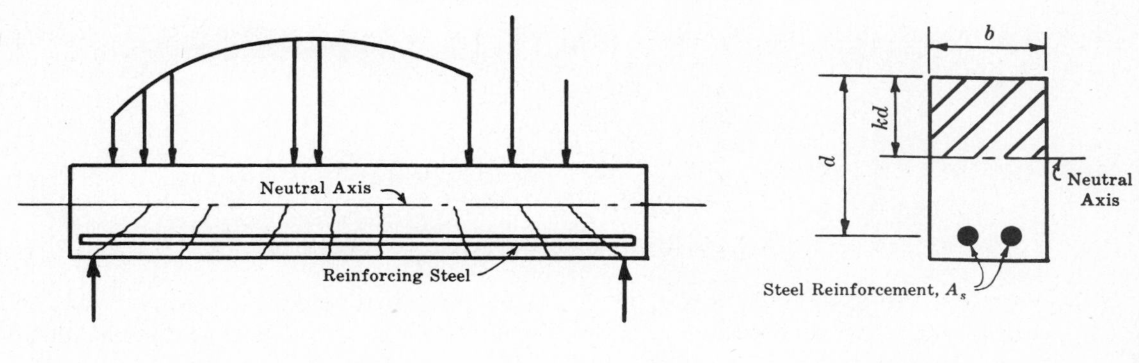

Fig. 4-1

Fig. 4-2

Obviously, when the concrete is cracked it is no longer capable of transmitting or resisting tensile forces. Then the tensile forces in the bottom of the beam must be resisted by the reinforcement, and the compressive forces at the top are resisted by the concrete. Thus the effective cross-section in resisting flexure is shown in Fig. 4-2. The shaded area above the neutral axis is the compression zone, and the steel reinforcement, A_s, is all that resists the tensile stresses below the neutral axis.

It is important to remember that because the tensile strength of concrete is so low, these cracks form at a load level such that the compressive stresses in the concrete and the tensile stresses in the reinforcement are still well below the ultimate or yield strengths. Also, the strains or deflections associated with the formation of these cracks are small enough that the appearance and serviceability of the structure are not impaired.

RECTANGULAR BEAMS

Fig. 4-3 shows the assumed distribution of strain and stress for a rectangular beam.

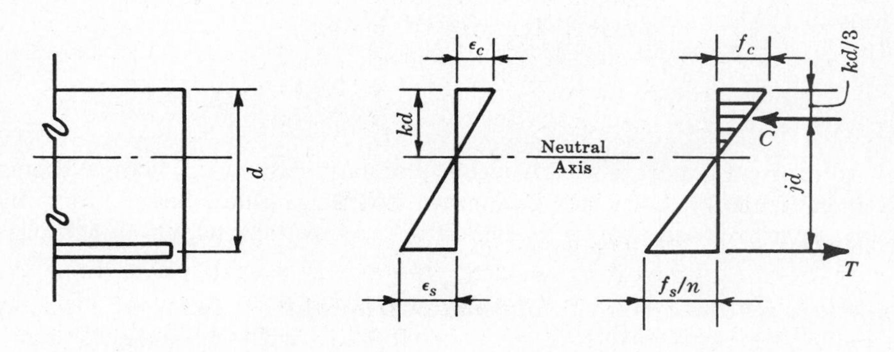

Fig. 4-3

As a matter of convenience, the principal expressions derived in Problem 4.1 are summarized below:

$$f_c = \frac{f_s k}{n(1-k)} \tag{4.1}$$

$$f_s = \frac{nf_c(1-k)}{k} \tag{4.2}$$

$$k = \sqrt{2pn + (pn)^2} - pn \tag{4.3}$$

$$j = 1 - k/3 \tag{4.4}$$

$$R = f_c jk/2 \tag{4.5}$$

$$M = Rbd^2 \tag{4.6}$$

$$d = \sqrt{M/Rb} \tag{4.7}$$

$$p = kf_c/2f_s \tag{4.8}$$

$$A_s = M/f_s jd \tag{4.9}$$

With the above equations it is possible to design a reinforced rectangular beam with tensile reinforcement. It will be shown later that aids can be developed for these equations which will greatly facilitate their solution.

TRANSFORMED SECTION

The foregoing expressions were developed assuming that the reinforced concrete beam was not homogeneous but was a composite of concrete and reinforcing steel. Although this method complies with the laws of mechanics, it can prove to be somewhat tedious when analyzing sections more complex in makeup than the previous simple example. A method will now be developed which will "transform" the composite section into a homogeneous beam. The transformed section is effected by substituting a certain cross-sectional area of concrete for the area of reinforcing steel. See Problem 4.2.

To transform the composite section of concrete and steel into an equivalent homogeneous section, the area of reinforcing steel is replaced by an area of concrete n times as great. The transformed section of Fig. 4-2 would then be as shown in Fig. 4-4.

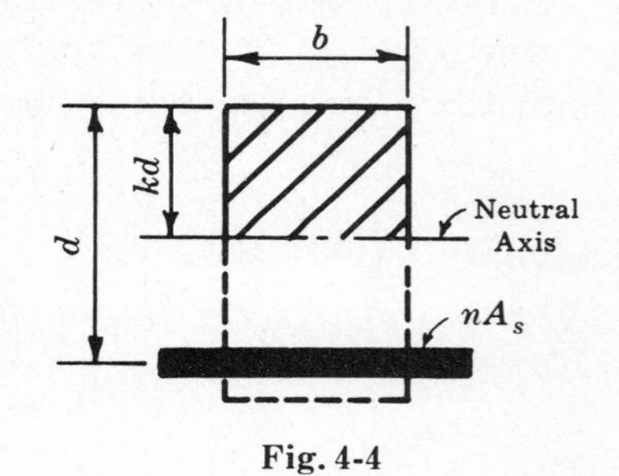

Fig. 4-4

Thus, with the section in Fig. 4-4, the standard formulas for flexure of a homogeneous beam may be used. And equations (4.1) through (4.9) may be derived beginning with the flexure formula $f = Mc/I$.

T-BEAMS

In much reinforced concrete construction and particularly in floor systems, a concrete slab is cast monolithically with and connected to rectangular beams forming a T-beam. Fig. 4-5 below is a cross-section of a concrete floor system which is composed of beams and slabs.

It is assumed, within certain limitations which will be discussed later, that the rectangular beam which has a width b' acts structurally with the slab, b wide and t thick. This then forms a T-beam shown shaded in the figure.

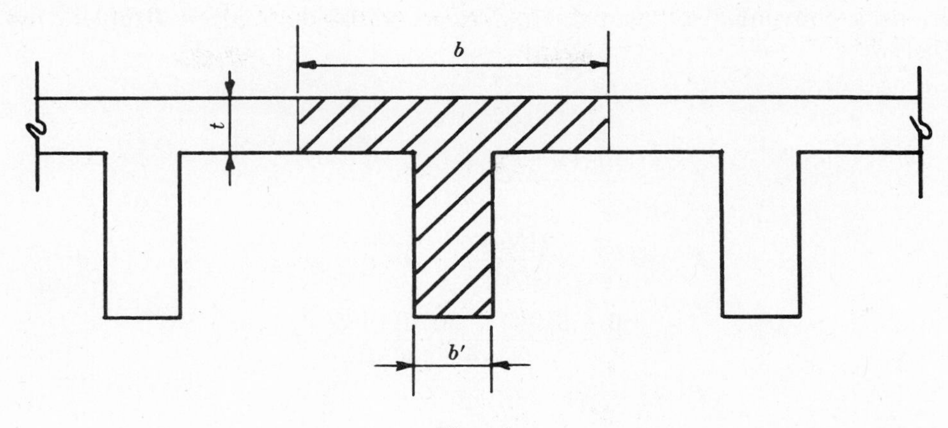

Fig. 4-5

If the neutral axis of the section falls within the slab or flange of the T-beam as shown in Fig. 4-6, then the analysis for the section is the same as for a rectangular beam. Because the effect of the concrete below the neutral axis has been neglected, the value for b' could be anything and equations (*4.1*) through (*4.9*) would still be valid. The beam would be analyzed as a simple rectangular one with width b.

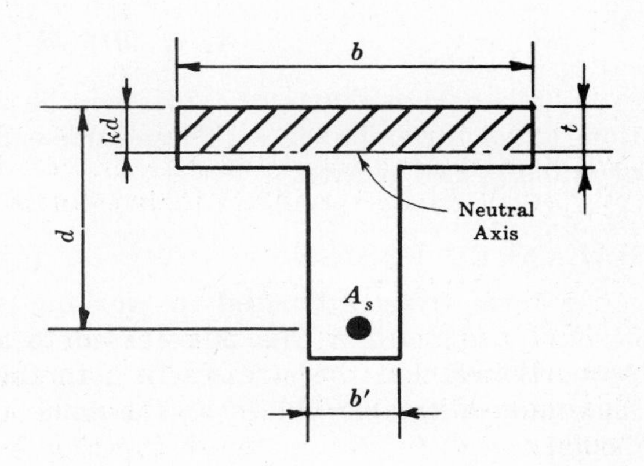

Fig. 4-6

If, however, the area of the flange is not capable of resisting the compressive force, then the neutral axis falls below the slab and the previous expressions are not applicable.

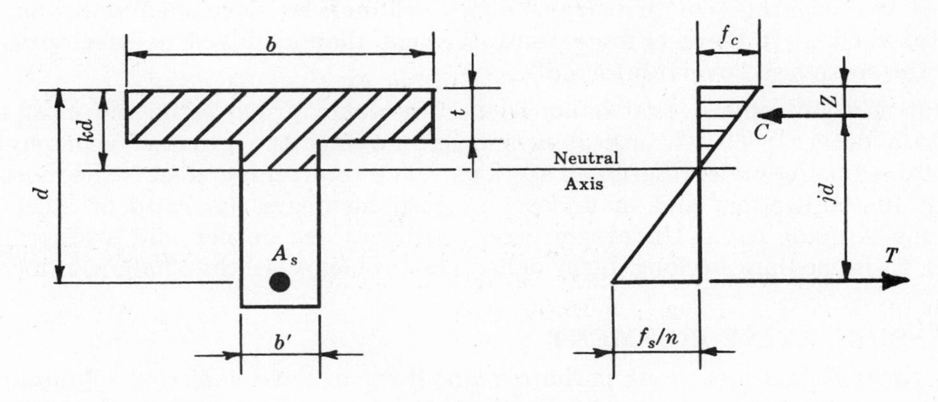

Fig. 4-7

Fig. 4-7 is similar to Fig. 4-3 except that the location of the centroid of the compressive force is not as readily determined. Hence the derivations are somewhat more complex.

The derivations of the formulas for a T-beam are similar to the rectangular beam. In the derivations contained in this chapter, the effect of the flange only is considered and then added to the rectangular portion or stem. See Problem 4.3.

Again, as a convenience, the principal expressions derived in Problem 4.3 are summarized below:

$$f_c = \frac{f_s k}{n(1-k)} \tag{4.10}$$

$$f_s = \frac{nf_c(1-k)}{k} \tag{4.11}$$

$$k = \frac{pn + \frac{1}{2}(t/d)^2}{pn + (t/d)} \tag{4.12}$$

$$j = \frac{6 - 6(t/d) + 2(t/d)^2 + (t/d)^3(\frac{1}{2}pn)}{6 - 3(t/d)} \tag{4.13}$$

$$R = f_c j(1 - t/2kd)(t/d) \tag{4.14}$$

$$M = Rbd^2 \tag{4.15}$$

$$d = \sqrt{M/Rb} \tag{4.16}$$

$$A_s = M/f_s jd \tag{4.17}$$

Later, design charts or tables will be developed which will aid in the solution of equations (*4.10*) through (*4.17*). These same equations could be developed using the homogeneous or transformed section.

BALANCED DESIGN

A term frequently used in working stress design of flexural members is "balanced design". Sometimes this is referred to as balanced reinforcement. If a member is so proportioned that the stresses in both the concrete and the reinforcing steel reach their maximum allowable values at the same time, then the section is said to have balanced design.

Balanced design means that there is exactly enough reinforcement to develop the maximum allowable compressive stress in the concrete. If there is a lesser amount of steel, then the concrete compressive strength cannot be developed and the section is "underreinforced". If there is more reinforcement than required to develop the concrete strength, the section is "overreinforced".

At first it would appear that balanced design would prove to be the most economical solution to a design. This is not always true. In fact the greatest economy is almost always attained using underreinforced sections. The reinforcing steel is the most expensive component in the section, and in underreinforced members the ratio of steel volume to total volume is made less. Underreinforced members are deeper and stiffer and are not as subject to immediate or long term deflection problems as the shallower members are.

COMPRESSION REINFORCEMENT

Many times it is desirable or perhaps mandatory to have a section of minimum depth in order to comply with some architectural or structural requirement. In fact, sometimes the section based on balanced design is too deep and the depth must be made even less.

If a section is arbitrarily made shallower than balanced design, there will be more than enough reinforcing steel to develop the concrete. Hence the concrete is overstressed when the steel is stressed to its allowable. When this condition exists, the allowable compressive force must be increased. And, it is increased by the addition of reinforcing steel in the compression zone. The reinforcing steel is capable of resisting compressive stresses several times greater than concrete can.

When compressive steel is added, the section is doubly reinforced and can withstand a greater moment and/or be made shallower. When this is done, the expressions previously derived are no longer valid because it was assumed that there was tension reinforcement only. Therefore with compressive reinforcement Fig. 4-2 would now look like Fig. 4-8.

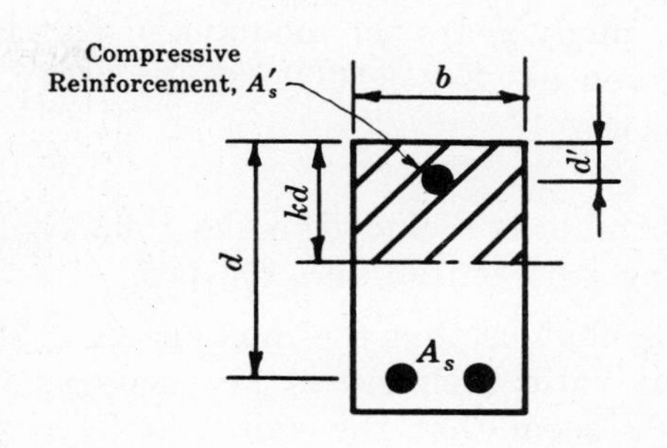

Fig. 4-8

Fig. 4-3 would now appear like Fig. 4-9. ϵ_c, ϵ_s', C_c, and C_s' are the unit strains and compressive forces in the concrete and steel respectively.

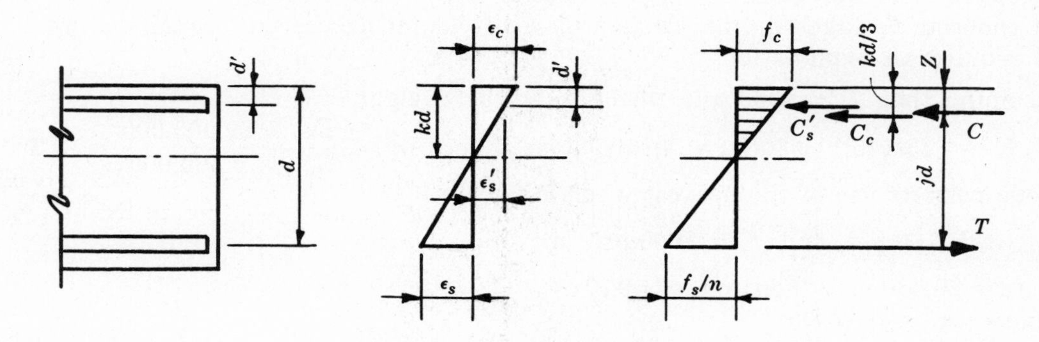

Fig. 4-9

Again as a summary, the expressions used in the working stress design of doubly reinforced beams as derived in Problem 4.4 are:

$$f_c = \frac{f_s k}{n(1-k)} \tag{4.18}$$

$$f_s = nf_c\left(\frac{1-k}{k}\right) \tag{4.19}$$

$$f_s' = 2nf_c\frac{k - d'/d}{k} \tag{4.20}$$

$$k = \sqrt{2n(p + 2p'd'/d) + n^2(p + 2p')^2} - n(p + 2p') \tag{4.21}$$

$$j = 1 - z/d \tag{4.22}$$

$$M_c = Rbd^2 \tag{4.23}$$

$$M_s = A_s' f_s'(d - d') \tag{4.24}$$

$$d = \sqrt{M/Rb} \tag{4.25}$$

$$A_s = M/f_s jd \tag{4.26}$$

$$A_s' = \frac{M - Rbd^2}{f_s'(d-d')} \tag{4.27}$$

There also will be design aids developed for the solution of equations (*4.18*) through (*4.27*).

MODULUS OF ELASTICITY OF CONCRETE

For many years the modulus of elasticity of concrete used in working stress design has been taken as $E_c = 1000 f'_c$. However, empirical data show that a much more precise formula is

$$E_c = 33 w^{1.5} \sqrt{f'_c} \tag{4.28}$$

This formula is included in the 1963 ACI Building Code and is assumed valid for concrete weighing between 90 and 155 pcf.

It is obvious that the modulus of elasticity of concrete varies as the weight. And, the modular ratio then varies inversely as the modulus of elasticity of the concrete. Hence it would seem that the value for the modular ratio for lightweight structural concrete would be much greater than that for normal weight concrete, 145 pcf, of equal cylinder strength.

Except in deflection calculations, the 1963 ACI Building Code specifies that n for lightweight concrete be taken as the same as normal weight concrete of equal strength. This should require an explanation.

Assuming that $f'_c = 3000$ psi, then for 145 pcf weight,

$$E_c = 33(145)^{1.5}\sqrt{3000} = 3{,}160{,}000 \text{ psi} \quad \text{and} \quad n = \frac{E_s}{E_c} = \frac{29{,}000{,}000}{3{,}160{,}000} = 9.2$$

The Code permits use of the nearest whole number.

For $w = 100$ pcf and $f'_c = 3000$ psi, $n = 16$.

If $f_s = 20{,}000$ psi, then for $n = 9.2$

$$k = \frac{1}{1 + f_s/nf_c} = \frac{1}{1 + 20{,}000/[9.2(3000)(0.45)]} = 0.383 \quad \text{and} \quad j = 1 - k/3 = 0.872$$

If $n = 16$,

$$k = \frac{1}{1 + 20{,}000/[16(3000)(0.45)]} = 0.52 \quad \text{and} \quad j = 1 - 0.52/3 = 0.827$$

The balanced steel ratios would be

$$p = f_c k/2f_s = 0.0129 \text{ for } n = 9.2, \text{ and } p = 0.0175 \text{ for } n = 16$$

This means that balanced reinforcement for the lightweight concrete would be approximately 35% greater than balanced reinforcement for the normal weight concrete. Ultimate strength design theory (Chapter 5) will show that the apparent ultimate capacity of the lightweight section would be approximately 30% greater than that of the normal weight section.

If the same values of n are compared for an underreinforced section, as an example for $p = 0.008$, then $k = \sqrt{2np + (np)^2} - np = \sqrt{2(9.2)(0.008) + (9.2)^2(0.008)^2} - 9.2(0.008) = 0.318$ for $n = 9.2$. And $k = 0.393$ for $n = 16$. Then $j = 0.894$ for $n = 9.2$, and $j = 0.869$ for $n = 16$. If $M = f_s jdA_s$, the moment capacity for the normal weight concrete would be approximately 3% greater than that for the lightweight section. The ultimate capacities would be identical.

Although the above comparison is merely an arbitrarily selected example, it does show what the results are when n is varied. Other examples would bear out this trend.

With balanced reinforcement in working stress design, the lightweight concrete section has ostensibly a much greater ultimate capacity if the two n values are used. For underreinforced sections the working stress resisting moments are almost equal and the ultimate resisting moments are identical regardless of what n is used. Therefore in flexural computations it is logical that the value of n should be taken as that determined for normal weight concrete regardless of the unit weight of the concrete.

DESIGN AIDS AND TABLES

Two methods have been presented with which the flexural design of reinforced concrete can be accomplished when the straight-line theory is used: the analysis (1) as a composite non-homogeneous section and (2) as a homogeneous transformed section. And with either of these two methods the laws of mechanics may be applied to a free body of the section, or values may be substituted directly into the derived equations (*4.1*) through (*4.27*). It is obvious that these methods are not suitable for rapid investigation or design and that they do not lend themselves to use as routine techniques for the design office.

As previously discussed, many design aids have been developed based on the derivations contained in this chapter. These have been in the form of charts, curves, tables, nomographs, etc. However, it seems that the solution of these equations by the use of tables has proved to be the most satisfactory for the design office.

The American Concrete Institute and others have developed series of tables which are universally accepted, and these will be the ones used herein.

Table 4.2, Page 71, which is reproduced with permission of the ACI is a series of coefficients developed for use in solving equations (*4.1*) through (*4.9*). The tables are divided into sections depending on the allowable steel stress f_s which varies from 16,000 psi to 33,000 psi. At the left hand column of each of the tables is the concrete cylinder strength f'_c and the modular ratio n. Also at the left is the allowable compressive stress in the concrete f_c. The body of the tables contain the appropriate values of the coefficient $K = R$; the ratio of the distance from the compression face to neutral axis to the distance from the compression face to centroid of the tension reinforcement, k; the ratio of the distance from the centroid of the compressive force to centroid of the tension reinforcement to the distance from the compression face to the centroid of the tension reinforcement, j; and the percentage of tension reinforcement, p.

Knowing f_s and f_c [equations (*4.1*) and (*4.2*)], Table 4.2 solves directly for k, j, R and p [equations (*4.3*), (*4.4*), (*4.5*) and (*4.8*)]. Knowing K or R, the value for the resisting moment M may then be determined by equation (*4.6*). Or knowing K or R and M, the effective depth d of the section is determined by equation (*4.7*). With the value of j and d, the area of tension reinforcement A_s is computed using equation (*4.9*).

The procedure outlined in the above paragraph is the one that would be followed when designing. In an investigation of an established section, this procedure may be reversed.

Table 4.2 indicates that for a given steel stress and over wide ranges of concrete stress, the value for j varies a small amount. If a new coefficient $a = f_s j/12{,}000$ is determined using the average of j for a given steel stress, then equation (*4.9*) is rewritten

$$A_s = M/ad \qquad (4.29)$$

where M is in foot-pounds, d in inches and A_s in square inches. Values of a are included at the top of each table and based on the average value of j.

As a matter of comparison, for $f_s = 20{,}000$ psi, the maximum value of j is 0.898 and the minimum is 0.828. Then

$$a(\text{max.}) = \frac{20{,}000}{12{,}000}(0.898) = 1.50, \qquad a(\text{min.}) = \frac{20{,}000}{12{,}000}(0.828) = 1.38$$

The average value of a is given as 1.44, which is approximately 4% different from both the maximum and minimum values. For a given concrete strength and modular ratio, the variation between the maximum and minimum values is even less. For $f_s = 20{,}000$ psi and $f'_c = 5000$ psi, the value of j varies from 0.872 to 0.828 and a would then vary from 1.45 to 1.38. Hence the assumption of an average value of a and j is well within the precision required for ordinary design practices.

In the flexural design of a slab, the moment capacity may be expressed in terms of a unit width. If in equation (*4.5*), $b = 12''$, then a table has been developed for the moment resisted by the concrete for various depths of slabs.

Knowing f'_c, f_c, f_s and d, the body of Table 4.3, Pages 72-74, contains the resisting moment. If the required moment capacity is known, then the effective depth is found at the top of the table. Table 4.3 is reproduced with permission of the ACI.

Equations (*4.13*) and (*4.14*) show that j and K or R are functions of the ratio t/d. Hence in the design of T-sections, the tables previously developed are not valid and a new set is needed which gives these coefficients for various values of t/d. Table 4.4 is such a table. The body of the table contains values for R, with the values for a at the top based on average values for j.

These tables neglect the compressive stress in the stem. If the stress in the stem is included, then Table 4.2 must be used along with Table 4.4. A solved problem will demonstrate this procedure.

Knowing f'_c, f_c, f_s and t/d, the values for R and a may be determined from Table 4.4, Pages 75-76. After R is known, then d may be found by equation (*4.16*). If the effective flange width b, effective depth d, and the required moment capacity are known, then R can be determined by equation (*4.15*) and the value for t/d is selected from Table 4.4. Table 4.4 is reproduced with permission of the ACI.

If the expression for f'_s given in equation (*4.20*) is substituted in equation (*4.27*), the expression for the area of compression reinforcement is

$$A'_s = \frac{M - Rbd^2}{cd} \tag{4.30}$$

If f'_s is equal to or less than f_s, then

$$c = \frac{f_s(2n-1)(1-d'/d)(k-d'/d)}{12{,}000n(1-k)}$$

If f'_s is restricted to f_s, then

$$c = \frac{f_s}{12{,}000}\left[1 - \frac{f_c}{f_s}\frac{(kd-d')}{kd}\right](1-d'/d)$$

In the above expressions for c, it is assumed that the effective modular ratio in compression is equal to $2n$.

Table 4.5, Pages 77-78, contains coefficients used in the design of doubly reinforced concrete beams and is reproduced with permission of the ACI.

If a doubly reinforced beam is to be designed by the tables, equation (*4.26*) is solved using Table 4.2 and equation (*4.30*) is solved using Table 4.5. Table 4.5 gives values of c for various ratios of d'/d.

If the required resisting moment M is known and if the resistance of the concrete Rbd^2 is determined by Table 4.2 and equation (*4.23*), the area of compressive reinforcement is then determined by equation (*4.30*).

The primary use for these tables is for the design of flexural members. However, these tables may also be used to investigate stresses in beams. Many times in an investigation it is convenient to use the transformed section technique rather than the tables.

In the solved problems that follow, the primary intent is to demonstrate the principles derived and discussed in this chapter. In the design of beams, factors other than flexural computations must be considered. Before a design is complete, such items as shear stresses, bond stresses, deflections, minimum reinforcement, etc., must be checked. All of these considerations are discussed elsewhere in this book.

TABLE 4.2
COEFFICIENTS (K, k, j, p) FOR RECTANGULAR SECTIONS

f'_c and n	f_c	K	k	j	p	K	k	j	p
		$f_s = 16{,}000$		$a = 1.13$		$f_s = 18{,}000$		$a = 1.29$	
	875.	137.	.356	.881	.0097	128.	.329	.890	.0080
2500	1000.	169.	.387	.871	.0121	158.	.359	.880	.0100
	1125.	201.	.415	.862	.0146	190.	.387	.871	.0121
10.1	1250.	235.	.441	.853	.0172	222.	.412	.863	.0143
	1500.	306.	.486	.838	.0228	291.	.457	.848	.0190
	1050.	173.	.376	.875	.0124	162.	.349	.884	.0102
3000	1200.	212.	.408	.864	.0153	199.	.380	.873	.0127
	1350.	252.	.437	.854	.0184	238.	.408	.864	.0153
9.2	1500.	294.	.463	.846	.0217	278.	.434	.855	.0181
	1800.	380.	.509	.830	.0286	362.	.479	.840	.0240
	1400.	249.	.412	.863	.0180	234.	.384	.872	.0149
4000	1600.	303.	.444	.852	.0222	286.	.416	.861	.0185
	1800.	359.	.474	.842	.0266	341.	.444	.852	.0222
8.0	2000.	417.	.500	.833	.0313	397.	.471	.843	.0261
	2400.	536.	.545	.818	.0409	513.	.516	.828	.0344
	1750.	327.	.437	.854	.0239	309.	.408	.864	.0199
5000	2000.	397.	.470	.843	.0294	376.	.441	.853	.0245
	2250.	468.	.500	.833	.0351	446.	.470	.843	.0294
7.1	2500.	542.	.526	.825	.0411	518.	.497	.835	.0345
	3000.	694.	.571	.810	.0535	666.	.542	.819	.0452

$$p = \frac{A_s}{bd}$$

$$k = \frac{1}{1 + f_s/nf_c} \qquad j = 1 - \tfrac{1}{3}k$$

$$p^* = \frac{f_c}{2f_s} \times k \qquad K = \frac{f_c}{2}kj$$

$$a = \frac{f_s}{12{,}000} \times (\text{av. } j\text{-value})$$

for use in

$$A_s = \frac{M}{ad} \quad \text{or} \quad A_s = \frac{NE}{adi}$$

f'_c and n	f_c	K	k	j	p	K	k	j	p	K	k	j	p
		$f_s = 20{,}000$		$a = 1.44$		$f_s = 22{,}000$		$a = 1.60$		$f_s = 24{,}000$		$a = 1.76$	
	875.	120.	.306	.898	.0067	113.	.287	.904	.0057	107.	.269	.910	.0049
2500	1000.	149.	.336	.888	.0084	141.	.315	.895	.0072	133.	.296	.901	.0062
	1125.	179.	.362	.879	.0102	170.	.341	.886	.0087	161.	.321	.893	.0075
10.1	1250.	211.	.387	.871	.0121	200.	.365	.878	.0104	191.	.345	.885	.0090
	1500.	277.	.431	.856	.0162	264.	.408	.864	.0139	253.	.387	.871	.0121
	1050.	152.	.326	.891	.0085	144.	.305	.898	.0073	136.	.287	.904	.0063
3000	1200.	188.	.356	.881	.0107	178.	.334	.889	.0091	169.	.315	.895	.0079
	1350.	226.	.383	.872	.0129	214.	.361	.880	.0111	204.	.341	.886	.0096
9.2	1500.	265.	.408	.864	.0153	252.	.385	.872	.0131	240.	.365	.878	.0114
	1800.	346.	.453	.849	.0204	331.	.429	.857	.0176	317.	.408	.864	.0153
	1400.	221.	.359	.880	.0126	210.	.337	.888	.0107	199.	.318	.894	.0093
4000	1600.	272.	.390	.870	.0156	258.	.368	.877	.0134	246.	.348	.884	.0116
	1800.	324.	.419	.860	.0188	309.	.396	.868	.0162	295.	.375	.875	.0141
8.0	2000.	379.	.444	.852	.0222	362.	.421	.860	.0191	347.	.400	.867	.0167
	2400.	492.	.490	.837	.0294	472.	.466	.845	.0254	454.	.444	.852	.0222
	1750.	292.	.383	.872	.0168	278.	.361	.880	.0144	265.	.341	.886	.0124
5000	2000.	358.	.415	.862	.0208	341.	.392	.869	.0178	326.	.372	.876	.0155
	2250.	426.	.444	.852	.0250	407.	.421	.860	.0215	390.	.400	.867	.0187
7.1	2500.	496.	.470	.843	.0294	475.	.447	.851	.0254	456.	.425	.858	.0221
	3000.	641.	.516	.828	.0387	617.	.492	.836	.0335	595.	.470	.843	.0294
		$f_s = 27{,}000$		$a = 2.00$		$f_s = 30{,}000$		$a = 2.24$		$f_s = 33{,}000$		$a = 2.48$	
	875.	99.	.247	.918	.0040	92.	.228	.924	.0033	86.	.211	.930	.0028
2500	1000.	124.	.272	.909	.0050	115.	.252	.916	.0042	108.	.234	.922	.0036
	1125.	150.	.296	.901	.0062	140.	.275	.908	.0052	132.	.256	.915	.0044
10.1	1250.	178.	.319	.894	.0074	167.	.296	.901	.0062	157.	.277	.908	.0052
	1500.	237.	.359	.880	.0100	224.	.336	.888	.0084	211.	.315	.895	.0072
	1050.	126.	.264	.912	.0051	117.	.244	.919	.0043	110.	.226	.925	.0036
3000	1200.	157.	.290	.903	.0064	147.	.269	.910	.0054	138.	.251	.916	.0046
	1350.	190.	.315	.895	.0079	178.	.293	.902	.0066	168.	.273	.909	.0056
9.2	1500.	225.	.338	.887	.0094	211.	.315	.895	.0079	199.	.295	.902	.0067
	1800.	299.	.380	.873	.0127	282.	.356	.881	.0107	267.	.334	.889	.0091
	1400.	185.	.293	.902	.0076	173.	.272	.909	.0063	162.	.253	.916	.0054
4000	1600.	230.	.322	.893	.0095	215.	.299	.900	.0080	203.	.279	.907	.0068
	1800.	277.	.348	.884	.0116	260.	.324	.892	.0097	246.	.304	.899	.0083
8.0	2000.	326.	.372	.876	.0138	308.	.348	.884	.0116	291.	.327	.891	.0099
	2400.	430.	.416	.861	.0185	407.	.390	.870	.0156	387.	.368	.877	.0134
	1750.	247.	.315	.895	.0102	231.	.293	.902	.0085	218.	.274	.909	.0073
5000	2000.	305.	.345	.885	.0128	287.	.321	.893	.0107	271.	.301	.900	.0091
	2250.	366.	.372	.876	.0155	346.	.347	.884	.0130	327.	.326	.891	.0111
7.1	2500.	430.	.397	.868	.0184	407.	.372	.876	.0155	386.	.350	.883	.0132
	3000.	564.	.441	.853	.0245	537.	.415	.862	.0208	511.	.392	.869	.0178

*"Balanced steel ratio" applies to problems involving bending only.

TABLE 4.3

RESISTING MOMENTS OF RECTANGULAR SECTIONS 1 FT. WIDE (SLABS)

Values of $\frac{Kd^2}{1000}$

Enter table with known M or NE (ft-kips)
Select effective depth (d: in.)

12″ — d — A_s

f'_c and n	f_c	Effective depth, d: 2	2.5	3	3.5	4	4.5	5	5.5	6	6.5	7	7.5	8	8.5	9	10	11	12	13	14
		f_s = 16,000																			
2500	875	.55	.86	1.23	1.68	2.20	2.8	3.4	4.2	4.9	5.8	6.7	7.7	8.8	9.9	11.1	13.7	16.6	19.8	23.2	26.9
	1000	.67	1.05	1.52	2.06	2.70	3.4	4.2	5.1	6.1	7.1	8.3	9.5	10.8	12.2	13.7	16.9	20.4	24.3	28.5	33.0
	1125	.81	1.26	1.81	2.47	3.22	4.1	5.0	6.1	7.2	8.5	9.9	11.3	12.9	14.5	16.3	20.1	24.4	29.0	34.0	39.4
10.1	1250	.94	1.47	2.12	2.88	3.76	4.8	5.9	7.1	8.5	9.9	11.5	13.2	15.0	17.0	19.0	23.5	28.5	33.9	39.7	46.1
	1500	1.22	1.91	2.75	3.74	4.89	6.2	7.6	9.2	11.0	12.9	15.0	17.2	19.6	22.1	24.8	30.6	37.0	44.0	51.7	59.9
3000	1050	.69	1.08	1.56	2.12	2.77	3.5	4.3	5.2	6.2	7.3	8.5	9.7	11.1	12.5	14.0	17.3	20.9	24.9	29.2	33.9
	1200	.85	1.32	1.90	2.59	3.39	4.3	5.3	6.4	7.6	8.9	10.4	11.9	13.5	15.3	17.1	21.2	25.6	30.5	35.8	41.5
	1350	1.01	1.58	2.27	3.09	4.03	5.1	6.3	7.6	9.1	10.6	12.3	14.2	16.1	18.2	20.4	25.2	30.5	36.3	42.6	49.4
9.2	1500	1.17	1.84	2.64	3.60	4.70	5.9	7.3	8.9	10.6	12.4	14.4	16.5	18.8	21.2	23.8	29.4	35.5	42.3	49.6	57.6
	1800	1.52	2.38	3.42	4.66	6.08	7.7	9.5	11.5	13.7	16.1	18.6	21.4	24.3	27.5	30.8	38.0	46.0	54.7	64.2	74.5
4000	1400	.99	1.55	2.24	3.05	3.98	5.0	6.2	7.5	9.0	10.5	12.2	14.0	15.9	18.0	20.1	24.9	30.1	35.8	42.0	48.7
	1600	1.21	1.89	2.73	3.71	4.85	6.1	7.6	9.2	10.9	12.8	14.8	17.0	19.4	21.9	24.5	30.3	36.6	43.6	51.2	59.4
	1800	1.44	2.24	3.23	4.40	5.74	7.3	9.0	10.9	12.9	15.2	17.6	20.2	23.0	25.9	29.1	35.9	43.4	51.7	60.7	70.4
8.0	2000	1.67	2.60	3.75	5.10	6.67	8.4	10.4	12.6	15.0	17.6	20.4	23.4	26.7	30.1	33.8	41.7	50.4	60.0	70.4	81.7
	2400	2.14	3.35	4.82	6.56	8.57	10.8	13.4	16.2	19.3	22.6	26.2	30.1	34.3	38.7	43.4	53.6	64.8	77.1	90.5	105
5000	1750	1.31	2.04	2.94	4.00	5.23	6.6	8.2	9.9	11.8	13.8	16.0	18.4	20.9	23.6	26.5	32.7	39.5	47.1	55.2	64.0
	2000	1.59	2.48	3.57	4.86	6.34	8.0	9.9	12.0	14.3	16.8	19.4	22.3	25.4	28.6	32.1	39.7	48.0	57.1	67.0	77.7
	2250	1.87	2.93	4.22	5.74	7.50	9.5	11.7	14.2	16.9	19.8	23.0	26.4	30.0	33.8	37.9	46.8	56.7	67.5	79.2	91.8
7.1	2500	2.17	3.39	4.88	6.64	8.67	11.0	13.6	16.4	19.5	22.9	26.6	30.5	34.7	39.2	43.9	54.2	65.6	78.1	91.6	106
	3000	2.77	4.33	6.24	8.50	1.10	14.0	17.3	21.0	25.0	29.3	34.0	39.0	44.4	50.1	56.2	69.4	83.9	99.9	117	136
		f_s = 18,000																			
2500	875	.51	.80	1.15	1.57	2.05	2.6	3.2	3.9	4.6	5.4	6.3	7.2	8.2	9.3	10.4	12.8	15.5	18.5	21.7	25.1
	1000	.63	.99	1.42	1.94	2.53	3.2	4.0	4.8	5.7	6.7	7.8	8.9	10.1	11.4	12.8	15.8	19.1	22.8	26.7	31.0
	1125	.76	1.19	1.71	2.32	3.03	3.8	4.7	5.7	6.8	8.0	9.3	10.7	12.1	13.7	15.4	19.0	22.9	27.3	32.0	37.2
10.1	1250	.89	1.39	2.00	2.72	3.56	4.5	5.6	6.7	8.0	9.4	10.9	12.5	14.2	16.1	18.0	22.2	26.9	32.0	37.6	43.6
	1500	1.16	1.82	2.61	3.56	4.65	5.9	7.3	8.8	10.5	12.3	14.2	16.3	18.6	21.0	23.5	29.1	35.2	41.8	49.1	56.9
3000	1050	.65	1.01	1.46	1.98	2.59	3.3	4.1	4.9	5.8	6.8	7.9	9.1	10.4	11.7	13.1	16.2	19.6	23.3	27.4	31.8
	1200	.80	1.25	1.79	2.44	3.19	4.0	5.0	6.0	7.2	8.4	9.8	11.2	12.7	14.4	16.1	19.9	24.1	28.7	33.7	39.0
	1350	.95	1.49	2.14	2.92	3.81	4.8	6.0	7.2	8.6	10.1	11.7	13.4	15.2	17.2	19.3	23.8	28.8	34.3	40.2	46.7
9.2	1500	1.11	1.74	2.51	3.41	4.45	5.6	7.0	8.4	10.0	11.8	13.6	15.7	17.8	20.1	22.6	27.8	33.7	40.1	47.0	54.6
	1800	1.45	2.26	3.26	4.44	5.80	7.3	9.1	11.0	13.0	15.3	17.8	20.4	23.2	26.2	29.4	36.2	43.8	52.2	61.2	71.0
4000	1400	.94	1.46	2.11	2.87	3.75	4.7	5.9	7.1	8.4	9.9	11.5	13.2	15.0	16.9	19.0	23.4	28.3	33.7	39.6	45.9
	1600	1.15	1.79	2.58	3.51	4.58	5.8	7.2	8.7	10.3	12.1	14.0	16.1	18.3	20.7	23.2	28.6	34.7	41.2	48.4	56.1
	1800	1.36	2.13	3.07	4.17	5.45	6.9	8.5	10.3	12.3	14.4	16.7	19.2	21.8	24.6	27.6	34.1	41.2	49.1	57.6	66.8
8.0	2000	1.59	2.48	3.57	4.86	6.35	8.0	9.9	12.0	14.3	16.8	19.4	22.3	25.4	28.7	32.1	39.7	48.0	57.1	67.1	77.8
	2400	2.05	3.21	4.62	6.28	8.20	10.4	12.8	15.5	18.5	21.7	25.1	28.8	32.8	37.1	41.5	51.3	62.0	73.8	86.7	101
5000	1750	1.23	1.93	2.78	3.78	4.94	6.3	7.7	9.3	11.1	13.0	15.1	17.4	19.8	22.3	25.0	30.9	37.4	44.5	52.2	60.5
	2000	1.50	2.35	3.39	4.61	6.02	7.6	9.4	11.4	13.5	15.9	18.4	21.2	24.1	27.2	30.5	37.6	45.5	54.2	63.6	73.7
	2250	1.78	2.79	4.01	5.46	7.14	9.0	11.2	13.5	16.1	18.8	21.9	25.1	28.5	32.2	36.1	44.6	54.0	64.2	75.4	87.4
7.1	2500	2.07	3.24	4.66	6.34	8.29	10.5	12.9	15.7	18.6	21.9	25.4	29.1	33.1	37.4	42.0	51.8	62.7	74.6	87.5	102
	3000	2.66	4.16	5.99	8.16	10.66	13.5	16.7	20.2	24.0	28.1	32.6	37.5	42.6	48.1	54.0	66.6	80.6	95.9	113	131

TABLE 4.3 (Cont.)

f'_c and n	f_c	Effective depth, d																			
		2	2.5	3	3.5	4	4.5	5	5.5	6	6.5	7	7.5	8	8.5	9	10	11	12	13	14
		f_s = 20,000																			
	875	.48	.75	1.08	1.47	1.93	2.4	3.0	3.6	4.3	5.1	5.9	6.8	7.7	8.7	9.8	12.0	14.6	17.3	20.3	23.6
2500	1000	.60	.93	1.34	1.83	2.38	3.0	3.7	4.5	5.4	6.3	7.3	8.4	9.5	10.8	12.1	14.9	18.0	21.5	25.2	29.2
	1125	.72	1.12	1.61	2.20	2.87	3.6	4.5	5.4	6.5	7.6	8.8	10.1	11.5	12.9	14.5	17.9	21.7	25.8	30.3	35.1
10.1	1250	.84	1.32	1.90	2.58	3.37	4.3	5.3	6.4	7.6	8.9	10.3	11.9	13.5	15.2	17.1	21.1	25.5	30.3	35.6	41.3
	1500	1.11	1.73	2.49	3.39	4.43	5.6	6.9	8.4	10.0	11.7	13.6	15.6	17.7	20.0	22.4	27.7	33.5	39.9	46.8	54.3
	1050	.61	.95	1.37	1.87	2.44	3.1	3.8	4.6	5.5	6.4	7.5	8.6	9.8	11.0	12.3	15.2	18.4	21.9	25.8	29.9
3000	1200	.75	1.18	1.69	2.30	3.01	3.8	4.7	5.7	6.8	7.9	9.2	10.6	12.0	13.6	15.2	18.8	22.8	27.1	31.8	36.9
	1350	.90	1.41	2.03	2.76	3.61	4.6	5.6	6.8	8.1	9.5	11.1	12.7	14.4	16.3	18.3	22.6	27.3	32.5	38.1	44.2
9.2	1500	1.06	1.65	2.38	3.24	4.23	5.4	6.6	8.0	9.5	11.2	13.0	14.9	16.9	19.1	21.4	26.5	32.0	38.1	44.7	51.9
	1800	1.38	2.16	3.12	4.24	5.54	7.0	8.7	10.5	12.5	14.6	17.0	19.5	22.2	25.0	28.0	34.6	41.9	49.8	58.5	67.8
	1400	.88	1.38	1.99	2.71	3.54	4.5	5.5	6.7	8.0	9.3	10.8	12.4	14.2	16.0	17.9	22.1	26.8	31.9	37.4	43.4
4000	1600	1.09	1.70	2.44	3.33	4.35	5.5	6.8	8.2	9.8	11.5	13.3	15.3	17.4	19.6	22.0	27.2	32.9	39.1	45.9	53.2
	1800	1.30	2.03	2.92	3.97	5.19	6.6	8.1	9.8	11.7	13.7	15.9	18.2	20.7	23.4	26.3	32.4	39.2	46.7	54.8	63.5
8.0	2000	1.51	2.37	3.41	4.64	6.06	7.7	9.5	11.5	13.6	16.0	18.6	21.3	24.2	27.4	30.7	37.9	45.8	54.5	64.0	74.2
	2400	1.97	3.07	4.43	6.02	7.87	10.0	12.3	14.9	17.7	20.8	24.1	27.7	31.5	35.5	39.8	49.2	59.5	70.8	83.1	96.4
	1750	1.17	1.83	2.63	3.58	4.68	5.9	7.3	8.8	10.5	12.4	14.3	16.5	18.7	21.1	23.7	29.2	35.4	42.1	49.4	57.3
5000	2000	1.43	2.24	3.22	4.38	5.72	7.2	8.9	10.8	12.9	15.1	17.5	20.1	22.9	25.8	29.0	35.8	43.3	51.5	60.5	70.1
	2250	1.70	2.66	3.83	5.21	6.81	8.6	10.6	12.9	15.3	18.0	20.9	23.9	27.2	30.8	34.5	42.6	51.5	61.3	71.9	83.4
7.1	2500	1.98	3.10	4.46	6.07	7.93	10.0	12.4	15.0	17.8	20.9	24.3	27.9	31.7	35.8	40.1	49.6	60.0	71.4	83.8	97.1
	3000	2.56	4.00	5.77	7.85	10.25	13.0	16.0	19.4	23.1	27.1	31.4	36.0	41.0	46.3	51.9	64.1	77.5	92.2	108	126
		f_s = 22,000																			
	875	.45	.71	1.02	1.39	1.81	2.3	2.8	3.4	4.1	4.8	5.6	6.4	7.3	8.2	9.2	11.3	13.7	16.3	19.2	22.2
2500	1000	.56	.88	1.27	1.73	2.25	2.9	3.5	4.3	5.1	5.9	6.9	7.9	9.0	10.2	11.4	14.1	17.0	20.3	23.8	27.6
	1125	.68	1.06	1.53	2.08	2.72	3.4	4.2	5.1	6.1	7.2	8.3	9.6	10.9	12.3	13.8	17.0	20.5	24.5	28.7	33.3
10.1	1250	.80	1.25	1.80	2.45	3.20	4.1	5.0	6.1	7.2	8.5	9.8	11.3	12.8	14.5	16.2	20.0	24.2	28.8	33.8	39.2
	1500	1.06	1.65	2.38	3.24	4.23	5.4	6.6	8.0	9.5	11.2	13.0	14.9	16.9	19.1	21.4	26.4	32.0	38.1	44.7	51.8
	1050	.58	.90	1.30	1.76	2.30	2.9	3.6	4.4	5.2	6.1	7.1	8.1	9.2	10.4	11.7	14.4	17.4	20.7	24.3	28.2
3000	1200	.71	1.11	1.60	2.18	2.85	3.6	4.5	5.4	6.4	7.5	8.7	10.0	11.4	12.9	14.4	17.8	21.6	25.7	30.1	34.9
	1350	.86	1.34	1.93	2.62	3.43	4.3	5.4	6.5	7.7	9.1	10.5	12.1	13.7	15.5	17.4	21.4	25.9	30.9	36.2	42.0
9.2	1500	1.01	1.57	2.27	3.09	4.03	5.1	6.3	7.6	9.1	10.6	12.3	14.2	16.1	18.2	20.4	25.2	30.5	36.3	42.6	49.4
	1800	1.32	2.07	2.98	4.06	5.30	6.7	8.3	10.0	11.9	14.0	16.2	18.6	21.2	23.9	26.8	33.1	40.1	47.7	56.0	64.9
	1400	.84	1.31	1.89	2.57	3.35	4.2	5.2	6.3	7.5	8.9	10.3	11.8	13.4	15.1	17.0	21.0	25.4	30.2	35.4	41.1
4000	1600	1.03	1.61	2.32	3.16	4.13	5.2	6.5	7.8	9.3	10.9	12.7	14.5	16.5	18.7	20.9	25.8	31.2	37.2	43.6	50.6
	1800	1.24	1.93	2.78	3.79	4.95	6.3	7.7	9.4	11.1	13.1	15.1	17.4	19.8	22.3	25.0	30.9	37.4	44.5	52.2	60.6
8.0	2000	1.45	2.26	3.26	4.43	5.79	7.3	9.0	10.9	13.0	15.3	17.7	20.4	23.2	26.2	29.3	36.2	43.8	52.1	61.2	70.9
	2400	1.89	2.95	4.25	5.79	7.56	9.6	11.8	14.3	17.0	20.0	23.1	26.6	30.2	34.1	38.3	47.2	57.2	68.0	79.8	92.6
	1750	1.11	1.74	2.50	3.40	4.45	5.6	6.9	8.4	10.0	11.7	13.6	15.6	17.8	20.1	22.5	27.8	33.6	40.0	47.0	54.5
5000	2000	1.36	2.13	3.07	4.18	5.46	6.9	8.5	10.3	12.3	14.4	16.7	19.2	21.8	24.6	27.6	34.1	41.3	49.1	57.6	66.8
	2250	1.63	2.54	3.66	4.98	6.51	8.2	10.2	12.3	14.6	17.2	19.9	22.9	26.0	29.4	33.0	40.7	49.2	58.6	68.8	79.8
7.1	2500	1.90	2.97	4.28	5.82	7.60	9.6	11.9	14.4	17.1	20.1	23.3	26.7	30.4	34.3	38.5	47.5	57.5	68.4	80.3	93.1
	3000	2.47	3.86	5.55	7.56	9.87	12.5	15.4	18.7	22.2	26.1	30.2	34.7	39.5	44.6	50.0	61.7	74.6	88.8	104	121
		f_s = 24,000																			
	875	.43	.67	.96	1.31	1.71	2.2	2.7	3.2	3.9	4.5	5.3	6.0	6.9	7.7	8.7	10.7	13.0	15.4	18.1	21.0
2500	1000	.53	.83	1.20	1.64	2.14	2.7	3.3	4.0	4.8	5.6	6.5	7.5	8.5	9.6	10.8	13.3	16.2	19.2	22.6	26.2
	1125	.65	1.01	1.45	1.98	2.58	3.3	4.0	4.9	5.8	6.8	7.9	9.1	10.3	11.7	13.1	16.1	19.5	23.2	27.3	31.6
10.1	1250	.76	1.19	1.72	2.34	3.05	3.9	4.8	5.8	6.9	8.1	9.3	10.7	12.2	13.8	15.4	19.1	23.1	27.5	32.2	37.4
	1500	1.01	1.58	2.28	3.10	4.04	5.1	6.3	7.6	9.1	10.7	12.4	14.2	16.2	18.3	20.5	25.3	30.6	36.4	42.7	49.5
	1050	.55	.85	1.23	1.67	2.18	2.8	3.4	4.1	4.9	5.8	6.7	7.7	8.7	9.8	11.0	13.6	16.5	19.6	23.0	26.7
3000	1200	.68	1.06	1.52	2.07	2.71	3.4	4.2	5.1	6.1	7.1	8.3	9.5	10.8	12.2	13.7	16.9	20.5	24.4	28.6	33.2
	1350	.82	1.28	1.84	2.50	3.26	4.1	5.1	6.2	7.3	8.6	10.0	11.5	13.1	14.7	16.5	20.4	24.7	29.4	34.5	40.0
9.2	1500	.96	1.50	2.16	2.95	3.85	4.9	6.0	7.3	8.7	10.2	11.8	13.5	15.4	17.4	19.5	24.0	29.1	34.6	40.6	47.1
	1800	1.27	1.98	2.86	3.89	5.08	6.4	7.9	9.6	11.4	13.4	15.6	17.9	20.3	22.9	25.7	31.7	38.4	45.7	53.6	62.2
	1400	.80	1.24	1.79	2.44	3.19	4.0	5.0	6.0	7.2	8.4	9.8	11.2	12.7	14.4	16.1	19.9	24.1	28.7	33.6	39.0
4000	1600	.98	1.54	2.21	3.01	3.94	5.0	6.2	7.4	8.9	10.4	12.1	13.8	15.7	17.8	19.9	24.6	29.8	35.4	41.6	48.2
	1800	1.18	1.85	2.66	3.62	4.73	6.0	7.4	8.9	10.6	12.5	14.5	16.6	18.9	21.3	23.9	29.5	35.7	42.5	49.9	57.9
8.0	2000	1.39	2.17	3.12	4.25	5.55	7.0	8.7	10.5	12.5	14.6	17.0	19.5	22.2	25.0	28.1	34.7	41.9	49.9	58.6	67.9
	2400	1.82	2.84	4.09	5.57	7.27	9.2	11.4	13.7	16.4	19.2	22.3	25.6	29.1	32.8	36.8	45.4	55.0	65.4	76.8	89.0
	1750	1.06	1.65	2.38	3.24	4.23	5.4	6.6	8.0	9.5	11.2	13.0	14.9	16.9	19.1	21.4	26.5	32.0	38.1	44.7	51.8
5000	2000	1.30	2.04	2.93	3.99	5.21	6.6	8.1	9.9	11.7	13.8	16.0	18.3	20.8	23.5	26.4	32.6	39.4	46.9	55.0	63.8
	2250	1.56	2.44	3.51	4.77	6.24	7.9	9.7	11.8	14.0	16.5	19.1	21.9	24.9	28.2	31.6	39.0	47.2	56.1	65.9	76.4
7.1	2500	1.82	2.85	4.11	5.59	7.30	9.2	11.4	13.8	16.4	19.3	22.4	25.7	29.2	33.0	36.9	45.6	55.2	65.7	77.1	89.4
	3000	2.38	3.72	5.35	7.29	9.52	12.0	14.9	18.0	21.4	25.1	29.1	33.5	38.1	43.0	48.2	59.5	72.0	85.6	101	117

TABLE 4.3 (Cont.)

f'_c and n	f_c	Effective depth, d																			
		2	2.5	3	3.5	4	4.5	5	5.5	6	6.5	7	7.5	8	8.5	9	10	11	12	13	14
		f_s = 27,000																			
	875	.40	.62	.89	1.21	1.58	2.0	2.5	3.0	3.6	4.2	4.9	5.6	6.3	7.2	8.0	9.9	12.0	14.3	16.7	19.4
2500	1000	.50	.77	1.11	1.52	1.98	2.5	3.1	3.7	4.5	5.2	6.1	7.0	7.9	8.9	10.0	12.4	15.0	17.8	20.9	24.3
	1125	.60	.94	1.35	1.84	2.40	3.0	3.8	4.5	5.4	6.3	7.4	8.4	9.6	10.8	12.2	15.0	18.2	21.6	25.4	29.4
10.1	1250	.71	1.11	1.60	2.18	2.85	3.6	4.4	5.4	6.4	7.5	8.7	10.0	11.4	12.9	14.4	17.8	21.5	25.6	30.1	34.9
	1500	.95	1.48	2.14	2.91	3.80	4.8	5.9	7.2	8.5	10.0	11.6	13.3	15.2	17.1	19.2	23.7	28.7	34.2	40.1	46.5
	1050	.50	.79	1.14	1.55	2.02	2.6	3.2	3.8	4.5	5.3	6.2	7.1	8.1	9.1	10.2	12.6	15.3	18.2	21.3	24.7
3000	1200	.63	.98	1.42	1.93	2.52	3.2	3.9	4.8	5.7	6.6	7.7	8.8	10.1	11.4	12.7	15.7	19.0	22.6	26.6	30.8
	1350	.76	1.19	1.71	2.33	3.05	3.9	4.8	5.8	6.9	8.0	9.3	10.7	12.2	13.8	15.4	19.0	23.0	27.4	32.2	37.3
9.2	1500	.90	1.41	2.03	2.76	3.60	4.6	5.6	6.8	8.1	9.5	11.0	12.7	14.4	16.3	18.2	22.5	27.2	32.4	38.0	44.1
	1800	1.20	1.87	2.69	3.66	4.78	6.1	7.5	9.0	10.8	12.6	14.6	16.8	19.1	21.6	24.2	29.9	36.2	43.0	50.5	58.6
	1400	.74	1.16	1.67	2.27	2.96	3.7	4.6	5.6	6.7	7.8	9.1	10.4	11.9	13.4	15.0	18.5	22.4	26.7	31.3	36.3
4000	1600	.92	1.44	2.07	2.81	3.68	4.7	5.7	6.9	8.3	9.7	11.3	12.9	14.7	16.6	18.6	23.0	27.8	33.1	38.8	45.0
	1800	1.11	1.73	2.49	3.39	4.43	5.6	6.9	8.4	10.0	11.7	13.6	15.6	17.7	20.0	22.4	27.7	33.5	39.9	46.8	54.2
8.0	2000	1.30	2.04	2.93	3.99	5.22	6.6	8.1	9.9	11.7	13.8	16.0	18.3	20.9	23.5	26.4	32.6	39.4	46.9	55.1	63.9
	2400	1.72	2.69	3.87	5.26	6.87	8.7	10.7	13.0	15.5	18.2	21.1	24.2	27.5	31.0	34.8	43.0	52.0	61.9	72.6	84.2
	1750	.99	1.54	2.22	3.02	3.95	5.0	6.2	7.5	8.9	10.4	12.1	13.9	15.8	17.8	20.0	24.7	29.9	35.5	41.7	48.4
5000	2000	1.22	1.91	2.75	3.74	4.88	6.2	7.6	9.2	11.0	12.9	14.9	17.2	19.5	22.0	24.7	30.5	36.9	43.9	51.6	59.8
	2250	1.47	2.29	3.30	4.49	5.86	7.4	9.2	11.1	13.2	15.5	18.0	20.6	23.4	26.5	29.7	36.6	44.3	52.8	61.9	71.8
7.1	2500	1.72	2.69	3.87	5.27	6.88	8.7	10.8	13.0	15.5	18.2	21.1	24.2	27.5	31.1	34.9	43.0	52.1	62.0	72.7	84.3
	3000	2.26	3.53	5.08	6.91	9.03	11.4	14.1	17.1	20.3	23.8	27.6	31.7	36.1	40.8	45.7	56.4	68.3	81.3	95.4	111
		f_s = 30,000																			
	875	.37	.58	.83	1.13	1.47	1.9	2.3	2.8	3.3	3.9	4.5	5.2	5.9	6.6	7.5	9.2	11.1	13.2	15.5	18.0
2500	1000	.46	.72	1.04	1.41	1.85	2.3	2.9	3.5	4.2	4.9	5.7	6.5	7.4	8.3	9.3	11.5	14.0	16.6	19.5	22.6
	1125	.56	.88	1.26	1.72	2.25	2.8	3.5	4.2	5.1	5.9	6.9	7.9	9.0	10.1	11.4	14.0	17.0	20.2	23.7	27.5
10.1	1250	.67	1.04	1.50	2.04	2.67	3.4	4.2	5.0	6.0	7.0	8.2	9.4	10.7	12.1	13.5	16.7	20.2	24.0	28.2	32.7
	1500	.89	1.40	2.01	2.74	3.58	4.5	5.6	6.8	8.0	9.4	11.0	12.6	14.3	16.1	18.1	22.4	27.0	32.2	37.8	43.8
	1050	.47	.73	1.06	1.44	1.88	2.4	2.9	3.6	4.2	5.0	5.8	6.6	7.5	8.5	9.5	11.7	14.2	16.9	19.9	23.0
3000	1200	.59	.92	1.32	1.80	2.35	3.0	3.7	4.4	5.3	6.2	7.2	8.3	9.4	10.6	11.9	14.7	17.8	21.2	24.8	28.8
	1350	.71	1.11	1.61	2.18	2.85	3.6	4.5	5.4	6.4	7.5	8.7	10.0	11.4	12.9	14.4	17.8	21.6	25.7	30.1	35.0
9.2	1500	.85	1.32	1.90	2.59	3.38	4.3	5.3	6.4	7.6	8.9	10.4	11.9	13.5	15.3	17.1	21.1	25.6	30.5	35.7	41.5
	1800	1.13	1.76	2.54	3.46	4.51	5.7	7.1	8.5	10.2	11.9	13.8	15.9	18.1	20.4	22.9	28.2	34.1	40.6	47.7	55.3
	1400	.69	1.08	1.56	2.12	2.77	3.5	4.3	5.2	6.2	7.3	8.5	9.7	11.1	12.5	14.0	17.3	20.9	24.9	29.2	33.9
4000	1600	.86	1.35	1.94	2.64	3.45	4.4	5.4	6.5	7.8	9.1	10.6	12.1	13.8	15.6	17.4	21.5	26.1	31.0	36.4	42.2
	1800	1.04	1.63	2.34	3.19	4.17	5.3	6.5	7.9	9.4	11.0	12.8	14.6	16.7	18.8	21.1	26.0	31.5	37.5	44.0	51.0
8.0	2000	1.23	1.92	2.77	3.77	4.92	6.2	7.7	9.3	11.1	13.0	15.1	17.3	19.7	22.2	24.9	30.8	37.2	44.3	52.0	60.3
	2400	1.63	2.55	3.67	4.99	6.52	8.2	10.2	12.3	14.7	17.2	20.0	22.9	26.1	29.4	33.0	40.7	49.3	58.7	68.8	79.8
	1750	.92	1.45	2.08	2.83	3.70	4.7	5.8	7.0	8.3	9.8	11.3	13.0	14.8	16.7	18.7	23.1	28.0	33.3	39.1	45.3
5000	2000	1.15	1.79	2.58	3.51	4.59	5.8	7.2	8.7	10.3	12.1	14.1	16.1	18.4	20.7	23.2	28.7	34.7	41.3	48.5	56.2
	2250	1.38	2.16	3.11	4.23	5.53	7.0	8.6	10.5	12.4	14.6	16.9	19.4	22.1	25.0	28.0	34.6	41.8	49.8	58.4	67.7
7.1	2500	1.63	2.54	3.66	4.99	6.51	8.2	10.2	12.3	14.7	17.2	19.9	22.9	26.1	29.4	33.0	40.7	49.3	58.6	68.8	79.8
	3000	2.15	3.35	4.83	6.57	8.59	10.9	13.4	16.2	19.3	22.7	26.3	30.2	34.3	38.8	43.5	53.7	64.9	77.3	90.7	105
		f_s = 33,000																			
	875	.34	.54	.77	1.05	1.37	1.7	2.1	2.6	3.1	3.6	4.2	4.8	5.5	6.2	7.0	8.6	10.4	12.4	14.5	16.8
2500	1000	.43	.68	.97	1.32	1.73	2.2	2.7	3.3	3.9	4.6	5.3	6.1	6.9	7.8	8.7	10.8	13.1	15.6	18.3	21.2
	1125	.53	.82	1.19	1.61	2.11	2.7	3.3	4.0	4.7	5.6	6.5	7.4	8.4	9.5	10.7	13.2	15.9	19.0	22.3	25.8
10.1	1250	.63	.98	1.41	1.92	2.51	3.2	3.9	4.7	5.7	6.6	7.7	8.8	10.0	11.3	12.7	15.7	19.0	22.6	26.5	30.8
	1500	.84	1.32	1.90	2.59	3.38	4.3	5.3	6.4	7.6	8.9	10.4	11.9	13.5	15.3	17.1	21.1	25.6	30.4	35.7	41.4
	1050	.44	.69	.99	1.35	1.76	2.2	2.7	3.3	4.0	4.6	5.4	6.2	7.0	7.9	8.9	11.0	13.3	15.8	18.6	21.5
3000	1200	.55	.86	1.24	1.69	2.21	2.8	3.4	4.2	5.0	5.8	6.8	7.8	8.8	10.0	11.2	13.8	16.7	19.8	23.3	27.0
	1350	.67	1.05	1.51	2.06	2.68	3.4	4.2	5.1	6.0	7.1	8.2	9.4	10.7	12.1	13.6	16.8	20.3	24.2	28.4	32.9
9.2	1500	.80	1.25	1.79	2.44	3.19	4.0	5.0	6.0	7.2	8.4	9.8	11.2	12.8	14.4	16.2	19.9	24.1	28.7	33.7	39.1
	1800	1.07	1.67	2.41	3.27	4.28	5.4	6.7	8.1	9.6	11.3	13.1	15.0	17.1	19.3	21.6	26.7	32.3	38.5	45.2	52.4
	1400	.65	1.02	1.46	1.99	2.60	3.3	4.1	4.9	5.8	6.9	8.0	9.1	10.4	11.7	13.2	16.2	19.7	23.4	27.4	31.8
4000	1600	.81	1.27	1.82	2.48	3.24	4.1	5.1	6.1	7.3	8.6	9.9	11.4	13.0	14.6	16.4	20.3	24.5	29.2	34.3	39.7
	1800	.98	1.54	2.21	3.01	3.93	5.0	6.1	7.4	8.8	10.4	12.0	13.8	15.7	17.8	19.9	24.6	29.7	35.4	41.5	48.2
8.0	2000	1.16	1.82	2.62	3.56	4.66	5.9	7.3	8.8	10.5	12.3	14.3	16.4	18.6	21.0	23.6	29.1	35.2	41.9	49.2	57.0
	2400	1.55	2.42	3.49	4.74	6.20	7.8	9.7	11.7	13.9	16.4	19.0	21.8	24.8	28.0	31.4	38.7	46.9	55.8	65.4	75.9
	1750	.87	1.36	1.96	2.66	3.48	4.4	5.4	6.6	7.8	9.2	10.7	12.2	13.9	15.7	17.6	21.8	26.3	31.3	36.8	42.6
5000	2000	1.08	1.69	2.44	3.32	4.33	5.5	6.8	8.2	9.7	11.4	13.3	15.2	17.3	19.6	21.9	27.1	32.8	39.0	45.7	53.1
	2250	1.31	2.04	2.94	4.01	5.23	6.6	8.2	9.9	11.8	13.8	16.0	18.4	20.9	23.6	26.5	32.7	39.6	47.1	55.3	64.1
7.1	2500	1.54	2.41	3.48	4.73	6.18	7.8	9.7	11.7	13.9	16.3	18.9	21.7	24.7	27.9	31.3	38.6	46.7	55.6	65.3	75.7
	3000	2.05	3.20	4.60	6.27	8.18	10.4	12.8	15.5	18.4	21.6	25.1	28.8	32.7	37.0	41.4	51.1	61.9	73.7	86.4	100

TABLE 4.4
COEFFICIENTS (a and K) FOR T-SECTIONS

$$a = \frac{f_s}{12{,}000} \times \text{average value of } j$$

Average values of j are taken from Table 4.2; a and K are used in:

$$A_s = \frac{M}{ad} \quad \text{or} \quad A_s = \frac{NE}{adi};$$

$$A'_s = \frac{M - KF}{cd} \quad \text{or} \quad A'_s = \frac{NE - KF}{cd}$$

$$K = \frac{f_c}{2} \times \frac{t}{d} \times \left(2 - \frac{t}{d} - \frac{t}{kd} + \frac{2t^2}{3kd^2}\right)$$

$f_s = 16{,}000$

f'_c and n	f_c	t/d = .10	.12	.14	.16	.18	.20	.24	.28	.32	.36	.40
	a	1.26	1.25	1.24	1.23	1.22	1.21	1.19	1.17	1.16	1.14	1.12
	K											
2500	875	72	82	92	101	108	115	125	132	136	137	136
	1000	83	96	107	118	127	135	149	158	165	168	168
	1125	94	109	122	135	146	156	172	185	193	199	201
10.1	1250	106	122	138	152	164	176	195	211	222	229	234
	1500	128	149	168	186	202	217	242	263	279	291	299
3000	1050	87	100	112	123	132	141	154	164	170	173	172
	1200	100	116	130	143	155	165	182	195	204	209	212
	1350	114	132	148	163	177	189	210	227	238	246	251
9.2	1500	127	148	167	184	200	214	238	258	273	283	290
	1800	154	180	203	224	244	263	295	321	341	357	368
4000	1400	117	135	152	167	181	193	213	229	239	246	249
	1600	135	157	176	194	211	226	251	271	285	295	301
	1800	153	178	201	222	241	258	288	312	331	344	353
8.0	2000	171	199	225	249	271	291	326	354	376	393	405
	2400	207	242	273	303	330	356	400	438	468	492	510
5000	1750	148	171	192	212	230	246	273	294	309	319	325
	2000	170	197	223	246	267	286	320	346	366	381	390
	2250	193	224	253	280	304	327	366	398	423	442	456
7.1	2500	215	251	283	314	342	368	413	450	481	504	521
	3000	260	304	344	382	417	449	507	555	595	627	652

$f_s = 18{,}000$

f'_c and n	f_c	t/d = .10	.12	.14	.16	.18	.20	.24	.28	.32	.36	.40
	a	1.42	1.41	1.40	1.39	1.38	1.37	1.34	1.32	1.30	1.28	1.26
	K											
2500	875	71	81	90	98	105	111	121	126	128	127	124
	1000	82	94	105	115	124	132	144	152	157	158	157
	1125	93	108	121	132	143	152	167	178	185	189	189
10.1	1250	105	121	136	149	162	172	191	204	214	220	222
	1500	127	147	166	183	199	213	237	257	271	281	287
3000	1050	86	99	110	120	129	137	149	157	161	162	160
	1200	99	114	128	141	152	161	177	188	195	199	199
	1350	113	130	146	161	174	186	205	220	230	236	238
9.2	1500	126	146	165	181	196	210	233	251	264	273	277
	1800	153	178	201	222	241	259	289	314	333	346	356
4000	1400	116	134	150	164	177	189	207	221	229	234	234
	1600	134	155	174	192	207	221	245	263	275	283	286
	1800	152	176	198	219	237	254	282	304	321	332	338
8.0	2000	170	197	223	246	267	286	320	346	366	381	391
	2400	206	240	271	300	327	351	394	430	458	479	495
5000	1750	146	169	190	209	226	241	266	285	298	306	309
	2000	169	196	220	243	263	281	313	337	355	367	374
	2250	191	222	250	276	300	322	359	389	412	429	439
7.1	2500	214	249	281	310	338	363	406	441	469	490	505
	3000	259	302	341	378	412	444	500	546	583	613	635

$f_s = 20{,}000$

f'_c and n	f_c	t/d = .10	.12	.14	.16	.18	.20	.24	.28	.32	.36	.40
	a	1.58	1.57	1.55	1.54	1.53	1.52	1.49	1.47	1.45	1.42	1.41
	K											
2500	875	70	80	89	96	103	108	116	120	120	118	112
	1000	81	93	104	113	121	128	139	146	149	148	145
	1125	92	106	119	130	140	149	162	172	177	179	178
10.1	1250	104	120	134	147	159	169	186	198	206	210	211
	1500	126	146	164	181	196	210	233	250	263	271	276
3000	1050	85	97	108	118	126	133	144	150	152	151	147
	1200	98	113	126	138	148	158	172	181	187	188	186
	1350	112	129	144	158	171	182	200	213	221	225	225
9.2	1500	125	145	163	179	193	206	228	244	255	262	264
	1800	152	177	199	220	238	255	284	307	324	336	343
4000	1400	115	132	148	161	174	184	201	213	219	221	219
	1600	133	153	172	189	204	217	239	255	265	270	271
	1800	151	175	196	216	234	249	276	296	311	320	324
8.0	2000	169	196	220	243	263	282	314	338	356	369	376
	2400	205	238	269	297	323	347	388	422	448	467	481
5000	1750	145	167	187	205	222	236	259	276	286	292	292
	2000	168	194	218	239	259	277	306	328	344	353	357
	2250	190	220	248	273	296	317	353	380	401	415	423
7.1	2500	213	247	278	307	334	358	399	432	458	476	488
	3000	258	300	339	375	408	439	493	537	572	599	619

$f_s = 22{,}000$

f'_c and n	f_c	t/d = .10	.12	.14	.16	.18	.20	.24	.28	.32	.36	.40
	a	1.74	1.72	1.71	1.70	1.69	1.67	1.64	1.62	1.59	1.56	1.55
	K											
2500	875	69	78	87	94	100	105	111	113	112	108	101
	1000	80	92	102	111	118	125	134	139	141	139	134
	1125	91	105	117	128	137	145	158	166	169	169	166
10.1	1250	103	118	132	145	156	166	181	192	198	200	199
	1500	125	145	163	179	193	206	228	244	255	262	264
3000	1050	84	96	106	115	123	129	139	143	144	140	134
	1200	97	112	124	136	145	154	167	174	178	177	173
	1350	111	128	143	156	168	178	195	206	212	214	213
9.2	1500	124	143	161	176	190	203	223	237	246	251	252
	1800	151	175	197	217	235	251	279	300	315	325	330
4000	1400	114	130	145	159	170	180	195	205	209	209	205
	1600	132	152	170	186	200	213	233	247	255	258	257
	1800	150	173	194	213	230	245	270	288	301	307	309
8.0	2000	168	194	218	240	260	278	307	330	346	356	361
	2400	204	237	267	294	320	343	382	414	438	455	466
5000	1750	144	165	185	202	218	231	252	267	275	278	276
	2000	166	192	215	236	255	272	299	319	332	339	341
	2250	189	218	245	270	292	312	346	371	389	401	406
7.1	2500	211	245	276	304	330	353	393	424	447	462	472
	3000	257	298	336	372	404	434	486	528	561	585	602

TABLE 4.4 (Cont.)

$f_s = 24{,}000$

f'_c and n	f_c	t/d = .10	.12	.14	.16	.18	.20	.24	.28	.32	.36	.40
	a	1.90	1.88	1.86	1.85	1.84	1.82	1.79	1.76	1.74	1.71	1.69
		K										
2500	875	68	77	85	92	97	101	106	107	104	98	89
	1000	79	90	100	109	116	121	130	133	133	129	122
	1125	91	104	115	126	134	142	153	159	161	160	155
10.1	1250	102	117	131	143	153	162	176	185	190	190	187
	1500	124	144	161	176	190	203	223	238	247	252	253
3000	1050	83	94	104	113	120	126	133	136	135	130	121
	1200	96	110	122	133	142	150	161	168	169	167	161
	1350	110	126	141	153	165	174	189	199	203	204	200
9.2	1500	123	142	159	174	187	199	217	230	238	240	239
	1800	150	174	195	215	232	248	274	293	306	314	317
4000	1400	112	129	143	156	167	176	189	197	199	197	190
	1600	131	150	167	183	197	208	227	239	245	246	242
	1800	149	171	192	210	226	241	264	280	291	295	294
8.0	2000	167	192	216	237	256	273	301	322	336	344	347
	2400	203	235	265	292	316	338	376	406	428	443	451
5000	1750	142	163	182	199	214	226	245	258	264	264	259
	2000	165	190	213	233	251	267	292	310	321	325	324
	2250	187	217	243	267	288	307	339	362	378	387	390
7.1	2500	210	243	273	301	326	348	386	415	435	448	455
	3000	255	296	334	369	400	429	479	519	549	571	586

$f_s = 27{,}000$

f'_c and n	f_c	t/d = .10	.12	.14	.16	.18	.20	.24	.28	.32	.36	.40
	a	2.13	2.11	2.10	2.09	2.07	2.05	2.02	1.98	1.95	1.92	1.90
		K										
2500	875	67	75	82	88	93	96	99	98	92	84	72
	1000	78	88	98	105	111	116	122	124	121	114	105
	1125	89	102	113	122	130	137	146	150	149	145	137
10.1	1250	100	115	128	139	149	157	169	176	178	176	170
	1500	123	142	158	173	186	198	216	228	235	237	235
3000	1050	81	92	101	109	115	120	125	126	122	114	102
	1200	95	108	120	129	138	144	153	157	156	151	141
	1350	108	124	138	150	160	169	181	188	190	188	181
9.2	1500	122	140	156	170	182	193	210	220	225	224	220
	1800	149	172	192	211	227	242	266	282	293	298	298
4000	1400	111	126	140	151	161	169	180	185	184	178	168
	1600	129	148	164	179	191	202	218	227	230	227	220
	1800	147	169	188	206	221	234	255	268	275	277	272
8.0	2000	165	190	213	233	251	267	292	310	321	326	325
	2400	201	232	261	287	311	332	367	394	413	424	429
5000	1750	140	161	179	194	207	219	235	244	247	243	234
	2000	163	187	209	228	245	259	282	297	304	305	300
	2250	186	214	239	262	282	300	329	349	361	366	365
7.1	2500	208	240	270	296	320	341	376	401	418	428	430
	3000	253	293	330	364	394	422	469	506	532	551	561

$f_s = 30{,}000$

f'_c and n	f_c	t/d = .10	.12	.14	.16	.18	.20	.24	.28	.32	.36	.40
	a	2.37	2.35	2.33	2.32	2.30	2.28	2.24	2.21	2.17	2.14	2.11
		K										
2500	875	65	73	80	85	89	91	92	88	80	69	54
	1000	76	87	95	102	107	111	115	114	109	100	87
	1125	88	100	110	119	126	132	139	140	137	130	120
10.1	1250	99	113	125	136	145	152	162	166	166	161	152
	1500	122	140	156	170	182	193	209	219	223	223	218
3000	1050	80	90	98	105	111	114	117	115	109	98	83
	1200	93	106	117	126	133	139	146	147	143	135	122
	1350	107	122	135	146	155	163	174	178	177	171	162
9.2	1500	120	138	153	166	178	187	202	209	211	208	201
	1800	147	170	189	207	223	236	258	272	280	282	279
4000	1400	109	124	137	147	156	163	171	173	169	160	146
	1600	127	145	161	174	186	195	208	215	215	209	198
	1800	145	166	185	202	216	228	246	256	260	258	250
8.0	2000	163	188	209	229	246	260	283	298	306	307	303
	2400	199	230	258	283	305	325	358	382	397	406	407
5000	1750	138	158	175	189	201	211	225	231	230	222	209
	2000	161	184	205	223	239	252	272	283	287	284	275
	2250	184	211	235	257	276	293	319	335	344	345	340
7.1	2500	206	237	266	291	314	333	365	388	401	407	405
	3000	251	291	326	359	388	415	459	492	515	530	536

$f_s = 33{,}000$

f'_c and n	f_c	t/d = .10	.12	.14	.16	.18	.20	.24	.28	.32	.36	.40
	a	2.61	2.59	2.56	2.55	2.53	2.51	2.46	2.43	2.39	2.35	2.32
		K										
2500	875	64	71	77	81	84	86	85	79	68	54	37
	1000	75	85	92	98	103	106	108	105	97	85	70
	1125	86	98	107	115	122	126	131	131	125	116	102
10.1	1250	98	111	123	132	140	147	155	157	154	147	135
	1500	120	138	153	166	178	187	201	209	211	208	200
3000	1050	78	88	96	102	106	109	110	105	95	82	64
	1200	92	104	114	122	128	133	138	136	130	118	103
	1350	105	120	132	142	151	157	166	168	164	155	142
9.2	1500	119	136	150	163	173	182	194	199	198	192	182
	1800	146	167	187	203	218	231	250	262	267	266	260
4000	1400	107	121	133	143	151	156	162	161	154	141	124
	1600	125	143	157	170	180	189	199	203	199	190	176
	1800	143	164	182	197	210	221	237	245	245	240	228
8.0	2000	161	185	206	224	240	254	274	286	291	289	281
	2400	198	228	255	279	300	319	349	370	382	387	385
5000	1750	136	155	171	184	195	204	215	217	213	202	185
	2000	159	182	201	218	233	245	262	270	270	263	250
	2250	182	208	232	252	270	285	308	322	327	324	315
7.1	2500	204	235	262	286	308	326	355	374	384	386	381
	3000	249	288	323	354	382	407	449	479	498	509	511

TABLE 4.5
COEFFICIENTS (c) FOR COMPRESSIVE REINFORCEMENT FOR RECTANGULAR AND T-SECTIONS

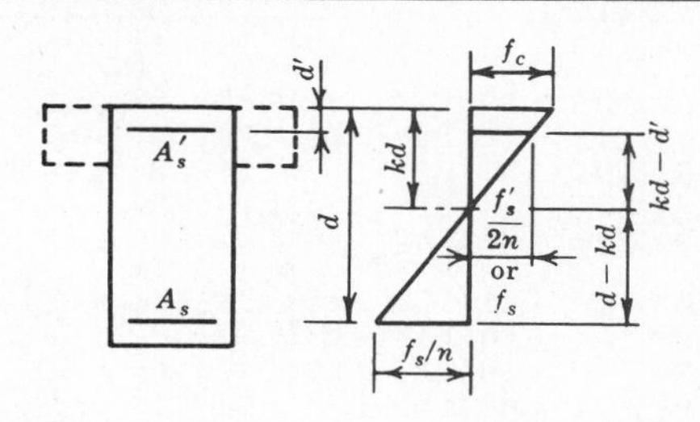

Enter table with known values of $\frac{d'}{d}$, f_s, n and f_c; select value of c.

Compute $A'_s = \dfrac{M - KF}{cd}$.

For values above heavy lines,

$$k \leqq 0.333 + 0.667\frac{d'}{d}$$

and

$$c = \frac{f_s(2n-1)\left(1-\frac{d'}{d}\right)\left(k-\frac{d'}{d}\right)}{12{,}000n(1-k)}$$

For values below heavy lines,

$$k > 0.333 + 0.667\frac{d'}{d}$$

and

$$c = \frac{f_s\left[1-\frac{f_c}{f_s}\left(1-\frac{d'}{kd}\right)\right]\left(1-\frac{d'}{d}\right)}{12{,}000}$$

Where f_s = allowable steel stress in tension.

$f_s = 16{,}000$

f'_c and n	f_c	.02	.04	.06	.08	.10	.12	.14	.16	.18	.20
	875	1.24	1.19	1.09	1.00	.91	.82	.73	.65	.57	.49
2500	1000	1.23	1.21	1.19	1.17	1.07	.97	.88	.79	.70	.62
	1125	1.22	1.20	1.18	1.16	1.14	1.11	1.03	.93	.84	.75
10.1	1250	1.21	1.19	1.17	1.15	1.13	1.11	1.09	1.06	.97	.87
	1500	1.19	1.17	1.15	1.13	1.11	1.09	1.07	1.05	1.03	1.01
	1050	1.23	1.20	1.18	1.10	1.01	.91	.82	.74	.65	.57
3000	1200	1.21	1.19	1.17	1.15	1.13	1.08	.98	.89	.80	.71
	1350	1.20	1.18	1.16	1.14	1.12	1.10	1.08	1.04	.94	.85
9.2	1500	1.19	1.17	1.15	1.13	1.11	1.09	1.07	1.05	1.03	.99
	1800	1.17	1.15	1.13	1.11	1.09	1.07	1.05	1.03	1.01	.99
	1400	1.20	1.18	1.16	1.14	1.12	1.09	.99	.90	.81	.72
4000	1600	1.18	1.16	1.14	1.13	1.11	1.09	1.07	1.05	.98	.88
	1800	1.17	1.15	1.13	1.11	1.09	1.07	1.06	1.04	1.02	1.00
8.0	2000	1.15	1.13	1.12	1.10	1.08	1.06	1.04	1.02	1.01	.99
	2400	1.12	1.10	1.09	1.07	1.05	1.04	1.02	1.00	.98	.97
	1750	1.17	1.15	1.14	1.12	1.10	1.08	1.06	1.03	.93	.84
5000	2000	1.15	1.13	1.12	1.10	1.08	1.06	1.05	1.03	1.01	.99
	2250	1.13	1.11	1.10	1.08	1.07	1.05	1.03	1.01	.99	.98
7.1	2500	1.11	1.10	1.08	1.06	1.05	1.03	1.02	1.00	.98	.96
	3000	1.07	1.06	1.04	1.03	1.01	1.00	.98	.97	.95	.94

$f_s = 18{,}000$

f'_c and n	f_c	.02	.04	.06	.08	.10	.12	.14	.16	.18	.20
	875	1.29	1.18	1.08	.98	.88	.78	.69	.60	.52	.44
2500	1000	1.39	1.37	1.25	1.14	1.04	.94	.84	.75	.65	.57
	1125	1.38	1.36	1.34	1.31	1.20	1.09	.99	.89	.79	.70
10.1	1250	1.37	1.35	1.33	1.30	1.28	1.25	1.14	1.03	.92	.82
	1500	1.35	1.33	1.31	1.29	1.26	1.24	1.22	1.19	1.17	1.08
	1050	1.39	1.29	1.19	1.08	.98	.88	.78	.69	.60	.52
3000	1200	1.38	1.35	1.33	1.26	1.15	1.05	.95	.85	.75	.66
	1350	1.37	1.34	1.32	1.30	1.27	1.22	1.11	1.00	.90	.80
9.2	1500	1.35	1.33	1.31	1.29	1.26	1.24	1.22	1.15	1.04	.94
	1800	1.33	1.31	1.29	1.27	1.24	1.22	1.20	1.18	1.15	1.13
	1400	1.36	1.34	1.32	1.27	1.16	1.06	.96	.86	.76	.67
4000	1600	1.35	1.32	1.30	1.28	1.26	1.24	1.14	1.03	.93	.83
	1800	1.33	1.31	1.29	1.27	1.25	1.22	1.20	1.18	1.10	.99
8.0	2000	1.31	1.29	1.27	1.25	1.23	1.21	1.19	1.17	1.15	1.12
	2400	1.28	1.26	1.24	1.22	1.20	1.18	1.16	1.14	1.12	1.10
	1750	1.33	1.31	1.29	1.27	1.25	1.20	1.09	.98	.88	.79
5000	2000	1.31	1.29	1.27	1.25	1.23	1.21	1.19	1.17	1.07	.96
	2250	1.29	1.28	1.26	1.24	1.22	1.20	1.18	1.16	1.14	1.11
7.1	2500	1.27	1.26	1.24	1.22	1.20	1.18	1.16	1.14	1.12	1.10
	3000	1.23	1.22	1.20	1.18	1.17	1.15	1.13	1.11	1.09	1.07

$f_s = 20{,}000$

f'_c and n	f_c	.02	.04	.06	.08	.10	.12	.14	.16	.18	.20
	875	1.28	1.17	1.06	.95	.85	.75	.65	.56	.47	.39
2500	1000	1.47	1.35	1.24	1.12	1.01	.90	.80	.70	.61	.52
	1125	1.55	1.52	1.41	1.29	1.17	1.06	.95	.84	.74	.65
10.1	1250	1.54	1.51	1.48	1.46	1.33	1.21	1.10	.99	.88	.77
	1500	1.52	1.49	1.47	1.44	1.41	1.39	1.36	1.27	1.15	1.03
	1050	1.40	1.28	1.17	1.06	.95	.85	.75	.65	.56	.47
3000	1200	1.54	1.48	1.36	1.24	1.13	1.01	.91	.80	.70	.61
	1350	1.53	1.50	1.48	1.42	1.30	1.18	1.07	.96	.85	.75
9.2	1500	1.52	1.49	1.47	1.44	1.42	1.35	1.23	1.11	1.00	.89
	1800	1.49	1.47	1.44	1.42	1.39	1.37	1.34	1.32	1.29	1.17
	1400	1.53	1.49	1.37	1.25	1.14	1.03	.92	.81	.72	.62
4000	1600	1.51	1.49	1.46	1.44	1.34	1.22	1.10	.99	.88	.78
	1800	1.49	1.47	1.45	1.42	1.40	1.37	1.29	1.17	1.05	.94
8.0	2000	1.48	1.45	1.43	1.41	1.38	1.36	1.34	1.31	1.22	1.10
	2400	1.45	1.42	1.40	1.38	1.36	1.33	1.31	1.29	1.26	1.24
	1750	1.50	1.47	1.45	1.40	1.28	1.16	1.05	.94	.84	.74
5000	2000	1.48	1.46	1.43	1.41	1.39	1.36	1.25	1.14	1.02	.91
	2250	1.46	1.44	1.41	1.39	1.37	1.35	1.32	1.30	1.21	1.09
7.1	2500	1.44	1.42	1.40	1.37	1.35	1.33	1.31	1.28	1.26	1.24
	3000	1.40	1.38	1.36	1.34	1.32	1.30	1.28	1.26	1.23	1.21

$f_s = 22{,}000$

f'_c and n	f_c	.02	.04	.06	.08	.10	.12	.14	.16	.18	.20
	875	1.28	1.16	1.04	.93	.82	.72	.62	.52	.43	.34
2500	1000	1.47	1.34	1.22	1.10	.98	.87	.76	.66	.56	.47
	1125	1.66	1.53	1.39	1.27	1.14	1.03	.91	.80	.70	.59
10.1	1250	1.70	1.67	1.57	1.44	1.31	1.18	1.06	.94	.83	.72
	1500	1.68	1.65	1.62	1.59	1.57	1.49	1.36	1.23	1.10	.98
	1050	1.39	1.27	1.15	1.03	.92	.81	.71	.61	.51	.42
3000	1200	1.60	1.47	1.34	1.22	1.10	.98	.87	.76	.66	.56
	1350	1.69	1.66	1.53	1.40	1.27	1.15	1.03	.92	.80	.70
9.2	1500	1.68	1.65	1.62	1.59	1.45	1.32	1.19	1.07	.95	.84
	1800	1.66	1.63	1.60	1.57	1.55	1.52	1.49	1.38	1.24	1.12
	1400	1.61	1.48	1.35	1.23	1.11	.99	.88	.77	.67	.57
4000	1600	1.67	1.65	1.57	1.44	1.31	1.19	1.07	.95	.84	.73
	1800	1.66	1.63	1.60	1.58	1.51	1.38	1.25	1.13	1.01	.89
8.0	2000	1.64	1.62	1.59	1.56	1.54	1.51	1.44	1.30	1.17	1.05
	2400	1.61	1.58	1.56	1.53	1.51	1.48	1.46	1.43	1.40	1.37
	1750	1.66	1.64	1.51	1.38	1.25	1.13	1.01	.90	.79	.69
5000	2000	1.64	1.62	1.59	1.56	1.48	1.34	1.22	1.09	.98	.86
	2250	1.62	1.60	1.57	1.55	1.52	1.50	1.42	1.29	1.16	1.04
7.1	2500	1.60	1.58	1.55	1.53	1.50	1.48	1.45	1.43	1.35	1.21
	3000	1.56	1.54	1.52	1.49	1.47	1.45	1.42	1.40	1.37	1.35

TABLE 4.5 (Cont.)

$f_s = 24{,}000$ (d'/d)

f'_c and n	f_c	.02	.04	.06	.08	.10	.12	.14	.16	.18	.20
	875	1.27	1.14	1.02	.91	.79	.68	.58	.48	.38	.29
2500	1000	1.46	1.33	1.20	1.07	.95	.84	.73	.62	.51	.42
	1125	1.65	1.51	1.38	1.24	1.12	.99	.87	.76	.65	.54
10.1	1250	1.85	1.70	1.55	1.41	1.28	1.15	1.02	.90	.78	.67
	1500	1.84	1.81	1.78	1.75	1.60	1.46	1.32	1.18	1.05	.93
	1050	1.39	1.26	1.13	1.01	.89	.78	.67	.57	.47	.37
3000	1200	1.60	1.46	1.32	1.19	1.07	.95	.83	.72	.61	.51
	1350	1.81	1.66	1.52	1.38	1.25	1.12	.99	.87	.76	.65
9.2	1500	1.84	1.81	1.71	1.56	1.42	1.28	1.15	1.03	.90	.79
	1800	1.82	1.79	1.76	1.73	1.70	1.62	1.47	1.33	1.20	1.07
	1400	1.61	1.47	1.33	1.21	1.08	.96	.84	.73	.62	.52
4000	1600	1.84	1.70	1.56	1.42	1.28	1.15	1.03	.91	.79	.68
	1800	1.82	1.79	1.76	1.63	1.49	1.35	1.21	1.08	.96	.84
8.0	2000	1.80	1.78	1.75	1.72	1.69	1.54	1.40	1.26	1.13	1.00
	2400	1.77	1.75	1.72	1.69	1.66	1.63	1.60	1.57	1.46	1.32
	1750	1.78	1.63	1.49	1.36	1.22	1.10	.98	.86	.75	.64
5000	2000	1.81	1.78	1.73	1.59	1.45	1.31	1.18	1.05	.93	.81
	2250	1.79	1.76	1.73	1.70	1.67	1.52	1.38	1.25	1.12	.99
7.1	2500	1.77	1.74	1.71	1.68	1.66	1.63	1.59	1.44	1.30	1.17
	3000	1.73	1.70	1.68	1.65	1.62	1.60	1.57	1.54	1.51	1.49

$f_s = 27{,}000$ (d'/d)

f'_c and n	f_c	.02	.04	.06	.08	.10	.12	.14	.16	.18	.20
	875	1.26	1.13	1.00	.87	.75	.63	.52	.41	.31	.21
2500	1000	1.45	1.31	1.17	1.04	.91	.79	.67	.55	.44	.34
	1125	1.64	1.49	1.35	1.21	1.07	.94	.82	.70	.58	.47
10.1	1250	1.84	1.68	1.53	1.38	1.24	1.10	.96	.84	.71	.60
	1500	2.09	2.05	1.88	1.72	1.56	1.41	1.26	1.12	.98	.85
	1050	1.38	1.24	1.11	.98	.85	.73	.61	.50	.40	.29
3000	1200	1.59	1.44	1.30	1.16	1.03	.90	.77	.66	.54	.43
	1350	1.80	1.64	1.49	1.34	1.20	1.07	.94	.81	.69	.57
9.2	1500	2.01	1.84	1.68	1.53	1.38	1.23	1.10	.96	.83	.71
	1800	2.07	2.03	2.00	1.90	1.73	1.57	1.42	1.27	1.13	.99
	1400	1.60	1.45	1.31	1.17	1.04	.91	.79	.67	.55	.45
4000	1600	1.84	1.68	1.53	1.38	1.24	1.10	.97	.84	.72	.61
	1800	2.07	1.91	1.75	1.59	1.44	1.30	1.16	1.02	.89	.77
8.0	2000	2.05	2.02	1.97	1.81	1.65	1.49	1.34	1.20	1.06	.93
	2400	2.02	1.99	1.95	1.92	1.89	1.85	1.71	1.55	1.39	1.25
	1750	1.77	1.61	1.47	1.32	1.18	1.05	.92	.80	.68	.56
5000	2000	2.03	1.87	1.71	1.55	1.41	1.26	1.12	.99	.86	.74
	2250	2.03	2.00	1.95	1.79	1.63	1.47	1.33	1.18	1.05	.91
7.1	2500	2.01	1.98	1.95	1.92	1.85	1.69	1.53	1.38	1.23	1.09
	3000	1.97	1.94	1.91	1.88	1.85	1.82	1.79	1.76	1.60	1.44

$f_s = 30{,}000$ (d'/d)

f'_c and n	f_c	.02	.04	.06	.08	.10	.12	.14	.16	.18	.20
	875	1.25	1.11	.97	.84	.71	.58	.46	.35	.24	.14
2500	1000	1.44	1.29	1.15	1.00	.87	.74	.61	.49	.37	.26
	1125	1.64	1.48	1.32	1.17	1.03	.89	.76	.63	.51	.39
10.1	1250	1.83	1.66	1.50	1.34	1.19	1.05	.91	.77	.64	.52
	1500	2.21	2.03	1.85	1.68	1.52	1.36	1.20	1.05	.91	.78
	1050	1.37	1.22	1.08	.94	.81	.68	.56	.44	.33	.22
3000	1200	1.58	1.42	1.27	1.12	.98	.85	.72	.59	.47	.36
	1350	1.79	1.62	1.46	1.31	1.16	1.02	.88	.75	.62	.50
9.2	1500	2.00	1.82	1.66	1.49	1.34	1.19	1.04	.90	.76	.64
	1800	2.31	2.22	2.04	1.86	1.69	1.52	1.36	1.21	1.06	.91
	1400	1.59	1.43	1.28	1.14	1.00	.86	.73	.60	.48	.37
4000	1600	1.83	1.66	1.50	1.35	1.20	1.05	.91	.78	.65	.53
	1800	2.07	1.89	1.72	1.56	1.40	1.25	1.10	.96	.82	.69
8.0	2000	2.30	2.12	1.94	1.77	1.60	1.44	1.28	1.13	.99	.85
	2400	2.26	2.23	2.19	2.15	2.01	1.83	1.65	1.49	1.33	1.17
	1750	1.76	1.60	1.44	1.29	1.14	1.00	.86	.73	.61	.49
5000	2000	2.02	1.85	1.68	1.52	1.36	1.21	1.07	.93	.79	.66
	2250	2.28	2.10	1.92	1.75	1.59	1.43	1.27	1.12	.98	.84
7.1	2500	2.26	2.22	2.17	1.99	1.81	1.64	1.47	1.32	1.16	1.02
	3000	2.22	2.18	2.15	2.11	2.08	2.04	1.88	1.70	1.53	1.37

$f_s = 33{,}000$ (d'/d)

f'_c and n	f_c	.02	.04	.06	.08	.10	.12	.14	.16	.18	.20
	875	1.24	1.09	.94	.80	.66	.53	.41	.29	.17	.06
2500	1000	1.43	1.27	1.12	.97	.83	.69	.55	.43	.30	.19
	1125	1.63	1.46	1.30	1.14	.99	.84	.70	.57	.44	.32
10.1	1250	1.82	1.64	1.47	1.31	1.15	1.00	.85	.71	.57	.44
	1500	2.20	2.01	1.83	1.65	1.47	1.31	1.15	.99	.84	.70
	1050	1.36	1.20	1.05	.91	.77	.63	.50	.38	.26	.14
3000	1200	1.57	1.40	1.24	1.09	.94	.80	.66	.53	.40	.28
	1350	1.78	1.60	1.44	1.27	1.12	.97	.82	.68	.55	.42
9.2	1500	1.99	1.80	1.63	1.46	1.29	1.14	.98	.84	.69	.56
	1800	2.40	2.21	2.01	1.83	1.65	1.47	1.30	1.14	.99	.84
	1400	1.58	1.41	1.26	1.10	.95	.81	.67	.54	.42	.30
4000	1600	1.82	1.65	1.48	1.31	1.16	1.00	.86	.72	.58	.46
	1800	2.06	1.88	1.70	1.52	1.36	1.20	1.04	.89	.75	.62
8.0	2000	2.30	2.11	1.92	1.74	1.56	1.39	1.23	1.07	.92	.78
	2400	2.51	2.47	2.36	2.16	1.97	1.78	1.60	1.42	1.26	1.10
	1750	1.75	1.58	1.41	1.25	1.10	.95	.81	.67	.54	.41
5000	2000	2.01	1.83	1.66	1.49	1.32	1.16	1.01	.87	.72	.59
	2250	2.28	2.08	1.90	1.72	1.54	1.38	1.21	1.06	.91	.77
7.1	2500	2.50	2.34	2.14	1.95	1.77	1.59	1.42	1.25	1.09	.94
	3000	2.46	2.42	2.39	2.35	2.21	2.02	1.83	1.64	1.46	1.29

Solved Problems

4.1. Derive the principal expressions used in the working stress design of rectangular reinforced concrete beams. Refer to Fig. 4-10.

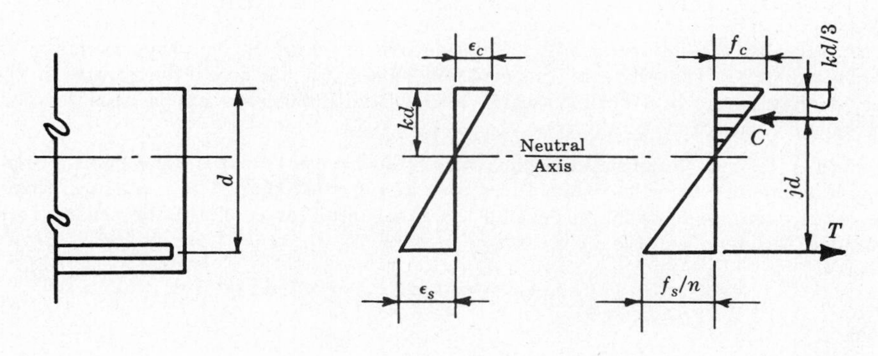

Fig. 4-10

For the section shown to be in equilibrium, the summation of horizontal forces must equal zero and the summation of moments must equal zero. Then $C = T$ or

$$\tfrac{1}{2}f_c\,kdb = A_s f_s \qquad (a)$$

And if the distance from the centroid of the compressive stress to the centroid of tensile stress is jd,

$$M_c = Cjd = \tfrac{1}{2}f_c\,kd^2bj \quad \text{and} \quad M_s = A_s f_s jd$$

From Fig. 4-10, $jd = d - kd/3$, or $j = 1 - k/3$. Hence to evaluate the expression for the resisting moment, the value of k must be determined.

If E = modulus of elasticity = f/ϵ, then $E_s = f_s/\epsilon_s$ and $E_c = f_c/\epsilon_c$. From the strain diagram,

$$\frac{\epsilon_c}{\epsilon_s} = \frac{kd}{d - kd} \quad \text{or} \quad \frac{f_c/E_c}{f_s/E_s} = \frac{kd}{d-kd} = \frac{k}{1-k}$$

And if $n = E_s/E_c$, where n is called the modular ratio,

$$\frac{nf_c}{f_s} = \frac{k}{1-k} \qquad (b)$$

from which

$$f_c = \frac{f_s k}{n(1-k)}, \qquad f_s = \frac{nf_c(1-k)}{k}$$

and

$$nf_c = f_s k + nf_c k = k(nf_c + f_s) \quad \text{or} \quad k = \frac{nf_c}{nf_c + f_s} = \frac{1}{1 + f_s/nf_c}$$

This is the usual form of the expression for k.

If the percentage of reinforcing steel is $p = A_s/bd$, then equation (a) becomes $\tfrac{1}{2}f_c\,kdb = pbdf_s$, or

$$f_c/f_s = 2p/k \qquad (c)$$

Substituting equation (b) into equation (c), we get

$$k^2 + 2pnk = 2pn \quad \text{and} \quad k = \sqrt{2pn + (pn)^2} - pn$$

Letting $R = \tfrac{1}{2}f_c\,jk$, we obtain

$$M_c = M_s = Rbd^2 \quad \text{and} \quad d = \sqrt{M/Rb}$$

These derived equations are developed using the straight-line theory and are applicable to the proportioning of rectangular reinforced concrete beams with tensile reinforcement only.

4.2. **Derive the expressions used in the transformed section method of analysis.**

Within the elastic range, stress is proportional to strain; hence $f_c = \epsilon_c E_c$ and $f_s = \epsilon_s E_s$. If the strain in concrete and steel are the same, or $\epsilon_c = \epsilon_s$, then

$$f_c/E_c = f_s/E_s \quad \text{or} \quad f_c = E_c f_s/E_s$$

If the modular ratio $n = E_s/E_c$, then

$$f_c = f_s/n \quad \text{or} \quad f_s = nf_c$$

In Fig. 4-3 the deformations or strains are assumed to be proportional to the distance from the neutral axis. Therefore at the centroid of the tensile force the strain in the concrete is the same as that in the reinforcing steel. (This assumes the concrete is capable of resisting tension.) Consequently the stress in the concrete is $f_c = f_s/n$.

If it is desired to substitute a quantity of concrete to act as the reinforcing steel, and if the stress in the concrete is less than the stress in the steel that it is replacing, then a larger area of concrete is required in order to develop the same total force. Or if the tensile force in the concrete is equal to the tensile force in the steel, then $T_c = T_s$ and $f_c A_c = f_s A_s$. Now since $f_c = f_s/n$, we have

$$f_s A_c/n = f_s A_s \quad \text{or} \quad A_c = nA_s$$

4.3. **Derive the principal expressions used in the working stress design of reinforced concrete T-beams. Refer to Fig. 4-11.**

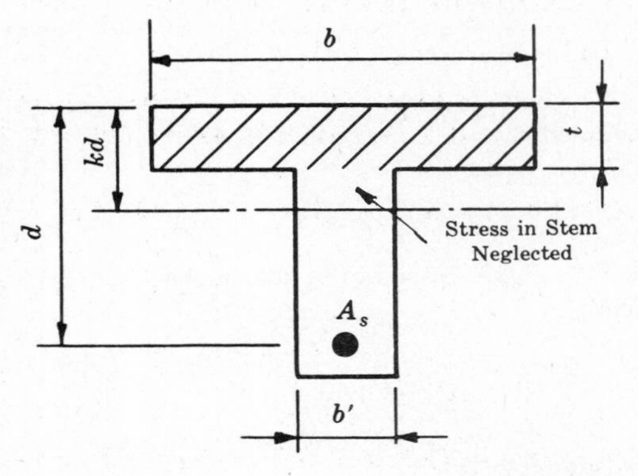

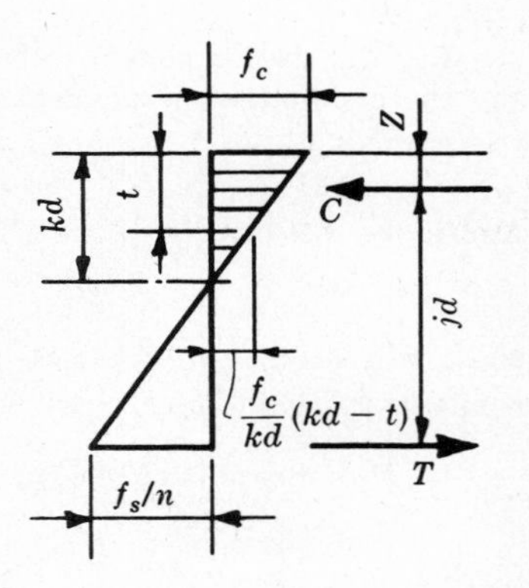

Fig. 4-11

If the compressive stress in the stem is neglected, the summation of horizontal forces is $C = T$. Since the compressive force is bounded by a trapezoid, $\frac{1}{2}[f_c + (f_c/kd)(kd-t)]bt = A_s f_s$ or

$$\frac{f_c(2kd-t)bt}{2kd} = A_s f_s \tag{a}$$

Rearranging (*a*),

$$\frac{f_c(2kd-t)bt}{f_s 2kd} = A_s = pbd \quad \text{or} \quad \frac{f_c}{f_s} = \frac{2kpd^2}{(2kd-t)t} \tag{b}$$

Substituting for f_c in (*b*), we obtain

$$\frac{1}{n(1-k)} = \frac{2pd^2}{(2kd-t)t} \quad \text{from which} \quad k = \frac{2pd^2n + t^2}{2(dt + pd^2n)} = \frac{pn + \frac{1}{2}(t/d)^2}{pn + t/d}$$

The distance z may be found by summing moments about the centroid of the compressive forces:

$$z = \frac{t(3kd-2t)}{3(2kd-t)} \tag{c}$$

From Fig. 4-11, $jd = d - z$. Then substituting (*c*) into $j = 1 - z/d$, we get

$$j = \frac{6 - 6(t/d) + 2(t/d)^2 + (t/d)^3(\frac{1}{2}pn)}{6 - 3(t/d)}$$

LEATHER TO BOOT

116 W. College Ave. State College, PA 16801

(814) 234-1022

NO CASH REFUNDS

Customer's Order No.________ Phone No.________ Date 7/17 19 78

Name________

Address________

SOLD BY	CASH	C. O. D.	CHARGE	ON ACCT.	MDSE. RETD.	PAID OUT	

QUAN.	DESCRIPTION	PRICE	AMOUNT	
	Levis Yokum		13	95
	7½			
	TOTAL			

ALL claims and returned goods MUST be accompanied by this bill.

0829 Rec'd by________

A B C HANGER & SUPPLY, INC., GARDEN CITY, N. Y.

The resisting moment of the concrete would be $M_c = C_c jd$ or $M_c = f_c(1 - t/2kd)btjd$

If $R = f_c j(1 - t/2kd)(t/d)$, then $M_c = Rbd^2$ and $d = \sqrt{M_c/Rb}$.

These expressions were developed assuming that the compressive stress in the stem portion is negligible and that the beam has tensile reinforcement only.

4.4. Derive the principal expressions used in the working stress design of doubly reinforced concrete beams. Refer to Fig. 4-9.

The expressions for a doubly reinforced beam will now be derived using the same basic assumptions as in Problems 4.1-4.3.

As with the section with tensile reinforcement only, $k = \dfrac{1}{1 + f_s/nf_c}$.

If the compressive stress in the compression steel is f_s', then $\dfrac{\epsilon_s'}{\epsilon_s} = \dfrac{kd - d'}{d - kd}$; and if $E\epsilon = f$, then

$$\frac{f_s'/E_s}{f_s/E_s} = \frac{kd - d'}{d - kd} \qquad \text{or} \qquad f_s' = \frac{f_s(kd - d')}{d - kd} \tag{a}$$

Also, $\dfrac{\epsilon_s'}{\epsilon_c} = \dfrac{kd - d'}{kd}$ or $\dfrac{f_s'/E_s}{f_c/E_c} = \dfrac{kd - d'}{kd}$. Letting $n = E_s/E_c$, $f_c' = nf_c\left(\dfrac{kd - d'}{kd}\right)$

In Fig. 4-9, if horizontal forces are summed, $T = C = C_s' + C_c$, where C_s' is the compressive force in the steel and C_c is the compressive force in the concrete. Then

$$\tfrac{1}{2}f_c kbd + A_s' f_s' = f_s A_s \tag{b}$$

It will be noted that the compressive force in the concrete should be a trifle less than that resulting from the above because the area occupied by the compressive steel was not deducted. This discrepancy will be negligible.

If equation (*a*) is substituted into equation (*b*), and if $A_s = pbd$ and $A_s' = p'bd$, then

$$\tfrac{1}{2}f_c kbd + p'bdf_s\left(\frac{kd - d'}{d - kd}\right) = f_s pbd \qquad \text{or} \qquad \frac{f_c}{f_s} = \frac{2}{k}\left[p - p'\left(\frac{kd - d'}{d - kd}\right)\right] \tag{c}$$

As before,

$$\frac{f_c}{f_s} = \frac{k}{n(1 - k)} \tag{d}$$

Equating f_c/f_s from (*c*) and (*d*), we obtain $k = \sqrt{2n(p + p'd'/d) + n^2(p + p')^2} - n(p + p')$.

The distance z can be determined by summing moments, $z = \dfrac{k^3d/3 + 2np'd'(k - d'/d)}{k^2 + 2np'(k - d'/d)}$; and $jd = d - z$.

The resisting moment due to the compressive force is

$$M_c = Cjd = (C_c + C_s')jd \qquad \text{or} \qquad M_c = jd(\tfrac{1}{2}f_c kbd + f_s' p'bd)$$

Substituting for f_s', $M_c = \tfrac{1}{2}f_c jbd^2[k + 2np'(1 - d'/kd)]$.

In the derivation of the above expressions for doubly reinforced beams, it was assumed that the compressive steel was n times as stiff or effective as concrete. However, because of observations of actual members in service and empirical data, the 1963 ACI Building Code along with most codes state that the effective modular ratio shall be taken as $2n$ for compressive steel in flexural design using the straight-line theory. But f_s' shall not exceed the allowable tensile stress of the steel. This means that in the previous expressions $2p'$ should be substituted for p'. Unless otherwise stated, a modular ratio of $2n$ shall always be used here.

In the design of doubly reinforced beams, it is assumed that the resisting moment is divided into two parts. One is the resisting moment of the beam section assuming there is no compressive reinforcement. The second is a couple formed by the additional compressive force in the compression steel and the tensile force in a like amount of additional tension steel.

If the resisting moment of the singly reinforced section is $M_1 = \tfrac{1}{2}f_c kjbd^2 = Rbd^2$ and the moment resisted by the compression steel is $M_2 = A_s' f_s'(d - d')$, then the total moment is $M = Rbd^2 + f_s' A_s'(d - d')$.

4.5. Fig. 4-12 shows a reinforced concrete beam that must resist a 50.0 ft-kip moment. If the concrete strength is 3000 psi, determine the flexural stresses in the concrete and steel by the transformed area method.

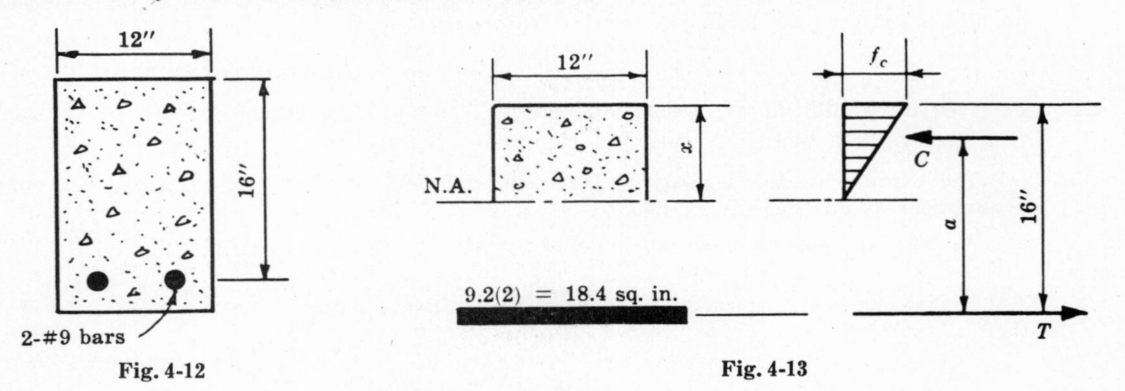

Fig. 4-12 Fig. 4-13

In order to determine the transformed section, the value of n must be known. Here $n = E_s/E_c$ where $E_s = 29{,}000{,}000$ psi and $E_c = 33w^{1.5}\sqrt{f'_c}$. If $w = 145$, then $E_c = 33(145)^{1.5}\sqrt{3000} = 3{,}160{,}000$ psi and $n = 29/3.16 = 9.2$.

The transformed section is as shown in Fig. 4-13.

Summing moments of the areas about the neutral axis, $12(x)(x/2) = 18.4(16 - x)$ or $x = 5.65$ in. The centroid of the compressive force would be $x/3$ below the top of the beam. Hence, $a = 16 - 5.65/3 = 14.12''$.

Then summing moments and equating the internal resisting moment to the external moment, $Ca = Ta = M = 50.0$ ft-kips or $C = T = 50.0(12{,}000)/14.12 = 42{,}800$ lb.

The steel stress would be $f_s = T/A_s = 42{,}800/2.0 = 21{,}400$ psi.

The compressive force in the concrete is $C = f_c(x)(12)/2$; then $42{,}800 = f_c(5.65)(12)/2$ or $f_c = 1260$ psi.

4.6. Solve Problem 4.5 using the classic flexure formula, $f = Mc/I$.

The moment of inertia $I = 12(5.65)^3/3 + 18.4(16 - 5.65)^2 = 2691 \text{ in}^4$.

$$f_s = Mcn/I = 50.0(12{,}000)(10.35)(9.2)/2691 = 21{,}300 \text{ psi}$$

$$f_c = Mc/I = 50.0(12{,}000)(5.65)/2691 = 1260 \text{ psi}$$

4.7. Solve Problem 4.5 using the formulas derived for rectangular beams.

If $p = A_s/bd = 2.0/(12)(16)$ and $n = 9.2$, using equation (*4.3*), $k = \sqrt{2pn + (pn)^2} - pn = 0.352$. And from equation (*4.4*), $j = 1 - k/3 = 1 - 0.352/3 = 0.883$. If $M = 50.0$ ft-kips the stress in the steel and concrete can be determined from equations (*4.9*) and (*4.1*) respectively

$$f_s = M/A_s jd = 21{,}300 \text{ psi} \qquad \text{and} \qquad f_c = f_s k/n(1-k) = 1260 \text{ psi}$$

4.8. Fig. 4-14 shows a reinforced concrete T-beam that must resist a 50.0 ft-kip moment. If $f'_c = 3000$ psi, determine the flexural stresses in the concrete and steel by the transformed area method.

(*a*) If $n = E_s/E_c = 9.2$ for $f'_c = 3000$ psi, the transformed area of steel $= 18.4 \text{ in}^2$.

If the neutral axis of the section is less than 3″ below the compression face, the section acts as a rectangular beam. If it is more than 3″, the section acts as a T-beam.

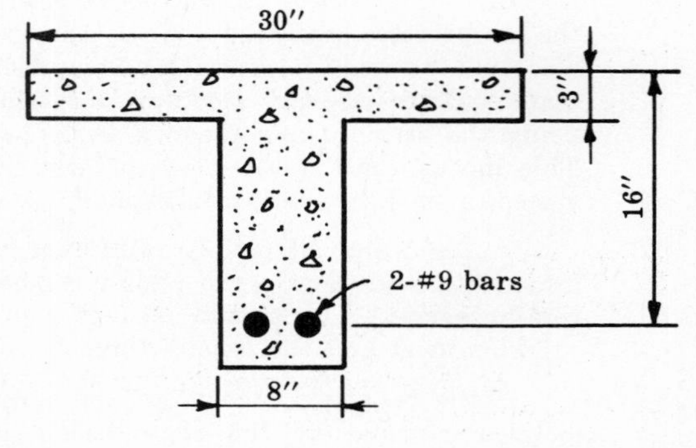

Fig. 4-14

To determine whether the neutral axis is above or below the slab, moments of the transformed areas are taken about the bottom of the flange. For the concrete flange, $3(30)(1.5) = 135\ \text{in}^3$; for the reinforcement, $18.4(13) = 239\ \text{in}^3$. Hence the neutral axis is below the flange.

(*b*) The summation of moments about the neutral axis is $8(x)(x/2) + (30-8)(3)(x-1.5) = 18.4(16-x)$ or $x = 3.92$ in.

(*c*) If the neutral axis is 0.92″ below the flange, the stress at the bottom of the flange is $0.92f_c/3.92 = 0.235f_c$.

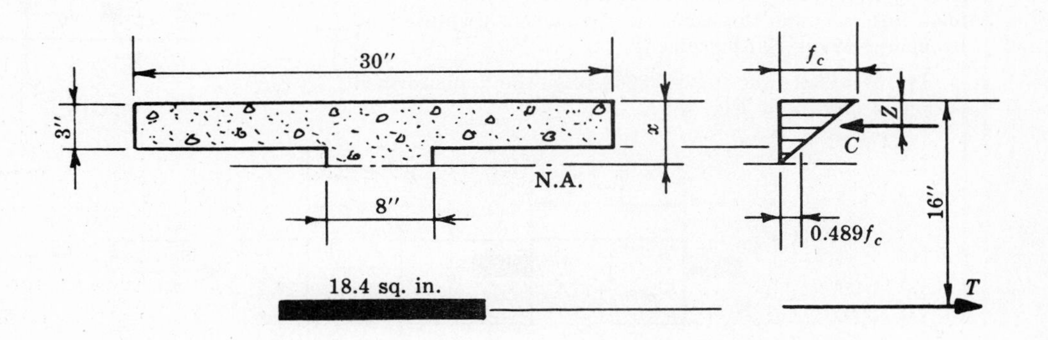

Fig. 4-15

The location of the centroid of the compressive force is not as readily determined here as in a rectangular beam. From Fig. 4-15, summing moments of the compressive forces about the top of the flange,

$$zC = [30.0(3.92)f_c/2](3.92/3) - [22.0(0.92)0.235f_c/2](3.0 + 0.92/3)$$
$$= 58.7f_c(1.31) - 2.38f_c(3.31) = 76.9f_c - 7.88f_c$$
$$z = 69.0f_c/(58.7f_c - 2.38f_c) = 1.22''$$

(*d*) The distance from the centroid of the compressive force to the tensile force is

$$jd = 16.0 - 1.22 = 14.78''$$

Summing internal resisting moments, $M = Cjd = Tjd$. Also, $Tjd = A_s f_s jd = 50.0(12{,}000)$. $f_s = 20{,}300$ psi and $C = 40{,}600$ lb. Then $(58.7 - 2.38)f_c = 40{,}600$ or $f_c = 720$ psi.

4.9. Solve Problem 4.8 using the classic flexure formula $f = Mc/I$.

Moment of inertia $I = 30(3)^3/12 + 3(30)(2.42)^2 + 8(0.92)^3/3 + 18.4(16-3.92)^2 = 3276\ \text{in}^4$.

$$f_s = Mcn/I = 50.0(12{,}000)(12.08)(9.2)/3276 = 20{,}400 \text{ psi}$$
$$f_c = Mc/I = 50.0(12{,}000)(3.92)/3276 = 720 \text{ psi}$$

4.10. Solve Problem 4.8 using the formulas derived for T-sections.

If $p = A_s/bd = 2.0/(30.0)(16) = 0.00417$, $n = 9.2$, $t = 3$ and $d = 16$, equation (*4.12*) gives

$$k = \frac{pn + \frac{1}{2}(t/d)^2}{pn + (t/d)} = 0.250$$

Then $z = \dfrac{t(3kd - 2t)}{3(2kd - t)} = 1.20$ and $j = 1 - z/d = 0.922$.

If $M = 50$ ft-kips, we can determine the steel and concrete stresses by equations (*4.17*) and (*4.10*) respectively:

$$f_s = M/A_s jd = 20{,}300 \text{ psi} \qquad \text{and} \qquad f_c = f_s k/n(1-k) = 736 \text{ psi}$$

4.11. **The beam shown in Fig. 4-16 resists a 50.0 ft-kip moment. If $f_c' = 3000$ psi, determine the flexural stresses in the concrete and steel by the transformed area method.**

The transformed area of the compression reinforcement is $(2n-1)A_s'$. This is in accordance with the assumption that the stress in the compression reinforcement is twice that in the concrete. The expression also takes into account the amount of concrete displaced by the compression reinforcement.

If $n = 9.2$ for $f_c' = 3000$ psi, the transformed section is shown in Fig. 4-17.

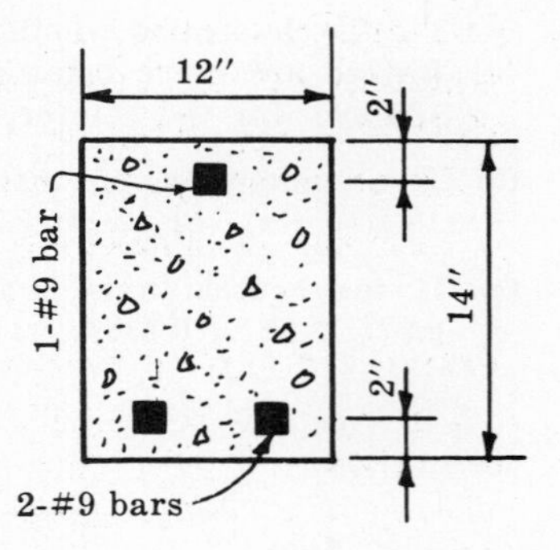

Fig. 4-16

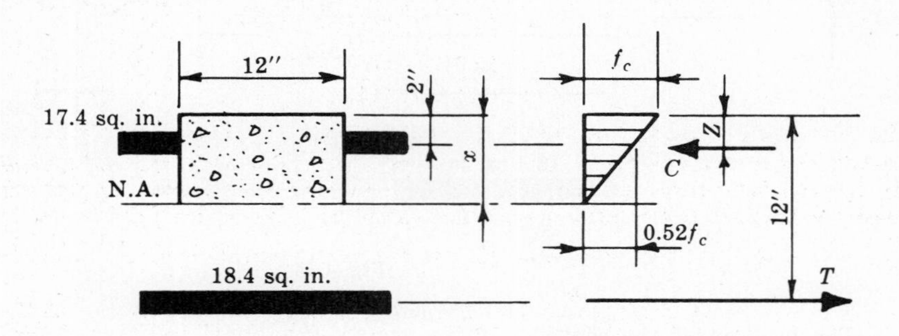

Fig. 4-17

Summing moments of the forces about the neutral axis, $12(x)(x/2) + 17.4(x-2) = 18.4(12-x)$ or $x = 4.17$ in. The centroid of the compressive force is located by taking moments about the top of the beam.

$$zC = (f_c/2)(12)(4.17)(4.17/3) + (0.52)(f_c)(18.4)(2) = 25.02f_c(1.39) + 9.56f_c(2)$$

$$z = (34.8f_c + 19.12f_c)/(25.02f_c + 9.56f_c) = 1.56''$$

Summing internal resisting moments, $M = Cjd = Tjd$. Also, $Tjd = A_s f_s jd = 50.0(12{,}000)$; and $f_s = 28{,}800$ psi and $C = 57{,}500$ lb.

Now $(25.02 + 9.56)f_c = 57{,}500$ or $f_c = 1670$ psi. Then $f_s' = 2n(0.52)f_c = 16{,}000$ psi.

4.12. **Solve Problem 4.11 using the classic flexure formula $f = Mc/I$.**

The moment of inertia $I = 12(4.17)^3/3 + 17.4(4.17-2.0)^2 + 18.4(12.0-4.17)^2 = 1502 \text{ in}^4$.

$$f_s = Mcn/I = 50.0(12{,}000)(7.83)(9.2)/1502 = 28{,}700 \text{ psi}$$

$$f_c = Mc/I = 50.0(12{,}000)(4.17)/1502 = 1670 \text{ psi}$$

$$f_s' = 2nMy/I = 2(9.2)(50.0)(12{,}000)(2.17)/1502 = 16{,}000 \text{ psi}$$

4.13. **Solve Problem 4.11 using the formulas derived for doubly reinforced beams.**

If $p = A_s/bd = 2.0/(12)(12) = 0.0139$, $p' = A_s'/bd = 1.0/(12)(12) = 0.00695$, and $n = 9.2$, using equation (*4.21*),

$$k = \sqrt{2n(p + 2p'd'/d) + n^2(p+2p')^2} - n(p+2p') = 0.348$$

and
$$z = \frac{k^3d/3 + 4np'd'(k - d'/d)}{k^2 + 4np'(k-d'/d)} = 1.56''$$

It should be noted that $2p'$ is substituted for p' in the above expressions in recognition of the doubled effectiveness of compression reinforcement.

From equation (*4.22*), $j = 1 - z/d = 0.87$. If $M = 50.0$ ft-kips, the stress in the steel and concrete can be determined by use of equations (*4.26*), (*4.18*) and (*4.20*):

$$f_s = \frac{M}{A_s jd} = 28{,}700 \text{ psi}, \quad f_c = \frac{f_s k}{n(1-k)} = 1670 \text{ psi}, \quad f_s' = 2nf_c\frac{k - d'/d}{k} = 16{,}000 \text{ psi}$$

4.14. Proportion a one-way continuous slab to support a maximum positive moment at midspan of 18.5 ft-kips/ft and a maximum negative moment at the support of 55 ft-kips/ft. Assume $f'_c = 3000$ psi, $f_s = 20{,}000$ psi and a minimum depth permitted with no compressive reinforcement.

Assume an allowable concrete stress $f_c = 0.45f'_c = 1350$ psi and $n = 9.2$. From (*4.1*),

$$k/(1-k) = 9.2(1350)/20{,}000 = 0.621 \quad \text{or} \quad k = 0.383$$

From equation (*4.4*), $j = 1 - k/3 = 0.872$; and from equation (*4.5*), $R = f_c jk/2 = 226$.

For a unit width of slab, $b = 12''$; and from equation (*4.7*) the minimum depth may be determined. The negative moment at the support is maximum. Hence,

$$d = \sqrt{M/Rb} = \sqrt{55(12{,}000)/[226(12)]} = 15.4''. \quad \text{Use } d = 15.0''.$$

From equation (*4.9*),

$$-A_s = \frac{M}{f_s jd} = \frac{55(12{,}000)}{20{,}000(0.872)(15.0)} = 2.52 \text{ in}^2$$

and

$$+A_s = \frac{18.5(12{,}000)}{20{,}000(0.872)(15.0)} = 0.85 \text{ in}^2$$

In the determination of the area of reinforcement at midspan, a value of $j = 0.872$ was used. This value for j was determined for a balanced section using the larger moment at the support. Hence the concrete stress at midspan will be less than $0.45f'_c$ and the section is underreinforced. Because j varies as the concrete stress, the value for $+A_s$ is not precise. If

$$M = Rbd^2 = f_c jkbd^2/2 \quad \text{and} \quad f_c = f_s k/n(1-k), \quad j = 1 - k/3$$

then

$$M = \frac{f_s k^2}{2n(1-k)}(1 - k/3)bd^2 \quad \text{or} \quad \frac{k^2(1-k/3)}{(1-k)} = \frac{2nM}{f_s bd^2}$$

And if the values in the problem are substituted into this expression,

$$\frac{k^2(1-k/3)}{(1-k)} = \frac{2(9.2)(18.5)(12{,}000)}{20{,}000(12)(15)^2} \quad \text{or} \quad k = 0.249$$

Then $j = 1 - k/3 = 1 - 0.083 = 0.917$. Substituting this value for j into the expression for the area of reinforcement at midspan,

$$+A_s = \frac{18.5(12{,}000)}{20{,}000(0.917)(15.0)} = 0.807 \text{ in}^2$$

Hence with the more precise value of j, a reduction of reinforcement required of approximately 5% was obtained. Normally, such a refinement is not justified in the usual design of reinforced concrete. In the design of underreinforced sections the generally accepted practice is to use the value of j at balanced design.

In underreinforced sections the value of j is greater than in balanced sections. Therefore the use of a lower value will yield a larger value for A_s and be conservative.

At midspan, $f_c = f_s k/n(1-k) = 722$ psi.

4.15. Solve Problem 4.14 using Tables 4.2 and 4.3.

Table 4.3 for $f_s = 20{,}000$, $f'_c = 3000$ and $f_c = 1350$ does not contain values as high as 55 ft-kips. Hence from Table 4.2, $R = 226$ and substituting into equation (*4.7*), $d = \sqrt{M/Rb} = 15.4''$. Use $d = 15.0''$.

From Table 4.2, $a = 1.44$; substituting in equation (*4.29*),

$$-A_s = M/ad = 2.54 \text{ in}^2 \quad \text{and} \quad +A_s = 18.5/1.44(15) = 0.86 \text{ in}^2$$

4.16. Proportion a balanced reinforced rectangular beam with tension reinforcement only to withstand a 250 ft-kip moment. Assume $f'_c = 3000$ psi and $f_y = 40{,}000$ psi. The length of the beam is 30′-0″.

At balanced design $f_c = 0.45(3000) = 1350$ psi; and from equations (*4.1*) and (*4.2*),

$$k = \frac{1}{1 + f_s/nf_c} = \frac{1}{1 + 20{,}000/9.2(1350)} = 0.383$$

From equations (*4.5*) and (*4.6*),

$$R = \tfrac{1}{2}f_c jk = \tfrac{1}{2}(1350)(1-0.383/3)(0.383) = 226, \quad bd^2 = M/R = 250(12{,}000)/226 = 13{,}300 \text{ in}^3$$

Try a ratio of $d/b = 1.5$. Then $b(1.5b)^2 = 13{,}300$ and $b = 18.1$; use $b = 18.5''$. Thus $d = 1.5(18.5) = 27.7$; use $d = 28.0''$. From equation (*4.9*), $A_s = M/f_s jd = 6.14 \text{ in}^2$.

Try a ratio of $d/b = 1.0$. Then $b^3 = 13{,}300$ and $b = 23.7$; use $b = 24.0''$. Thus $d = 24.0''$ and $A_s = 7.17 \text{ in}^2$.

Try a ratio of $d/b = 0.67$. Then $b(0.67b)^2 = 13{,}300$; $b^3 = 29{,}600$ and $b = 30.9$; use $b = 31.0''$. Thus $d = 0.67(31.0) = 20.8$; use $d = 21.0''$ and $A_s = 8.20 \text{ in}^2$.

The 1963 ACI Building Code requires the following minimum amount of reinforcement in beams. It specifies that for flexural members, except constant thickness slabs, if positive reinforcement is required, the minimum ratio p shall not be less than $200/f_y$ unless the area of reinforcement at every section, positive or negative, is at least one-third greater than that required. If $f_y = 40{,}000$ psi, $200/f_y = 200/40{,}000 = 0.005 < 6.14/28.0(18.5)$.

The deflections of beams must be computed in accordance with the 1963 ACI Building Code if the depth is less than a specified value. For simply supported beams, this minimum depth is $d = l/20 = 30.0(12)/20 = 18.0''$. The deflection check is not mandatory.

This problem could be solved using Table 4.2 by finding the value of $R = 226$ and a value of $a = 1.44$.

4.17. Determine the area of reinforcing steel required for the section shown in Fig. 4-18. Assume $f'_c = 4000$ psi, $f_s = 24{,}000$ psi and $M = 300$ ft-kips.

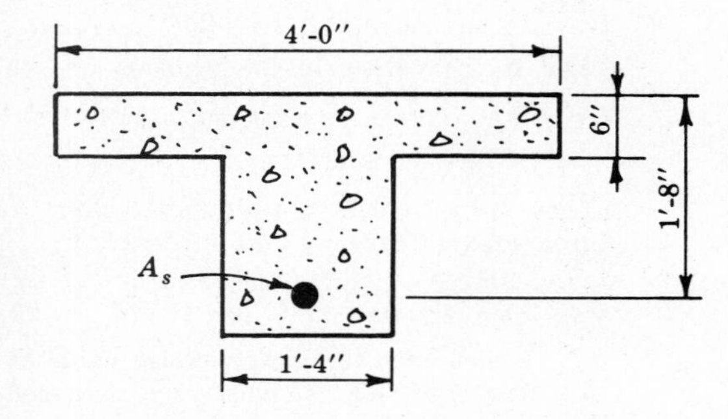

Fig. 4-18

(*a*) The stress in the stem will be neglected. Because f_c is not known, the value of k must be determined by substitution into equations (*4.10*) through (*4.17*). This would be quite cumbersome due to the complex expression for j. Hence a trial and error procedure will be used.

A value $j = 0.9$ will be tried first. Substituting into equation (*4.17*),

$$A_s = M/f_s jd = 300(12{,}000)/24{,}000(0.9)(20) = 8.33 \text{ in}^2$$

Now $p = 8.33/48(20) = 0.00866$, $n = 8.0$, and $t/d = 6.0/20.0 = 0.30$.

From equation (*4.12*), $k = \dfrac{pn + \frac{1}{2}(t/d)^2}{pn + (t/d)} = 0.310$. As previously derived, $z = \dfrac{t(3kd-2t)}{3(2kd-t)}$ and $j = 1 - z/d$. Then

$$j = 1 - \frac{t(3k - 2t/d)}{3(2kd-t)} = 0.897$$

This value for j is near enough to the original value assumed that another trial is not necessary. For T-beams of normal proportions with the t/d ratio varying between 0.10 and 0.40, the value of j will vary from approximately 0.95 to 0.84. Regardless of what value of j is assumed for the initial trial, if it is within this range it will result in a design that would be close to that resulting from the "exact" value.

(*b*) The area of reinforcement is $A_s = 8.33 \text{ in}^2$. The 1963 ACI Building Code requires a minimum amount of reinforcement in beams. The specification states that for flexural members, except constant thickness slabs, if positive reinforcement is required, the minimum ratio p shall not be less than $200/f_y$ unless the area of reinforcement at every section, positive or negative, is at least one third greater than that required. Therefore if $f_y = 50{,}000$ psi, $200/f_y = 0.004 < 8.33/16(20)$. The minimum positive reinforcement requirement is met.

(*c*) If this is a simple span beam, the depth must be $d \geqq l/20$ so that a deflection check is not mandatory.

4.18. Solve Problem 4.17 using Table 4.4.

From Table 4.4, for $f_s = 24{,}000$ psi, $f'_c = 4000$ psi, $f_c = 1800$ psi and $t/d = 0.30$: $a = 1.75$ and $R = 285$.

The area of reinforcement is $A_s = M/ad = 300/1.75(20) = 8.58\ \text{in}^2$.

As a check, the resisting moment is $M = Rbd^2 = 456$ ft-kips > 300.

4.19. Determine the area of reinforcement required for the section shown in Fig. 4-19 if the compressive stress in the stem is included. Assume $f'_c = 3000$ psi, $f_s = 20{,}000$ psi, and $M = 150$ ft-kips. Use Tables 4.2 and 4.4.

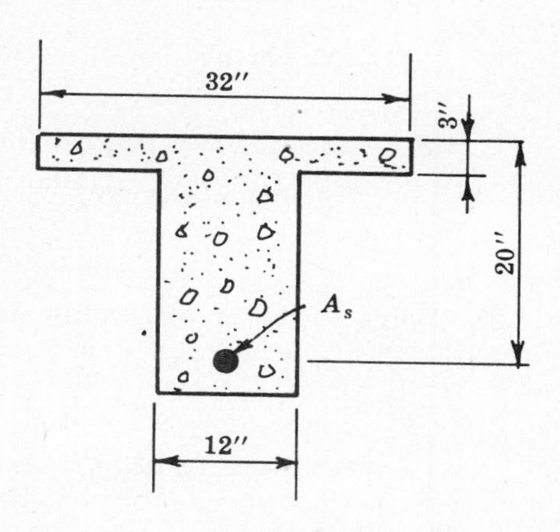

Fig. 4-19

(a) Table 4.4 neglects the effect of the stress in the stem. Hence the effect of the web must be determined using Table 4.2 and then added to the effect of the flange which has a net width equal to $32'' - 12'' = 20''$.

From Table 4.2, for $f_s = 20{,}000$ psi, $f'_c = 3000$ psi, $f_c = 1350$ psi: $R_w = 226$ (Balanced Design) and $a_w = 1.44$. Table 4.4, for $t/d = 0.15$: $R_f = 150$ (Balanced Design) and $a_f = 1.54$.

(b) The resisting moments are $M_w = R_w bd^2 = 90.3$ ft-kips and $M_f = R_f bd^2 = 100.0$ ft-kips. $M_w + M_f = 190$ ft-kips > 150. So the section will be under-reinforced.

(c) The area of reinforcement required to develop the web is $A_{sw} = M_w/a_w d = 3.13\ \text{in}^2$.

The area of reinforcement required to develop the flange is $A_{sf} = (M - M_w)/a_f d = 1.94\ \text{in}^2$.

The total area of tension reinforcement is $A_s = A_{sw} + A_{sf} = 5.07\ \text{in}^2$.

(d) $p = 5.07/20(12) = 0.0211 > 200/f_y$. Hence the minimum positive reinforcement requirement is met.

(e) If the beam is simply supported and $l/20 \leqq 20''$, a deflection check is not mandatory.

4.20. Compare the area of reinforcement required in Problem 4.19 to that required if the compressive stress in the stem is neglected.

From Table 4.4, for $t/d = 0.15$: $R = 150$ and $a = 1.54$. The resisting moment is $M = Rbd^2 = 160$ ft-kips > 150.

The area of reinforcement is $A_s = M/ad = 4.87\ \text{in}^2$.

Because the stress in the stem is neglected, the centroid of the compressive force is located higher than in Problem 4.19. Therefore the internal moment arm jd is increased and the required area of reinforcement is decreased. The resisting moment of the concrete is decreased. The difference in reinforcement required is $(5.07 - 4.87)(100)/5.07 = 3.94\%$.

4.21. Determine the area of reinforcing steel required for the section shown in Fig. 4-20. Assume $f'_c = 3000$ psi, $f_s = 20{,}000$ psi, and $M = 200$ ft-kips.

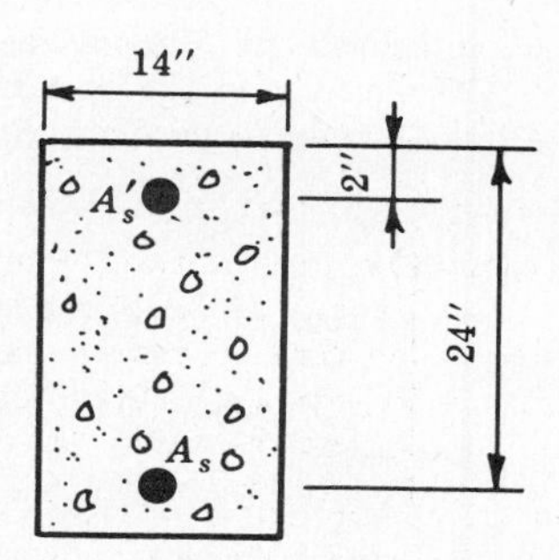

Fig. 4-20

A check should be made to determine if compressive reinforcement is required. If $f_c = 1350$ psi, $f_s = 20{,}000$ psi and $n = 9.2$, $k = 1/(1 + f_s/nf_c) = 0.383$; then $j = 1 - k/3 = 0.872$.

From equation (4.5), $R = \frac{1}{2} f_c jk = 226$. The resisting moment is

$$M = Rbd^2 = 226(14)(24)^2/12{,}000$$
$$= 152 \text{ ft-kips} < 200$$

Hence the section requires compressive reinforcement.

The resisting moment of a balanced design is 152 ft-kips. The compressive reinforcement must resist a moment $M' = 200 - 152 = 48$ ft-kips.

From equation (*4.27*), $A'_s = M'/f'_s(d-d')$. The value of f'_s is a function of f_c and the ratio d'/d. If $d'/d = 0.0833$, from (*4.20*),

$$f'_s = 2nf_c(k - d'/d)/k = 19{,}400 \text{ psi} < 20{,}000$$

Substituting into equation (*4.27*), $A'_s = 48(12{,}000)/(19{,}400)(22) = 1.35 \text{ in}^2$.

From equation (*4.26*), the total area of tension reinforcement is

$$A_s = M/f_s jd = 5.73 \text{ in}^2 \qquad \text{and} \qquad p = 5.73/14(24) = 0.0170 > 200/f_y$$

Hence the minimum positive reinforcement requirement is met.

If the beam is simply supported, and $l/20 \leqq 24''$, a deflection check is not mandatory.

4.22. Solve Problem 4.21 using Tables 4.2 and 4.5.

From Table 4.2 for $f'_c = 3000$ psi, $f_c = 1350$ psi, and $f_s = 20{,}000$ psi: $R = 226$, $a = 1.44$, $j = 0.872$, and $k = 0.383$. The resisting moment is $M = Rbd^2 = 152$ ft-kips < 200. Hence $M' = 200 - 152 = 48$ ft-kips.

From Table 4.5 for $f'_c = 3000$ psi, $f_c = 1350$ psi, $f_s = 20{,}000$ psi and $d'/d = 0.0833$: $c = 1.41$. From equations (*4.30*) and (*4.29*),

$$A'_s = M'/cd = 1.41 \text{ in}^2 \qquad \text{and} \qquad A_s = M/ad = 5.79 \text{ in}^2$$

The other design checks would be the same as in Problem 4.21.

4.23. Determine the area of tension and compression reinforcement required in the section shown in Fig. 4-21 if $M = 167$ ft-kips. Assume $f'_c = 3000$ psi and $f_s = 20{,}000$ psi.

This is the same section as in Problem 4.18. Hence $R = 226$, $a = 1.44$, $j = 0.872$, $k = 0.383$, and $c = 1.41$. Then the resisting moment is $M = 152$ ft-kips < 167. Hence $M' = 15$ ft-kips.

$$A'_s = M'/cd = 0.444 \text{ in}^2$$

and

$$A_s = M/ad = 4.84 \text{ in}^2$$

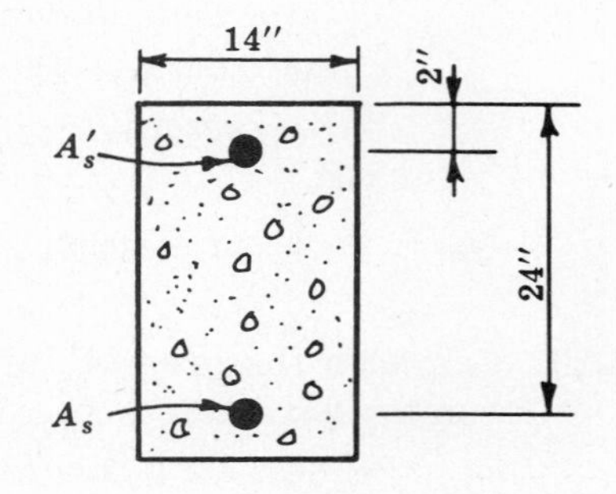

Fig. 4-21

4.24. Determine the area of tension reinforcement required for the section in Problem 4.23 if there is no compression reinforcement and $f_c = 1350$ psi.

If the compressive force as shown in Fig. 4-3 is made larger, the section can resist greater moments. If f_c must remain fixed, the only manner in which to accomplish this is by increasing k. By increasing the area of tension reinforcement, the neutral axis may be lowered. From equations (*4.5*) and (*4.6*), $M = Rbd^2 = \frac{1}{2}f_c jkbd^2$; and from equation (*4.4*), $j = 1 - k/3$. Hence

$$M = \tfrac{1}{2}f_c(1 - k/3)kbd^2 \qquad \text{or} \qquad (1 - k/3)k = 2M/f_c bd^2 = 0.368, \qquad \text{and} \qquad k = 0.429$$

From equations (*4.2*) and (*4.9*), $f_s = nf_c\left(\frac{1-k}{k}\right) = 16{,}500$ psi and $A_s = \frac{M}{f_s jd} = 5.90 \text{ in}^2$.

In Problem 4.23 the total area of reinforcing steel is $4.84 + 0.44 = 5.28$. Comparing the answers,

$$(5.90 - 5.28)(100)/5.28 = 12\% \text{ increase}$$

in reinforcement required when tension steel only is used. This is required for a section that must resist a moment that is merely 10% greater than the balanced resisting moment.

4.25. Determine the area of reinforcement required for the edge or spandrel beam shown in Fig. 4-22. Assume $f'_c = 4000$ psi and $f_s = 24{,}000$ psi. The beam has a span of 10′-0″ and the 5″ slab has a clear span of 16′-0″. The beam is continuous and must resist a negative moment at the support of 170 ft-kips and a positive moment at midspan of 55 ft-kips.

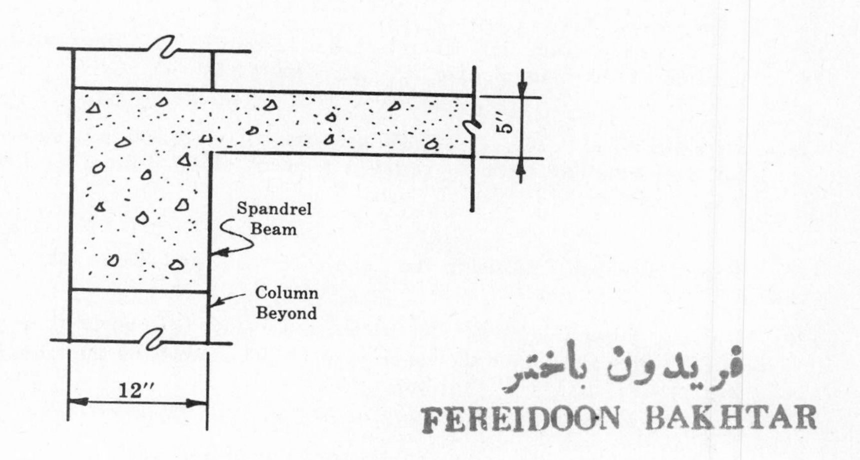

Fig. 4-22

(*a*) The negative moment at the support is the largest absolute and controls the size of the member. If the section is to be designed using balanced reinforcement, $f_c = 1800$ psi, $f_s = 24{,}000$ psi, and $n = 8.0$.

If no compression reinforcement is to be used at the support, then $Rbd^2 \geqq 170$ ft-kips. From Table 4.2, for $f_c = 1800$ psi and $f_s = 24{,}000$ psi: $R = 295$ and $a = 1.76$. Hence $bd^2 = M/R = 6900$ in^3. Because the width of the beam has been set equal to the width of the column, 12″, $d^2 = 6900/12 = 575$. Use $d = 24''$.

The area of negative reinforcement at the support is $-A_s = M/ad = 4.02$ in^2.

(*b*) At the supports, the beam acts as a rectangular section. However, in the positive moment region the beam acts as an unsymmetrical T-section.

The 1963 ACI Building Code specifies that in beams having a flange on one side only the effective flange shall not be greater than 1/12 the beam span length and that the effective overhanging flange shall not be greater than six times the slab thickness nor greater than one-half the slab span. Applying these rules,

$$b = (1/12)(10)(12) + 12 = 22'', \quad b = 12 + 6(5) = 42'', \quad \text{or} \quad b = 12 + \tfrac{1}{2}(16)(12) = 108''$$

Hence the assumed effective flange width is 22″.

(*c*) From Table 4.4, for $f_c = 1800$ psi, $f_s = 24{,}000$ psi and $t/d = 5/24 = 0.208$: $R = 246$ and $a = 1.82$. Then $Rbd^2 = 260$ ft-kips > 55 and the area of positive reinforcement is $+A_s = M/ad = 1.26$ in^2.

Checking for the minimum area of reinforcement, $200/f_y = 0.0033 < 1.26/12(24)$.

(*d*) The beam is continuous at both ends. Consequently the 1963 ACI Building Code would not require a deflection check because $l/26 < 24''$.

Supplementary Problems

4.26. A 10″ wide rectangular concrete beam has an effective depth of 14″ and is reinforced with 1.32 in^2 of steel. If $n = 15$, determine the moment of inertia of the transformed section.
Ans. $I = 1980 \text{ in}^4$

4.27. A 12″ wide rectangular concrete beam is reinforced with 4-#8 bars. If $n = 12$, determine the location of the neutral axis. *Ans.* $kd = 8.9''$

4.28. If $b = 12''$, $d = 16''$, $f_c = 1350$ psi, and $f_s = 20{,}000$ psi, determine the required reinforcement for a rectangular beam to resist a moment of 60 ft-kips. Use the transformed section method.
Ans. 2.59 in^2

4.29. Repeat Problem 4.28 using the flexure formula and Table 4.2.

4.30. Given a T-beam with $b = 50''$, $b' = 10''$, $t = 6''$, $d = 20''$, $A_s = 2.50 \text{ in}^2$, and an applied moment of 65 ft-kips, determine the concrete and steel stresses by the transformed section method if $n = 10$.
Ans. $f_c = 460$ psi, $f_s = 16{,}800$ psi

4.31. Repeat Problem 4.30 using the flexure formula and Table 4.4.

4.32. Given a rectangular doubly reinforced beam with $b = 14''$, $d = 16''$, $f_c = 1350$ psi, $f_s = 20{,}000$ psi, and an applied moment of 140 ft-kips, determine the area of reinforcement required. Assume $d' = 2''$ and $n = 9.2$. *Ans.* $A_s = 6.08 \text{ in}^2$, $A'_s = 2.91 \text{ in}^2$

4.33. Proportion a one-way slab with simple supports to resist an applied moment of 12.0 ft-kips/ft. Assume balanced design, $f'_c = 3000$ psi, and $f_s = 24{,}000$ psi. Use equations (*4.1*) through (*4.9*).
Ans. $d = 8''$, $A_s = 0.85 \text{ in}^2\text{/ft}$

4.34. Repeat Problem 4.33 but use Tables 4.2 and 4.3.

4.35. Proportion by use of equations (*4.1*) through (*4.9*) a balanced reinforced rectangular beam to resist a moment of 300 ft-kips. Assume $f'_c = 5000$ psi and $f_s = 24{,}000$ psi.
Ans. $b = 15''$, $d = 25''$, $A_s = 6.82 \text{ in}^2$

4.36. Repeat Problem 4.35 but use Table 4.2.

4.37. If $b = 35''$, $b' = 12''$, $d = 20''$, $t = 3''$, $f'_c = 3000$ psi, $f_s = 20{,}000$ psi and $M = 65$ ft-kips, determine the reinforcement required in a T-beam. Use equations (*4.10*) through (*4.17*). Assume balanced design. *Ans.* $A_s = 2.14 \text{ in}^2$

4.38. Using equations (*4.18*) through (*4.27*), proportion a doubly reinforced beam if $b = 13''$, $d = 26''$, $d' = 2.5''$, $f'_c = 2500$ psi, $f_s = 20{,}000$ psi, and $M = 200$ ft-kips.
Ans. $A_s = 5.30 \text{ in}^2$, $A'_s = 2.16 \text{ in}^2$

4.39. Repeat Problem 4.38 but use Table 4.5.

Chapter 5

Ultimate Strength Design

GENERAL PROVISIONS AND FLEXURAL COMPUTATIONS

NOTATION

A_s = area of tension reinforcement
A'_s = area of compression reinforcement
A_{sf} = area of reinforcement to develop compressive strength of overhanging flanges in I- and T-sections
a = depth of equivalent rectangular stress block = $k_1 c$
b = width of compression face of flexural member
b' = width of web in I- and T-sections
c = distance from extreme compression fiber to neutral axis at ultimate strength
D = dead load
d = distance from extreme compression fiber to centroid of tension reinforcement
d' = distance from extreme compression fiber to centroid of compression reinforcement
E = earthquake load
f'_c = compressive strength of concrete
f_y = yield strength of reinforcement
k_1 = a factor defined elsewhere
L = specified live load plus impact
M_u = ultimate resisting moment
p = A_s/bd
p' = A'_s/bd
p_b = reinforcement producing balanced conditions at ultimate strength
p_f = $A_{sf}/b'd$
p_w = $A_s/b'd$
q = $A_s f_y/bdf'_c$
t = flange thickness in I- and T-sections
U = required ultimate load capacity of section
W = wind load
ϕ = capacity reduction factor

(The notation used here is the same as in the 1963 ACI Building Code.)

INTRODUCTION

Modern day American practice of structural design of reinforced concrete has been based almost exclusively on elastic design or working stress design techniques with building codes not recognizing ultimate strength design procedures. However, in October 1955 the American Society of Civil Engineers issued Proceedings Paper No. 809 entitled "Report of ASCE-ACI Joint Committee on Ultimate Strength Design" that presented a comprehensive historical background of ultimate strength design of reinforced concrete as well as recommendations for design by this technique.

In the 1956 edition of the ACI Building Code, USD was given its introduction to the American structural engineers because in the Appendix of this Code many of the provisions contained in the ASCE-ACI Report were included, and the Code permitted their use in the design of reinforced concrete.

Ultimate strength design procedures differ from working stress design procedures. In the former it is recognized that at high stress levels in concrete, stress is not proportional to strain and, secondly, in ultimate strength design procedures design loads are multiples of anticipated service loads. In working stress design procedures stress is assumed to be proportional to strain and design loads are equal to service loads.

As pointed out in the Report, there are several advantages or reasons for using ultimate strength design:

(1) Ultimate strength design better predicts the ultimate strength of a section because of the recognition of the non-linearity of the stress-strain diagram at high stress levels.

(2) Because the dead loads to which a structure is subjected are more certainly determined than the live loads, it is unreasonable to apply the same factor of safety to both.

(3) Elastic column design is a modification of ultimate strength design and therefore is not compatible with working stress design of flexural members. Hence a consistent design technique is desirable.

(4) A more certain evaluation of the critical moment-thrust ratio for columns is possible with ultimate strength design rather than working stress design.

(5) Ultimate strength design must be used when determining the ultimate capacity of prestressed concrete.

Perhaps of the above, the first two advantages listed are the most significant. The first has been discussed thoroughly elsewhere. The second warrants some explanation.

Assuming that it is desired to have a reasonable and common margin of safety of one (or a factor of safety of two) and assuming that the dead load effects are almost certain, then with the working stress design some adjustment must be made in order to predict the relative capacity of any structure. If the margin of safety for dead load is one, for various dead load-live load ratios of 0.5, 1, 2 and 4 we obtain:

For $f_c = f'_c/2$ and $f_s = f_y/2$: Margin of safety for live load $= \dfrac{2(D+L)-D}{L} - 1$

For D.L./L.L. = 0.5, M.S. for L.L. $= \dfrac{2(0.5+1.0)-0.5}{1.0} - 1 = 1.5$

For D.L./L.L. = 1.0, M.S. for L.L. $= \dfrac{2(1.0+1.0)-1.0}{1.0} - 1 = 2.0$

$$\text{For D.L./L.L.} = 2.0, \ \text{M.S. for L.L.} = \frac{2(2.0+1.0)-2.0}{1.0} - 1 = 3.0$$

$$\text{For D.L./L.L.} = 4.0, \ \text{M.S. for L.L.} = \frac{2(4.0+1.0)-4.0}{1.0} - 1 = 5.0$$

It is obvious that the factor of safety of a structure varies as the dead load-live load ratio varies, if the ultimate capacity is based on a multiple of the sum of the dead and live load effects.

Before derivation of the basic equations and relationships it is necessary to delineate the basic assumptions, which are:

(1) Plane sections before bending remain plane after bending.

(2) At ultimate capacity strain and stress are not proportional.

(3) Strain in the concrete is proportional to the distance from the neutral axis [see (1) above].

(4) Tensile strength of concrete is neglected in flexural computations.

(5) The ultimate concrete strain is 0.003.

(6) The modulus of elasticity of the reinforcing steel is 29,000,000 psi.

(7) The maximum compressive stress in the concrete is $0.85f'_c$.

(8) The ultimate tensile stress in the reinforcement does not exceed f_y.

In addition to these eight assumptions, the assumed compressive stress distribution in the concrete is most important (Fig. 5-1). Any distribution such as a rectangle, trapezoid, parabola, sine wave, or any other shape is permitted if the predicted ultimate capacity is in close agreement with test data.

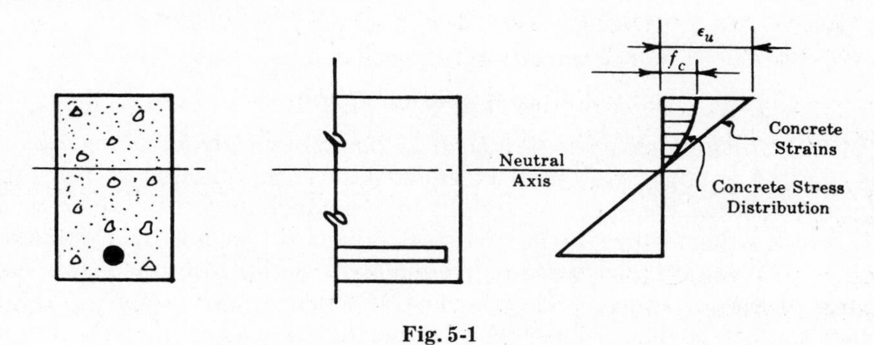

Fig. 5-1

Many stress distributions have been proposed, the three most common being the parabola, trapezoid, and rectangle, Fig. 5-2. All yield reasonable results.

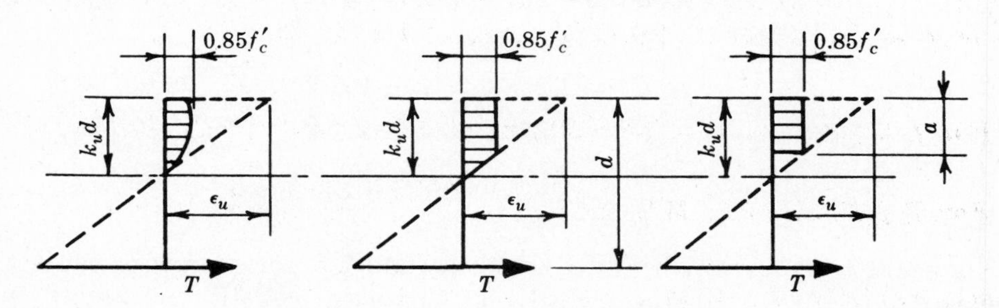

Fig. 5-2

The mechanics and derivations using the rectangular distribution are somewhat simpler. This is the assumed distribution in the ACI Building Code and will be the one used herein. It is further assumed that $a = k_1 c$ and that k_1 be taken as 0.85 for concrete strengths of 4000 psi and less, and that k_1 will be diminished by the value 0.05 for each 1000 psi of concrete strength in excess of 4000 psi.

LOAD FACTORS

In the ASCE-ACI Report and in the 1956 edition of the ACI Building Code, all of the factor of safety in ultimate strength design was provided for ostensibly in the load factor. That is, the service loads were increased by some multiple and the idealized capacity, or ultimate strength, of the member had to be equal to or greater than this assumed loading. However, in the 1963 edition of the ACI Building Code, the factor of safety is applied both to the load and to the idealized capacity sides of the equation. The 1956 version would be

$$\text{idealized capacity } (U) = \text{ultimate load } (C_1 D + C_2 L)$$

And the 1963 version would be

$$\phi \times \text{idealized capacity } (\phi U) = \text{ultimate load } (C_3 D + C_4 L)$$

In so doing, the 1963 Code provides for possible overloads (right hand side of equation) and possible under-capacity (left hand side of equation).

This idealized capacity reduction factor provides for the possibility of the concrete or reinforcing steel being of less strength than required and for the possibility of members being understrength due to inaccuracies or mistakes in construction. This reduction factor ϕ does vary with the importance of the member and with the mode of anticipated failure. In design the following should be used:

$$\begin{aligned} \phi &= 0.90 \text{ for flexure} \\ &= 0.85 \text{ for diagonal tension, bond and anchorage} \\ &= 0.75 \text{ for spirally reinforced compression members} \\ &= 0.70 \text{ for tied compression members} \end{aligned}$$

It is noted that for flexural members such as beams and girders, the capacity reduction factor is the highest (0.90). As it will be noted later, the amount of reinforcing steel in flexure is limited so that the steel will yield in tension before the concrete fails in compression. Therefore the failure in flexure will normally be due to a gradual yielding of the reinforcing. In columns, however, the mode of failure could be an explosive compression failure of the concrete. Too, in most structures, the action of columns is more important than that of beams. Consequently the idealized capacity of columns is reduced more than that for beams, 0.75 or 0.70 vs. 0.90.

The load factor equations contained in the ASCE-ACI Report are

$$U = 1.2D + 2.4L \tag{5.1}$$

$$U = K(D + L) \tag{5.2}$$

$$U = 1.2D + 2.4L + 0.6W \tag{5.3}$$

$$U = 1.2D + 0.6L + 2.4W \tag{5.4}$$

$$U = K(D + L + 0.5W) \tag{5.5}$$

$$U = K(D + 0.5L + W) \tag{5.6}$$

where K is equal to 2 for axially load members and 1.8 for members subject to flexure. When seismic forces (E) must be considered, these are substituted for W in the above equations.

The load factor equations now proposed are

$$U = 1.5D + 1.8L \tag{5.7}$$

$$U = 1.25(D + L + W) \tag{5.8}$$

$$U = 0.9D + 1.1W \tag{5.9}$$

Again E is substituted for W when seismic loads are involved.

In equations (*5.7*) through (*5.9*) it must be remembered that the ultimate capacity U of the member is a function of the capacity reduction factor ϕ; and hence these equations may have four different values depending on the value of ϕ.

The strength of a member must be such that all of the load factor equations must be satisfied. Hence the design must be checked to determine which one of the equations is critical. Assuming no aids are available, it is necessary to establish an adequate and efficient system so that all load factors are tested. See Problem 5.1.

In actual design practice, the tables shown in Problem 5.1 could be abbreviated as the designer becomes familiar with the procedure. In a given structure with a relatively constant ratio of dead, live and wind loads, the critical load factor may be determined by inspection after a few cases are checked. Even so, it can become quite tedious when many elements of the structure must be analyzed. It is desirable that some design aid be developed to relieve the designer of this burden. Such design aids have been developed and will be discussed here. If we let T.L. = total load = D.L. + L.L. + W.L. and operate on equations (*5.3*) through (*5.6*) as given in the ASCE-ACI Report for flexural members, we obtain

$$U = 0.6D + 1.8L + 0.6T \qquad U = 0.9D + 0.9L + 0.9T$$

$$U = -1.2D - 1.8L + 2.4T \qquad U = -0.9L + 1.8T$$

By equating the various equations, rearranging, and solving for L/T as a function of D/T, we can obtain

$$L/T = -1.33D/T + 0.67 \qquad L/T = 0.33D/T + 0.33$$

$$L/T = -0.5D/T + 0.5 \qquad L/T = -0.5D/T + 0.5$$

The graph of each of these equations is shown in Fig. 5-3.

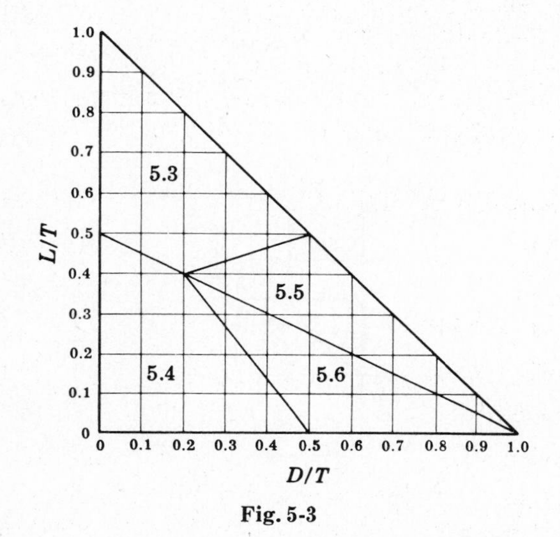

Fig. 5-3

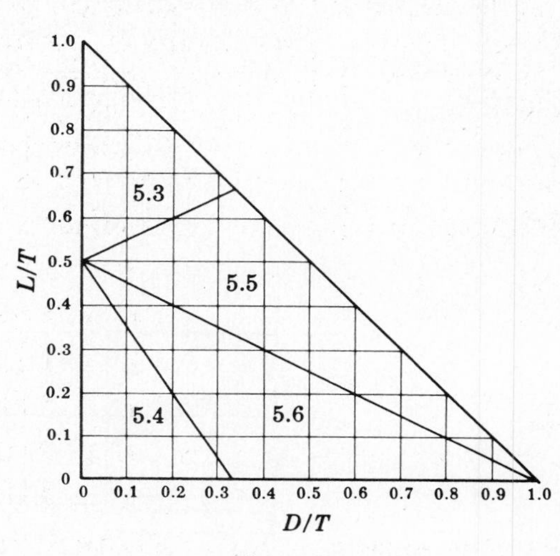

Fig. 5-4

The same procedure can be utilized for columns and the result is graphed in Fig. 5-4.

If the element sustains no wind effects, then the critical equation may be readily determined. If the ratio L/D is greater than 1.0, equation (*5.1*) is critical. If L/D is less than 1.0, equation (*5.2*) governs.

It is obvious that the absolute value of equation (*5.9*) will be less than that of equations (*5.7*) or (*5.8*). Equation (*5.9*) is used to minimize gravity loads and maximize horizontal loads so that possible stress or moment reversals will be accounted for. Hence the choice is between equations (*5.7*) and (*5.8*) in determining the maximum absolute values.

Following the same procedure as before, we obtain $L/T = -0.833D/T + 0.694$. This equation is plotted in Fig. 5-5.

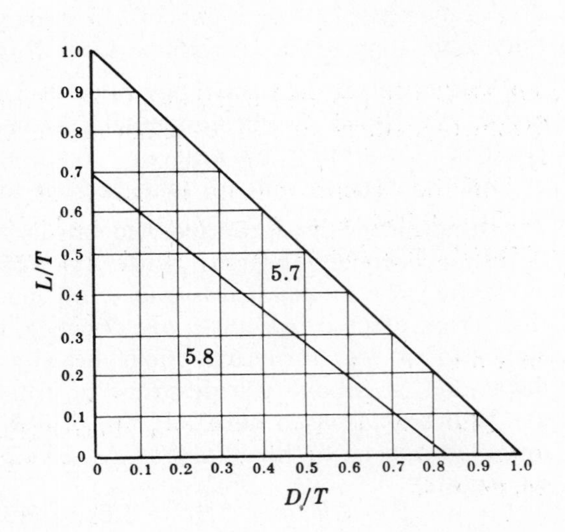

Fig. 5-5

With the analysis based on the service loads and with a simple arithmetic operation, it is possible with Fig. 5-3, 5-4 or 5-5 to readily determine which loading condition is critical. See Problem 5.2. Fig. 5-3 and 5-4 are used for the equations in the 1956 ACI Building Code and Fig. 5-5 for the 1963 edition. Equation (*5.9*) of the 1963 edition must be considered also.

Stresses due to the effect of creep, elastic deformation, support settlement, temperature variations, etc., if critical, must be considered as a dead load and must be modified by the appropriate load factor.

The general requirements for such as allowable steel stresses, deflections, etc., that are particularly applicable to ultimate strength design are discussed elsewhere in the appropriate sections.

FLEXURAL COMPUTATIONS BY ULTIMATE STRENGTH DESIGN

Test data show that the assumed ultimate concrete strain of 0.003 is conservative but yet reasonable. Further, the maximum allowable concrete stress of $0.85f'_c$ is compatible with the ultimate strain. If the free-body diagram given in the ASCE-ACI Report is used (Fig. 5-6), the summation of horizontal forces shows that

$$C = T \quad \text{or} \quad 0.85f'_c\, k_1 k_u bd = A_s f_s \qquad (a)$$

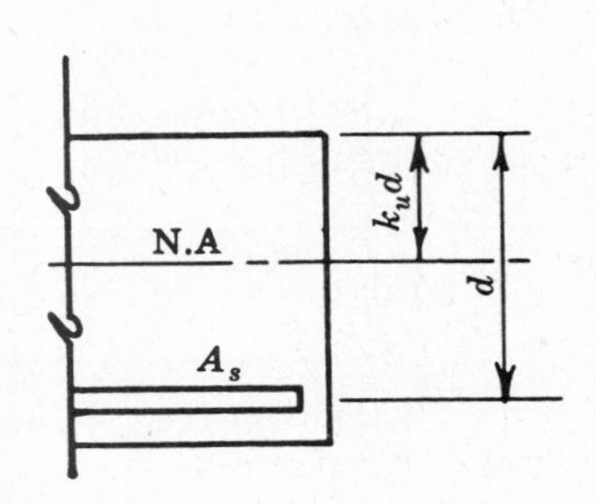

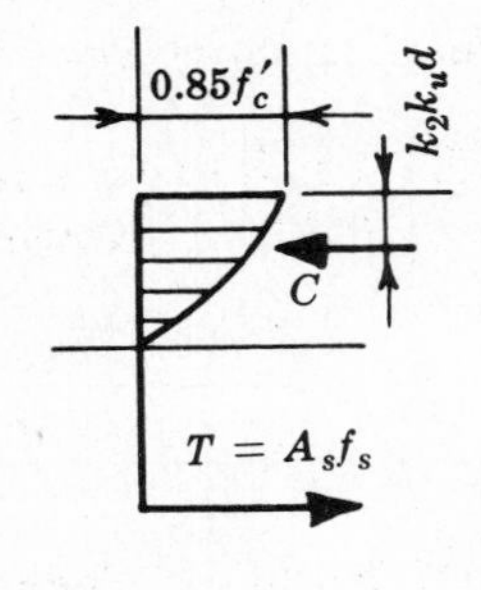

Fig. 5-6

k_1 is the average compressive stress divided by $0.85f'_c$. The summation of moments about T yields

$$M = 0.85f'_c k_1 k_u(1 - k_2 k_u)bd^2 \tag{b}$$

where $k_2 k_u d$ = distance from compression face to centroid of compressive force.

If $f_s = f_y$, $p = A_s/bd$ and $m = f_y/0.85f'_c$; combining (a) and (b) above, we obtain the expression for the ultimate resisting moment

$$M_u = A_s f_y d\left(1 - \frac{k_2}{k_1}pm\right) \tag{5.10}$$

which is the general expression for the ultimate resisting moment for an underreinforced beam regardless of the assumed shape of the compressive force diagram.

If the second degree parabola of the ASCE-ACI Report is used, it is determined $k_2/k_1 = 0.554$ and $k_1 = 0.925 - 0.438f'_c/E_c\epsilon_u$. If $\epsilon_u = 0.0035$ and $E_c = 1000f'_c$, $k_1 = 0.80$. Substituting for k_1 and k_2 in the expression for the ultimate resisting moment

$$M_u = f'_c bd^2 q(1 - 0.65q) \tag{5.11}$$

in which $q = pf_y/f'_c$.

Likewise, if the trapezoidal distribution contained in the ASCE-ACI Report is assumed, $k_2/k_1 = 0.50$ and $k_1 = 0.879$. Substituting values of k_1 and k_2, the ultimate resisting moment is

$$M_u = f'_c bd^2 q(1 - 0.59q) \tag{5.12}$$

If the beam is underreinforced so that the failure in flexure is due to a yielding of the steel, the expression for the ultimate resisting moment assuming a rectangular stress distribution is

$$M_u = f'_c bd^2 q(1 - 0.59q) \tag{5.13}$$

This is the expression for the idealized flexural capacity of an underreinforced beam recommended by the ASCE-ACI Report and included in the ACI Building Code and will be the one generally used here. Methods of solution of this expression will be discussed and illustrated later. See Problem 5.3.

If the condition is such that the concrete and reinforcing steel are stressed to the allowable ultimate at the same time, then the section or beam has balanced reinforcement. Tests show that when $q = 0.456$ the condition of balanced reinforcement exists. Because of assumed underreinforcement in the derivations and because of the desirability of tension control of failure, it is necessary to limit the value of p or q in order to assure this condition.

In the ASCE-ACI Report and in the 1956 ACI Building Code, q was limited to a maximum value of 0.40. However, in the 1963 ACI Building Code a different expression is presented

$$p_b = \frac{0.85f'_c k_1}{f_y}\,\frac{87{,}000}{87{,}000 + f_y} \tag{5.14}$$

The 1963 ACI Building Code defines balanced reinforcement by this expression and requires for beams with tension reinforcement only that the maximum percentage of reinforcement be limited to 0.75 times this value. See Problem 5.4.

Fig. 5-7 shows a series of curves obtained by plotting values of p_{max} versus f_y for various concrete strengths.

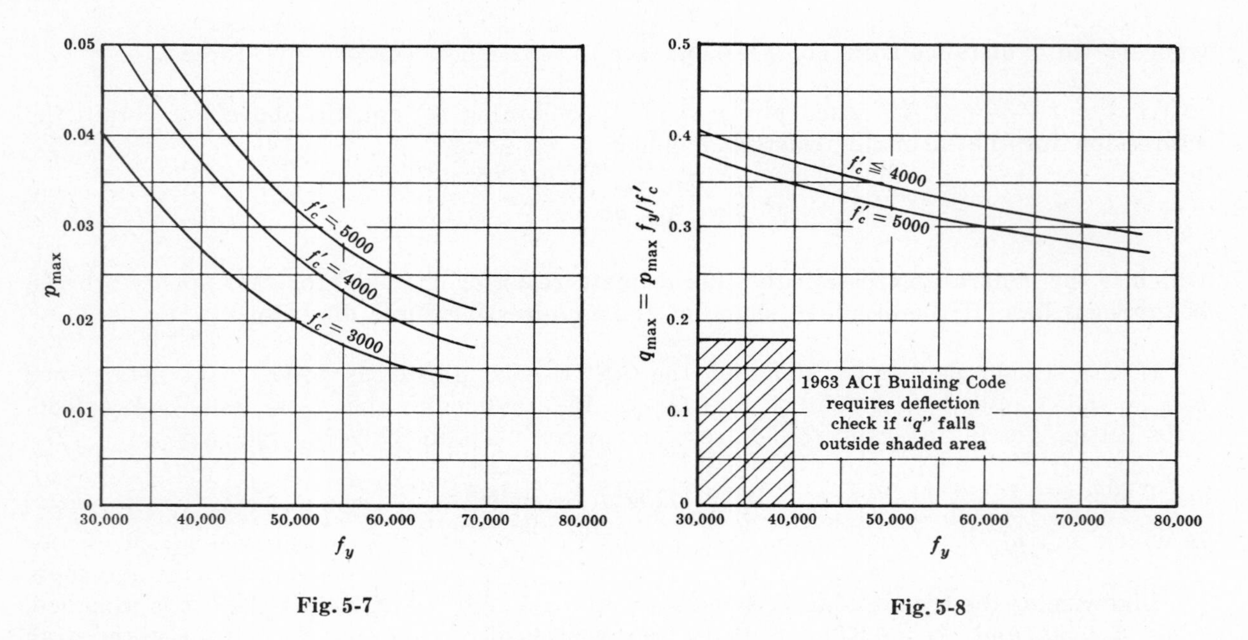

Fig. 5-7

Fig. 5-8

Fig. 5-8 is a plot of the maximum value of q which would be equal to $0.75p_b f_y/f'_c$.

COMPRESSION REINFORCEMENT

Compression reinforcement in flexural members designed by ultimate strength design procedures will very seldom be required. Because of the much increased value given to the concrete in compression due to the shape and increased depth of the compression zone, it will be practically unsound to place enough reinforcement in a beam or girder to attain the balanced condition. If $f'_c = 3000$ psi and $f_y = 40{,}000$ psi, then $p = 0.18f'_c/f_y$ for a beam of balanced design using working stress design. For the same conditions in ultimate strength design, $p_b = 0.37f'_c/f_y$. It is generally impractical, if not impossible, to use twice as much reinforcement in a beam as that which would be required by balanced condition in working stress design. Hence the discussion and derivation of flexural formulas for members requiring compression reinforcement is somewhat academic rather than practical.

In ultimate strength design of double reinforced beams, beams requiring compression reinforcement, it is assumed that the resisting moment is composed of two separate moments. First, there is the resisting capacity of the concrete and the balanced reinforcement. Second, there is the resisting capacity of the compression steel and a like amount of tension steel.

If the expression for the ultimate resisting moment of a beam with tensile reinforcement is rearranged,

$$M_u = A_s f_y (d - a/2) \tag{5.12a}$$

The moment of the additional compression steel would be $M'_u = A'_s f_y (d - d')$ if the steel is stressed to the yielding. Adding these two moments and deducting the effect of the concrete in compression occupied by the compressive reinforcement,

$$M_u = (A_s - A'_s) f_y (d - a/2) + A'_s f_y (d - d') \tag{5.15}$$

For the expression to be valid, it is necessary that the compressive reinforcement reach its yield strength at the ultimate strength of the member. This is satisfied if

$$p - p' \geqq \frac{0.85 f'_c k_1 d'}{f_y d} \frac{87{,}000}{87{,}000 - f_y} \tag{5.16}$$

This is the expression presented in the 1963 ACI Building Code. If the total tensile reinforcement is greater than this value for $(p - p')$, then there is more than enough tensile force, $A_s f_y$, to develop the concrete and any compression reinforcement. Therefore the tensile steel is not at yield and the failure is not controlled by tension. See Problem 5.5.

The quantity $(p - p')$ is also limited to $0.75p_b$ in order to protect against brittle or compression failures of the concrete.

I- AND T-SECTIONS

In flanged sections such as an I- or T-section in ultimate strength design, it is again assumed that the resisting moment is composed of two separate moments. First, there is the moment capacity of the rectangular portion of the concrete, b' times d, and a corresponding amount of tensile reinforcement. Second, there is the moment capacity of the overhang portion of the concrete, $(b - b')t$, and another amount of tensile reinforcement. The idealized ultimate flexural capacity of an I- or T-section is

$$M_u = (A_s - A_{sf}) f_y (d - 0.5a) + A_{sf} f_y (d - 0.5t) \tag{5.17}$$

This expression assumes that the section acts as an I- or T-section and that the neutral axis of the sections falls without the flanged section. The 1963 ACI Building Code requires that the neutral axis falls within the flange if

$$t \geqq 1.18qd/k_1 \tag{5.18}$$

Then the member acts as a rectangular beam and it can be proportioned according to requirements of rectangular beams with tension reinforcement.

In an investigation of a member, it is easy to substitute in equation (*5.18*). However, in a design it might be more practical to determine if the flange of the section is capable of resisting the applied moment. If not, then the neutral axis falls outside of the flange and the section does not act as a rectangular member.

If the moment capacity of the flange,

$$M_F/\phi = 0.85 f'_c bt(d - 0.5t)/\phi$$

is greater than the applied moment, then the member is proportioned as a rectangular beam.

In I- and T-sections, the area of tension steel available to develop the concrete of the web in compression is limited to $0.75p_b$, that is,

$$A_s/b'd - A_{sf}/b'd \leqq 0.75p_b \quad \text{or} \quad p_w - p_f \leqq 0.75p_b$$

This, too, is to guard against possible brittle or compressive failures of the concrete.

All of the expressions derived in this chapter are the idealized capacities. It should be remembered that these expressions must be modified by the appropriate capacity reduction factor ϕ. For flexure, $\phi = 0.90$.

Design specifications require that in reinforced concrete structures proportioned by ultimate strength design techniques the analysis of the frame to determine shears, moments, and thrusts be based on elastic behavior. Usually, in reinforced concrete no redistribution of moments or limit design is considered. Therefore, no matter how the structure is to be proportioned (working stress design or ultimate strength design), the analysis will have the same basis.

SOLUTION OF EQUATIONS AND DESIGN AIDS

The solution and application of the expressions derived in this chapter will be demonstrated in the solved problems. However, it is appropriate to discuss the solution of the expression for the ultimate moment capacity of a rectangular beam with tension reinforcement only. In the solution of the expression $M_u = bd^2 f'_c q(1-0.59q) = A_s f_y (d-a/2)$, there are the variables b, d and A_s which are not known before the design is completed. It is obvious that there are an infinite number of solutions to the equations that would yield the same value of M_u. Hence the designer must assume the value of one or more of these variables before he can determine the unique solution he is seeking.

The designer may assume the overall dimensions of the member; or may assume a value of p; or may assume that the member shall have minimum depth or maximum p permitted by specifications. As previously shown, the last assumption will seldom be used because of the extremely high quantity of reinforcing steel required which is not an economical solution. Often the overall width or depth, or both, of a beam are established by some other structural or architectural consideration.

If the minimum depth of the overall member size is not determined or assumed, then the designer must assume a value for p or q. The value $q = 0.18$ is one of the limiting values above which the 1963 ACI Building Code requires a deflection computation for the member. Although this value of $q = 0.18$ is comparable to working stress design, it should be remembered that balanced reinforcement usually does not result in the most economical proportions. Therefore, if a value for q is assumed, it should be equal to or less than 0.18.

There are four basic techniques for the solution of single reinforced beams. They would include:

(1) Direct substitution into formulas

(2) Use of tables for the formulas

(3) Use of curves or charts for the formulas

(4) Derivation of constants analogous to working stress design procedures

All of these four methods will be used in the solved problems.

Table 5.1 is a solution of a form of equation *(5.13)*,

$$M_u/f'_c bd^2 = q(1-0.59q)$$

with the first column being the first two decimal places and the first row being the third decimal place for the value of q; and the body of the table gives the corresponding values for $M_u/f'_c bd^2$. With Table 5.1, the design may be accomplished by assuming a value of q or p and solving for $M_u/f'_c bd^2$ and subsequently for b and d; or, a value for bd^2 may be assumed and the value of q determined. Note that this is the solution of the idealized capacity. Therefore, M_u must be modified by $\phi = 0.90$.

TABLE 5.1

ULTIMATE MOMENT OF RECTANGULAR SECTIONS

q	.000	.001	.002	.003	.004	.005	.006	.007	.008	.009
	$M_u/f'_c bd^2$									
.0	0	.0010	.0020	.0030	.0040	.0050	.0060	.0070	.0080	.0090
.01	.0099	.0109	.0119	.0129	.0139	.0149	.0159	.0168	.0178	.0188
.02	.0197	.0207	.0217	.0226	.0236	.0246	.0256	.0266	.0275	.0285
.03	.0295	.0304	.0314	.0324	.0333	.0343	.0352	.0362	.0372	.0381
.04	.0391	.0400	.0410	.0420	.0429	.0438	.0448	.0457	.0467	.0476
.05	.0485	.0495	.0504	.0513	.0523	.0532	.0541	.0551	.0560	.0569
.06	.0579	.0588	.0597	.0607	.0616	.0625	.0634	.0643	.0653	.0662
.07	.0671	.0680	.0689	.0699	.0708	.0717	.0726	.0735	.0744	.0753
.08	.0762	.0771	.0780	.0789	.0798	.0807	.0816	.0825	.0834	.0843
.09	.0852	.0861	.0870	.0879	.0888	.0897	.0906	.0915	.0923	.0932
.10	.0941	.0950	.0959	.0967	.0976	.0985	.0994	.1002	.1011	.1020
.11	.1029	.1037	.1046	.1055	.1063	.1072	.1081	.1089	.1098	.1106
.12	.1115	.1124	.1133	.1141	.1149	.1158	.1166	.1175	.1183	.1192
.13	.1200	.1209	.1217	.1226	.1234	.1243	.1251	.1259	.1268	.1276
.14	.1284	.1293	.1301	.1309	.1318	.1326	.1334	.1342	.1351	.1359
.15	.1367	.1375	.1384	.1392	.1400	.1408	.1416	.1425	.1433	.1441
.16	.1449	.1457	.1465	.1473	.1481	.1489	.1497	.1506	.1514	.1522
.17	.1529	.1537	.1545	.1553	.1561	.1569	.1577	.1585	.1593	.1601
.18	.1609	.1617	.1624	.1632	.1640	.1648	.1656	.1664	.1671	.1679
.19	.1687	.1695	.1703	.1710	.1718	.1726	.1733	.1741	.1749	.1756
.20	.1764	.1772	.1779	.1787	.1794	.1802	.1810	.1817	.1825	.1832
.21	.1840	.1847	.1855	.1862	.1870	.1877	.1885	.1892	.1900	.1907
.22	.1914	.1922	.1929	.1937	.1944	.1951	.1959	.1966	.1973	.1981
.23	.1988	.1995	.2002	.2010	.2017	.2024	.2031	.2039	.2046	.2053
.24	.2060	.2067	.2075	.2082	.2089	.2096	.2103	.2110	.2117	.2124
.25	.2131	.2138	.2145	.2152	.2159	.2166	.2173	.2180	.2187	.2194
.26	.2201	.2208	.2215	.2222	.2229	.2236	.2243	.2249	.2256	.2263
.27	.2270	.2277	.2284	.2290	.2297	.2304	.2311	.2317	.2324	.2331
.28	.2337	.2344	.2351	.2357	.2364	.2371	.2377	.2384	.2391	.2397
.29	.2404	.2410	.2417	.2423	.2430	.2437	.2443	.2450	.2456	.2463
.30	.2469	.2475	.2482	.2488	.2495	.2501	.2508	.2514	.2520	.2527
.31	.2533	.2539	.2546	.2552	.2558	.2565	.2571	.2577	.2583	.2590
.32	.2596	.2602	.2608	.2614	.2621	.2627	.2633	.2639	.2645	.2651
.33	.2657	.2664	.2670	.2676	.2682	.2688	.2694	.2700	.2706	.2712
.34	.2718	.2724	.2730	.2736	.2742	.2748	.2754	.2760	.2766	.2771
.35	.2777	.2783	.2789	.2795	.2801	.2807	.2812	.2818	.2824	.2830
.36	.2835	.2841	.2847	.2853	.2858	.2864	.2870	.2875	.2881	.2887
.37	.2892	.2898	.2904	.2909	.2915	.2920	.2926	.2931	.2937	.2943
.38	.2948	.2954	.2959	.2965	.2970	.2975	.2981	.2986	.2992	.2997
.39	.3003	.3008	.3013	.3019	.3024	.3029	.3035	.3040	.3045	.3051
.40	.3056									

Fig. 5-9 is a chart or series of curves that accomplish the solution similar to Table 5.1. This is the same chart contained in the ASCE-ACI Joint Committee Report and is reproduced with permission of the ACI.

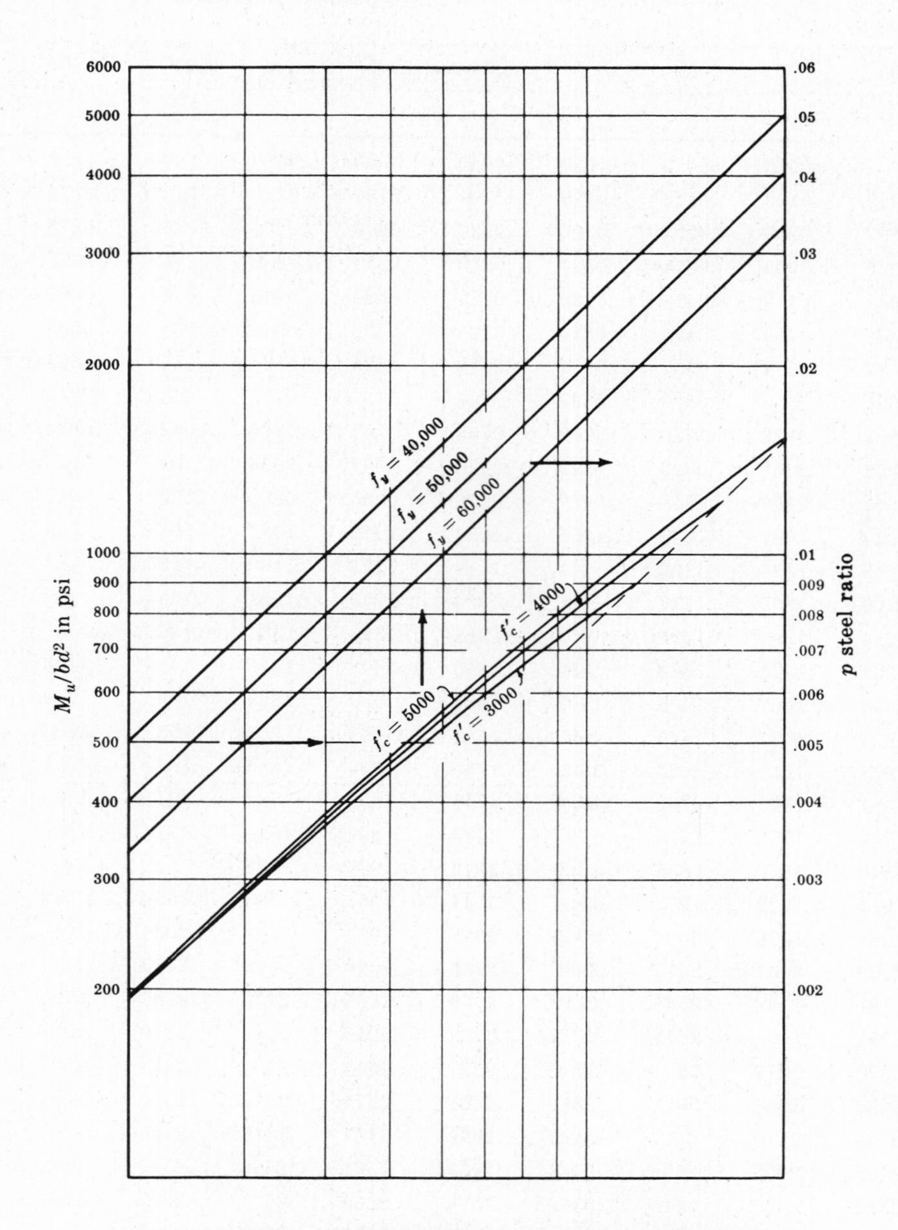

Fig. 5-9. Ultimate Bending Moment for Rectangular Beams

The chart may be entered with a value of $M_u/f'_c bd^2$; then traverse horizontally to the concrete strength, then vertically to the yield strength of the steel, then horizontally to the value of p. If p is assumed, this procedure is reversed.

In Chapter 4 it was shown that the basic equations using the transformed section method could be reduced to simpler forms, and with the help of design aid tables or charts the flexural computations could be made much easier. A method similar to this can be employed in ultimate strength design procedures.

If $a = A_s f_y/0.85f'_c b$ or $a/d = pm$, where $m = f_y/0.85f'_c$, the distance from the centroid of the compression to the tensile reinforcement $j_u d = d - a/2$ or $j_u = 1 - a/2d$. The resisting moment $M_u/\phi = Tj_u d = A_s f_y j_u d$. In working stress design, $R = M/bd^2 = f_c/2kj$. In ultimate strength design if the member is underreinforced, then the analogous R is $R_u = f'_c q(1 - 0.59q) = pf_y(1 - 0.59pf_y/f'_c)$ and $M_u = \phi R_u bd^2$. In working stress design for a given steel stress, the values of j were averaged and a value was computed so that $a = f_s j/12{,}000$ and hence $A_s = M/ad$.

In working stress design for a given steel stress, typical values of j would not vary more than 5 to 10 percent regardless of the concrete stress. However, in ultimate strength design because of the wider range of possible reinforcing steel percentages, there obviously would be a greater variation in the values of j as the concrete stress varies. Therefore, if it is considered that a_u (this is not the depth of the stress block) $= f_y j_u$, the area of steel must be $A_s = M_u/\phi f_y j_u d$.

Table 5.2 lists the various constants for ultimate strength design that are similar to those developed for working stress design.

In Table 5.2, $\phi a_u = f_y j_u/12{,}000$. Therefore, if the moment is in foot-kips, then the area of reinforcement in square inches is $A_s = M_u/\phi a_u d$. Note that this table *does* include the effect of ϕ, 0.90.

TABLE 5.2

COEFFICIENTS FOR BEAMS WITH TENSION REINFORCEMENT

f'_c	$f_y = 30{,}000$				
	ϕR_u	c/d	ϕj_u	ϕa_u	p
3000	53.4	.0277	.8894	2.223	.0020
	105.5	.0554	.8788	2.197	.0040
	156.3	.0830	.8681	2.170	.0060
	205.8	.1107	.8575	2.144	.0080
	254.1	.1384	.8469	2.117	.0100
	301.1	.1661	.8363	2.091	.0120
	346.8	.1938	.8257	2.064	.0140
	391.2	.2215	.8150	2.038	.0160
	434.4	.2491	.8044	2.011	.0180
	476.3	.2768	.7938	1.985	.0200
	516.9	.3045	.7832	1.958	.0220
	556.2	.3322	.7726	1.931	.0240
	594.3	.3599	.7619	1.905	.0260
	631.1	.3875	.7513	1.878	.0280
	666.6	.4152	.7407	1.852	.0300
	700.9	.4429	.7301	1.825	.0320
	733.8	.4706	.7195	1.799	.0340
	765.5	.4983	.7088	1.772	.0360
	796.0	.5260	.6982	1.746	.0380
	825.1	.5536	.6876	1.719	.0400
3/4 × Balanced Steel Ratio					
	829.3	.5577	.6860	1.715	.0403

For values below horizontal line, q exceeds 0.18.

TABLE 5.2 (Cont.)

f'_c	$f_y = 30{,}000$				
	ϕR_u	c/d	ϕj_u	ϕa_u	p
4000	53.5	.0208	.8920	2.230	.0020
	106.1	.0415	.8841	2.210	.0040
	157.7	.0623	.8761	2.190	.0060
	208.4	.0830	.8681	2.170	.0080
	258.1	.1038	.8602	2.150	.0100
	306.8	.1246	.8522	2.131	.0120
	354.6	.1453	.8442	2.111	.0140
	401.4	.1661	.8363	2.091	.0160
	447.3	.1869	.8283	2.071	.0180
	492.2	.2076	.8204	2.051	.0200
	536.2	.2284	.8124	2.031	.0220
	579.2	.2491	.8044	2.011	.0240
	621.2	.2699	.7965	1.991	.0260
	662.3	.2907	.7885	1.971	.0280
	702.5	.3114	.7805	1.951	.0300
	741.7	.3322	.7726	1.931	.0320
	779.9	.3529	.7646	1.911	.0340
	817.2	.3737	.7566	1.892	.0360
	853.5	.3945	.7487	1.872	.0380
	888.8	.4152	.7407	1.852	.0400
	923.2	.4360	.7327	1.832	.0420
	956.7	.4567	.7248	1.812	.0440
	989.2	.4775	.7168	1.792	.0460
	1020.7	.4983	.7088	1.772	.0480
	1051.3	.5190	.7009	1.752	.0500
	1080.9	.5398	.6929	1.732	.0520
3/4 × Balanced Steel Ratio					
	1105.7	.5577	.6860	1.715	.0537

For values below horizontal line, q exceeds 0.18.

TABLE 5.2 (Cont.)

f'_c	$f_y = 30{,}000$				
	ϕR_u	c/d	ϕj_u	ϕa_u	p
5000	53.6	.0176	.8936	2.234	.0020
	106.5	.0353	.8873	2.218	.0040
	158.6	.0529	.8809	2.202	.0060
	209.9	.0706	.8745	2.186	.0080
	260.4	.0882	.8681	2.170	.0100
	310.2	.1059	.8618	2.154	.0120
	359.3	.1235	.8554	2.138	.0140
	407.5	.1412	.8490	2.123	.0160
	455.0	.1588	.8427	2.107	.0180
	501.8	.1765	.8363	2.091	.0200
	547.7	.1941	.8299	2.075	.0220
	592.9	.2118	.8235	2.059	.0240
	637.4	.2294	.8172	2.043	.0260
	681.1	.2471	.8108	2.027	.0280
	724.0	.2647	.8044	2.011	.0300
	766.1	.2824	.7980	1.995	.0320
	807.5	.3000	.7917	1.979	.0340
	848.1	.3176	.7853	1.963	.0360
	888.0	.3353	.7789	1.947	.0380
	927.1	.3529	.7726	1.931	.0400
	965.4	.3706	.7662	1.915	.0420
	1003.0	.3882	.7598	1.900	.0440
	1039.8	.4059	.7534	1.884	.0460
	1075.8	.4235	.7471	1.868	.0480
	1111.1	.4412	.7407	1.852	.0500
	1145.6	.4588	.7343	1.836	.0520
	1179.3	.4765	.7280	1.820	.0540
	1212.3	.4941	.7216	1.804	.0560
	1244.5	.5118	.7152	1.788	.0580
	1275.9	.5294	.7088	1.772	.0600
	1306.6	.5471	.7025	1.756	.0620
3/4 × Balanced Steel Ratio	1324.7	.5577	.6986	1.747	.0632

For values below horizontal line, q exceeds 0.18.

TABLE 5.2 (Cont.)

f'_c	$f_y = 40{,}000$				
	ϕR_u	c/d	ϕj_u	ϕa_u	p
3000	70.9	.0369	.8858	2.953	.0020
	139.5	.0738	.8717	2.906	.0040
	205.8	.1107	.8575	2.858	.0060
	269.9	.1476	.8434	2.811	.0080
	331.7	.1845	.8292	2.764	.0100
	391.2	.2215	.8150	2.717	.0120
	448.5	.2584	.8009	2.670	.0140
	503.5	.2953	.7867	2.622	.0160
	556.2	.3322	.7726	2.575	.0180
	606.7	.3691	.7584	2.528	.0200
	654.9	.4060	.7442	2.481	.0220
	700.9	.4429	.7301	2.434	.0240
	744.6	.4798	.7159	2.386	.0260
3/4 × Balanced Steel Ratio					
	782.8	.5138	.7029	2.343	.0278

For values below horizontal line, q exceeds 0.18.

f'_c	$f_y = 40{,}000$				
	ϕR_u	c/d	ϕj_u	ϕa_u	p
4000	71.2	.0277	.8894	2.965	.0020
	140.6	.0554	.8788	2.929	.0040
	208.4	.0830	.8681	2.894	.0060
	274.4	.1107	.8575	2.858	.0080
	338.8	.1384	.8469	2.823	.0100
	401.4	.1661	.8363	2.788	.0120
	462.4	.1938	.8257	2.752	.0140
	521.6	.2215	.8150	2.717	.0160
	579.2	.2491	.8044	2.681	.0180
	635.0	.2768	.7938	2.646	.0200
	689.2	.3045	.7832	2.611	.0220
	741.7	.3322	.7726	2.575	.0240
	792.4	.3599	.7619	2.540	.0260
	841.5	.3875	.7513	2.504	.0280
	888.8	.4152	.7407	2.469	.0300
	934.5	.4429	.7301	2.434	.0320
	978.5	.4706	.7195	2.398	.0340
	1020.7	.4983	.7088	2.363	.0360
3/4 × Balanced Steel Ratio					
	1043.7	.5138	.7029	2.343	.0371

For values below horizontal line, q exceeds 0.18.

TABLE 5.2 (Cont.)

f'_c	$f_y = 40,000$				
	ϕR_u	c/d	ϕj_u	ϕa_u	p
5000	71.3	.0235	.8915	2.972	.0020
	141.3	.0471	.8830	2.943	.0040
	209.9	.0706	.8745	2.915	.0060
	277.1	.0941	.8660	2.887	.0080
	343.0	.1176	.8575	2.858	.0100
	407.5	.1412	.8490	2.830	.0120
	470.7	.1647	.8405	2.802	.0140
	532.5	.1882	.8320	2.773	.0160
	592.9	.2118	.8235	2.745	.0180
	652.0	.2353	.8150	2.717	.0200
	709.8	.2588	.8065	2.688	.0220
	766.1	.2824	.7980	2.660	.0240
	821.1	.3059	.7896	2.632	.0260
	874.8	.3294	.7811	2.604	.0280
	927.1	.3529	.7726	2.575	.0300
	978.0	.3765	.7641	2.547	.0320
	1027.6	.4000	.7556	2.519	.0340
	1075.8	.4235	.7471	2.490	.0360
	1122.6	.4471	.7386	2.462	.0380
	1168.1	.4706	.7301	2.434	.0400
	1212.3	.4941	.7216	2.405	.0420
3/4 × Balanced Steel Ratio					
	1248.1	.5138	.7145	2.382	.0437

For values below horizontal line, q exceeds 0.18.

f'_c	$f_y = 50,000$ *				
	ϕR_u	c/d	ϕj_u	ϕa_u	p
3000	88.2	.0461	.8823	3.676	.0020
	172.9	.0923	.8646	3.603	.0040
	254.1	.1384	.8469	3.529	.0060
	331.7	.1845	.8292	3.455	.0080
	405.8	.2307	.8115	3.381	.0100
	476.3	.2768	.7938	3.308	.0120
	543.3	.3230	.7761	3.234	.0140
	606.7	.3691	.7584	3.160	.0160
	666.6	.4152	.7407	3.086	.0180
	723.0	.4614	.7230	3.013	.0200
3/4 × Balanced Steel Ratio					
	740.5	.4763	.7173	2.989	.0206

For values below horizontal line, q exceeds 0.18.

*Always check deflection when f_y exceeds 40.0 ksi.

TABLE 5.2 (Cont.)

f'_c	f_y = 50,000 *				
	ϕR_u	c/d	ϕj_u	ϕa_u	p
4000	88.7	.0346	.8867	3.695	.0020
	174.7	.0692	.8735	3.639	.0040
	258.1	.1038	.8602	3.584	.0060
	338.8	.1384	.8469	3.529	.0080
	416.8	.1730	.8336	3.473	.0100
	492.2	.2076	.8204	3.418	.0120
	565.0	.2422	.8071	3.363	.0140
	635.0	.2768	.7938	3.308	.0160
	702.5	.3114	.7805	3.252	.0180
	767.3	.3460	.7673	3.197	.0200
	829.4	.3806	.7540	3.142	.0220
	888.8	.4152	.7407	3.086	.0240
	945.7	.4498	.7274	3.031	.0260
3/4 × Balanced Steel Ratio					
	987.3	.4763	.7173	2.989	.0275

For values below horizontal line, q exceeds 0.18.
*Always check deflection when f_y exceeds 40.0 ksi.

f'_c	f_y = 50,000 *				
	ϕR_u	c/d	ϕj_u	ϕa_u	p
5000	88.9	.0294	.8894	3.706	.0020
	175.8	.0588	.8788	3.662	.0040
	260.4	.0882	.8681	3.617	.0060
	343.0	.1176	.8575	3.573	.0080
	423.5	.1471	.8469	3.529	.0100
	501.8	.1765	.8363	3.485	.0120
	578.0	.2059	.8257	3.440	.0140
	652.0	.2353	.8150	3.396	.0160
	724.0	.2647	.8044	3.352	.0180
	793.8	.2941	.7938	3.308	.0200
	861.5	.3235	.7832	3.263	.0220
	927.1	.3529	.7726	3.219	.0240
	990.5	.3824	.7619	3.175	.0260
	1051.8	.4118	.7513	3.131	.0280
	1111.1	.4412	.7407	3.086	.0300
	1168.1	.4706	.7301	3.042	.0320
3/4 × Balanced Steel Ratio					
	1178.9	.4763	.7280	3.033	.0324

For values below horizontal line, q exceeds 0.18.
*Always check deflection when f_y exceeds 40.0 ksi.

TABLE 5.2 (Cont.)

f'_c	$f_y = 60{,}000$ *				
	ϕR_u	c/d	ϕj_u	ϕa_u	p
3000	105.5	.0554	.8788	4.394	.0020
	205.8	.1107	.8575	4.288	.0040
	301.1	.1661	.8363	4.181	.0060
	391.2	.2215	.8150	4.075	.0080
	476.3	.2768	.7938	3.969	.0100
	556.2	.3322	.7726	3.863	.0120
	631.1	.3875	.7513	3.757	.0140
	700.9	.4429	.7301	3.650	.0160
3/4 × Balanced Steel Ratio					
	702.1	.4439	.7297	3.649	.0161

For values below horizontal line, q exceeds 0.18.
*Always check deflection when f_y exceeds 40.0 ksi.

f'_c	$f_y = 60{,}000$ *				
	ϕR_u	c/d	ϕj_u	ϕa_u	p
4000	106.1	.0415	.8841	4.420	.0020
	208.4	.0830	.8681	4.341	.0040
	306.8	.1246	.8522	4.261	.0060
	401.4	.1661	.8363	4.181	.0080
	492.2	.2076	.8204	4.102	.0100
	579.2	.2491	.8044	4.022	.0120
	662.3	.2907	.7885	3.942	.0140
	741.7	.3322	.7726	3.863	.0160
	817.2	.3737	.7566	3.783	.0180
	888.8	.4152	.7407	3.704	.0200
3/4 × Balanced Steel Ratio					
	936.1	.4439	.7297	3.649	.0214

For values below horizontal line, q exceeds 0.18.
*Always check deflection when f_y exceeds 40.0 ksi.

TABLE 5.2 (Cont.)

f'_c	$f_y = 60{,}000$ *				
	ϕR_u	c/d	ϕj_u	ϕa_u	p
	106.5	.0353	.8873	4.436	.0020
	209.9	.0706	.8745	4.373	.0040
	310.2	.1059	.8618	4.309	.0060
	407.5	.1412	.8490	4.245	.0080
	501.8	.1765	.8363	4.181	.0100
5000	592.9	.2118	.8235	4.118	.0120
	681.1	.2471	.8108	4.054	.0140
	766.1	.2824	.7980	3.990	.0160
	848.1	.3176	.7853	3.927	.0180
	927.1	.3529	.7726	3.863	.0200
	1003.0	.3882	.7598	3.799	.0220
	1075.8	.4235	.7471	3.735	.0240
3/4 × Balanced Steel Ratio					
	1116.4	.4439	.7397	3.699	.0252

For values below horizontal line, q exceeds 0.18.
*Always check deflection when f_y exceeds 40.0 ksi.

Note the variation in j_u as p varies. Table 5.2 is the solution of the reduced capacity, equation (*5.13*), and does include the effects of the capacity reduction factor ϕ.

In the solved problems that follow, the primary intent is to demonstrate the principles derived and discussed in this chapter. In the design of beams, other than flexural computations must be considered. Before a design is complete, such things as shear stresses, bond stresses, deflections, minimum reinforcement, etc., must be checked. All of these considerations are discussed elsewhere in this book.

Solved Problems

5.1. Assuming a typical floor section as shown in Fig. 5-10; the analysis yields the service load effects, $1.0D + 1.0L + 1.0W$, which are tabulated in the following table. Determine the ultimate load effects.

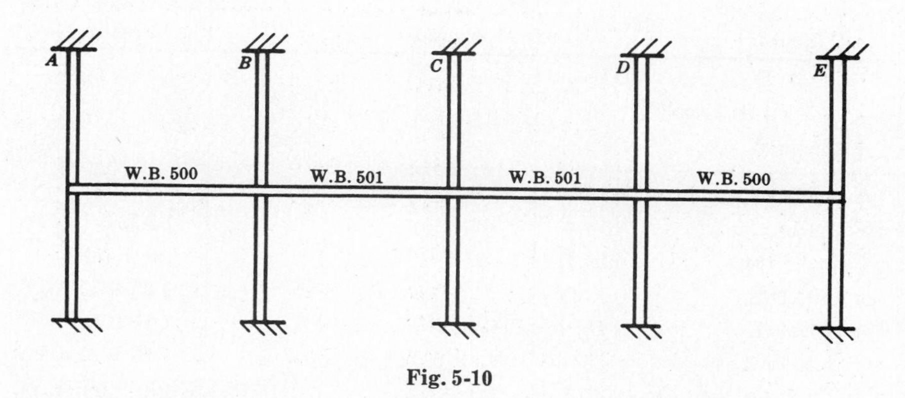

Fig. 5-10

(*a*) With the load effects tabulated, a system as shown in the table is adequate for determining the critical loading conditions for beams marked W.B. 500.

TABULATION OF BEAM MOMENTS AND SHEARS

W.B. 500

	Moment			Shear		
Load Factor	Col. A	Pos. Mom.	Col. B	Load Factor	Col. A	Col. B
Basic D.L.	−40	+40	−50	Basic D.L.	+15	+20
Basic L.L.	−40	+50	−45	Basic L.L.	+10	+15
Basic W.L.	±120		±100	Basic W.L.	±15	±15
(*a*) 1.5 D.L.	−60	+60	−75	1.5 D.L.	+22.5	+30
(*b*) 1.8 L.L	−72	+90	−81	1.8 L.L.	+18	+27
(*c*) (*a*) + (*b*)	−132	+150	−156	(*a*) + (*b*)	+40.5	+57
(*d*) 1.25 D.L.	−50		−62.5	1.25 D.L.	+18.8	+25
(*e*) 1.25 L.L.	−50		−56.2	1.25 L.L.	+12.5	+18.8
(*f*) 1.25 W.L.	±150		±125	1.25 W.L.	±18.8	±18.8
(*g*) (*d*) + (*e*) + (*f*)	−250		−243.7	(*d*) + (*e*) + (*f*)	+50.1	+62.6
(*h*) 0.9 D.L.	−36		−45	0.9 D.L.	+13.5	+18
(*i*) 1.1 W.L.	±132		±110	1.1 W.L.	±16.5	±16.5
(*j*) (*h*) + (*i*)	−168		−155	(*h*) + (*i*)	+30.0	+34.5
(*k*) (*h*) + (*i*)	+96		+65	(*h*) + (*i*)	−3.0	+1.5
Critical Loading	(*g*) & (*k*)	(*c*)	(*g*) & (*k*)		(*g*)	(*g*)

The various load effects are multiplied by the appropriate factors and summed on lines (*c*), (*g*), (*j*) and (*k*). Each of these lines represents the results of one of the three basic load factor equations. After the summation, the critical condition becomes apparent.

(*b*) Likewise, in the following table, the basic load effects are tabulated for columns A and B. These forces are then operated on by the appropriate multipliers representing the various load factor equations.

TABULATION OF COLUMN THRUSTS, SHEARS AND MOMENTS

Columns A and B					
Load Factor	Moment	Load Factor	Shear	Load Factor	Axial
Basic D.L.	+10				+400
Basic L.L.	+8				+100
Basic W.L.	±100		±15		±5
(*a*) 1.5 D.L.	+15	1.5 D.L.		1.5 D.L.	+600
(*b*) 1.8 L.L.	+14.4	1.8 L.L.		1.8 L.L.	+180
(*c*) (*a*) + (*b*)	+29.4			(*a*) + (*b*)	+780
(*d*) 1.25 D.L.	+12.5	1.25 D.L.		1.25 D.L.	+500
(*e*) 1.25 L.L.	+10	1.25 L.L.		1.25 L.L.	+125
(*f*) 1.25 W.L.	±125	1.25 W.L.	±18.7	1.25 W.L.	±6
(*g*) (*d*) + (*e*) + (*f*)	+147.5	(*d*) + (*e*) + (*f*)	±18.7	(*d*) + (*e*) + (*f*)	+631
(*h*) 0.9 D.L.	+9	0.9 D.L.		0.9 D.L.	+360
(*i*) 1.1 W.L.	±110	1.1 W.L.	±16.5	1.1 W.L.	±6
(*j*) (*h*) + (*i*)	+119	(*h*) + (*i*)	±16.5	(*h*) + (*i*)	+366
(*k*) (*h*) + (*i*)	−101	(*h*) + (*i*)	±16.5	(*h*) + (*i*)	−354
Critical Loading	(*g*)		(*g*)		(*c*)

Again after the summation on lines (*c*), (*g*), (*j*) and (*k*), the critical loading becomes obvious.

It should be recalled that this is the "right hand side of the equation" only and does not take into account the capacity reduction factor.

5.2. Using the values given in Problem 5.1 for the load effects on beam W.B. 500, determine the critical ultimate load factor equation. Also check columns A and B.

(*a*)

Moment			Shear	
Col. A	Positive	Col. B	Col. A	Col. B
D.L. − 40	+40	−50	+15	+20
L.L. − 40	+50	−45	+10	+15
W.L. ± 120		±100	±15	±15
T.L. − 200	+90	−195	+40	+50
D/T 0.20	0.45	0.26	0.38	0.40
L/T 0.20	0.55	0.23	0.25	0.30

Using Fig. 5-5, the results obtained in Problem 5.1 are readily verified.

(*b*) And from the table for columns A and B:

	Moment	Shear	Axial
D.L.	+10		+400
L.L.	+8		+100
W.L.	±100	±15	±5
T.L.	+118	±15	+505
D/T	0.08	0	0.79
L/T	0.07	0	0.20

Again with Fig. 5-5, the results of Problem 5.1 can be verified.

(c) With charts or aids such as these, the operation with the load factors may be made less burdensome. Fig. 5-5 is valid regardless of the value of ϕ.

5.3. Derive the expression for the idealized ultimate flexural capacity of an underreinforced rectangular beam assuming a rectangular stress distribution.

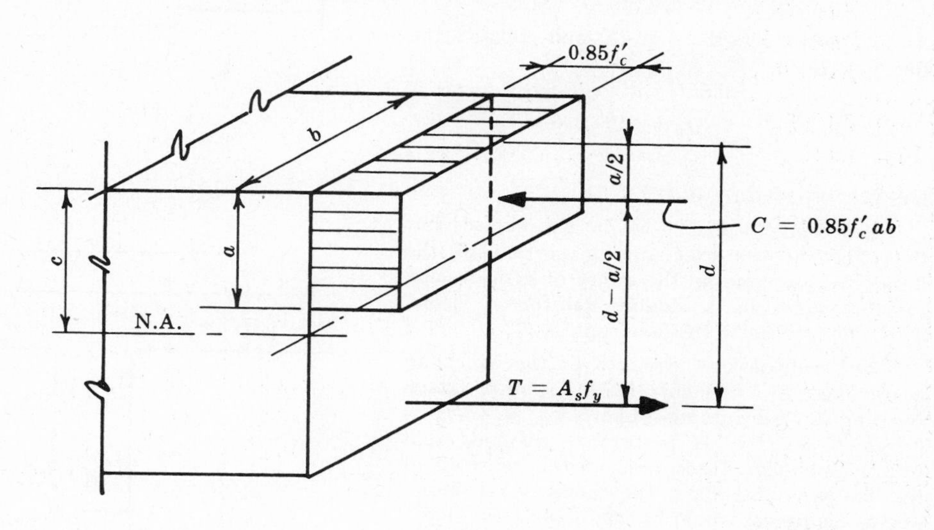

Fig. 5-11

Summing horizontal forces, $T = C$. Then

$$A_s f_y = 0.85 f'_c ab \qquad \text{and} \qquad a = A_s f_y / 0.85 f'_c b$$

Summing moments about the centroid of the compressive force, $M_u = A_s f_y(d - a/2)$. Substituting for a,

$$M_u = A_s f_y d(1 - 0.59 p f_y / f'_c) \qquad \text{or} \qquad M_u = f'_c bd^2 q(1 - 0.59q)$$

5.4. Derive the expression for balanced reinforcement as given by equation (*5.14*).

Referring to Fig. 5-12, by geometry:

$$\epsilon_y c = \epsilon_u(d - c) \qquad \text{or} \qquad c = \epsilon_u d/(\epsilon_u + \epsilon_y)$$

But, $c = a/k_1 = A_s f_y / 0.85 f'_c bk_1$. Then

$$\epsilon_u d/(\epsilon_u + \epsilon_y) = A_s f_y / 0.85 f'_c bk_1$$

or

$$\frac{A_s}{bd} = \frac{0.85 f'_c k_1}{f_y} \frac{\epsilon_u}{\epsilon_u + \epsilon_y} = p$$

If p is defined as the balanced reinforcement, p_b, and if we let $\epsilon_u = 0.003$, then

$$p_b = \frac{0.85 f'_c k_1}{f_y} \frac{0.003 E_s}{(0.003 + \epsilon_y) E_s}$$

If $E_s = 29{,}000{,}000$ psi and $f_y = \epsilon_y E_s$, then

$$p_b = \frac{0.85 f'_c k_1}{f_y} \frac{87{,}000}{87{,}000 + f_y}$$

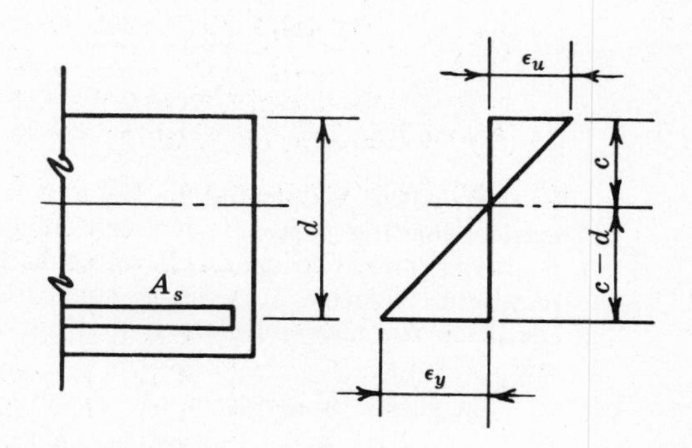

Fig. 5-12

5.5. Derive equation (*5.16*).

Referring to Fig. 5-13, by geometry,

$$\epsilon_s' c = \epsilon_u(c - d') \quad \text{or} \quad c = \epsilon_u d'/(\epsilon_u - \epsilon_s')$$

But $c = a/k_1 = A_s f_y / 0.85 f_c' b k_1$. Then

$$\epsilon_u d'/(\epsilon_u - \epsilon_s') = A_s f_y / 0.85 f_c' b k_1$$

or

$$\frac{A_s}{bd} = \frac{0.85 f_c' k_1 d'}{f_y d} \frac{\epsilon_u}{\epsilon_u - \epsilon_s'}$$

At ultimate strength $\epsilon_s' = \epsilon_y$, and substituting for $f_y = E_s \epsilon_y$,

$$p - p' = \frac{0.85 f_c' k_1 d'}{f_y d} \frac{87{,}000}{87{,}000 - f_y}$$

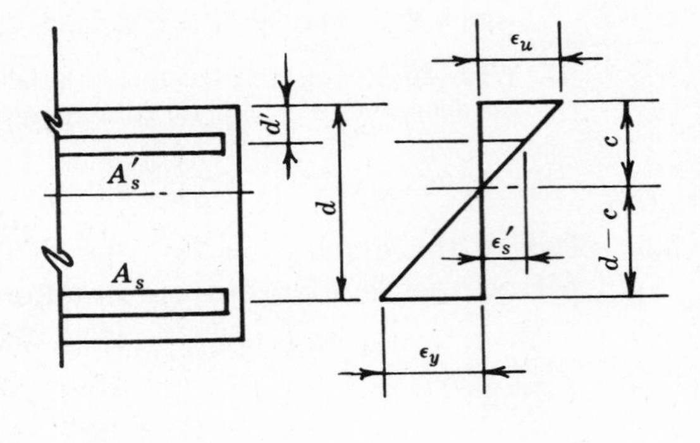

Fig. 5-13

5.6. Derive equation (*5.17*).

Referring to Fig. 5-14, it is assumed that the effect of the overhanging portion of the flange is the same as the effect of compression reinforcement in a double reinforced beam. This "additional" reinforcement is A_{sf}.

The compressive strength of the overhang would be $C' = 0.85 f_c'(b - b')t$, and transforming to reinforcing steel, $A_{sf} = C'/f_y = 0.85(b - b')t f_c'/f_y$. It is further assumed that this compressive force acts at the centroid of the flange. Therefore, the moment resistance due to the flange would be $M_F = A_{sf} f_y (d - 0.5t)$. Substituting these values in the expression for a double reinforced beam,

$$M_u = (A_s - A_{sf}) f_y (d - 0.5a) + A_{sf} f_y (d - 0.5t)$$

where $a = (A_s - A_{sf}) f_y / 0.85 f_c' b'$.

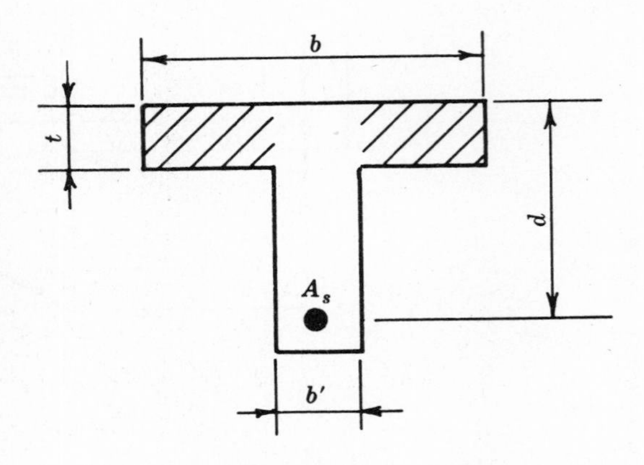

Fig. 5-14

5.7. The three span beam shown in Fig. 5-15 is subjected to a dead load of 1.0 kip/ft and a live load of 2.0 kips/ft. Determine the ultimate positive and negative moments and shears. Use the load factors of the 1963 ACI Building Code.

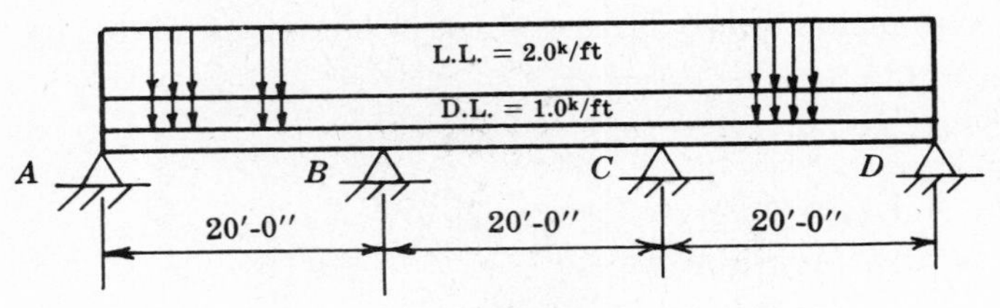

Fig. 5-15

Since the L.L. could be on any number of the spans, it will be necessary to determine the critical loading pattern. The critical loading condition for maximum shear at A and for maximum positive moment in span AB would be live load in spans AB and CD only. The critical loading for maximum positive moment in span BC would be live load in span BC only. The critical loading condition for maximum shear and maximum negative moment at B would be live load in spans AB and BC only.

An elastic analysis of the structure yields the following shears and moments. For dead load:

$V_{AB} = 0.4wl = 0.4(1.0)(20) = 8.0$ kips $\qquad +M_{AB} = 0.08wl^2 = 0.08(1.0)(20)^2 = 32.0$ ft-kips

$V_{BA} = 0.6wl = 0.6(1.0)(20) = 12.0$ kips $\qquad -M_B = 0.10wl^2 = 0.10(1.0)(20)^2 = 40.0$ ft-kips

$V_{BC} = 0.5wl = 0.5(1.0)(20) = 10.0$ kips $\qquad +M_{BC} = 0.025wl^2 = 0.025(1.0)(20)^2 = 10.0$ ft-kips

For live load on spans AB and CD only:

$$V_{AB} = 0.45wl = 0.45(2)(20) = 18.0 \text{ kips}$$
$$+M_{AB} = 0.101wl^2 = 0.101(2)(20)^2 = 80.8 \text{ ft-kips}$$

For live load on span BC only:

$$+M_{BC} = 0.075wl^2 = 0.075(2)(20)^2 = 60.0 \text{ ft-kips}$$

For live load on spans AB and BC only:

$$V_{BA} = 0.617wl = 0.617(2)(20) = 24.7 \text{ kips}$$
$$V_{BC} = 0.583wl = 0.583(2)(20) = 23.3 \text{ kips}$$
$$-M_B = 0.117wl^2 = 0.117(2)(20)^2 = 93.6 \text{ ft-kips}$$

Summarizing,

		D.L.	Max. L.L.
V_{AB}	=	8.0 kips	18.0 kips
V_{BA}	=	12.0	24.7
V_{BC}	=	10.0	23.3
$+M_{AB}$	=	32.0 ft-kips	80.8 ft-kips
$-M_B$	=	40.0	93.6
$+M_{BC}$	=	10.0	60.0

The load factor equations (*5.7*) and (*5.8*) are

$$U = 1.5D + 1.8L \quad \text{and} \quad U = 1.25(D + L + W)$$

It is apparent that the first equation governs, since there are no wind load effects. Therefore:

$$\text{Ult. } V_{AB} = 1.5(8.0) + 1.8(18) = 44.4 \text{ kips}$$
$$\text{Ult. } V_{BA} = 1.5(12.0) + 1.8(24.7) = 62.5 \text{ kips}$$
$$\text{Ult. } V_{BC} = 1.5(10.0) + 1.8(23.3) = 56.9 \text{ kips}$$
$$\text{Ult. } +M_{AB} = 1.5(32.0) + 1.8(80.8) = 194 \text{ ft-kips}$$
$$\text{Ult. } -M_B = 1.5(40.0) + 1.8(93.6) = 229 \text{ ft-kips}$$
$$\text{Ult. } +M_{BC} = 1.5(10.0) + 1.8(60.0) = 123 \text{ ft-kips}$$

5.8. Given the beam section in Fig. 5-16 with the concrete cylinder strength equal to 3500 psi and the yield strength of the steel equal to 50,000 psi. Determine the idealized ultimate flexural capacity of the beam assuming the parabolic and rectangular compressive stress distributions.

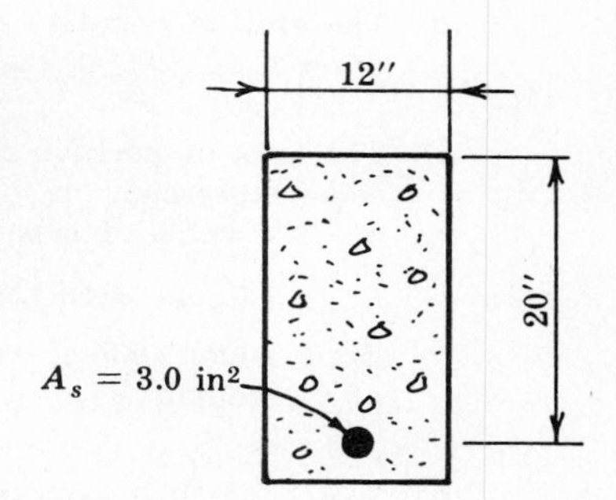

Fig. 5-16

The general expression for the ultimate capacity of the beam is equation (*5.10*),

$$M_u = A_s f_y d\left(1 - \frac{k_2}{k_1} pm\right)$$

It was shown that for the parabolic distribution, $k_2/k_1 = 0.554$ when $f_c = 0.85f'_c$ and $f_s = f_y$. Then from equation (*5.10*),

$$M_u = 3.0(50{,}000)(20)\left[1 - \frac{0.554(3.0)(50)}{12(20)(0.85)(3.5)}\right] = 2{,}650{,}000 \text{ in-lb}$$

If the rectangular distribution is assumed, $k_2/k_1 = 0.50$ and $M_u = 2{,}690{,}000$ in-lb.

A comparison of the results yields $2{,}690{,}000/2{,}650{,}000 = 1.016$, or less than 2% difference.

5.9. Design a one way continuous slab to support the moments tabulated. Assume $f'_c = 3000$ psi, $f_y = 40{,}000$ psi, and design for minimum depth permitted with no compressive reinforcement.

Dead Load Moment = +9.5 ft-kips/ft at midspan
= −15.0 ft-kips/ft at supports

Live Load Moment = +9.0 ft-kips/ft at midspan
= −10.0 ft-kips/ft at supports

Wind Load Moment = ±30.0 ft-kips/ft at the supports. (Wind Load produces no moment at midspan.)

(*a*) The first step is to determine the ultimate moments, both positive and negative, by use of load factor equations:

$$U = 1.5D + 1.8L, \qquad U = 1.25(D + L + W), \qquad U = 0.9D + 1.1W$$

For positive moments,

$$U = 1.5(9.5) + 1.8(9.0) = 30.5 \text{ ft-kips/ft}$$

This will be the ultimate positive moment because wind has no effect at midspan and the last load factor equation will be critical only at the supports.

For negative moment,

$$U = 1.5(-15.0) + 1.8(-10.0) = -40.5 \text{ ft-kips/ft}$$

$$U = -1.25(15.0 + 10.0 + 30.0) = -68.7 \text{ ft-kips/ft}$$

$$U = 0.9(15.0) \pm 1.1(30.0) = -19.5 \text{ ft-kips/ft} \quad \text{or} \quad = +46.5 \text{ ft-kips/ft}$$

(*b*) The maximum ultimate negative moment of −68.7 ft-kips/ft is the largest absolute value and governs the thickness of the slab. It is necessary to determine the maximum percentage of reinforcement permitted. Then

$$p_{\max} = 0.75\left[\frac{0.85k_1 f'_c}{f_y}\,\frac{87{,}000}{87{,}000 + f_y}\right] = 0.75\left[\frac{0.85(0.85)(3000)}{40{,}000}\,\frac{87{,}000}{127{,}000}\right] = 0.0279$$

(*c*) With $p_{\max}$ determined, the depth of the section is determined by substitution into the expression for the ultimate moment capacity; assuming the rectangular distribution,

$$M_u = \phi[bd^2 f'_c\, q(1 - 0.59q)]$$

where $\phi = 0.90$ for flexure and $q = p(f_y/f'_c) = 0.0279(40{,}000/3000) = 0.372$. Then $d = 9.38''$. Use $d = 9.5''$.

(*d*) The area of negative reinforcement at the support is

$$-A_s = (bd)(qf'_c/f_y) = 0.0279bd = 0.0279(12)(9.5) = 3.18 \text{ in}^2/\text{ft}$$

(*e*) The area of positive reinforcement at the support would be determined by substitution in the same expression for ultimate moment capacity. However, now the depth of the section is known and the value of q is unknown.

$$M_u = \phi[bd^2 f'_c\, q(1 - 0.59q)] \quad \text{or} \quad 46.5(12{,}000) = 0.9[12(9.5)^2(3000)q(1 - 0.59q)]$$

from which $0.59q^2 - q = -0.191$ and $q = 0.219$. Hence the area of positive reinforcement at the support is

$$+A_s = (bd)(qf'_c/f_y) = 1.87 \text{ in}^2/\text{ft}$$

(*f*) The area of positive reinforcement at midspan may be determined by modifying the constant in the quadratic in (*e*) by the ratio of moments.

$$558{,}000(30.5/46.5) = 2{,}920{,}000q(1 - 0.59q) \quad \text{or} \quad q = 0.136 \quad \text{and} \quad A_s = (bd)(qf'_c/f_y) = 1.16 \text{ in}^2/\text{ft}$$

5.10. Solve Problem 5.9 by using charts or tables for proportioning the member.

(*a*) From Fig. 5-7, $p_{\max} = 0.028$; then $q_{\max} = p(f_y/f'_c) = 0.372$. Or from Fig. 5-8, $q_{\max} = 0.37$.

(*b*) Solving for $M_u/bd^2f'_c$ by Table 5.1 for $q = 0.37$,

$$\frac{M_u}{\phi f'_c\,bd^2} = 0.2892 \quad \text{or} \quad d^2 = \frac{68.7(12{,}000)}{0.9(3000)(12)(0.2892)} = 88.0$$

from which $d = 9.4''$. Use $d = 9.5''$.

(*c*) Negative reinforcement at the support is $-A_s = 0.028(12)(9.5) = 3.19$ in²/ft.

(*d*) Positive reinforcement at support is determined by solving for q. We have

$$\frac{M_u}{\phi f'_c\,bd^2} = \frac{46.5(12{,}000)}{(0.9)(3000)(12)(9.5)^2} = 0.191$$

From Table 5.1, $q = 0.219$. Then $+A_s = 0.219(3000/40{,}000)(12)(9.5) = 1.88$ in²/ft.

(*e*) Similarly, the positive reinforcement at midspan is determined.

$$\frac{M_u}{\phi f'_c\,bd^2} = \frac{30.5(12{,}000)}{(0.9)(3000)(12)(9.5)^2} = 0.125$$

From Table 5.1, $q = 0.136$. Then $A_s = 1.16$ in²/ft.

5.11. Solve Problem 5.10 using Fig. 5-9.

After determining maximum p and minimum d as before, enter Fig. 5-9 with values of M_u/bd^2.

$$\frac{M_u}{\phi bd^2} = \frac{46.5(12{,}000)}{(0.9)(12)(9.5)^2} = 572$$

From Fig. 5-14, $p = 0.018$. Then $+A_s = 0.018(12)(9.5) = 2.05$ in²/ft.

$$\frac{M_u}{\phi bd^2} = \frac{30.5(12{,}000)}{(0.9)(12)(9.5)^2} = 376$$

and $p = 0.011$. Then $A_s = 0.011(12)(9.5) = 1.25$ in²/ft.

5.12. Solve Problem 5.10 using Table 5.2.

(*a*) Using Table 5.2, the maximum resisting moment at balanced conditions is $M_u/\phi = R_u bd^2$. From the table, for $f'_c = 3000$ and $f_y = 40{,}000$, $\phi R_u = 783$.

Then $d^2 = \dfrac{68.7(12{,}000)}{0.9(12)(871)} = 87.7$, $d = 9.38$. Use $d = 9.5''$.

(*b*) From Table 5.2, $p_{max} = 0.0278$. Then $-A_s = 0.0278(12)(9.5) = 3.18$ in²/ft.

(*c*) Using an average value for ϕj_u selected from Table 5.2,

$$+A_s = \frac{M_u}{\phi f_y\,j_u\,d} = \frac{46.5(12{,}000)}{0.80(40{,}000)(9.5)} = 1.83 \text{ in}^2/\text{ft}$$

and $p = 1.83/(9.5)(12) = 0.016$. This value checks very close to the value for p given in the table. If greater refinement is required, the next trial may be made using $p = 0.016$ and the corresponding $\phi a_u = 2.622$. The value of ϕa_u is determined so that the moment is in foot-kips.

$$+A_s = \frac{M}{\phi a_u d} = \frac{46.5}{2.622(9.5)} = 1.86 \text{ in}^2/\text{ft}$$

In normal design procedures, the first trial yielding an area of 1.83 in²/ft would be sufficiently accurate. This is comparable to the accuracy attained in working stress design.

(*d*) Because the value of the midspan moment is less than that in (*c*), a value for ϕa_u somewhat greater than that given above will be used. Hence try $\phi a_u = 2.85$:

$$A_s = \frac{M}{\phi a_u d} = \frac{30.5}{2.85(9.5)} = 1.13 \text{ in}^2/\text{ft} \quad \text{and} \quad p = \frac{1.13}{9.5(12)} = 0.0099$$

Try $\phi a_u = 2.80$:

$$A_s = \frac{30.5}{2.80(9.5)} = 1.15 \text{ in}^2/\text{ft} \quad \text{and} \quad p = \frac{1.15}{9.5(12)} = 0.0101$$

(*e*) In (*c*) and (*d*) it is shown that a refinement in the selection of ϕj_u or ϕa_u results in a very small change in the value of the area of reinforcement. With experience, a "feel" for such variations will be developed by the designer. Large discrepancies may be adjusted with only two trials in the selection of ϕa_u or ϕj_u.

5.13. Design a minimum depth rectangular beam with tension reinforcement only to carry a D.L. moment of 50 ft-kips and a L.L. moment of 200 ft-kips. Assume $f'_c = 3000$ psi and $f_y = 50{,}000$ psi.

(*a*) The ultimate moment must be determined by use of the load factor equations (*5.7*), (*5.8*) and (*5.9*). $U = 1.5D + 1.8L = 1.5(50) + 1.8(200) = 435$ ft-kips. It is clear that equation (*5.7*) governs because there are no wind load effects.

(*b*) A minimum depth section requires maximum percentage of steel reinforcement. Therefore substituting in the expression for p_b,

$$p_b = \frac{0.85 f'_c k_1}{f_y}\,\frac{87{,}000}{87{,}000 + f_y} = 0.0276 \quad \text{and} \quad p_{\max} = 0.75(0.0276) = 0.0207$$

(*c*) With $p_{\max}$ determined, the value for bd^2 is found by substitution in the expression for the ultimate moment capacity.

$$M_u = \phi[bd^2 f'_c\, q(1 - 0.59q)]$$

where $q = p(f_y/f'_c) = 0.0207(50{,}000/3000) = 0.345$. Substituting,

$$435(12{,}000) = (0.9)(bd^2)(3000)(0.345)(1 - 0.204) \quad \text{or} \quad bd^2 = 7050 \text{ in}^3$$

There would be many combinations of the beam width and depth that would satisfy the above expression. A width or a depth is often set by some other architectural or structural consideration.

Try a ratio $d/b = 1.5$. Then $b(1.5b)^2 = 7050$, $b^3 = 3120$, $b = 14.6$; use $b = 14.5''$. Thus $d = 1.5(14.5) = 21.8$; use $d = 22''$. Then $A_s = pbd = 0.0207(14.5)(22) = 6.60$ in^2.

(*d*) Try a ratio of $d/b = 1.0$. Then $b^3 = 7050$, $b = 19.2$; use $b = 19.5''$. Use $d = 19.5''$. Then $A_s = pbd = 0.0207(19.5)^2 = 7.86$ in^2.

(*e*) Try a ratio of $d/b = 0.67$. Then $b(0.67b)^2 = 7050$, $b^3 = 15{,}800$, $b = 25.1$; use $b = 25''$. Thus $d = 0.67(25) = 16.7$; use $d = 17''$. Then $A_s = pbd = 0.0207(25)(17) = 8.80$ in^2.

(*f*) The 1963 ACI Building Code requires a minimum amount of reinforcement in beams. The specification states that for flexural members, except constant thickness slabs, if positive reinforcement is required, the minimum reinforcement ratio p shall be not less than $200/f_y$ unless the area of reinforcement at every section, positive or negative, is at least one third greater than that required.

Of course in a section of maximum p this requirement will not control. However, for $f_y = 50{,}000$ psi, $200/f_y = 0.004$.

(*g*) For a beam such as this, the deflections at service loads should be checked. The 1963 ACI Building Code requires that the deflection of a beam must be checked if it is proportioned according to ultimate strength design procedures and if p exceeds $0.18 f'_c/f_y$ or if f_y exceeds 40,000 psi.

In this example, $q = 0.345 > 0.18$ and $f_y = 50{,}000 > 40{,}000$ psi. Therefore the deflection check would be required.

5.14. Solve Problem 5.13 using Table 5.2.

If $f'_c = 3000$ psi, $f_y = 50{,}000$ psi, $M_u = 435$ ft-kips, from Table 5.2, the maximum values for ϕR_u and p are $\phi R_u = 740.5$, $p = 0.0206$. Since in ultimate strength design $\phi R_u = M_u/bd^2$, then $bd^2 = 435(12{,}000)/740.5 = 7060$ in^3.

Using the same ratios of d/b and selecting ϕa_u from Table 5.2, $\phi a_u = 2.989$.

For $d/b = 1.5$, $d = 22''$ and $A_s = 435/(2.99)(22) = 6.61\text{ in}^2$.

For $d/b = 1.0$, $d = 19.5''$ and $A_s = 435/(2.99)(19.5) = 7.48\text{ in}^2$.

For $d/b = 0.67$, $d = 17''$ and $A_s = 435/(2.99)(17) = 8.56\text{ in}^2$.

It will be noted that the areas of reinforcement above for ratios of $d/b = 1.0$ and 0.67 do not check with those of Problem 5.13. This discrepancy is due to the fact that the values determined for d and b were rounded off.

In Problem 5.13, if the actual values were used for $d/b = 1.0$, $d = 19.2''$, $b = 19.2''$ and $A_s = 0.0207(19.2)^2 = 7.62\text{ in}^2$. For $d/b = 0.67$, $d = 16.7''$, $b = 25.1''$ and $A_s = 0.0207(16.7)(25.1) = 8.68\text{ in}^2$.

In this Problem 5.14, if these same values for d are used, for $d/b = 1.0$, $d = 19.2''$ and $A_s = 7.59\text{ in}^2$; and for $d/b = 0.67$, $d = 16.7''$ and $A_s = 8.70\text{ in}^2$.

Thus when the "exact" values for d and b are used, both methods yield essentially the same results.

Regardless of which method is used, the rounded-off values result in areas of reinforcement within 3% of the results determined using the more precise values for b and d. This degree of accuracy would be normally acceptable in ordinary design procedures.

5.15. Determine the area of reinforcing steel required for the section shown in Fig. 5-17. Assume $f'_c = 3500$ psi, $f_y = 50{,}000$ psi and $M_u = 150$ ft-kips.

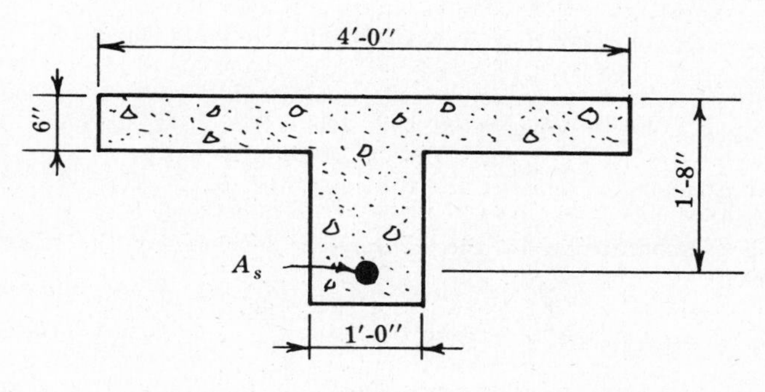

Fig. 5-17

(*a*) The first step in the solution is to determine if the neutral axis of the section falls within the stem and the section acts as a T-beam. If not, it is proportioned as a rectangular beam. If the moment capacity of the flange is greater than 150 ft-kips, then the section acts as a rectangular beam.

The moment capacity of the flange is

$$M_{uf} = 0.85f'_c\,bt(d-0.5t) = 0.85(3500)(48)(6)(20-3)/12{,}000 = 1215\text{ ft-kips} > 150/0.9$$

Hence the section does not act as a T-beam and should be designed as a rectangular section.

(*b*) Table 5.2 does not contain a series of coefficients for $f'_c = 3500$ psi. However, Table 5.1 may be used.

$$\frac{M_u}{f'_c\,bd^2} = \frac{150(12{,}000)}{(0.90)(3500)(48)(20)^2} = 0.0298$$

Now using Table 5.1, we find $q = 0.030$. Then $p = qf'_c/f_y = 0.030(3500)/50{,}000 = 0.00210$ and $A_s = pbd = 0.00210(48)(20) = 2.02\text{ in}^2$.

(*c*) As discussed in Problem 5.13, the 1963 ACI Building Code requires that under certain conditions the minimum area of tension reinforcement shall not be less than $200/f_y$. $200/f_y = 200/50{,}000 = 0.004 < 2.02/(12)(20)$.

It should be noted that here p is defined as the ratio of the area of tension reinforcement to the effective area of concrete in the web of a flanged member.

(*d*) $t = 6'' > 1.18qd/k_1 = 1.18(0.030)(20)/0.85 = 0.83''$

(*e*) Although q is less than 0.18, f_y is greater than 40,000 psi. Hence the 1963 ACI Building Code would require a deflection check.

5.16. Determine the area of reinforcement for a T-section with the following given: $t = 2.0''$, $b = 30''$, $b' = 14''$, $d = 30''$, $f'_c = 3000$ psi, $f_y = 40{,}000$ psi, $M_u = 600$ ft-kips.

(*a*) As in Problem 5.15, we must determine if the neutral axis of the section falls within the stem.

The idealized moment capacity of the concrete flange is

$$M_{uf} = 0.85f'_c\,bt(d-0.5t) = 370 \text{ ft-kips} < 600/0.90 = M_u/\phi$$

Hence the section acts as a T-beam.

(*b*) In the design of T-beams, the overhanging portion of the flange is considered the same as an equivalent amount of compression steel.

$$A_{sf} = 0.85(b-b')tf'_c/f_y = 2.04 \text{ in}^2$$

The idealized moment capacity of the overhanging portion is $M'_{uf} = A_{sf}f_y(d-0.5t) =$ 197 ft-kips.

(*c*) The idealized moment capacity of the rectangular section, b' times d, must be

$$M_w = M_u/\phi - M'_{uf} = 667 - 197 = 470 \text{ ft-kips}$$

And the idealized moment capacity of the web portion is $M_w = b'd^2f'_c\,q(1-0.59q)$. Then

$$\frac{M_w}{f'_c\,b'd^2} = \frac{470(12{,}000)}{3000(14)(30)^2} = 0.149 = q(1-0.59q)$$

From Table 5.1, $q = 0.165$ or the reinforcement available to develop the web is $A_{sw} = p_{sw}b'd$. Then

$$A_{sw} = 0.165(3000)(14)(30)/40{,}000 = 5.19 \text{ in}^2$$

(*d*) The idealized moment capacity of the overhanging portion of the section is 197 ft-kips, and the idealized moment capacity of the web is 470 ft-kips. Thus $197 + 470 = 667 = 600/0.9$.

(*e*) The total area of the tension reinforcement is $A_s = A_{sf} + A_{sw} = 2.04 + 5.19 = 7.23 \text{ in}^2$.

(*f*) As a check, substituting in the expression derived for the ultimate moment capacity of a T-section,

$$M_u = \phi[(A_s - A_{sf})f_y(d-a/2) + A_{sf}f_y(d-0.5t)]$$

where $a = A_{sw}f_y/0.85f'_c\,b' = 5.80$. Then $M_u = 7{,}200{,}000$ in-lb $= 600$ ft-kips.

(*g*) The 1963 ACI Building Code requires that in I- and T-sections the area of the tension reinforcement available to develop the web portion be limited to prevent the possibility of brittle or compression failures of the concrete. Therefore $p_w - p_f \leqq 0.75p_b$ or

$$\frac{A_s}{b'd} - \frac{A_{sf}}{b'd} \leqq 0.75p_b \quad \text{or} \quad \frac{A_{sw}}{b'd} \leqq 0.75p_b$$

In this example, $$\frac{A_{sw}}{b'd} = \frac{5.19}{14(30)} = 0.0124 < 0.75p_b = 0.75(0.0373)$$

(*h*) Checking for minimum reinforcement requirements:

$$\text{Min. } p = 200/40{,}000 = 0.005 \quad \text{and} \quad \text{Min. } A_s = 0.005(14)(30) = 2.10 \text{ in}^2 < 7.23$$

(*i*) $t = 2.0'' < 1.18qd/k_1 = 1.18(0.165)(30)/0.85 = 6.9''$

(*j*) The 1963 ACI Building Code requires that a deflection check must be made if

$$p_w - p_f > 0.18(f'_c/f_y) \quad \text{or} \quad 0.0124 < 0.18(3000)/40{,}000 = 0.0135$$

and $f_y = 40{,}000$ psi. Consequently, no check of deflections is necessary.

5.17. With the given data, determine the area of reinforcement required to resist an ultimate moment of 900 ft-kips: $b = 12''$, $d = 30''$, $f'_c = 3000$ psi, $f_y = 40{,}000$ psi.

(*a*) It is necessary to determine the area of tension reinforcement. Examining Table 5.2, a value for $\phi a_u = 2.4$ is selected. Then

$$A_s = \frac{M_u}{\phi a_u d} = \frac{900}{2.4(30)} = 12.5 \text{ in}^2, \qquad p = \frac{A_s}{bd} = \frac{12.5}{12(30)} = 0.0348$$

This value is greater than the maximum value of $p = 0.0278$ given in Table 5.2. Hence compression reinforcement will be required.

Using the maximum value of $p = 0.0278$, $A_s = 0.0278(12)(30) = 10.0 \text{ in}^2$. Then the idealized moment capacity of this area of reinforcement is

$$M_u/\phi = A_s a_u d = 10.0(2.343)(30)/0.9 = 782 \text{ ft-kips}$$

(*b*) The second portion of the ultimate resisting moment is the couple formed by the compression steel and a like amount of tension steel. Or,

$$M'_u/\phi = 900/\phi - 782 = 218$$

Assuming the depth of the compression reinforcement $d' = 2.0''$, the resisting moment is

$$M'_u/\phi = A'_s f_y (d - d'); \quad 218(12{,}000) = A'_s(40{,}000)(28); \quad A'_s = 2.34 \text{ in}^2$$

Checking to see if the compression steel is at yield,

$$p - p' \geqq \frac{0.85k_1 f'_c d'}{f_y d} \, \frac{87{,}000}{87{,}000 - f_y} = 0.00668$$

$$p - p' = \frac{10.0}{12(30)} = 0.0278 > 0.00668$$

Therefore the condition is satisfied.

(*c*) The total area of tension reinforcement is $A_s = 10.0 + 2.34 = 12.34 \text{ in}^2$. The area of compression reinforcement is $A'_s = 2.34 \text{ in}^2$.

(*d*) The minimum steel requirement of $p \geqq 200/f_y$ is satisfied.

(*e*) It is obvious that $q > 0.18$. Hence a deflection check of the section would be required.

(*f*) The requirement that $p - p' \leqq 0.75p_b$ is met.

5.18. Because Table 5.2 is limited to a certain combination of f_y and f'_c, solve Problem 5.17 without the use of this table.

(*a*) First, to determine the area of reinforcement required,

$$\frac{M_u}{\phi f'_c bd^2} = \frac{900(12{,}000)}{0.9(3000)(12)(30)^2} = 0.370 = q(1 - 0.59q)$$

Entering Table 5.1, it is seen that there is no value of $M_u/\phi f'_c bd^2 = 0.370$. Therefore, it will be necessary to solve for q,

$$q(1 - 0.59q) = 0.370, \qquad q = 0.551$$

It is seen in Fig. 5-8, that this value exceeds the maximum value for q. Hence $p > p_b$ and compression reinforcement is required.

$$p_{\max} = \frac{0.75(0.85)f'_c k_1}{f_y} \, \frac{87{,}000}{87{,}000 + f_y} = 0.0278 \quad \text{and} \quad q_{\max} = \frac{0.0278 f_y}{f'_c} = 0.371$$

Entering Table 5.1 with $q = 0.371$,

$$\frac{M_u}{f'_c bd^2} = 0.2898 \quad \text{or} \quad M_u = \frac{0.2898(3000)(12)(30)^2}{12{,}000} = 782 \text{ ft-kips}$$

This value is the idealized moment capacity. If the applied moment is divided by ϕ so that $M_u/\phi = 900/0.90 = 1000$ ft-kips, then it will be unnecessary to consider ϕ further in the calculations.

(*b*) The second portion of the resisting moment,

$$M'_u = 1000 - 782 = 218 \text{ ft-kips} \quad \text{and} \quad M'_u = A'_s f_y (d - d');$$

$$218(12{,}000) = A'_s(40{,}000)(30 - 2), \quad A'_s = 2.34 \text{ in}^2$$

(*c*) The check of the yielding of the compression steel would be the same as in Problem 5.17.

(*d*) The total area of tension reinforcement is $A_s = 0.0278(12)(30) + 2.34 = 12.34 \text{ in}^2$. The area of compression reinforcement is $A'_s = 2.34 \text{ in}^2$.

(*e*) The other checks would be the same as Problem 5.17.

5.19. Given the floor section shown in Fig. 5-18 and $f'_c = 4000$ psi and $f_y = 50{,}000$ psi, select the area of flexural reinforcement required to resist the moments indicated in the beams.

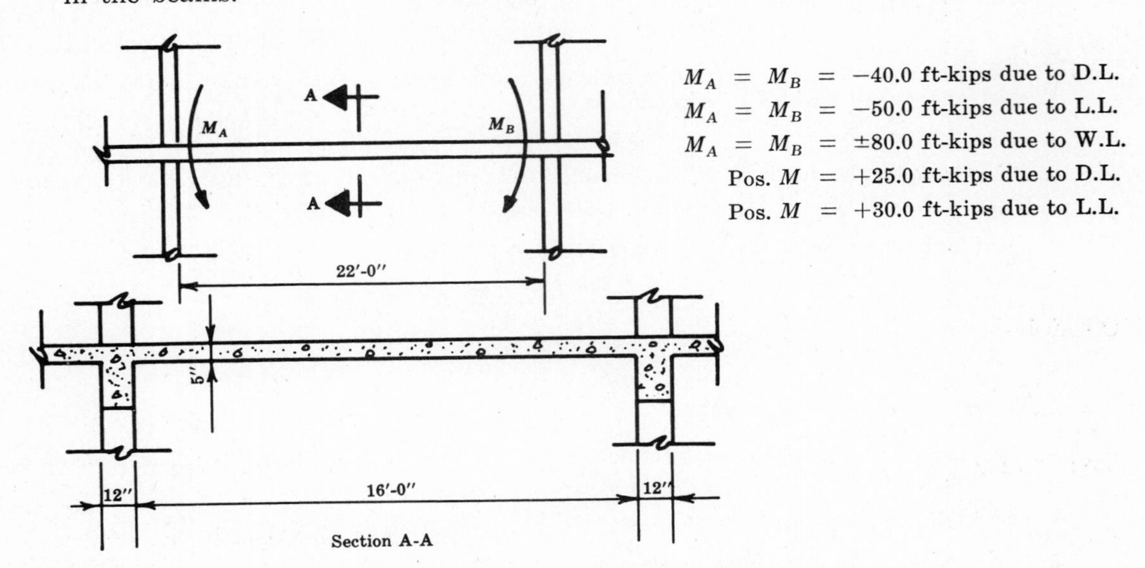

$M_A = M_B = -40.0$ ft-kips due to D.L.
$M_A = M_B = -50.0$ ft-kips due to L.L.
$M_A = M_B = \pm 80.0$ ft-kips due to W.L.
Pos. $M = +25.0$ ft-kips due to D.L.
Pos. $M = +30.0$ ft-kips due to L.L.

Fig. 5-18

(*a*) It is first necessary to determine the critical load factor equations. Using Fig. 5-5, for M_A: $T = 170$, $L/T = 0.294$, $D/T = 0.234$. Hence equation (*5.8*) is critical. For Pos. M, there are no wind effects.

The ultimate design moments are:

$$-M_{Au} = 1.25(170) = -213 \text{ ft-kips}$$
$$+M_{Au} = 0.9(-40.0) + 1.1(80.0) = +52.0 \text{ ft-kips}$$
$$\text{Pos. } M_u = 1.5(25.0) + 1.8(30.0) = +91.5 \text{ ft-kips.}$$

(*b*) The negative moment at the support is the largest absolute and controls the size of the member.

Rather than design a minimum depth member, a member with $q = 0.18$ will be used. This will result in a member comparable to working stress design with balanced reinforcement.

For $q = 0.18$, $p = 0.18(f'_c/f_y) = 0.0144$. From Table 5.2, $\phi j_u = 0.805$ and $\phi a_u = 3.35$. And if $M_u = \phi R_u bd^2 = A_s f_y \phi j_u d$, then $\phi R_u = p f_y \phi j_u$. Hence

$$M_u/bd^2 = 0.0144(50{,}000)(0.805) = 580$$

Then

$$bd^2 = 213(12{,}000)/(580) = 4410, \quad d^2 = 4410/12 = 367, \quad d = 19.2. \quad \text{Use } d = 19.5''.$$

(*c*) The area of steel reinforcement at the support in the top is

$$-A_s = M_u/\phi a_u d = 3.26 \text{ in}^2$$

The area of steel reinforcement at the support at the bottom is

$$+A_s = 52.0/3.35(19.5) = 0.80 \text{ in}^2$$

It is unnecessary to check for maximum reinforcement requirements because $q = 0.18$.

(*d*) At the supports the beam acts as a rectangular section. However, in the positive moment region the beam is a T-section.

The 1963 ACI Building Code specifies that in T-sections the effective flange width shall not be greater than one-fourth the beam span length and that the overhang on either side of the web shall not be greater than eight times the slab thickness or greater than one-half the slab span. Applying these rules:

$$b = 1/4(22)(12) = 66'', \quad b = 12 + 8(5)(2) = 92'', \quad b = 12 + (16)(12) = 204''$$

Therefore the effective flange width is 66″.

(e) It is necessary to evaluate the capacity of the flange in order to determine if the section acts as a T-beam.

$M_{uf} = 0.85f'_c bt(d - 0.5t) = 0.85(4000)(66)(5)(19.5 - 2.5) = 19{,}000{,}000$ in-lb $= 1590$ ft-kips > 85.2

Therefore the section acts as a rectangular beam.

(f) Using the same value of $d = 19.5''$,

$$A_s = \frac{M_u}{\phi a_u d} = \frac{91.5}{3.6(19.5)} = 1.30 \text{ in}^2 \quad \text{and} \quad p = \frac{1.30}{66(19.5)} = 0.00101$$

Try $\phi a_u = 3.7$: $A_s = \dfrac{91.5}{3.7(19.5)} = 1.27 \text{ in}^2$.

(g) Checking for minimum area of reinforcement:

$$200/f_y = 200/50{,}000 = 0.004; \quad 0.004 < 1.27/(19.5)(12)$$

(h) It is obvious that the positive moment section is underreinforced.

(i) Although $q \leqq 0.18$, $f_y > 40{,}000$ psi, and deflections must be checked.

Supplementary Problems

5.20. The basic D.L., L.L. and W.L. moments in a beam are 20.0, 20.0 and 80.0 ft-kips respectively. The corresponding shears are 35.0, 35.0 and 5.0 kips. Determine the controlling load factor equation according to the 1963 ACI Building Code.

Ans. Equation (*5.8*) for moment, equation (*5.7*) for shear.

5.21. A rectangular beam that has a width of 12″ and an effective depth of 13″ is reinforced at the bottom with 2-#10 bars. If $f'_c = 3000$ psi and $f_y = 40{,}000$ psi, determine the idealized ultimate flexural capacity of the beam assuming the parabolic distribution. *Ans.* 94.5 ft-kips

5.22. Determine the capacity of the beam in Problem 5.21 according to the 1963 ACI Building Code.

Ans. 95.8 ft-kips

5.23. A simply supported one way slab has a span of 12′-0″ and must resist service uniform dead and live loads of 750 lb/ft² and 1000 lb/ft² respectively. Design by USD a minimum depth slab if $f'_c = 3000$ psi and $f_y = 50{,}000$ psi. *Ans.* $d = 9''$, $A_s = 2.05$ in²/ft

5.24. A simply supported rectangular beam must resist service dead load and live load moments of 60.0 ft-kips and 30.0 ft-kips respectively. If $f'_c = 4000$ psi and $f_y = 50{,}000$ psi, proportion a minimum depth beam with tension reinforcement only. *Ans.* $b = 12''$, $d = 12.5''$, $A_s = 3.85$ in²

5.25. Repeat Problem 5.24 but assume $q = 0.18$. *Ans.* $b = 12''$, $d = 16''$, $A_s = 2.70$ in²

5.26. A concrete T-beam must resist an ultimate moment of 125 ft-kips. If $b = 55''$, $b' = 12''$, $d = 20''$, $t = 6''$, $f'_c = 4000$ psi, and $f_y = 50{,}000$ psi, determine the required area of tension reinforcement.

Ans. $A_s = 1.66$ in²

5.27. Repeat Problem 5.26 but let $t = 3''$ and the ultimate moment = 900 ft-kips.

Ans. $A_{sw} = 4.53$ in², $A_{sf} = 8.77$ in²

5.28. A rectangular beam with a width of 11″ and an effective depth of 25″ must resist an ultimate moment of 700 ft-kips. If $f'_c = 3000$ psi and $f_y = 40{,}000$ psi, determine the reinforcement required. *Ans.* If $d' = 2''$, $A_s = 11.30$ in², $A'_s = 3.65$ in²

5.29. A fixed end rectangular beam must support uniform service dead and live loads of 15.0 kips/ft and 10.0 kips/ft. If $l = 20.0'$, $f'_c = 4000$ psi and $f_y = 60{,}000$ psi, determine the flexural reinforcement requirements so that $q < 0.18$.

Ans. $b = 24''$, $d = 35''$, $+A_s = 4.15$ in², $-A_s = 9.60$ in²

Chapter 6

Shear and Diagonal Tension

WORKING STRESS DESIGN AND ULTIMATE STRENGTH DESIGN

NOTATION

A_g = gross area of section
A_s = area of tension reinforcement
A_v = total area of web reinforcement in tension within a distance, s, measured in a direction parallel to the longitudinal reinforcement
α = angle between inclined web bars and longitudinal axis of member
b = width of compression face of flexural member
b' = width of web in I and T sections
b_o = periphery of critical section for slabs and footings
d = distance from extreme compression fiber to centroid of tension reinforcement
d' = distance from the concrete face to the centroid of the near reinforcement
f'_c = compressive strength of concrete
f_v = tensile stress in web reinforcement
f_y = yield strength of reinforcement, including web reinforcement
F_{sp} = ratio of splitting tensile strength to the square root of the compressive strength
M = bending moment
M' = modified bending moment (due to axial force)
N = load normal to the cross section, to be taken as positive for compression, negative for tension, and to include the effects of tension due to shrinkage and creep
p_w = $A_s/b'd$
s = spacing of stirrups or bent bars in a direction parallel to the longitudinal reinforcement
t = total depth of section
v_c = shear stress carried by concrete
v_u = nominal ultimate shear stress as a measure of diagonal tension
V = total shear at section
V_u = total ultimate shear
V' = shear carried by web reinforcement
V'_u = ultimate shear carried by web reinforcement
ϕ = capacity reduction factor = 0.85 for shear

SHEAR STRESS IN HOMOGENEOUS BEAMS

In the study of Strength of Materials it is shown that, for *homogeneous elastic materials*, the unit horizontal shear stress at a section of a beam may be calculated using the equation

$$v = VA\bar{y}/Ib \tag{6.1}$$

where v = unit horizontal shear stress, psi
A = cross-section area above element in question, in^2
$\bar{y}$ = distance from the centroid of A to the neutral axis, in.
I = moment of inertia of the beam, in^4
b = thickness of the element, in.

This shear stress develops due to the change in bending moment from one side of the element to the other.

Concrete is neither truly elastic nor homogeneous, and results of experiments related to shear stress in concrete cannot be correlated directly with equation (*6.1*). At the present time, statistical correlation of test data must be utilized in order to provide rational equations for use in designing concrete elements to successfully resist shear stress.

Shear stress does not work alone to cause failure of concrete beams. The complex mechanism of concrete and reinforcing bars provides for an equally complex resistance to stress. Although no precise theory has been developed to properly explain shear as related to failure, it is known that *diagonal tension stresses* develop and that failure is due to tension in the concrete, rather than shear stress.

In order to develop a rational method of design, the shear stress distribution is assumed as shown in Fig. 6-1. Tension is neglected in the concrete, causing a constant shear stress distribution to be assumed below the neutral axis.

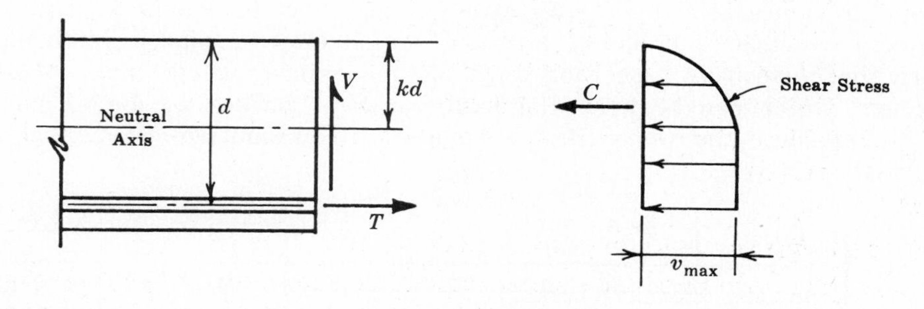

Fig. 6-1

Since the equations for shear design have been empirically devised using test data obtained for *first-cracking* of the concrete, the equations are, of necessity, *ultimate strength design* equations. There is in actuality no true *working stress method* for shear design. Suitable safety factors have been applied to the ultimate strength equations in order to provide working stress equations for those designers who prefer to use equations which already contain safety factors.

Since the ultimate strength design equations for shear stress differ from those for working stress design only in the constant terms, it is convenient to study both methods simultaneously. This process avoids needless duplication of effort in many respects.

SHEAR STRESS ON UNREINFORCED WEBS OF CONCRETE BEAMS

The actual *shear stress* on a section is computed nominally as force per unit area:

$$v = V/bd \tag{6.2}$$

for working stress design, and

$$v_u = V_u/bd \tag{6.3}$$

for ultimate strength design. All terms have been defined in the Notation.

The allowable shear stress on a section, obtained experimentally, is calculated using

$$v_c = \sqrt{f'_c} + 1300 p_w Vd/M \leqq 1.75\sqrt{f'_c} \tag{6.4}$$

for working stress design, and

$$v_{cu} = \phi(1.9\sqrt{f'_c} + 2500 p_w Vd/M) \leqq 3.5\phi\sqrt{f'_c} \tag{6.5}$$

for ultimate strength design. In all cases, V and M are taken as the ultimate values for ultimate strength design. Further, for both methods of design, Vd/M is always considered to be positive and may never be greater than 1.0.

Instead of the comprehensive equations (*6.4*) and (*6.5*), less liberal but more simple expressions may be used:

$$v_c = 1.1\sqrt{f'_c} \tag{6.6}$$

for working stress design, and

$$v_{cu} = 2\phi\sqrt{f'_c} \tag{6.7}$$

for ultimate strength design.

Equations (*6.4*), (*6.5*), (*6.6*) and (*6.7*) refer to the stress permitted on the web of a beam when special shear reinforcement is not provided.

WEB REINFORCEMENT

When the concrete cross section has insufficient area to maintain shear stresses below the permissible values, additional resistance to shear may be provided. One form of shear reinforcement consists of hoops or *stirrups,* which may be placed vertically or at some angle with the horizontal. Another form of web reinforcement may consist of flexural reinforcement which can be bent diagonally upward (where no longer needed to resist bending) to reinforce the web. Fig. 6-2 shows a combination of various types of web reinforcement.

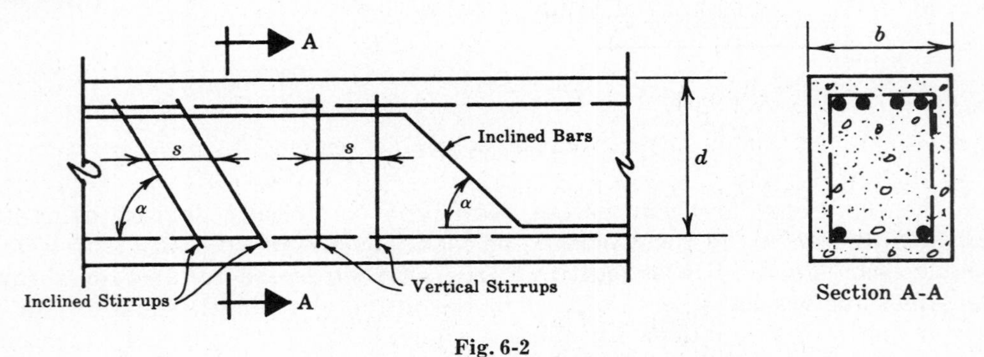

Fig. 6-2

Vertical or Inclined Stirrups

Vertical or inclined stirrups must be designed to resist the excess *shear force* (V' or V'_u) that cannot be taken by the concrete alone. For working stress design,

$$V' = (v - v_c)bd = v'bd \tag{6.8}$$

and for ultimate strength design,

$$V'_u = (v_u - v_{cu})bd = v'_u bd \tag{6.9}$$

The longitudinal spacing (s) of the stirrups depends on the intensity of V' or V'_u. For *vertical stirrups,*

$$s = A_v f_v d/V' \tag{6.10}$$

for working stress design, and

$$s = \phi A_v f_y d/V'_u \tag{6.11}$$

for ultimate strength design.

For *inclined stirrups,*

$$s = A_v f_v d(\sin\alpha + \cos\alpha)/V' \tag{6.12}$$

for working stress design, and

$$s = \phi A_v f_y d(\sin\alpha + \cos\alpha)/V'_u \tag{6.13}$$

for ultimate strength design. In equations (*6.12*) and (*6.13*) the angle α is measured as the small angle between the stirrup and the horizontal.

When stirrups are required by the analysis, under no circumstances may the spacing exceed $d/2$. Further, for working stress design, s may not exceed $d/4$ when v exceeds $3\sqrt{f'_c}$. Similarly, for ultimate strength design, s may not exceed $d/4$ when v_u exceeds $6\phi\sqrt{f'_c}$.

Bent Up Bars

Bent up bars may be used in lieu of, or in combination with stirrups. When a single bar or a single group of parallel bars is bent-up at the same distance from the support, the total area of steel so bent must be at least

$$A_v = V'/(f_v \sin\alpha) \leqq 1.5bd\sqrt{f'_c}/(f_v \sin\alpha) \tag{6.14}$$

for working stress design, and

$$A_v = V'_u/(\phi f_y \sin\alpha) \leqq 3bd\sqrt{f'_c}/(f_y \sin\alpha) \tag{6.15}$$

for ultimate strength design.

If a series of groups of parallel bars is bent up as web reinforcement, the spacing between groups must be calculated using equation (*6.12*) or (*6.13*).

Restrictions on Web Reinforcement

When web reinforcement is required the following restrictions must be met:

(1) A_v may not be less than $(0.0015)bs$ for any interval.

(2) For working stress design no single type of web reinforcement may resist a total shear force greater than $2V'/3$, nor may the total shear stress exceed $5\sqrt{f'_c}$.

(3) For ultimate strength design no single type of web reinforcement may resist a total shear force greater than $2V'_u/3$, nor may the total shear stress exceed $10\phi\sqrt{f'_c}$.

Stresses in Web Reinforcement

For working stress design the value of f_v depends on the type of steel being used for stirrups.

(1) For billet steel or structural grade axle steel bars, $f_v = 18{,}000$ psi.

(2) For *deformed bars* for which $f_y \geqq 60{,}000$ psi, $f_v = 24{,}000$ psi.

(3) For all other types of steel, $f_v = 20{,}000$ psi.

For ultimate strength design of stirrups the ultimate value f_y replaces f_v in the equations, but may not exceed 60,000 psi even when higher strength steel is used.

SHEAR IN SLABS AND FOOTINGS

Special consideration must be given to slabs and footings. For working stress design v_c may not exceed $2\sqrt{f'_c}$ without shear reinforcement nor $3\sqrt{f'_c}$ with web reinforcement. For ultimate strength design v_{cu} may not exceed $4\phi\sqrt{f'_c}$ without shear reinforcement nor $6\phi\sqrt{f'_c}$ with web reinforcement. Beam action should also be checked in slabs and footings.

Reinforcing bars may not be used to resist shear in slabs or footings less than 10 inches thick. Further, stresses in shear reinforcement may not exceed $0.5f_y$ in footings for either method of design. (Footing design is discussed in Chapter 12.)

SHEAR FOR LIGHTWEIGHT STRUCTURAL CONCRETE

The provisions stated heretofore for shear refer generally to normal weight concrete made using sand and gravel as aggregates. All of these provisions apply to lightweight structural concrete, with several exceptions.

For working stress design, the maximum shear stress in beams (without web reinforcement) may not exceed

$$v_c = 0.15F_{sp}\sqrt{f'_c} + 1300p_wVd/M \tag{6.16}$$

or the approximate value

$$v_c = 0.17F_{sp}\sqrt{f'_c} \tag{6.17}$$

For slabs and footings,

$$v_c = 0.3F_{sp}\sqrt{f'_c} \tag{6.18}$$

Similarly, for ultimate strength design of beams, the shear stress on an unreinforced web may not exceed

$$v_{cu} = \phi(0.28F_{sp}\sqrt{f'_c} + 2500p_wVd/M) \tag{6.19}$$

or the approximate value

$$v_{cu} = 0.3\phi F_{sp}\sqrt{f'_c} \tag{6.20}$$

For slabs and footings,

$$v_{cu} = 0.6\phi F_{sp}\sqrt{f'_c} \tag{6.21}$$

In the absence of comprehensive test data, F_{sp} is taken as 4.0 for both methods of design. This is very conservative, a more realistic value being about 5.5.

AXIAL LOAD COMBINED WITH SHEAR AND FLEXURE

Because of the reduction in tensile stress due to compressional axial loads and the increase in tensile stress due to tensile axial loads, the ACI Code provides equations for use when axial load is present in combination with shear and flexure.

In all of the equations in which the term Vd/M appears, the variable M is replaced by

$$M' = M - N(4t - d)/8 \tag{6.22}$$

and the values of v_c and v_{cu} (on the unreinforced web) are limited to

$$v_c = 1.75\sqrt{f'_c(1 + 0.004N/A_g)} \tag{6.23}$$

for working stress design, and

$$v_{cu} = 3.5\phi\sqrt{f'_c(1 + 0.002N/A_g)} \tag{6.24}$$

for ultimate strength design.

It is important to note that equations (*6.23*) and (*6.24*) do not apply to lightweight aggregate concrete.

CRITICAL SECTIONS FOR SHEAR STRESS

In computing shear stresses, the critical sections for shear are assumed to exist at the following locations:

(1) For beams and slabs, at a distance d from the face of the support and across the width of the section, for flexural shear stress.

(2) For slabs and footings, at a distance $d/2$ from the face of the column, and around the perimeter of the column.

Solved Problems

WORKING STRESS DESIGN

In all of the problems for working stress design, unless stated otherwise, $f'_c = 3000$ psi, $f_s = 20{,}000$ psi and $b = 12''$.

6.1. A beam is 20′ long and must resist shear and moment at the critical section for shear as follows:

Dead load: $V = 10$ kips, $M = -6$ ft-kips

Live load: $V = 5$ kips, $M = -12$ ft-kips

The tension reinforcement, A_s, consists of 3 No. 9 bars. $d = 18''$. Use the *approximate method* to check shear stress.

$$v_c = 1.1\sqrt{f'_c} = (1.1)\sqrt{3000} = 60.3 \text{ psi allowed}$$

Actual $v = V/bd = (10+5)(1000)/(12)(18) = 69.5 \text{ psi} > 60.3$. Web reinforcement is required.

6.2. Solve Problem 6.1 using the more detailed analysis.

$$v_c = \sqrt{f'_c} + 1300\, p_w Vd/M = \sqrt{3000} + (1300)(0.0139)(1.0) = 72.9 \text{ psi}$$

where $p_w = A_s/bd = 3/(12)(18) = 0.0139$ and $Vd/M = (15)(18)/(18)(12) = 1.25$ (use 1.0).

From Problem 6.1, $v = 69.5$ psi; stirrups are not required.

This problem illustrates the fact that stirrups may often be eliminated by using the more detailed analysis.

6.3. A beam is reinforced with 4 No. 9 tension bars having $b = 13''$, $d = 20''$. At a distance 20″ from the face of the support the following conditions exist:

Dead load: $V = 12$ kips, $M = -20$ ft-kips, $N = 6$ kips

Live load: $V = 6$ kips, $M = -30$ ft-kips, $N = 10$ kips

Check the beam for shear using the approximate method.

$$v = V/bd = (12+6)(1000)/(13)(20) = 69.3 \text{ psi}$$

Allowable $v_c = 1.1\sqrt{f'_c} = 1.1\sqrt{3000} = 60.3 \text{ psi} < 69.3$; stirrups are required.

6.4. Solve Problem 6.3 using the detailed analysis, but neglecting the axial load.

$p_w = A_s/bd = 4/(13)(20) = 0.0154$. $Vd/M = (18)(20)/[(30+20)(12)] = 0.6$.

$$v_c = \sqrt{f'_c} + 1300 p_w Vd/M = 54.8 + (1300)(0.0154)(0.6) = 66.8 \text{ psi allowed}$$

Actual $v = V/bd = 69.3$ psi > 66.8 psi; stirrups are required.

6.5. Solve Problem 6.3 considering the effects of axial load.

$M' = M - N(4t-d)/8 = (50)(12) - (10+6)[(4)(22)-20]/8 = 464$ inch-kips

$Vd/M' = (18)(20)/464 = 0.778$. $v_c = 54.8 + (1300)(0.0154)(0.778) = 70.4$ psi.

Actual $v = 69.3$ psi < 70.4 psi allowed; stirrups are not required.

It is necessary to insure that $v_c \leqq 1.75\sqrt{f'_c(1+0.004N/A_g)}$ or

$$1.75\sqrt{3000[1+(0.004)(16{,}000)/(13)(22)]} = 106.1 \text{ psi} > 70.1 \text{ psi}$$

6.6. Noting that an increase in the steel ratio p_w also increases the allowable shear stress v_c, using the detailed analysis determine A_s required to eliminate stirrups in Problem 6.4 if the axial forces do not exist.

$v = 69.3 = v_c = 54.8 + 1300 p_w(0.6)$ or $p_w = 14.5/780 = 0.0186$.

$$A_s = p_w bd = (0.0186)(13)(20) = 4.83 \text{ in}^2$$

6.7. A beam has dimensions $b = 12''$ and $d = 18''$. A_s consists of 4 No. 9 bars, and $f_v = 24{,}000$ psi. The conditions at the critical section are as follows:

Dead load: $V = 16$ kips, $M = -28$ ft-kips

Live load: $V = 12$ kips, $M = -18$ ft-kips

Check the section for shear, and design vertical stirrups (if required) at the critical section.

Using the approximate method, $v_c = 1.1\sqrt{f'_c} = 60.2$ psi. $v = V/bd = (16+12)(1000)/(12)(18) = 130$ psi; stirrups are required.

Using detailed analysis, $p_w = 4/(12)(18) = 0.0185$ and $Vd/M = (28)(18)/[(46)(12)] = 0.91$. $v_c = \sqrt{f'_c} + 1300\,p_w Vd/M = 54.8 + (1300)(0.0185)(0.91) = 76.7$ psi; stirrups are required.

With stirrups, v_c max. $= 5\sqrt{f'_c} = (5)(54.5) = 272.5$ psi > 130 psi.

Dimensions (b, d) need not be increased. Try No. 3 bars with 2 vertical legs, $A_v = 0.22$ in^2.

$$V' = (v - v_c)bd = (130.0 - 76.7)(12)(18) = 11{,}500 \text{ lb}$$

V' may not exceed $1.5bd\sqrt{f'_c} = (1.5)(12)(18)(54.8) = 17{,}700$ lb.

Stirrup spacing $= s = A_v f_v d/V' = (0.22)(24{,}000)(18)/11{,}500 = 8.25''$.

Minimum $A_v = (0.0015)bs = (0.0015)(12)(8.25) = 0.1485$ in^2.

A_v provided $= 0.22$ in^2 > 0.1485 in^2 minimum.

Note. Stirrups must be placed between the critical section and the face of the support. For practical purposes, use 8″ spacing. Beyond the critical section the shear decreases, so the stirrup spacing may be increased.

ULTIMATE STRENGTH DESIGN PROBLEMS

6.8. A beam has been satisfactorily designed to resist flexure. At a distance d from the support face the following conditions exist:

Dead load: $V = 10$ kips, $M = -6$ ft-kips

Live load: $V = 5$ kips, $M = -12$ ft-kips

A previous check shows that wind forces are not critical. The clear span is 20′, $d = 18''$, $b = 12''$, $f'_c = 3000$ psi. A_s consists of 3 No. 9 bars. Check the beam for shear.

(*a*) Using the approximate method, $v_{cu} = 2\phi\sqrt{f'_c} = 2(0.85)(54.8) = 93.2$ psi.

(*b*) Using the more detailed method, $v_{cu} = \phi(1.9\sqrt{f'_c} + 2500p_w Vd/M)$.

$p_w = A_s/bd = 3/(12)(18) = 0.0139$.

Load factor, $U = 1.5D + 1.8L$; then $M_u = (1.5)(6) + (1.8)(12) = 30.6$ ft-kips.

$V_u = (1.5)(10) + (1.8)(5) = 24.0$ kips. $Vd/M = (24)(18)/[(30.6)(12)] = 1.175$ (use 1.0).

$$v_{cu} = 0.85[(1.9)(54.77) + (2500)(0.0139)(1)] = 118.0 \text{ psi}$$

Actual shear stress: $v_u = V_u/bd = 24{,}000/(12)(18) = 111.1$ psi.

Note that stirrups are required using the approximate analysis, whereas stirrups are not required using the more detailed method.

6.9. A beam has been designed for flexure using $f'_c = 3000$ psi, $b = 13''$, $d = 20''$, $d' = 2''$ and $A_s = 4.0$ in^2. The conditions at a point 20″ from the face of the support are as follows:

Dead load: $V = 12$ kips, $M = -20$ ft-kips, $N = 6$ kips

Live load: $V = 6$ kips, $M = -30$ ft-kips, $N = 10$ kips

Check the beam for shear. Wind is not critical.

$M_u = 1.5D + 1.8L = 1.5(20) + 1.8(30) = 84$ ft-kips

$V_u = 1.5(12) + 1.8(6) = 28.8$ kips. $N_u = 1.5(6) + 1.8(10) = 27.0$ kips.

(*a*) By the approximate method, $v_{cu} = 2\phi\sqrt{f'_c} = 93.0$ psi.

(*b*) By more detailed analysis, $p_w = A_s/bd = 4/(13)(20) = 0.0154$.

Neglecting axial load, $Vd/M = 28.8(20)/[(84)(12)] = 0.573$.

$$v_{cu} = \phi(1.9\sqrt{f'_c} + 2500p_w Vd/M) = 0.85[(1.9)(54.77) + (2500)(0.0154)(0.573)] = 107.0 \text{ psi}$$

(*c*) Considering axial load, $M' = M - N(4t - d)/8 = 84(12) - 27[(4)(22) - 20]/8 = 778.9$ inch-kips. $Vd/M' = 28.8(20)/778 = 0.742$.

$$v_{cu} = 0.85[(1.9)(54.77) + (2500)(0.0154)(0.742)] = 112.7 \text{ psi}$$

Check maximum permissible v_{cu}: $N/A_g = 27(1000)/(13)(22) = 9.45$ and

$$v_{cu} = 3.5\phi\sqrt{f'_c(1 + 0.002N/A_g)} = 3.5(0.85)\sqrt{3000[1 + 0.002(9.45)]} = 178 \text{ psi}$$

Actual $v_u = V_u/bd = 28.8(1000)/(13)(20) = 111.0$ psi.

Note that stirrups are required unless the axial load is considered in the analysis.

6.10. A beam has dimensions $b = 12''$, $d = 18''$, is reinforced with 4 No. 9 bars, $f'_c = 3000$ psi and $f_y = 60{,}000$ psi. The following conditions exist at a cross-section 18″ from the support:

Dead load: $V = 16$ kips, $M = -28$ ft-kips

Live load: $V = 12$ kips, $M = -18$ ft-kips

Check the shear stresses at the critical section and design stirrups if required. Consider No. 3 bars for stirrups.

$V_u = (1.5)(16) + (1.8)(12) = 45.6$ kips. $M_u = (1.5)(28) + (1.8)(18) = 74.4$ ft-kips.

The actual shear stress is $v_u = V_u/bd = (45.6)(1000)/(12)(18) = 210.0$ psi.

The allowable shear stress is:

(*a*) by the approximate method, $v_{cu} = 2\phi\sqrt{f'_c} = 1.7\sqrt{3000} = 93.0$ psi.

(*b*) by the more detailed analysis, $v_{cu} = \phi\,[1.9\sqrt{f'_c} + 2500 p_w Vd/M] = 124.5$ psi.

where $p_w = 4/(12)(18) = 0.0185$ and $Vd/M = (45.6)(18)/[(74.4)(12)] = 0.923$.

The allowable shear stress is exceeded, so stirrups are required.

Using the detailed analysis, the excess shear stress will be $v'_u = v_u - v_{cu} = 210.0 - 124.5 = 85.5$ psi.

The total shear force for stirrup design will be $V'_u = v'_u bd = (85.5)(12)(18) = 18{,}500$ lb, which does not exceed the maximum permissible value: $V'_u = 3\phi bd\sqrt{f'_c}$ or 30,170 lb.

The stirrup spacing required will be

$$s = \phi A_v f_y d/V'_u = (0.85)(0.22)(60{,}000)(18)/18{,}500 = 10.9''$$

The code does not permit the spacing to exceed $d/2$, so $s = 18.0/2 = 9.0''$.

The stirrups should therefore be spaced at 9″ on centers in the vicinity of the support. The spacing may increase toward the center of the span where the shear force decreases.

Supplementary Problems

WORKING STRESS DESIGN

For all working stress design problems use $d = 20''$, $b = 12''$, $f'_c = 3000$ psi, $f_v = 20{,}000$ psi, $A_s = 4$ No. 10 bars. Loads include the weight of the beam.

6.11. For the beam shown in Fig. 6-3, $w_{DL} = 1.0$ kip/ft, $w_{LL} = 1.5$ kip/ft. Determine (*a*) the allowable shear stress on the concrete using the approximate method, (*b*) the allowable shear stress using the detailed analysis, (*c*) the actual shear stress.
Ans. (*a*) 60.3 psi (*b*) 73.97 psi (*c*) 113.0 psi

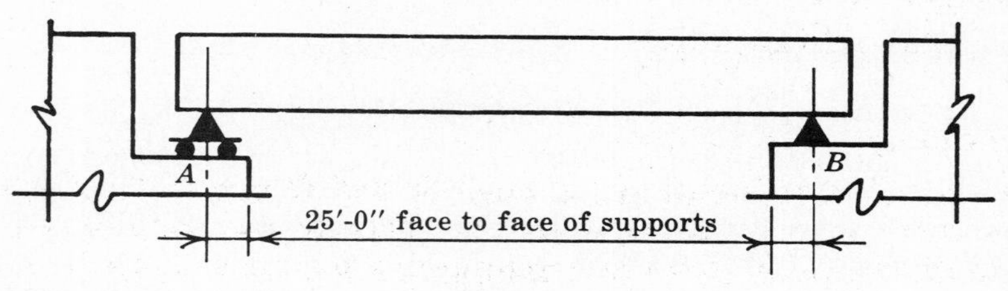

Fig. 6-3

6.12. For the critical section shear in Problem 6.11, determine the theoretical spacing required for No. 3 vertical stirrups having 2 legs. *Ans.* $s = 6.96''$ by approximate analysis

6.13. Consider the addition of a concentrated load of 4.0 kips (dead load) applied at the $\frac{1}{4}$ point and the $\frac{3}{4}$ point in Problem 6.11. Use the approximate analysis to determine the theoretical spacing for No. 3 vertical stirrups at the critical section. *Ans.* $s = 5.27''$

ULTIMATE STRENGTH DESIGN

For all ultimate strength design problems use $d = 20''$, $b = 12''$, $f'_c = 3000$ psi, $f_y = 60{,}000$ psi. (Normal weight aggregates unless otherwise stated.)

6.14. Solve Problem 6.11 using ultimate strength design.
Ans. (*a*) 93.2 psi (*b*) 120.0 psi (*c*) 190.0 psi

6.15. Solve Problem 6.12 using ultimate strength design. *Ans.* $s = 11.15''$

6.16. Solve Problem 6.13 using ultimate strength design and the detailed analysis. *Ans.* $s = 9.15''$

6.17. Solve Problem 6.11 using ultimate strength design and lightweight aggregate concrete. (Do not make deductions for change in the weight of the concrete). Use $F_{sp} = 4.0$.
Ans. (*a*) 55.9 psi (*b*) 83.3 psi (*c*) 190.0 psi

Chapter 7

Bond, Anchorage, and Splices

NOTATION

d = distance from extreme compression fiber to centroid of tension reinforcement

D = nominal diameter of bar, inches

f'_c = compressive strength of concrete

j = ratio of distance between centroid of compression and centroid of tension to the depth, d

j_u = ratio of distance between centroid of compression and centroid of tension to the depth, d, at ultimate strength

Σo = sum of perimeters of all effective bars crossing the section on the tension side if of uniform size; for mixed sizes, substitute $4A_s/D$, where A_s is the total steel area and D is the largest bar diameter. For bundled bars use the sum of the exposed portions of the perimeters.

u = bond stress

u_u = ultimate bond stress

V = total shear

V_u = total ultimate shear

INTRODUCTION

One of the basic assumptions in the analysis and design of reinforced concrete is that there is absolutely no slippage between the concrete and reinforcing steel. Whenever there is a rate of change of stress in a reinforcing bar there must be some interchange of stress or shear flow between the concrete and the reinforcement. The resistance to slippage is termed bond. And the intensity of this bonding force is termed bond stress or unit bond stress.

There are three basic components to bond. They are adhesion, friction, and mechanical anchorage. If the steel bar in Fig. 7-1 is subjected to a tensile force P which is resisted by bond stresses in the concrete, the pull-out is first resisted by an adhesion between the concrete and the steel bar. The bond stress is highest at the face of the concrete and

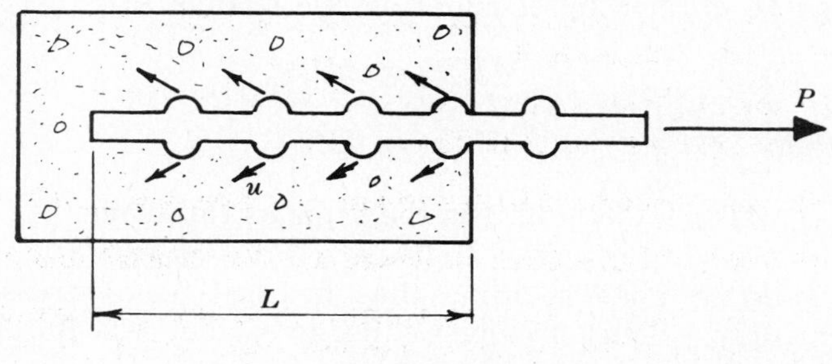

Fig. 7-1

gradually diminishes along the embedded length. In fact, if L is great enough the tensile force in the bar may be zero some distance from the end. As P is increased, longer segments of the bar are subjected to tensile and bond stresses. If P is increased sufficiently so that the bond stress exceeds the adhesion between the concrete and the bar, then a friction force is developed which helps resist pull-out. This frictional force is caused by an attempted wedging action which is resisted by the confining force of the concrete.

If the tensile force in the bar is still increased, the adhesion and friction forces along all or part of the bar are exceeded and the bond is due to mechanical lug action of the deformations on the bar.

The bar may be pulled out by stripping the concrete lugs or by splitting of the concrete due to formation of tension forces caused by the wedging action or inclined friction forces.

There are two basic types of bond stresses that are a concern of the designer. They are termed flexural bond and anchorage bond stresses.

FLEXURAL BOND

Flexural bond stresses are the bond stresses developed between tension reinforcement and concrete in flexural members. A short length of a beam is taken out and made a free body in Fig. 7-2.

The tensile stress in the reinforcing steel varies as the moment. Therefore the change in the total tensile force dT must be resisted by bond between the concrete and reinforcing steel if there is to be no slippage. Or if the unit bond stress per unit surface area is designated as u, and the unit area of contact between the bar per unit length is designated as Σo:

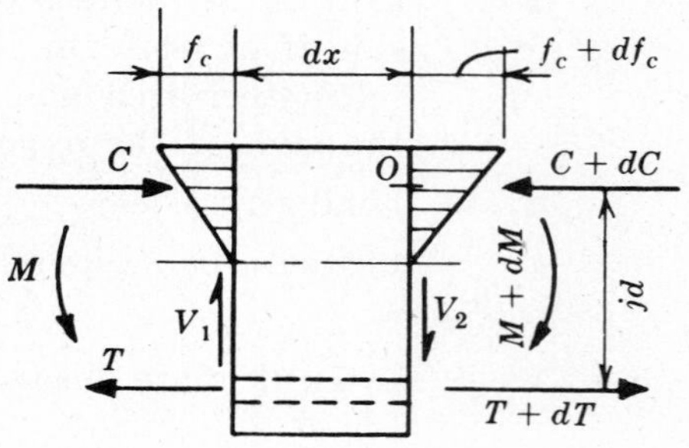

Fig. 7-2

$$u(\Sigma o)\,dx = dT \qquad (a)$$

If dx is so small that $V_1 = V_2 = V$, and summing moments about point O:

$$V\,dx + Tjd - (T + dT)jd = 0 \quad \text{or} \quad dT = \frac{V\,dx}{jd} \qquad (b)$$

Substituting (b) in (a),

$$u(\Sigma o)\,dx = \frac{V\,dx}{jd} \quad \text{or} \quad u = \frac{V}{\Sigma o\, jd} \qquad (7.1)$$

This is the classic formula for computing the flexural bond stresses. Normally, u is in pounds per square inch, Σo is in square inches per linear inch, and d is in inches.

Equation (7.1) can be derived by considering the change in resisting moment. If

$$dT\,jd = dM \qquad (c)$$

and if dx is small, then

$$V\,dx = dM \qquad (d)$$

Substituting (c) in (d), $dT = \dfrac{V\,dx}{jd}$. This is the same as (b) above.

The 1963 ACI Building Code requires that flexural bond stresses be computed for tension reinforcement only. Some previous editions of the ACI Building Code as well as other codes required bond calculation for compression steel.

As shown in the derivation of equation (*7.1*), the tensile reinforcement must resist all of the change in moment from section to section. This is not true for compression reinforcement. In the compression zone, the concrete shares in resisting dC (Fig. 7-2). If the shear is proportioned between the concrete and compression reinforcement according to their compressive strengths or stiffnesses, then the shear proportioned to the compression reinforcement is

$$V_s = \frac{VA'_s f'_s}{A'_s f'_s + f_c bkd/2}$$

The second term in the denominator is the compressive force in the concrete, and $A'_s f'_s$ is the compressive force in the reinforcement. Therefore V_s will always be less than V, and f'_s nearly always less than f_s. Consequently, if the compression reinforcement has the same ratio $\Sigma o/A_s$ as the tension reinforcement, then the bond stress in the former will be the lesser of the two. As previously stated, the 1963 ACI Building Code requires flexural bond stress calculations for tension reinforcement only. It will be shown later that the allowable bond stress for top bars is sometimes less than that for bottom bars. Hence in positive moment regions, even though the actual bond stress in the compression reinforcement is less than that for the tension reinforcement, the allowable for the former may be less and may occasionally be critical.

ANCHORAGE BOND

The second type of bond stress is anchorage bond stress, which is sometimes called development bond.

Referring again to Fig. 7-1, if it is assumed that the bond stress is uniform along the length l the average stress is $u = P/(\Sigma o\, l)$. Then

$$u = \frac{A_s f_s}{\Sigma o\, l} \tag{7.2}$$

This then is the formula used to calculate anchorage bond stresses. Although the stress varies along the embedded length of the bar, an average value is computed. It is possible that the extreme outer end of the bar has slipped while the innermost end may sustain no stress. Therefore the allowable development bond stress is based on the results of pull-out tests.

Empirical data indicate that regardless what the flexural bond stresses in tension reinforcement may be, if the bars are sufficiently anchored or developed, there will be no slippage between the concrete and the bars. Some authorities believe that if the anchorage bond requirements are met, then the flexural bond stresses are not critical. In fact, the 1963 ACI Building Code specifies that flexural bond stresses need not be checked if the anchorage bond stress is less than 0.8 of that permitted.

Many times in the design of a reinforced concrete member, it is impractical, if not impossible, to extend a straight bar a sufficient distance to develop the anchorage requirements. A common remedy to this problem is to bend the bar to form a hook which aids in the resistance to pull-out. The hooked or bent portion attempts to slip out along the bend and there is a mechanical anchorage developed by the bearing of the bar on the concrete.

As defined in the 1963 ACI Building Code a standard hook for flexural reinforcement is shown in Fig. 7-3 below.

In addition, the minimum radius of bend shall be that given in Table 7.1.

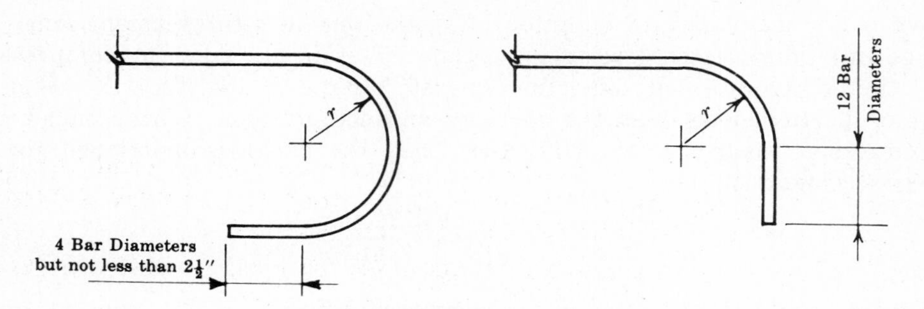

Fig. 7-3

TABLE 7.1

MINIMUM BAR BENDING RADII

Bar Size	Minimum Bend Radius	
	Intermediate and Structural Grades	All Others
#3 thru #5	2½ Bar Diameters	2½ Bar Diameters
#6 thru #8	2½ Bar Diameters	3 Bar Diameters
#9 thru #11	2½ Bar Diameters	4 Bar Diameters
#14S & #18S	5 Bar Diameters	5 Bar Diameters

The requirements for a hook for anchorage of stirrups or ties are a 90° or 135° bend plus an extension of six bar diameters or at least 2½ inches.

LAPS AND SPLICES

It is frequently required in reinforced concrete to lap or splice bars so that the stress in one bar may be transmitted from one bar to the next through the surrounding concrete. The length required for a bar to "unload" its stress into the concrete is given by equation (*7.2*), or

$$l = \frac{A_s f_s}{u\,\Sigma o} \tag{7.3}$$

It will be discussed later that the allowable bond stress in working stress design for other than top bars is

$$u_{\text{All}} = 4.8\sqrt{f'_c}/D$$

If $f'_c = 3000$ psi and $f_s = 20{,}000$ psi, equation (*7.3*) gives

$$l = \frac{\pi D^2 f_s}{4(4.8)\sqrt{f'_c}(\pi D)/D} = 18.9D^2$$

If $f'_c = 5000$ psi and $f_s = 24{,}000$ psi,

$$l = \frac{D^2(24{,}000)}{19.2\sqrt{5000}} = 17.6D^2$$

If $D = 1''$ or a #8 bar, then the required l would be approximately 20″ to develop the bar stress. The 1963 ACI Building Code requires that in splices the bond stress allowable must be reduced to 75% of the allowable and that the minimum lap of deformed bars for a tension splice shall be 24, 30 and 36 bar diameters for yield strengths of 40,000, 50,000 and 60,000 psi respectively. If $f'_c \geqq 3000$ psi, the requirements for compression splices are 20, 24 and 30 bar diameters for yield strengths of 50,000 and less, 60,000 and 75,000 psi respectively.

All of the above splices should have a minimum length of 12″, and splices of plain bars should be twice that specified for deformed bars.

ALLOWABLE BOND STRESSES

The 1963 ACI Building Code contains two chapters devoted to bond and anchorage. One chapter is for working stress design and the other is for ultimate strength design. They are essentially identical except for the allowable stresses.

In working stress design, the allowable bond stress for tension bars with deformations according to ASTM-A 305 are

$$u = 3.4\sqrt{f'_c}/D \leqq 350 \text{ psi} \qquad \text{For top bars}$$

$$u = 4.8\sqrt{f'_c}/D \leqq 500 \text{ psi} \qquad \text{For other than top bars}$$

For tension bars with deformations according to ASTM-A 408, the permissible bond stresses are

$$u = 2.1\sqrt{f'_c} \qquad \text{For top bars}$$

$$u = 3\sqrt{f'_c} \qquad \text{For other than top bars}$$

For all deformed compression bars the permissible bond stress is

$$u = 6.5\sqrt{f'_c} \leqq 400 \text{ psi}$$

For plain bars the allowable stress is one half of that permitted for ASTM-A 305 bars but should not exceed 160 psi.

In ultimate strength design, the corresponding values for ASTM-A 305 bars are

$$u_u = 6.7\sqrt{f'_c}/D \leqq 560 \text{ psi} \qquad \text{For top bars}$$

$$u_u = 9.5\sqrt{f'_c}/D \leqq 800 \text{ psi} \qquad \text{For other than top bars}$$

For ASTM-A 408 bars,

$$u_u = 4.2\sqrt{f'_c} \qquad \text{For top bars}$$

$$u_u = 6\sqrt{f'_c} \qquad \text{For other than top bars}$$

For deformed compression bars

$$u_u = 13\sqrt{f'_c} \leqq 800 \text{ psi}$$

And for plain bars, u_u shall be one-half the value given for ASTM-A 305 bars but should not exceed 250 psi.

Tables 7.2 and 7.3 give the allowable bond stress values for working stress design and ultimate strength design respectively.

TABLE 7.2

ALLOWABLE BOND VALUES — WORKING STRESS DESIGN

f'_c	Bar Size	#3	#4	#5	#6	#7	#8	#9	#10	#11	#14S	#18S
2500	$3.4\sqrt{f'_c}/D$	350	340	272	227	194	170	151	134	121	—	—
	$4.8\sqrt{f'_c}/D$	500	480	384	320	274	240	213	189	170	—	—
	$2.1\sqrt{f'_c}$	—	—	—	—	—	—	—	—	—	105	105
	$3\sqrt{f'_c}$	—	—	—	—	—	—	—	—	—	150	150
	$6.5\sqrt{f'_c}$	325	325	325	325	325	325	325	325	325	325	325

TABLE 7.2 (Cont.)

f'_c	Bar Size	#3	#4	#5	#6	#7	#8	#9	#10	#11	#14S	#18S
3000	$3.4\sqrt{f'_c}/D$	350	350	299	249	214	187	166	147	133	—	—
	$4.8\sqrt{f'_c}/D$	500	500	421	352	302	264	234	208	187	—	—
	$2.1\sqrt{f'_c}$	—	—	—	—	—	—	—	—	—	115	115
	$3\sqrt{f'_c}$	—	—	—	—	—	—	—	—	—	165	165
	$6.5\sqrt{f'_c}$	356	356	356	356	356	356	356	356	356	356	356
4000	$3.4\sqrt{f'_c}/D$	350	350	345	287	246	215	191	170	153	—	—
	$4.8\sqrt{f'_c}/D$	500	500	487	405	348	304	270	239	216	—	—
	$2.1\sqrt{f'_c}$	—	—	—	—	—	—	—	—	—	133	133
	$3\sqrt{f'_c}$	—	—	—	—	—	—	—	—	—	190	190
	$6.5\sqrt{f'_c}$	400	400	400	400	400	400	400	400	400	400	400
5000	$3.4\sqrt{f'_c}/D$	350	350	350	320	275	240	213	189	170	—	—
	$4.8\sqrt{f'_c}/D$	500	500	500	453	388	340	301	267	241	—	—
	$2.1\sqrt{f'_c}$	—	—	—	—	—	—	—	—	—	148	148
	$3\sqrt{f'_c}$	—	—	—	—	—	—	—	—	—	212	212
	$6.5\sqrt{f'_c}$	400	400	400	400	400	400	400	400	400	400	400

TABLE 7.3

ALLOWABLE BOND VALUES — ULTIMATE STRENGTH DESIGN

f'_c	Bar Size	#3	#4	#5	#6	#7	#8	#9	#10	#11	#14S	#18S
2500	$6.7\sqrt{f'_c}/D$	560	560	536	447	383	335	297	264	238	—	—
	$9.5\sqrt{f'_c}/D$	800	800	760	633	543	475	421	374	337	—	—
	$4.2\sqrt{f'_c}$	—	—	—	—	—	—	—	—	—	210	210
	$6\sqrt{f'_c}$	—	—	—	—	—	—	—	—	—	300	300
	$13\sqrt{f'_c}$	650	650	650	650	650	650	650	650	650	650	650
3000	$6.7\sqrt{f'_c}/D$	560	560	560	490	420	368	326	290	261	—	—
	$9.5\sqrt{f'_c}/D$	800	800	800	695	596	521	462	411	370	—	—
	$4.2\sqrt{f'_c}$	—	—	—	—	—	—	—	—	—	231	231
	$6\sqrt{f'_c}$	—	—	—	—	—	—	—	—	—	330	330
	$13\sqrt{f'_c}$	714	714	714	714	714	714	714	714	714	714	714

TABLE 7.3 (Cont.)

f'_c	Bar Size	#3	#4	#5	#6	#7	#8	#9	#10	#11	#14S	#18S
4000	$6.7\sqrt{f'_c}/D$	560	560	560	560	485	424	376	334	301	—	—
	$9.5\sqrt{f'_c}/D$	800	800	800	800	688	602	534	473	427	—	—
	$4.2\sqrt{f'_c}$	—	—	—	—	—	—	—	—	—	266	266
	$6\sqrt{f'_c}$	—	—	—	—	—	—	—	—	—	380	380
	$13\sqrt{f'_c}$	800	800	800	800	800	800	800	800	800	800	800
5000	$6.7\sqrt{f'_c}/D$	560	560	560	560	542	473	420	373	336	—	—
	$9.5\sqrt{f'_c}/D$	800	800	800	800	768	671	596	528	476	—	—
	$4.2\sqrt{f'_c}$	—	—	—	—	—	—	—	—	—	297	297
	$6\sqrt{f'_c}$	—	—	—	—	—	—	—	—	—	424	424
	$13\sqrt{f'_c}$	800	800	800	800	800	800	800	800	800	800	800

TOP BAR BOND

After concrete is placed and consolidated, the water in the mix tends to migrate upward due to capillary action. And even after thorough consolidation, the concrete continues to subside. The water gain which can accumulate on the underside of reinforcing steel coupled with voids formed by subsidence of the concrete away from rigidly fixed bars can cause a weakened bond between the concrete and the underneath side of the bars. As the depth of concrete below the bars is increased, the water gain and subsidence increases, and the probability of a weakened bond increases.

The 1963 ACI Building Code states that if more than 12″ of concrete is placed beneath the reinforcing steel, the allowable bond must be decreased. This reduction is approximately 30%. Some authorities believe that this is too severe a penalty and a reduction of perhaps 10 to 20% is more reasonable. Because of modern construction techniques, concrete with lesser slumps and water content can now be properly placed and consolidated and which should be much less subject to excessive water gain and subsidence. Therefore in future versions of building codes, this penalty placed on top bars may be made more liberal.

WORKING STRESS VS. ULTIMATE STRENGTH

The only differences in bond and anchorage stresses in working stress design and ultimate strength design are the differences in allowable stresses, the different load levels (service loads vs. ultimate loads), and the capacity reduction factor, ϕ. For working stress design,

$$u = \frac{V}{\Sigma o\, jd}$$

And in ultimate strength design,

$$u_u = \frac{V_u}{\phi\, \Sigma o\, j_u d}$$

The 1963 ACI Building Code specifies that a standard hook as previously defined may be considered capable of developing sufficient anchorage to withstand a bar stress of 10,000 psi in working stress design or 19,000 psi in ultimate strength design.

BAR TERMINATIONS

When a bar is terminated in a concrete beam, the possibility of stress concentrations are cause for some concern, particularly if the bar is in a tension zone. Empirical data indicate that these stress concentrations in a tension zone can create a tension type crack at the end of the bar. This tension crack when extended can become inclined to become ostensibly the same as a diagonal tension crack. Thus the shear capacity of the beam may be materially reduced. Hence the 1963 ACI Building Code does not permit the termination of reinforcement in tension zones unless:

(1) The reinforcement that continues beyond the point provides an area or a perimeter that is at least twice that required to resist stresses due to flexure. Or,

(2) The shear stress at the point of termination is less than one-half of that permitted. Or,

(3) Additional stirrups in excess of those normally required to resist shearing stresses are provided.

This additional shear reinforcement in (3) above shall be at least equal to the minimum discussed in Chapter 6. These stirrups shall be placed on each side of the termination point for a distance at least equal to 3/4 of the effective depth of the beam and shall have a maximum spacing of $d/8r_b$. r_b is the ratio of the area of bars terminated to the total area of bars at the section.

There are several other requirements concerning bar cutoffs which will be discussed and illustrated in the solved problems.

ANCHORAGE OF SHEAR REINFORCEMENT

Thus far, the discussion of tension reinforcement has been concerned with principal or longitudinal flexure reinforcement. Shear reinforcement or stirrups as discussed herein are also subjected to tensile stresses and, therefore, must be sufficiently anchored into the concrete to prevent slippage.

The 1963 ACI Building Code requires that stirrups should be anchored by at least one of four different ways:

(1) A standard hook plus sufficient embedment, with the effective embedment length being the distance between mid-depth of the member and the center of radius of the hook.

(2) Welding to longitudinal reinforcing steel.

(3) Bending at least 180° around longitudinal reinforcement.

(4) Embedment in the compression zone as in (1) above to develop bar stress with a minimum of 24 bar diameters.

Of the above, usually (1) or (4) are the most desirable from a construction standpoint. In (2) the fitting and welding is a problem and can best be accomplished outside the formwork. In (3) it is difficult, if not sometimes impossible, to bend a stirrup tightly enough around another bar to insure anchorage.

Solved Problems

7.1. If 2-#8 bars are lapped 24 diameters and if the allowable bond stress is 300 psi, what is the allowable tensile stress in the bars? Assume a working stress analysis.

It is assumed that the stress in the bars is transferred to the concrete through the lap length. Using equation (*7.3*) and l = 24 diameters = 24″, A_s = 0.79 in^2, and Σo = 3.1, then $f_s = (u\,\Sigma o\,l)/A_s$ = 28,200 psi.

7.2. A #6 bar with deformations according to ASTM-A 305 is a top bar in a beam and is subjected to a 20,000 psi stress. What length is required to develop the anchorage for the bar if f'_c = 3000 psi? Assume structural grade steel.

From Table 7.2, assuming a straight-line theory design, the allowable bond stress is $u = 3.4\sqrt{f'_c}/D$ = 249 psi.

Then using equation (*7.3*), $l = A_s f_s/(u\,\Sigma o)$ = 14.7″. Say 15″.

Assuming an isolated bar, the 1963 ACI Building Code requires a minimum anchorage length of 12 bar diameters. Or, 12(0.75) = 9″ < 15″.

7.3. Solve Problem 7.2 assuming that a standard 180° hook is used.

In working stress design, a standard hook is assumed to develop 10,000 psi tensile stress in the bar. The embedment length required to develop 10,000 psi is $l = 0.44(10{,}000)/(249)(2.4)$ = 7.4″. Say 7.5″.

The length of bar required for a standard hook is approximately $l' = 2.5(0.75)\pi + 4(0.75)$ = 8.9″. The total length of bar required is 16.5″ as compared to 15″ determined in Problem 7.2.

7.4. Assuming an ultimate strength design, repeat Problem 7.2. Assume f_y = 40,000 psi.

From Table 7.3, the allowable bond stress is $u_u = 6.7\sqrt{f'_c}/D = 6.7\sqrt{3000}/0.75$ = 491 psi.

Assuming $\phi = 0.85$ and substituting in equation (*7.3*),

$$l = \frac{A_s f_s}{\phi u_u\,\Sigma o} = \frac{0.44(40{,}000)}{0.85(491)(2.4)} = 17.6''. \quad \text{Say } 17.5''.$$

7.5. Solve Problem 7.4 assuming that a standard 180° hook is used.

In ultimate strength design, a standard hook is assumed to develop 19,000 psi tensile stress in the bar. The embedment length required to develop the remaining stress is

$$l = \frac{0.44(40{,}000/0.85 - 19{,}000)}{491(2.4)} = 10.5''$$

The length of bar required for a standard hook is approximately $l' = 2.5(0.75)\pi + 4(0.75)$ = 8.9″. The total length of bar required is 19.5″ as compared to 17.5″ determined in Problem 7.4.

7.6. Compare total bottom bar lengths required to fully develop the tensile anchorage for concrete strengths of 3000, 4000, and 5000 psi and for reinforcing with yield strengths of 40,000, 50,000 and 60,000 psi using straight bars, 180° hooks and 90° hooks. Check for bar sizes #4, #8 and #14S. Assume working stress design.

(*a*) It will be assumed that in working stress design the allowable steel stress will be f_s = 20,000 psi for f_y = 40,000 and 50,000 psi and for #14S bar, and f_s = 24,000 psi for f_y = 60,000 psi.

(*b*) From the 1963 ACI Building Code, the allowable bond stresses are

$$u = 4.8\sqrt{f'_c}/D < 500 \text{ psi for ASTM-A 305} \quad \text{and} \quad u = 3\sqrt{f'_c} \text{ for ASTM-A 408}$$

f'_c	#4	#8	#14S
3000	500	264	165
4000	500	304	190
5000	500	340	212

(*c*) Substituting in equation (*7.2*), $l = A_s f_s/(u\,\Sigma o)$. For $f_s = 20{,}000$ psi, the straight embedment length in inches is in the following table.

f'_c	#4	#8	#14S
3000	5.1	19.0	51.4
4000	5.1	16.5	44.6
5000	5.1	14.8	40.0

For $f_s = 24{,}000$ psi, the lengths are

f'_c	#4	#8	#14S
3000	6.1	22.8	61.6
4000	6.1	19.8	53.6
5000	6.1	17.8	48.0

(*d*) A standard hook develops 10,000 psi tension. Therefore the straight bar portion when a hook is used must develop 10,000 and 14,000 psi respectively.

For $f_s = 10{,}000$ psi, the embedment lengths are

f'_c	#4	#8	#14S
3000	2.6	9.5	25.7
4000	2.6	8.2	22.3
5000	2.6	7.4	20.0

For $f_s = 14{,}000$ psi, the embedment lengths are

f'_c	#4	#8	#14S
3000	3.6	13.3	36.0
4000	3.6	11.5	31.2
5000	3.6	10.4	28.0

(*e*) The minimum radii of bend given in the 1963 ACI Building Code are:

$r = 2.5$ diameters for #4

$r = 2.5$ diameters for #8, $f_s = 20{,}000$ psi

$r = 3$ diameters for #8, $f_s = 24{,}000$ psi

$r = 5$ diameters for #14S

(*f*) The length of bar required for a 180° hook is approximately:

$l' = \pi(2.5)(0.5) + 2.5 = 6.43''$ for #4 bar

$l' = \pi(2.5)(1.0) + 4.0 = 11.86''$ for #8 bar, $f_s = 20{,}000$ psi

$l' = \pi(3)(1.0) + 4.0 = 13.44''$ for #8 bar, $f_s = 24{,}000$ psi

$l' = \pi(5)(1.69) + 6.76 = 33.36''$ for #14S

(g) The length of bar required for a 90° hook is approximately:

$$l' = \tfrac{1}{2}\pi(2.5)(0.5) + 12(0.5) = 7.96'' \text{ for \#4 bar}$$

$$l' = \tfrac{1}{2}\pi(2.5)(1.0) + 12(1.0) = 15.93'' \text{ for \#8 bar}, \quad f_s = 20{,}000 \text{ psi}$$

$$l' = \tfrac{1}{2}\pi(3)(1.0) + 12(1.0) = 16.71'' \text{ for \#8 bar}, \quad f_s = 24{,}000 \text{ psi}$$

$$l' = \tfrac{1}{2}\pi(5)(1.69) + 12(1.69) = 33.6'' \text{ for \#14S bar}$$

(h) Summarizing for a #4 bar, the required lengths are

f'_c	$f_s = 20{,}000$			$f_s = 24{,}000$		
	Straight	180°	90°	Straight	180°	90°
3000	5.1	9.0	10.6	6.1	10.0	11.6
4000	5.1	9.0	10.6	6.1	10.0	11.6
5000	5.1	9.0	10.6	6.1	10.0	11.6

(i) Summarizing for a #8 bar, the required lengths are

f'_c	$f_s = 20{,}000$			$f_s = 24{,}000$		
	Straight	180°	90°	Straight	180°	90°
3000	19.0	21.4	25.4	22.8	26.7	30.0
4000	16.5	20.0	24.1	19.8	24.9	28.2
5000	14.8	19.3	23.3	17.8	23.8	27.1

(j) Summarizing for a #14S bar, the required lengths are

f'_c	$f_s = 20{,}000$			$f_s = 24{,}000$		
	Straight	180°	90°	Straight	180°	90°
3000	51.4	59.1	59.3	61.6	69.4	69.6
4000	44.6	55.7	55.9	53.6	64.6	64.8
5000	40.0	53.4	53.6	48.0	61.4	61.6

(k) Reviewing the previous tabulations, it is seen that as the concrete and steel strengths increase the hooks require more and more bar lengths. Therefore the hooks for the #8 and #14S bars should be shortened and considered as bar extensions only. It should be remembered that these values are the theoretical anchorage requirements based on the allowable stress values in the 1963 ACI Building Code.

7.7. The beam shown in Fig. 7-4 is subjected to a shear of 10 kips. If $f'_c = 3000$ psi and $f_s = 24{,}000$ psi, compute the flexural bond stresses. Assume working stress design.

The perimeter of a #8 bar is 3.14 inches. Therefore using equation (7.1),

$$u = \frac{V}{\Sigma o\, jd} = \frac{10{,}000}{3(3.14)(0.88)(12)} = 100 \text{ psi}$$

The allowable bond according to the 1963 ACI Building Code is

$$u = 4.8\sqrt{f'_c}/D = 4.8\sqrt{3000}/1.0 = 264 \text{ psi}$$

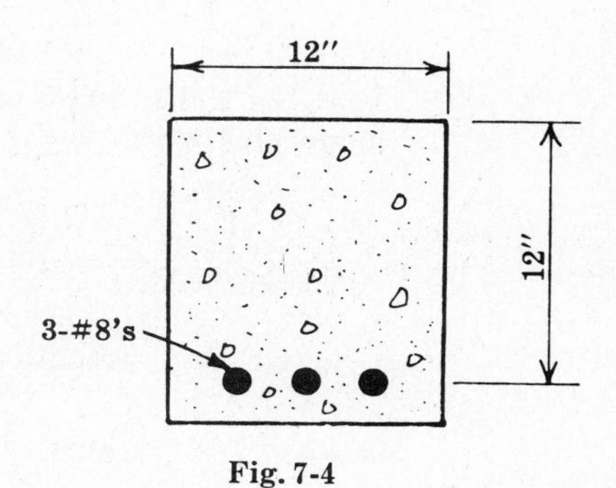

Fig. 7-4

7.8. The beam shown in Fig. 7-5 is subjected to a uniform dead load of 1.0 kip per foot and a uniform live load of 2.0 kips per foot. Check the bond stresses and compute the anchorage requirements. Assume $f'_c = 4000$ psi, $f_s = 24{,}000$ psi, and working stress design according to the 1963 ACI Building Code.

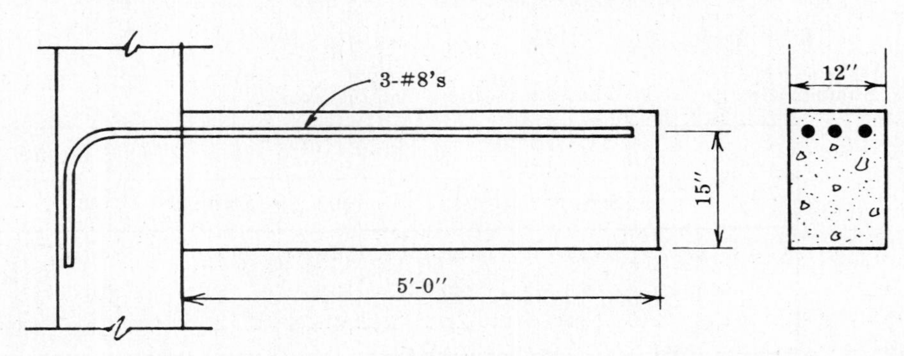

Fig. 7-5

(*a*) The maximum shear at the face of the support is $V = 3.0(5.0) = 15.0$ kips. The perimeter of 3-#8's is $\Sigma o = 3(3.14) = 9.42$ in.

The flexural bond stress at the face of the support is $u = V/(\Sigma o\, jd) = 121$ psi.

The allowable top bar bond is $u_{\text{All}} = 3.4\sqrt{4000}/1.0 = 215 \text{ psi} > 121;\ 121/215 = 0.563.$

(*b*) The straight embedment required to develop the bar is

$$l = \frac{A_s f_s}{u\,\Sigma o} = \frac{3(0.79)(14{,}500)}{215(9.42)} = 17.0''$$

where $f_s = \dfrac{M}{A_s\, jd} = \dfrac{15.0(2.5)(12{,}000)}{3(0.79)(0.875)(15)} = 14{,}500$ psi.

This assumes no anchorage due to the hook except that it is an extension of the bar.

(*c*) It is required that the calculated tension or compression in the bars at any section must be developed on each side of the section. This means that the length of the bar in the beam must be at least equal to 17″. If, in a beam such as this example, the embedment length is greater than the beam length, then it would be necessary to hook the bars or reduce the length requirements in some other manner.

7.9. If $f_y = 60{,}000$ psi, solve Problem 7.8 using ultimate strength design. (*Note.* This beam is overreinforced according to USD but is used as comparison only.)

(*a*) Because there are no wind load effects, the critical load factor equation is obviously $U = 1.5D + 1.8L$. Hence $V_u = 1.5(1.0)(5.0) + 1.8(2.0)(5.0) = 25.5$ kips.

The ultimate flexural bond stress at the face of the support is

$$u_u = \frac{V_u}{\phi\,\Sigma o\, j_u d}$$

At the higher stress levels in ultimate strength design j_u should be somewhat higher than j in working stress design. Hence it is assumed here that $j_u = 0.90$. Then

$$u_u = \frac{25{,}500}{0.85(9.42)(0.85)(15)} = 250 \text{ psi}$$

The allowable top bar bond is $u_{\text{All}} = 6.7\sqrt{4000}/1.0 = 424 \text{ psi} > 250;\ 250/424 = 0.590.$

In ultimate strength design, the steel stress is $f_s = \dfrac{M_u}{A_s \phi j_u d}$ and $M_u = V_u(2.5) = 25.5(2.5) = 63.7$ ft-kips. Then $f_s = 25{,}400$ psi. Hence $l = \dfrac{A_s f_s}{u_u\,\Sigma o} = 15.1''$.

(*b*) This embedment of 15.1″ compares with 17.0″ determined in Problem 7.8. This same variation obviously would not always be true because in ultimate strength design the stresses are a function of the ratio of dead load to live load effects. This is not so in working stress design.

7.10. For the beam shown in Fig. 7-6 determine the working stress bond and anchorage requirements according to the 1963 ACI Building Code. Assume $f'_c = 3000$ psi, $f_s = 20{,}000$ psi, and complete fixity of the beam at the ends. D.L. = 2.0 kips per foot, L.L. = 2.8 kips per foot.

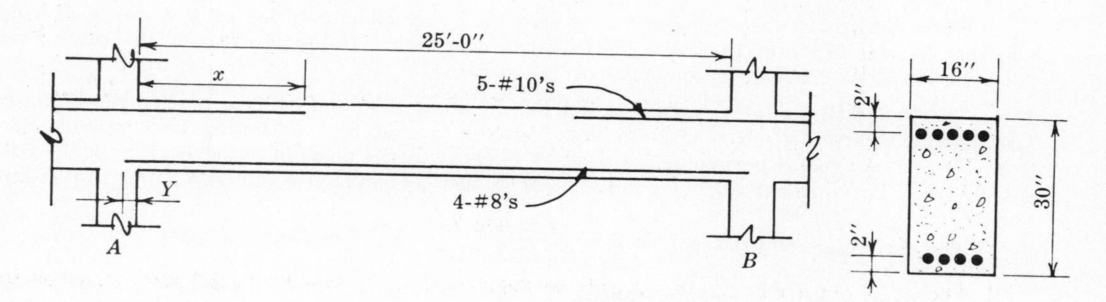

Fig. 7-6

(*a*) The beam shears and moments are

$$V_A = V_B = 25.0(4.8)/2 = 60.0 \text{ kips}, \qquad M_A = M_B = 4.8(25.0)^2/12 = 250 \text{ ft-kips},$$
$$+M = 4.8(25.0)^2/24 = 125 \text{ ft-kips}$$

The point of inflection is located at a distance of 5.28 ft from the face of the support.

(*b*) The allowable bond stresses are

$$u_{\text{All}} = 3.4\sqrt{3000}/1.27 = 147 \text{ psi for \#10 bars}, \qquad u_{\text{All}} = 4.8\sqrt{3000}/1.0 = 264 \text{ psi for \#8 bars}$$

(*c*) The 1963 ACI Building Code states that critical sections for the computation of flexural bond stresses occur at the face of the support, at a point of inflection, and where tension bars terminate.

The shear at the point of inflection is $V = 60.0 - 5.28(4.8) = 34.6$ kips.

The flexural bond stresses are

$$u = \frac{60.0(1000)}{5(3.99)(0.872)(28)} = 123 \text{ psi for \#10's} < 147$$

$$u = \frac{34.6(1000)}{4(3.14)(0.872)(28)} = 113 \text{ psi for \#8's at P.I.} < 264$$

The maximum tensile stresses in the reinforcement are

$$f_s = \frac{250(12{,}000)}{5(1.27)(0.872)(28)} = 19{,}400 \text{ psi for \#10's at face of support}$$

$$f_s = \frac{125(12{,}000)}{4(0.79)(0.872)(28)} = 19{,}500 \text{ psi for \#8's at midspan}$$

(*d*) The bar developments for the above stresses are

$$l = \frac{(1.27)(19{,}400)}{(3.99)(147)} = 42.0'' \text{ for \#10's}, \qquad l = \frac{(0.79)(19{,}500)}{(3.14)(264)} = 18.5'' \text{ for \#8's}$$

(*e*) The 1963 ACI Building Code requires that at least one-third of the negative moment steel shall be carried past the extreme point of inflection at least 1/16 of the clear span or the effective depth of the member. Also, every bar must be extended past where it is no longer needed at least 12 bar diameters or the effective depth of the member.

$$\frac{\text{Span}}{16} = \frac{25.0}{16} = 1.56', \quad \text{or} \quad d = \frac{28.0}{12} = 2.33', \quad \text{or} \quad 12 \text{ bar diameters} = 1.27'$$

Hence if all of the #10's are cut off at the same point the distance x would be $5.28 + 2.33 = 7.61$ ft.

(*f*) The 1963 ACI Building Code requires that in continuous beams, at least one-fourth of the positive moment reinforcement shall extend straight into the support at least 6 inches. Therefore if all of the #8's are cut off at the same point, the distance Y would be 0.5 feet.

7.11. In Problem 7.10 all the bars were cut off at the same time. Cut off three of the #10's where they are no longer needed and check the bond and anchorage requirements.

The resisting moment of 3-#10's is $M = A_s f_s jd$. Then $M = 155$ ft-kips. The three bars could be cut off at a distance of 2.9 ft from the face of the support.

These bars must be extended past where they are no longer needed a distance at least equal to the effective depth of the member or 12 bar diameters. Hence the end of the three bars would be $2.9 + 2.33 = 5.23'$ from the face of the support.

The 1963 ACI Building Code does not permit a bar to be cut off in a tension zone unless certain requirements are met. One of these requirements is that the continuing bars provide double the cross-sectional area or perimeter required. The shear at the point of cutoff is $V = 60.0 - 5.23(4.8) = 34.9$ kips. The moment at the cutoff is $M = 250 - 60.0(5.23) + (4.8/2)(5.23)^2 = 1.6$ ft-kips.

The perimeter required is $\Sigma o = \dfrac{(2)34{,}900}{147(0.872)(28)} = 19.46$ inches $> 2(3.99)$.

The area of reinforcement required is $A_s = 2(1.6)/1.44(28) = 0.079$ in^2 $< 2(1.27)$. Therefore the area requirement is met and the bars may be terminated.

7.12. The beam shown in Fig. 7-7 is proportioned using ultimate strength procedures. The tension reinforcement consists of 4-#7's. Check the bond and anchorage requirements at the cutoff point if 2-#7's are terminated where they are no longer needed. Assume $b = 15''$, $d = 20''$, $f'_c = 3000$ psi and $f_y = 60{,}000$ psi.

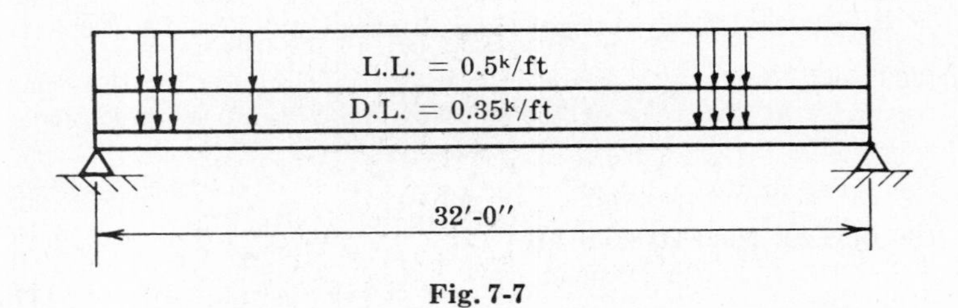

Fig. 7-7

The ultimate moment capacity of 2-#7's is

$$M_u = A_s f_s \phi j_u d = 2(0.60)(60{,}000)(0.84)(20)/12{,}000 = 101 \text{ ft-kips}$$

The ultimate reaction is

$$R_u = 1.5(0.35)(32.0)/2 + 1.8(0.50)(32.0)/2 = 22.8 \text{ kips}$$

The ultimate uniform load is $w_u = 1.5(0.35) + 1.8(0.50) = 1.425$ kips/ft. The 2-#7's are theoretically no longer needed at a distance from the reaction $x = 5.2$ ft.

The #7's must be extended an additional 12 diameters or the effective depth of the beam, whichever is greater. Or $12(0.875) = 10.5'' < 20''$. Hence the bars are cut off at a distance of 3'-6" from the reaction.

The shear at 3'-6" from the support is $V = 22.8 - (3.5)(1.425) = 17.8$ kips. The moment at 3'-6" from the support is

$$M = 22.8(3.5) - 1.425(3.5)^2/2 = 71.2 \text{ ft-kips}$$

The allowable bond stress is $u = 9.5\sqrt{f'_c}/D = 9.5\sqrt{3000}/0.875 = 597$ psi.

Because the bars are terminated in a tension zone, the perimeter requirements for the continuing bars are $\Sigma o = \dfrac{2(17{,}800)}{597(0.84)(20)} = 3.54 < 2(2.79)$. Hence the bond requirement of the continuing bars is met.

7.13. The beam of Problem 7.12 is now subjected to two concentrated live loads at the third points in lieu of the uniform live load. The service live loads are 15.0 kips each. Check the bond and anchorage requirements at the cutoff point if 2-#8's are terminated where they are no longer needed. The total tensile reinforcement at midspan is now 2-#10's and 2-#8's.

(*a*) The ultimate reaction is

$$R_u = 1.5(0.35)(32.0)/2 + 1.8(15.0) = 35.4 \text{ kips}$$

The ultimate moment capacity of 2-#10's is

$$M_u = 2(1.27)(60{,}000)(0.84)(20)/12{,}000 = 214 \text{ ft-kips}$$

The 2-#8's are theoretically no longer needed at a distance from the reaction $x = 6.3$ ft.

The 2-#10's must be extended an additional 20″. Hence the cutoff point is 4′-7″ from the support.

The shear at the cutoff point is $V_u = 35.4 - 1.5(0.35)(4.58) = 33.0$ kips.

The moment at the cutoff point is $M_u = 35.4(4.83) - 1.5(0.35)(4.83)^2/2 = 164$ ft-kips.

(*b*) Because the bars are terminated in a tension zone, the perimeter requirements for the continuing bars are

$$\Sigma o = \frac{2(33{,}000)}{411(0.84)(20)} = 9.56 > 2(3.99)$$

If the perimeter of the extended bars is not sufficient, the area of these bars should be checked to determine if it is twice that required. Twice the area required is

$$A_s = \frac{2(164)(12{,}000)}{60{,}000(0.84)(20)} = 3.90 \text{ in}^2 > 2(1.27)$$

The next check should be that of unit shear stress. A bar may be terminated in a tension zone if the shear stress is not more than half that normally permitted. The shear stress at the cutoff point is

$$v = V/bd = 33{,}000/(15)(20) = 110 \text{ psi}$$

If the web is unreinforced, the allowable shear is

$$v_u = 2\phi\sqrt{f_c'} = 2(0.85)\sqrt{3000} = 93.5 \text{ psi} < 110$$

(*c*) Because the requirements of the perimeter and area of the continuing bars and the unit shear stress were not met, it is necessary to furnish additional stirrups at the cutoff point. The 1963 ACI Building Code specifies that the area of the web reinforcement shall not be less than $A_v = 0.0015bs$, where b is the web width and s is the stirrup spacing.

The maximum spacing shall be $s = d/8r_b$, where r_b is the ratio of the area of the bars terminated to the total area of the bars at the section. The maximum spacing is $s = 20(4.12)/(8)(1.58) = 6.53''$ Use $s = 6.5''$. Then $A_v = 0.0015(15)(6.5) = 0.146$ in². This area would require a U shaped #3 stirrup.

These additional stirrups should be placed on both sides of the cutoff for a distance equal to $\frac{3}{4}d$, or $\frac{3}{4}(20) = 15''$. The stirrup spacing would be as shown in Fig. 7-8.

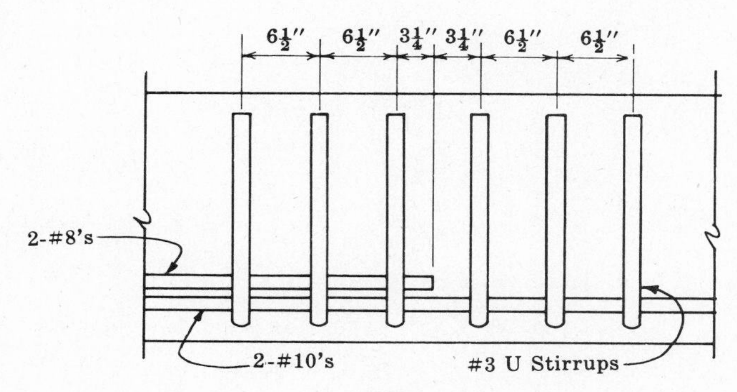

Fig. 7-8

Supplementary Problems

7.14. If 2 isolated #10 top bars which are stressed in tension to 20,000 psi must be lap spliced, what is the lap length required by the 1963 ACI Building Code? Assume $f_c' = 3000$ psi and WSD.
Ans. 57.8″

7.15. A #8 compression bar must be anchored by a straight extension in 3000 psi concrete. If $f_s = 24{,}000$ psi, determine the anchorage length required by the 1963 ACI Building Code. Assume WSD. *Ans.* 17″

7.16. Repeat Problem 7.15 using a standard 90° hook. *Ans.* Total extension must be at least 17″.

7.17. Repeat Problem 7.15 but assume $f_y = 60{,}000$ psi and USD. *Ans.* 25″

7.18. Repeat Problem 7.16 but assume $f_y = 60{,}000$ psi and USD.
Ans. Total extension must be at least 25″.

7.19. Compare top bar lengths required by the 1963 ACI Building Code to develop compression anchorage for concrete strengths of 3000, 4000 and 5000 psi and reinforcing with yield strengths of 40,000 and 60,000 psi using straight bars, 180° hooks and 90° hooks. Check bar sizes #4 and #11 and assume USD.

7.20. Repeat Problem 7.10 but assume $f_y = 60{,}000$ psi and USD.

7.21. A beam similar to that shown in Fig. 7-7 must resist a total uniform service load of 600 lb/ft. The tension reinforcement consists of 4-#7's. If $f_s = 24{,}000$ psi, check by WSD the bond and anchorage requirements if 2-#7's are cut off where they are no longer needed.

7.22. Assume $f_c' = 3000$, 4000 and 5000 psi, $f_s = 20{,}000$ and 24,000 psi, and $f_y = 40{,}000$ and 60,000 psi. Compare the straight bar anchorage requirements by both WSD and USD.

Chapter 8

General Provisions for Columns

NOTATION AND DEFINITIONS

a = depth of equivalent rectangular stress block = k_1c, inches

a_b = depth of equivalent rectangular stress block for balanced conditions = k_1c_b, inches

A_c = area of core of spirally reinforced column, measured to the outside diameter of the spiral, in^2

A_g = gross area of section, in^2

A_s = area of tension reinforcement, in^2

A'_s = area of compression reinforcement, in^2

A_{st} = total area of longitudinal reinforcement, in^2

b = width of compression face of flexural member, inches

c = distance from extreme compression fiber to neutral axis, inches

c_b = distance from extreme compression fiber to neutral axis for balanced conditions = $d(87{,}000)/(87{,}000+f_y)$, inches

d = distance from extreme compression fiber to centroid of tension reinforcement, inches

d' = distance from extreme compression fiber to centroid of compression reinforcement, inches

d'' = distance from plastic centroid to centroid of tension reinforcement, inches

D = overall diameter of circular section, inches

D_s = diameter of the circle through centers of reinforcement arranged in a circular pattern, inches (also called gD or gt)

e = eccentricity of axial load at end of member measured from plastic centroid of the section, calculated by conventional methods of frame analysis, ($e = M/N$), inches

e' = eccentricity of axial load at end of member measured from the centroid of the tension reinforcement, calculated by conventional methods of frame analysis, inches

e_b = eccentricity of load P_b measured from plastic centroid of section, inches

E_c = modulus of elasticity of concrete = $(33)W^{1.5}\sqrt{f'_c}$, psi

E_s = modulus of elasticity of steel = 29×10^6 psi

F_a = allowable axial stress that would be permitted for axial load alone, or $0.34(1+p_gm)f'_c$, psi

F_b = allowable bending stress that would be permitted for bending alone, or $0.45f'_c$, psi

f_a = axial load divided by area of member, A_g, psi

f'_c = compressive strength of concrete, psi

f_r = allowable stress in the metal core of a composite column, psi

f'_r = allowable stress on unencased metal columns and pipe columns, psi
f_s = yield stress of reinforcement, psi
f_y = yield stress for reinforcing steel
g = the ratio, D_s/t, D_s/D or $(d-d')/t$
h = actual unsupported length of column, inches
h' = effective length of column, inches
I = moment of inertia of beam or column, in^4
j = ratio of distance between centroid of compression force and centroid of tension force to the depth, d
K = stiffness factor $= EI/L$ in^4; also, $P_u/(f'_c bt)$ or $P_u/(f'_c D^2)$
K_c = radius of gyration of concrete in pipe columns, inches; or, stiffness of any column, in^3
K_s = radius of gyration of metal pipe in pipe columns, inches
k_1 = a factor defined as $[0.85-(0.05)(f'_c-4.0)]$ for $f'_c > 4.0$ ksi and 0.85 for $f'_c \leqq 4.0$ ksi
L = span length of slab or beam, ft
L' = clear span for positive moment and shear and the average of the two adjacent clear spans for negative moment, ft
m = $f_y/0.85f'_c$
m' = $m-1$
M = bending moment, ft-kips
M_b = ultimate moment capacity at simultaneous crushing of concrete and yielding of tension steel (balanced conditions, $M_b = P_b e_b$), ft-kips
M_o = bending moment when pure bending exists, ft-kips
M_u = ultimate moment capacity under combined axial load and bending, ft-kips
n = ratio of modulus of elasticity of steel to that of concrete, or E_s/E_c
N = eccentric load normal to the cross section of a column, kips
N_b = the value of N below which the allowable load is controlled by tension, and above which the allowable load is controlled by compression, kips
p = ratio of area of tension reinforcement to effective area of concrete
p' = ratio of area of compression reinforcement to effective area of concrete
p_g = ratio of area of vertical reinforcement to the gross area, A_g
p_s = ratio of volume of spiral reinforcement to total volume of core (out to out of spirals) of a spirally reinforced concrete or composite column
p_t = A_{st}/A_g
P_b = axial load capacity at simultaneous crushing of concrete and yielding of tension steel (balanced conditions), kips
P_o = ultimate axial load capacity of actual member when concentrically loaded, kips
P_u = ultimate axial load capacity under combined axial load and bending, kips
r = radius of gyration of gross concrete area of a column, inches
r' = the ratio of ΣK of columns to ΣK of floor members in a plane at one end of a column (stiffness ratio)
R = a reduction factor for long columns
s = pitch or distance between successive turns of a spiral bar, inches

t = overall depth of a rectangular section or diameter of a circular section, inches

ϕ = capacity reduction factor

GENERAL PROVISIONS FOR CAST-IN-PLACE COLUMNS

A number of general requirements are stipulated in the ACI Code relative to columns. Some of the more important factors which should be remembered are listed in the paragraphs which follow.

Principal columns which support a floor or roof are subject to the following limitations:

(*a*) *Rectangular columns* must have an area of at least 96 square inches and the smallest dimension may not be less than 8 inches.

(*b*) *Circular columns* may not have an overall diameter less than 10 inches.

(*c*) The main *vertical reinforcement* must consist of bars having a diameter equal to or greater than $\frac{5}{8}$ inch. For columns having *single ties* (rectangular or round), the minimum number of bars is 4. For columns in which *continuous spirals* are used, at least 6 vertical bars must be used. The area of the vertical bars must be at least 1 percent and not more than 8 percent of the gross area of the column; that is, $(0.01)A_g \leqq p_g \leqq (0.08)A_g$.

(*d*) When *spirals* are used, the minimum *ratio of spiral steel* (p_s) shall be given by the equation:

$$p_s = 0.45[(A_g/A_c) - 1](f'_c/f_y) \tag{8.1}$$

in which $f_y \leqq 60{,}000$ psi.

Spirals may not be made of bars less than $\frac{1}{4}$ inch in diameter for rolled bars or No. 4 AS & W gage for drawn wire.

The *pitch*, or center to center spacing of turns in the spiral may not exceed (*a*) $\frac{1}{6}$ of the outer diameter of the *spiral core* or (*b*) 3″ + spiral bar diameter, whichever is the lesser dimension.

The *minimum pitch* shall be such that the *clear distance* between spiral turns shall be the *greater* of (*a*) $1\frac{3}{8}$″ or (*b*) $1\frac{1}{2}$ times the maximum size of the coarse aggregate.

For *tied columns*, the tie bars shall not be less than $\frac{1}{4}$″ in diameter. The tie spacing (center to center) shall be the *lesser* of (*a*) 16 diameters of the vertical bars, (*b*) 48 diameters of the tie bars or (*c*) the least dimension of the column.

Note. Circular ties may be used when the vertical bars are arranged in a circle, but shall conform to the conditions stated above for ties. (It is usually assumed that tied columns refer to rectangular ties surrounding the vertical bars. However, round ties are sometimes used.)

A column may be designed as a circular section of diameter D and built as a section of *any* shape having a least dimension equal to D. The steel ratio shall be based on the assumed circular section diameter.

In columns, the clear distance between vertical bars shall be the *greater* of (*a*) $1\frac{1}{2}$ times the bar diameter, (*b*) $1\frac{1}{2}$ times the maximum size of coarse aggregate or (*c*) $1\frac{1}{2}$″. (This also applies to the distance between contact splices and adjacent bars or splices.)

Length of splice for *deformed bars* is governed by the vertical steel strength, as shown in Table 8.1, which applies when $f'_c \geqq 3000$ psi. For $f'_c < 3000$ psi, multiply the tabular values by 1.33.

TABLE 8.1

f_y, psi	Splice, inches (not less than 12 inches)
50,000	20 bar diameters
60,000	24 bar diameters
75,000	30 bar diameters

For *plain bars,* the splice length shall be *twice* the values stated for deformed bars.

In tied columns, the amount of reinforcement spliced by lapping shall *not* exceed $(0.04)A_g$ in any 3 foot length.

Column spirals or ties shall have a *clear cover* of concrete at least equal to the *greater* of (*a*) $1\frac{1}{2}''$, (*b*) $1\frac{1}{2}$ times the maximum coarse aggregate size or (*c*) the diameter of the vertical bars.

For columns formed *below grade* and in contact with earth after removal of the forms or for columns *in a corrosive atmosphere,* the cover shall be *at least* 2 inches.

HEIGHT OF COLUMNS

As the height of a column increases, the allowable load decreases because of the tendency of long columns to *buckle*. The height of a column is therefore an important factor in the design. Fig. 8-1 indicates the height that must be used in the design for different situations.

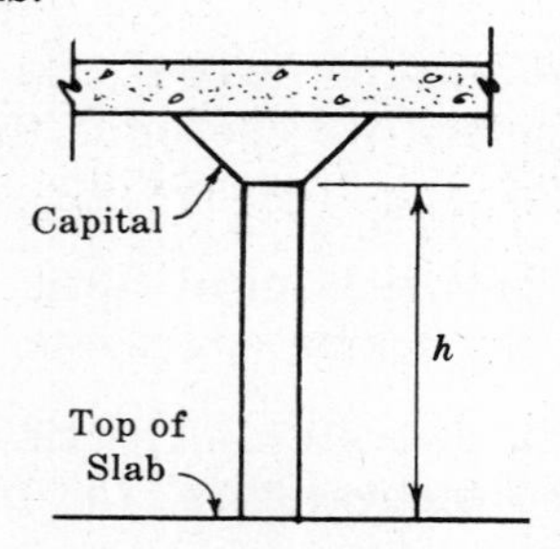

(*a*) Flat Slab Construction with Column Capital

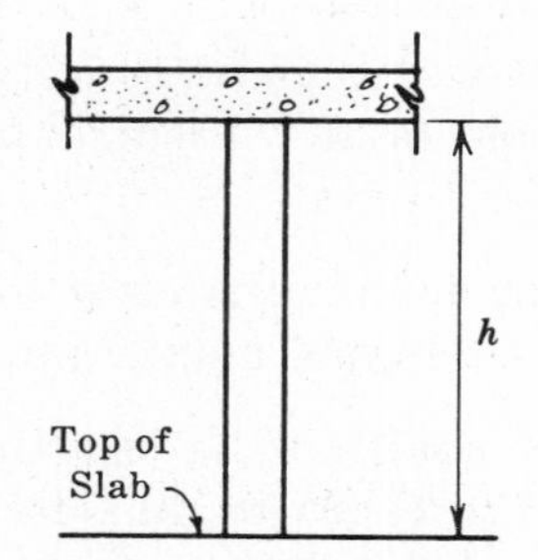

(*b*) Flat Slab (or Plate) Construction without Column Capital

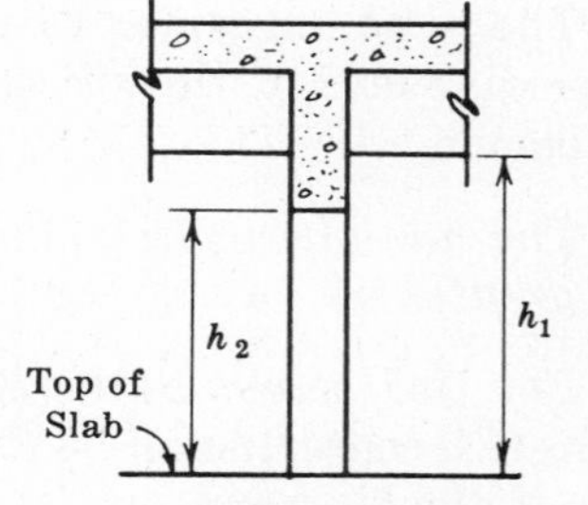

(*c*) Beam and Girder Construction

Note: In beam and girder construction, use the height in the plane of bending. When the design is made for *minimum eccentricity*, use the particular value of h which will provide the *largest* h/r *ratio*. (Minimum eccentricity and h/r are discussed in subsequent paragraphs.)

Fig. 8-1

VARIATION IN CONCRETE STRENGTH

For column concrete having 40% or higher strength than that of the connecting floor, either:

(1) place the higher strength concrete in the slab concentrically around the column, covering an area equal to *4 times* that of the column area, A_g, or

(2) compute the column strength at the floor using the weaker strength concrete, adding *vertical dowels* and spirals at the floor as required to provide the required additional strength, or

(3) for *interior columns,* (supported on four sides by beams of approximately equal depth, or by slabs), compute the column capacity at the floor level using an *equivalent concrete strength* equal to 75% of the column concrete strength plus 35% of the floor concrete strength.

DESIGN OF COLUMNS

Prior to 1951 reinforced concrete columns were designed exclusively by the *elastic theory.* Laboratory tests have proven that this theory provides erroneous results and that the ultimate strength theory must be used in order to obtain reasonable correlation between theory and reality.

Since 1951 a gradual transition has been made from the elastic theory to the ultimate strength theory for column design. Although the working stress method provided in the 1963 ACI Code contains terms usually related to the elastic theory, this method is actually a derivative of the ultimate strength design method. There is, in fact, no truly elastic method for designing reinforced concrete columns. Consequently, the basic working stress design equations which appear in the 1963 ACI Code cannot be derived from the elastic theory.

Fundamentally, this method consists of the ultimate strength method with a safety factor of approximately 2.5, and the resulting equations formulated in such a manner as to have the appearance of being related to the elastic theory.

Because the ultimate strength theory is based primarily on test data, its usefulness is limited to the data from which it was devised. It is important to note then, that the equations are truly valid for steel ratios of 0.04 or less, and for steel having yield point stress values of 50,000 psi or less. Use of the equations for larger values involves extrapolation beyond the test data range and can often provide results which are grossly in error. As a consequence, the working stress theory is also limited in usefulness to this range.

RADIUS OF GYRATION

The radius of gyration is defined as $\sqrt{I/A}$. The following close approximations are satisfactory:

(1) for *rectangular columns,* $r = (0.3)t$, where t is measured perpendicular to the axis about which bending occurs, and

(2) for *round columns,* $r = (0.25)D$, where D is the column diameter.

AXIAL LOADS AND MOMENTS

The 1963 ACI Code requires that the axial loads and moments should be obtained using *acceptable elastic design* methods. The axial loads or moments for dead load and live load are added directly when using the working stress design method. The axial loads and moments obtained from elastic analysis must be increased by appropriate load factors in order to obtain the ultimate values for use in ultimate strength design.

Solved Problems

8.1. A round spirally reinforced concrete column has an overall diameter of 24″ and a core diameter of 20″. Determine the required pitch s for a No. 5 spiral bar. Maximum aggregate size is 1″, $f'_c = 3000$ psi and $f_y = 40{,}000$ psi.

The spiral steel ratio p_s is defined as

$$p_s = \frac{\text{volume of spiral bar}}{\text{volume of concrete core per turn of spiral}}$$

and may not be less than

$$p_s \geqq 0.45[(A_g/A_c) - 1](f'_c/f_y)$$

Here $A_g = \pi(24)^2/4 = 452.0$ in², $A_c = \pi(20)^2/4 = 314.2$ in² and $f'_c/f_y = 3/40 = 0.075$. Hence

$$p_s \geqq 0.45[(452.0/314.2) - 1](0.075) \geqq 0.0149$$

The bar volume per turn is equal to $(\pi d_s/4)A_b$, so

$$V_s = \pi(20.0 - 0.625)(0.31) = 18.9 \text{ in}^3\text{/turn}$$

The concrete volume per turn of spiral will be $V_c = A_c(\text{pitch}) = 314.2s$. Thus $V_s/V_c = 18.9/(314.2s) \geqq 0.0149$, from which $s \leqq 4.03''$.

Other requirements to be satisfied include (*a*) $s \leqq D_c/6 = 20/6 = 3.67''$, (*b*) $s \leqq 3'' +$ bar diameter $\leqq 3.625''$, (*c*) $s \geqq 1.5$(maximum aggregate size) + bar diameter $= 1.5(1) + 0.625 = 2.13''$, (*d*) $s \geqq 1\frac{3}{8}''$ + bar diameter $= 1.375 + 0.625 = 2.0''$.

All requirements will be satisfied if the pitch is 3.0″.

8.2. A rectangular tied column measures 10″ x 20″. The ties will be $\frac{1}{4}''$ bars and the main vertical bars will be No. 9 size. Determine the required tie spacing.

The tie spacing may not exceed the least of (*a*) 16 vertical bar diameters $= (16)(1.128) = 18.05''$, (*b*) 48 tie bar diameters $= (48)(\frac{1}{4}) = 12.0''$, (*c*) the least dimension of the column, 10.0″. Thus the tie spacing may not exceed 10.0″. Use $s = 10.0''$.

8.3. A column is subjected to direct force and flexure. The analysis indicates that a steel ratio $p_t = 0.036$ will satisfy load requirements. Determine whether or not this design will satisfy the Code.

$p_{\min} = 0.01$, $p_{\max} = 0.08$

$0.01 < 0.036 < 0.08$. Code is satisfied.

8.4. A 20″ square column is to be constructed above the ground. The vertical steel consists of No. 9 bars and the maximum aggregate size is 1″. Determine the cover required over the tie bars.

The minimum clear cover over the ties must be the largest of: (*a*) $1\frac{1}{2}''$; (*b*) the diameter of the vertical bars, 1.128″; (*c*) $1\frac{1}{2} \times$ the maximum aggregate size $= (1.5)(1) = 1.5''$. Thus a minimum clear cover of $1\frac{1}{2}''$ must be used.

8.5. A 23″ square column contains 12 No. 9 bars arranged in a circle within a No. 5 spiral bar. The clear cover over the spiral is $1\frac{1}{2}''$ Check the clearance between the vertical bars for Code requirements.

The diameter of the circle through the centers of the vertical bars will be

$$D_s = gD = 23.0 - (2)(1.5 + 0.625) - 1.0 = 17.75''$$

The circumference of that circle will be $\pi gD = \pi(17.75) = 55.76''$.

For the 12 bars the center to center spacing will be $55.76/12 = 4.646''$. The clear spacing will then be $4.646 - 1.128 = 3.518''$.

The spacing is governed by the largest of (a) 1.5 × vertical bar diameter = 1.692″, (b) 1.5 × maximum aggregate size = 1.5″, (c) $1\frac{1}{2}''$.

Therefore the 3.518″ clear spacing provided will be satisfactory.

8.6. **Check the clear cover provided for the spiral in Problem 8.5 for compliance with Code requirements.**

The minimum permissible cover will be the greater of (a) $1\frac{1}{2}''$, (b) 1.5 × maximum aggregate size = 1.5″, (c) the diameter of the vertical bars, 1.128″.

The $1\frac{1}{2}''$ clear cover provided is therefore satisfactory.

8.7. **Determine the spiral pitch s required for the column described in Problem 8.5. Use a No. 5 spiral bar.**

The governing equation is

$$p_s \geqq 0.45[(A_g/A_c) - 1](f_c'/f_y) = V_s/V_c$$

where $A_g = (23)(23) = 529$ in², $A_c = (\pi/4)(20)^2 = 314.2$ in² and $f_c'/f_y = 5/60 = 0.0833$. By direct substitution of these values, obtain $p_s \geqq 0.0257$.

The length of the spiral bar per turn will be $L_s = \pi d_s = \pi(20.0 - 0.625) = 60.87''$.

The volume of the spiral bar per turn will be $V_s = A_b L_s = (0.31)(60.87) = 18.87$ in³.

The volume of the concrete core within the outer limits of the spiral will be $V_c = A_c \times \text{pitch} = 314.2s$. Thus $s \leqq 2.36''$.

The pitch must also satisfy the requirements: (a) $s \geqq 1\frac{3}{8}''$ + bar diameter = 2.0″, (b) $s \leqq 3''$ + bar diameter $\leqq 3.625''$, (c) $s \geqq 1.5$(maximum aggregate size) + bar diameter = 2.13″, and (d) $s \leqq \frac{1}{6}$(core diameter) = 3.33″.

Therefore a practical pitch of $2\frac{1}{4}''$ will be satisfactory.

8.8. **Derive a general expression for determining the spiral pitch s in terms of the appropriate variables to satisfy the equation**

$$p_s \geqq 0.45[(A_g/A_c) - 1](f_c'/f_y)$$

Let L_s = length of the spiral bar per turn
A_b = area of the spiral bar
V_s = volume of the spiral bar per turn
V_c = volume of the concrete core per turn of the spiral bar
d_c = core diameter
d_s = mean diameter of the spiral hoop
s = vertical pitch of the spiral bar
d_b = diameter of the spiral bar

By definition, and to satisfy the Code equation stated,

$$p_s = V_s/V_c = \pi d_s A_b/(A_c s) \geqq 0.45[(A_g/A_c) - 1](f_c'/f_y)$$

from which it follows that

$$A_b/s \geqq 0.45[(A_g/A_c) - 1](f_c'/f_y)[A_c/(\pi d_s)]$$

Now, since $d_s = d_c - d_b$, the final equation is

$$s \leqq A_b(d_c - d_b)(\pi f_y)/[0.45 f_c'(A_g - A_c)] \tag{8.2}$$

The resulting spiral pitch must also satisfy the clearance requirements and the maximum spacing previously discussed.

8.9. Use equation (*8.2*) to solve for the theoretical minimum spiral pitch s required in Problem 8.7.

Referring to Problem 8.7 and substituting into equation (*8.2*), obtain

$$s \leqq (0.31)(20.0 - 0.625)(60\pi)/[(0.45)(5)(529.0 - 314.2)]$$

from which $s \leqq 2.36''$, as previously determined.

8.10. Calculate the tie spacing for a rectangular section $18'' \times 24''$ reinforced with No. 9 vertical bars and No. 4 *circular hoops*.

The circular hoops must satisfy the same requirements as those for rectangular ties. Thus the tie spacing must be the lesser dimension of (*a*) 16 diameters of the vertical bars = 16″, (*b*) 48 diameters of the tie bars = 24″, (*c*) the least dimension of the column = 18″. Therefore the spacing of the circular hoops may not exceed 16″.

Supplementary Problems

In the following problems, consider the maximum aggregate size to be 1″.

8.11. Solve Problem 8.1 using a 26″ diameter column having a 20″ diameter core, all other data remaining the same. Check all clearances and minimum requirements. *Ans.* Pitch = 3.0″

8.12. Using $\frac{1}{4}''$ ties, determine the tie spacing for a 20″ square column reinforced with No. 11 vertical bars. *Ans.* $s = 12.0''$

8.13. Solve Problem 8.4 considering No. 11 vertical bars. *Ans.* Cover = 1.5″

8.14. Solve Problem 8.7 using $f'_c = 3000$ psi and $f_y = 50{,}000$ psi. *Ans.* $s = 3.0''$

8.15. Solve Problem 8.10 using No. 11 vertical bars. *Ans.* $s = 18.0''$

Chapter 9

Short Columns — Working Stress Design

NOTATION

The notation and definitions pertaining to this chapter are identical to those listed for Chapter 8. Additional definitions are provided as necessary in connection with the appropriate material.

TYPES OF COLUMNS AND METHODS OF LOADING

Column sections most often used include round or square sections with vertical bars arranged in a circle, bound together with closely spaced spirals, and rectangular sections with vertical bars arranged parallel to the sides, bound together with rectangular shaped tie bars. These sections will be considered in this chapter.

Columns may be loaded only with axially applied forces, but in most cases bending moments are also present. The former are called axially loaded columns, and the latter are called eccentrically loaded columns.

Because of imperfections which occur during construction, columns are almost always eccentrically loaded. To account for this fact, the ACI Code requires that columns must be designed for minimum eccentricity, even when the axial loads exist without bending. The Code presents equations for axially loaded columns, however, to be used as an index and as a reference point for use in deriving equations for eccentrically loaded columns.

It is important to emphasize that the working stress design method is merely a modified method of ultimate strength design, using a safety factor of approximately 2.5.

Short Columns and Long Columns

The equations presented in this chapter are valid when there is no danger of failure due to buckling because the column is slender. When the length of a column is great compared to its least lateral dimension, failure due to buckling is possible. For such conditions the basic equations for short columns must be modified using methods covered in Chapter 11, Long Columns.

Axially Loaded Columns

The maximum axial load permitted on columns subject to axial load without bending is

$$P = A_g(0.25f'_c + f_s p_g) \tag{9.1}$$

for a spirally reinforced column, and

$$P = A_g(0.2125f'_c + 0.85f_s p_g) \tag{9.2}$$

for a tied column. In equations (*9.1*) and (*9.2*),

$$f_s = 0.4f_y \leqq 30{,}000 \text{ psi}, \quad p_g = \text{steel ratio } A_{st}/A_g, \quad A_{st} = \text{total steel area, in}^2,$$

$$A_g = \text{gross area of the concrete cross-section, in}^2$$

Eccentrically Loaded Columns. Definitions and explanatory notes.

(1) *Principal centroidal axes* are those axes which are 90 degrees apart, intersect at the centroid of the section, and about which the product of inertia is zero. (Any axis of symmetry passing through the centroid is a principal centroidal axis.) Further, the moments of inertia I_x and I_y are the maximum and minimum (or vice versa) moments of inertia about the centroid.

(2) *Product of inertia,* by definition is

$$I_{xy} = \int_A xy\, dA \qquad (9.3)$$

or

$$I_{xy} = \sum_1^n A_n \bar{x}_n \bar{y}_n \qquad (9.4)$$

(3) *Balanced conditions* are those conditions for which, simultaneously, tension steel is fully stressed to the maximum permissible value of f_s and the concrete is stressed in compression to the maximum value of f_c.

(4) *Tension controls* the design when the *eccentricity* ($e = M/N$) is greater than the balanced eccentricity, e_b.

(5) *Compression controls* the design when the eccentricity is less than the balanced eccentricity, e_b.

(6) *Eccentricity* ($e = M/N$) refers to the coordinate of the applied eccentric load, or

$$e_x = M_x/N \qquad (9.5)$$

and

$$e_y = M_y/N \qquad (9.6)$$

where, as shown in Fig. 9-1, M_x = bending moment about the x-axis, inch-kips; M_y = bending moment about the y-axis, inch-kips; N = axial load, kips.

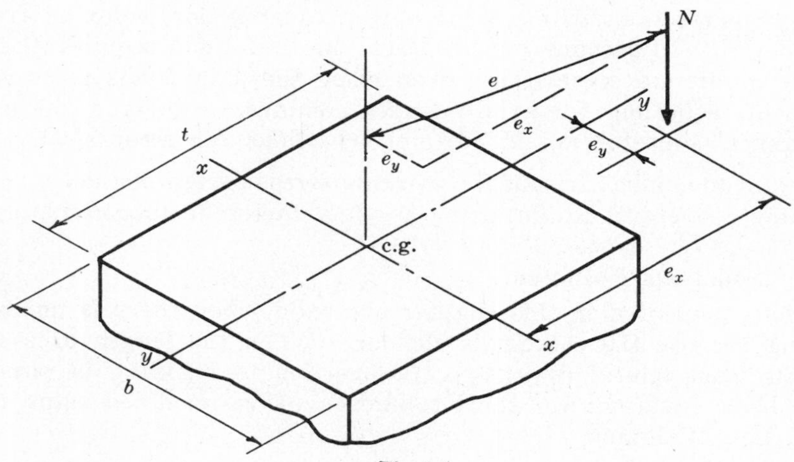

Fig. 9-1

Further, the resultant eccentricity in bi-axial bending is

$$e = \sqrt{(e_x)^2 + (e_y)^2} \qquad (9.7)$$

and the resultant moment is

$$M = \sqrt{(M_x)^2 + (M_y)^2} \qquad (9.8)$$

(7) *Balanced eccentricity* e_{bx} and e_{by} for which balanced stress conditions exist, i.e. f_c maximum occurs when f_s maximum develops, may be calculated using empirical equations (*9.9*) through (*9.14*); for these equations, $m = f_y/0.85f'_c$, $p' = A'_s/bd$ and $p = A_s/bd$.

The dimension terms b, d, and d' are illustrated on Fig. 9-2.

BALANCED ECCENTRICITY EQUATIONS

For symmetrical spiral columns

(*a*) Square sections

$$e_{bx} = e_{by} = 0.43(p_g m)D_s + 0.14t \tag{9.9}$$

(*b*) Round sections

$$e_{bx} = e_{by} = 0.43(p_g m)D_s + 0.14D \tag{9.10}$$

For symmetrical tied columns

(*a*)
$$e_{bx} = [0.67(p_g m) + 0.17](t - d') \tag{9.11}$$

(*b*)
$$e_{by} = [0.67(p_g m) + 0.17](b - d') \tag{9.12}$$

For unsymmetrical tied columns

$$e_{bx} = \frac{p'm(t-2d') + 0.1(t-d')}{(p'-p)m + 0.6} \tag{9.13}$$

$$e_{by} = \frac{p'm(b-2d') + 0.1(b-d')}{(p'-p)m + 0.6} \tag{9.14}$$

It is important to note that the dimension d' may differ in the x and y directions.

Allowable Stresses

(*a*) The steel stress may not exceed $0.4f_y$ for spiral columns, but is limited to 30,000 psi. For tied columns f_s may not exceed $0.34f_y$, but is limited to 25,500 psi.

(*b*) Under pure axial load ($M = 0$) the maximum stress in the concrete may not exceed

$$F_a = 0.34(1 + p_g m)f'_c \tag{9.15}$$

(*c*) Under pure bending ($N = 0$) the maximum stress in the concrete may not exceed

$$F_b = 0.45f'_c \tag{9.16}$$

Computation of Actual Bending Stresses

(*a*) The stress in the concrete due to bending about the x-axis is

$$f_{bx} = M_x c_y / I_x \tag{9.17}$$

(*b*) The stress in the concrete due to bending about the y-axis is

$$f_{by} = M_y c_x / I_y \tag{9.18}$$

(*c*) The axial stress due to axial load (N) is

$$f_a = N/A_g \tag{9.19}$$

The moments of inertia refer to the gross moments of inertia of the concrete and the transformed steel area, using as the transformed area of steel the quantity $A_t = (2n-1)(A_{st})$, in which $n = E_s/E_c$.

For the bending stresses about any *principal axis*, the section modulus principle may be used, or

$$f_b = M/S \tag{9.20}$$

where the section modulus S replaces the term I/c.

The moments of inertia, c distances and gross areas of most commonly used sections are calculated using the appropriate equations of the group (*9.21*) through (*9.36*). The corresponding sections are shown in Fig. 9-2.

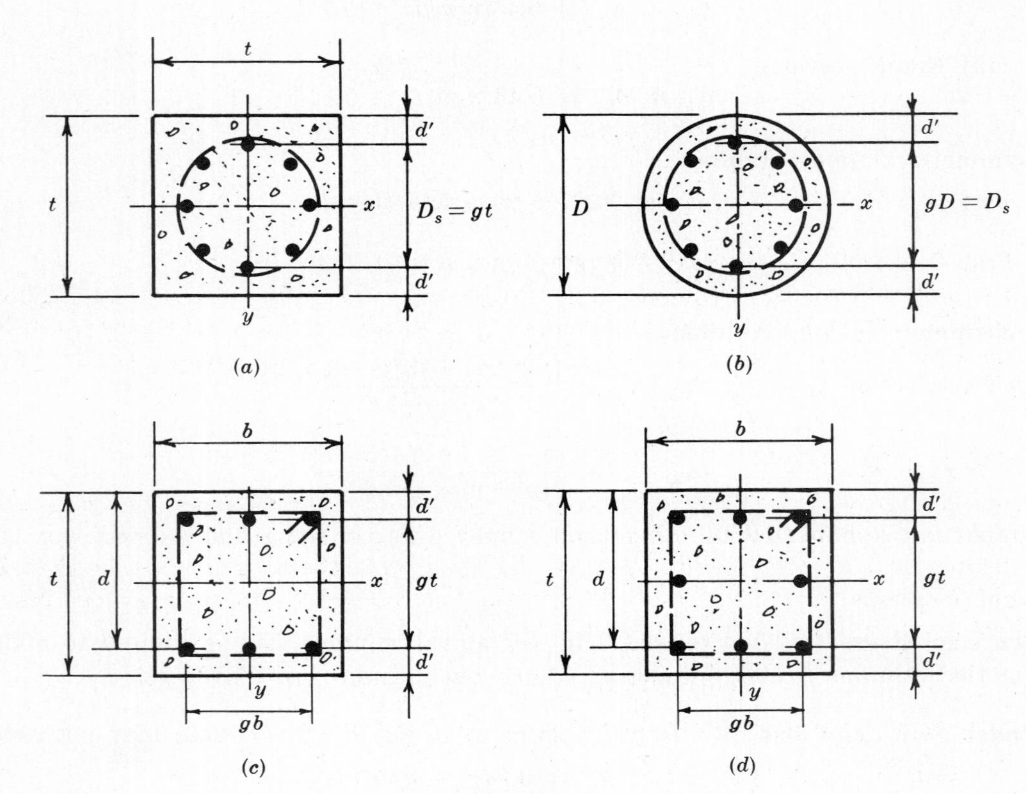

Fig. 9-2

PROPERTIES OF SECTIONS COMMONLY USED

Square sections, steel arranged in a circle

$$I_x = I_y = t^4/12 + A_{st}(2n-1)D_s^2/8 \tag{9.21}$$

$$A_g = t^2 \tag{9.22}$$

$$c_x = c_y = t/2 \tag{9.23}$$

Round sections, steel arranged in a circle

$$I_x = I_y = (\pi D^4/64) + A_{st}(2n-1)D_s^2/8 \tag{9.24}$$

$$c_x = c_y = D/2 \tag{9.25}$$

$$A_g = \pi D^2/4 \tag{9.26}$$

Rectangular sections, symmetrical steel parallel to two faces

$$I_x = bt^3/12 + (2n-1)(A_{st})(gt)^2/4 \tag{9.27}$$

$$I_y = tb^3/12 + (2n-1)(A_{st})(gb)^2/4 \tag{9.28}$$

$$c_x = b/2 \tag{9.29}$$

$$c_y = t/2 \tag{9.30}$$

$$A_g = bt \tag{9.31}$$

Note that g may be different in the x and y directions.

Rectangular sections, symmetrical steel equally distributed along four faces

$$I_x = bt^3/12 + (2n-1)(A_{st})(gt)^2/6 \tag{9.32}$$

$$I_y = tb^3/12 + (2n-1)(A_{st})(gb)^2/6 \tag{9.33}$$

$$c_x = b/2 \tag{9.34}$$

$$c_y = t/2 \tag{9.35}$$

$$A_g = bt \tag{9.36}$$

Note that g may be different in the x and y directions.

Other shapes and other reinforcement patterns

For column shapes and patterns of reinforcement other than the common sections for which equations have been stated, it is necessary to use the *transformed gross section* with the transformed steel equal to $A_{st}(2n-1)$, and to sum the moments of inertia about the principal centroidal axes.

COMBINED STRESSES

When a section is subjected to the combined effects of a compressive axial load and bending moment, an *interaction equation* is used. The allowable stresses due to bending differ from those allowed for axial load, and it is not permissible to add the stresses directly. The ratios of the actual stresses to the allowable stresses for combined conditions are added in accord with the general equation

$$f_a/F_a + f_{bx}/F_{bx} + f_{by}/F_{by} \leqq 1.0 \tag{9.37}$$

which applies for bending about two axes (i.e. bi-axial bending).

In the usual case, there will be bending about only one axis.

LIMITATIONS OF THE ACI CODE EQUATIONS

In order to conform to the ultimate strength equations from which the working stress equations were devised, the same limitations must be considered for both methods. Accordingly, the interaction diagram for working stress design is divided into three regions:

Region I: When the eccentricity in either principal direction is less than the minimum eccentricity e_a, the section must be designed for a *pure axial load* using the appropriate equation (*9.1* or *9.2*).

Region II: When the eccentricity in both principal directions exceeds the minimum eccentricity e_a but is less than the balanced eccentricity e_b (see equations (*9.9*) through (*9.14*)), the section is proportioned using equation (*9.37*) which appears to be a form of the elastic theory, but is empirical.

Region III: When the eccentricity exceeds the balanced eccentricity e_b, an empirically derived *pure moment* is used for the design. It is assumed that this pure moment varies linearly from the moment capacity of the section when axial load does not exist, to the moment which corresponds to balanced eccentricity and balanced failure. The values of pure moment for $N=0$ are obtained from equations (*9.38*) through (*9.42*) and are approximately 4/10 of the values of pure moment for ultimate strength design. The equations which apply are as follows:

(a) ***Spirally reinforced columns, round or square***

$$M_{ox} = M_{oy} = 0.12A_{st}f_yD_s \tag{9.38}$$

(b) ***Symmetrical tied columns***

$$M_{ox} = 0.40A_sf_y(t - 2d') \tag{9.39}$$

$$M_{oy} = 0.40A_sf_y(b - 2d') \tag{9.40}$$

(c) ***Unsymmetrical tied columns***

$$M_{ox} = 0.40A_sf_y(j_x)(t - d') \tag{9.41}$$

$$M_{oy} = 0.40A_sf_y(j_y)(b - d') \tag{9.42}$$

In equations (*9.38*) through (*9.42*), the following definitions apply:

A_{st} = total steel area

A_s = area of tension steel only

j = coefficient of the depth (jd or jb) which describes the lever arm of the resultant tension and compression forces. The subscript indicates the axis about which bending occurs.

(d) ***Bending about two principal axes, all types of columns***

$$M_x/M_{ox} + M_y/M_{oy} \leqq 1.0 \tag{9.43}$$

When bi-axial bending occurs, the pure moment for $N = 0$ must conform to equation (*9.43*). The properties of a section which conforms to Region III must be changed until equation (*9.43*) is satisfied.

DESIGN METHODS

Fig. 9-3 shows the three regions described by the foregoing equations for working stress design. The definitions and terminology are shown on the figure.

Design aids consist of plotted diagrams which conform to the three regions for various types of columns. When such aids are not available it is a simple matter to prepare the diagrams; this is illustrated in the solved problems.

Since the intent of the working stress design method is to provide a safety factor of 2.5 for the ultimate strength design method, it is satisfactory to use the ultimate strength charts with this safety factor and still comply with this chapter.

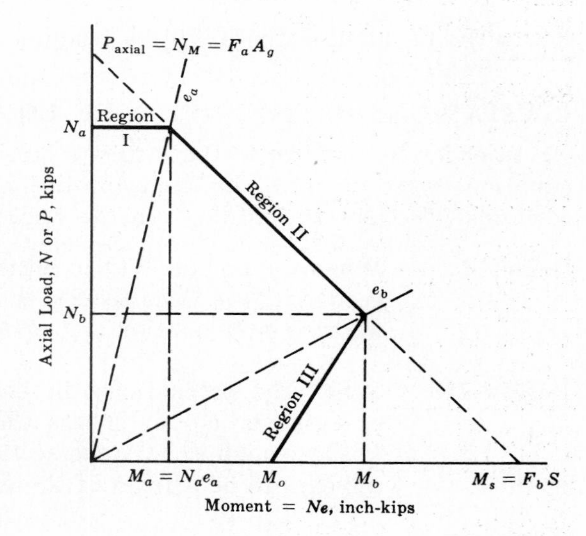

Fig. 9-3

Criteria for Each Design Region

(*a*) Region I, $N > N_b$, $e < e_a$, *compression controls design*

(*b*) Region II, $N > N_b$, $e_a \leqq e \leqq e_b$, *compression controls design*

(*c*) Region III, $N < N_b$, $e > e_b$, *tension controls design*

Design Equations

Region I:

$$e = M/N \leqq e_a$$

$$e_a = M_s(1/N_a - 1/N_m) \tag{9.44}$$

$$N_m = F_a A_g \tag{9.45}$$

Obtain N_a using equation (*9.1*) or (*9.2*).

Region II:

$$e_a \leqq e \leqq e_b, \quad e = M/N$$

Obtain e_a using equation (*9.44*) to determine if this region applies. When applicable, design for an equivalent axial load equal to

$$N_q = G[N + CD_1(12M/t)] \tag{9.46}$$

in which $C = F_a/F_b$, $D_1 = A_g t/S$, and $G = N_a/(F_a A_g)$.

Equation (*9.1*) or (*9.2*) is used with the equivalent load for design. It is necessary to prove that the resulting section conforms to equation (*9.37*). For preliminary designs, approximate values of CD_1 and G are very useful. Typical values of those variables are listed in Tables 9.1 and 9.2. In general, small values of CD_1 apply to large values of p_g and g, whereas small values of G apply to small values of p_g. Note that the variable g refers to gt or gD, corresponding to D_s for spiral columns and $(t - 2d')$ for tied columns.

The variable D_1 is referenced in the *Working Strength Design Handbook* as D. However, it is desirable to make a distinction between this variable and the column diameter D. Thus D_1 is used herein.

Comprehensive tables for CD_1 and G may be obtained from the *Working Stress Design Handbook,* available from the Portland Cement Association of the United States, or from the American Concrete Institute.

Region III:

$$e > e_b, \quad e = M/N$$

Design for an equivalent pure moment, M_{oe},

$$M_{oe} = 12(M - D'Nt/12) \tag{9.47}$$

in which

$$D' = (M'_b - M'_o)/N'_b \tag{9.48}$$

$$M'_b = M_b/f'_c t A_g \tag{9.49}$$

$$M'_o = M_o/f'_c A_g \tag{9.50}$$

$$N'_b = N_b/f'_c A_g \tag{9.51}$$

$$M_b = N_b e_b \tag{9.52}$$

$$N_b = F_a A_g/[1 + (CD_1 e/t)] \tag{9.53}$$

(It should be noted that N'_b is designated as J in the *Working Stress Handbook* of the American Concrete Institute.)

Equations (*9.47*) through (*9.53*) may be solved in inverse order to obtain the equivalent pure moment M_{oe}. The section must then be checked to insure that

$$M_{oe}/S \leqq F_b \tag{9.54}$$

in which $F_b = 0.45f'_c$. Equation (*9.54*) is used as the *equivalent pure moment,* and the column steel is designed using the equations for M_o, *substituting therein* M_{oe} for M_o.

$f'_c = 3{,}000$

TABLE 9.1
TIED COLUMNS: DESIGN COEFFICIENTS (e_a/t, J or N'_b, CD_1, D')

Reinf.	g	Percent of reinforcement ($100p_g$) .0	.5	1.0	1.5	2.0	2.5	3.0	3.5	4.0	4.5	5.0	5.5	6.0	6.5	7.0	7.5	8.0	C'
$f_y = 40{,}000$; $f_s = 16{,}000$		Values of e_a/t																	↓
	.95	.13	.14	.15	.15	.15	.16	.16	.16	.16	.16	.16	.16	.16	.16	.16	.16	.16	
	.90	.13	.14	.14	.15	.15	.15	.15	.15	.15	.15	.15	.15	.15	.14	.14	.14	.14	
	.85	.13	.14	.14	.14	.14	.14	.14	.14	.14	.14	.14	.14	.13	.13	.13	.13	.13	
	.80	.13	.13	.13	.13	.13	.13	.13	.13	.13	.13	.13	.12	.12	.12	.12	.12	.12	
	.75	.13	.13	.13	.13	.13	.12	.12	.12	.12	.12	.12	.11	.11	.11	.11	.11	.11	
	.70	.13	.13	.13	.12	.12	.12	.12	.11	.11	.11	.11	.11	.10	.10	.10	.10	.10	
	.65	.13	.13	.12	.12	.11	.11	.11	.11	.10	.10	.10	.10	.10	.09	.09	.09	.09	
	.60	.13	.13	.12	.12	.11	.11	.11	.10	.10	.09	.09	.09	.09	.09	.08	.08	.08	
		Values of J or N'_b																	
	.95	.19	.20	.20	.21	.21	.21	.22	.22	.22	.23	.23	.23	.24	.24	.24	.24	.25	
	.90	.20	.20	.20	.20	.21	.21	.21	.22	.22	.22	.22	.23	.23	.23	.23	.23	.24	
	.85	.20	.20	.20	.20	.20	.21	.21	.21	.21	.21	.22	.22	.22	.22	.22	.23	.23	
	.80	.20	.20	.20	.20	.20	.20	.20	.21	.21	.21	.21	.21	.21	.21	.21	.22	.22	
	.75	.20	.20	.20	.20	.20	.20	.20	.20	.20	.20	.20	.20	.20	.20	.21	.21	.21	
	.70	.21	.20	.20	.20	.20	.20	.20	.20	.20	.20	.20	.20	.20	.20	.20	.20	.20	
	.65	.21	.20	.20	.20	.19	.19	.19	.19	.19	.19	.19	.19	.19	.19	.19	.19	.19	
	.60	.21	.20	.20	.20	.19	.19	.19	.19	.19	.18	.18	.18	.18	.18	.18	.18	.18	
		Values of CD_1																	
	.95	4.5	4.0	3.6	3.3	3.1	2.9	2.8	2.7	2.6	2.5	2.4	2.4	2.3	2.3	2.2	2.2	2.1	
	.90	4.5	4.0	3.7	3.4	3.2	3.1	2.9	2.8	2.7	2.7	2.6	2.5	2.5	2.4	2.4	2.4	2.3	
	.85	4.5	4.1	3.8	3.6	3.4	3.2	3.1	3.0	2.9	2.9	2.8	2.7	2.7	2.7	2.6	2.6	2.5	
	.80	4.5	4.2	3.9	3.7	3.6	3.4	3.3	3.2	3.2	3.1	3.0	3.0	2.9	2.9	2.8	2.8	2.8	
	.75	4.5	4.3	4.1	3.9	3.8	3.6	3.5	3.5	3.4	3.3	3.3	3.2	3.2	3.1	3.1	3.1	3.1	
	.70	4.5	4.3	4.2	4.0	3.9	3.8	3.8	3.7	3.6	3.6	3.5	3.5	3.5	3.4	3.4	3.4	3.4	
	.65	4.5	4.4	4.3	4.2	4.1	4.1	4.0	4.0	3.9	3.9	3.8	3.8	3.8	3.8	3.7	3.7	3.7	
	.60	4.5	4.5	4.4	4.4	4.3	4.3	4.3	4.2	4.2	4.2	4.2	4.2	4.1	4.1	4.1	4.1	4.1	
		Values of D'																	
	.95	.17	.15	.14	.13	.13	.12	.12	.12	.12	.12	.13	.13	.14	.14	.15	.16	.16	1.6
	.90	.16	.15	.14	.13	.13	.12	.12	.12	.12	.12	.12	.13	.13	.13	.14	.14	.15	1.7
	.85	.16	.15	.14	.13	.13	.12	.12	.12	.12	.12	.12	.12	.12	.13	.13	.13	.14	1.8
	.80	.15	.15	.14	.13	.13	.13	.12	.12	.12	.12	.12	.12	.12	.12	.12	.12	.12	1.9
	.75	.15	.14	.14	.14	.13	.13	.12	.12	.12	.12	.11	.11	.11	.11	.11	.11	.11	2.0
	.70	.14	.14	.14	.14	.13	.13	.13	.12	.12	.12	.11	.11	.11	.11	.11	.10	.10	2.1
	.65	.14	.14	.14	.14	.14	.13	.13	.13	.12	.12	.12	.11	.11	.11	.10	.10	.10	2.3
	.60	.14	.14	.14	.14	.14	.14	.13	.13	.13	.12	.12	.11	.11	.11	.10	.10	.09	2.5
$f_y = 50{,}000$; $f_s = 20{,}000$		Values of e_a/t																	C' ↓
	.95	.13	.14	.14	.14	.14	.14	.14	.14	.14	.13	.13	.13	.13	.13	.13	.13	.13	
	.90	.13	.13	.13	.13	.13	.13	.13	.13	.13	.13	.12	.12	.12	.12	.12	.12	.12	
	.85	.13	.13	.13	.13	.13	.12	.12	.12	.12	.12	.12	.11	.11	.11	.11	.11	.11	
	.80	.13	.13	.13	.12	.12	.12	.11	.11	.11	.11	.11	.10	.10	.10	.10	.10	.10	
	.75	.13	.13	.12	.12	.11	.11	.11	.10	.10	.10	.10	.10	.09	.09	.09	.09	.09	
	.70	.13	.12	.12	.11	.11	.10	.10	.10	.10	.09	.09	.09	.09	.09	.08	.08	.08	
	.65	.13	.12	.11	.11	.10	.10	.10	.09	.09	.09	.08	.08	.08	.08	.08	.08	.07	
	.60	.13	.12	.11	.10	.10	.09	.09	.09	.08	.08	.08	.08	.07	.07	.07	.07	.07	
		Values of J or N'_b																	
	.95	.19	.19	.20	.20	.20	.20	.20	.21	.21	.21	.21	.22	.22	.22	.22	.22	.22	
	.90	.20	.19	.19	.20	.20	.20	.20	.20	.20	.20	.21	.21	.21	.21	.21	.21	.21	
	.85	.20	.20	.19	.19	.19	.19	.20	.20	.20	.20	.20	.20	.20	.20	.20	.20	.20	
	.80	.20	.20	.19	.19	.19	.19	.19	.19	.19	.19	.19	.19	.19	.19	.19	.19	.19	
	.75	.20	.20	.19	.19	.19	.19	.19	.19	.19	.19	.19	.19	.19	.19	.19	.19	.19	
	.70	.21	.20	.19	.19	.19	.18	.18	.18	.18	.18	.18	.18	.18	.18	.18	.18	.18	
	.65	.21	.20	.19	.19	.18	.18	.18	.18	.18	.17	.17	.17	.17	.17	.17	.17	.17	
	.60	.21	.20	.19	.19	.18	.18	.18	.17	.17	.17	.17	.16	.16	.16	.16	.16	.16	
		Values of CD_1																	
	.95	4.5	4.0	3.7	3.4	3.2	3.1	3.0	2.9	2.8	2.7	2.7	2.6	2.6	2.5	2.5	2.5	2.4	
	.90	4.5	4.1	3.8	3.6	3.4	3.3	3.2	3.1	3.0	2.9	2.9	2.8	2.8	2.8	2.7	2.7	2.7	
	.85	4.5	4.2	3.9	3.7	3.6	3.5	3.4	3.3	3.2	3.2	3.1	3.1	3.0	3.0	3.0	2.9	2.9	
	.80	4.5	4.3	4.1	3.9	3.8	3.7	3.6	3.5	3.5	3.4	3.4	3.3	3.3	3.3	3.2	3.2	3.2	
	.75	4.5	4.3	4.2	4.1	4.0	3.9	3.8	3.8	3.7	3.7	3.6	3.6	3.6	3.5	3.5	3.5	3.5	
	.70	4.5	4.4	4.3	4.2	4.2	4.1	4.1	4.0	4.0	4.0	3.9	3.9	3.9	3.9	3.9	3.8	3.8	
	.65	4.5	4.5	4.4	4.4	4.4	4.4	4.3	4.3	4.3	4.3	4.3	4.3	4.2	4.2	4.2	4.2	4.2	
	.60	4.5	4.6	4.6	4.6	4.6	4.6	4.6	4.6	4.6	4.6	4.6	4.6	4.6	4.6	4.6	4.6	4.7	
		Values of D'																	
	.95	.17	.15	.13	.12	.10	.09	.09	.08	.07	.07	.06	.06	.06	.06	.06	.06	.06	1.3
	.90	.16	.15	.13	.12	.11	.10	.09	.08	.07	.06	.06	.05	.05	.05	.04	.04	.04	1.3
	.85	.16	.15	.13	.12	.11	.10	.09	.08	.07	.06	.05	.05	.04	.04	.03	.02	.02	1.4
	.80	.15	.14	.13	.12	.11	.10	.09	.08	.07	.06	.05	.04	.03	.03	.02	.01	.00	1.5
	.75	.15	.14	.13	.12	.11	.10	.09	.08	.07	.06	.05	.04	.03	.02	.01	.00	−.01	1.6
	.70	.14	.14	.14	.13	.12	.11	.10	.09	.07	.06	.05	.04	.03	.01	.00	−.01	−.02	1.7
	.65	.14	.14	.14	.13	.12	.11	.10	.09	.08	.07	.05	.04	.03	.01	.00	−.02	−.03	1.8
	.60	.14	.14	.14	.13	.13	.12	.11	.10	.09	.07	.06	.04	.03	.01	.00	−.02	−.04	2.0

Simplification in design procedure may be achieved by using equation (*9.47*) directly with *approximate values of* D' and the actual values of N and M, then checking for satisfaction of equation (*9.54*). Approximate values of D' may be obtained from Table 9.1. More complete tables appear in the *Working Stress Design Handbook*.

TABLE 9.2

TIED COLUMNS: DESIGN COEFFICIENTS (G)

f'_c	f_y	f_s	Percentage of Reinforcement ($100\ p_g$)								
			.0	1.0	2.0	3.0	4.0	5.0	6.0	7.0	8.0
3000	40,000	16,000	.63	.66	.68	.70	.71	.72	.73	.74	.75
	50,000	20,000	.63	.66	.69	.71	.72	.74	.75	.76	.76
	60,000	24,000	.63	.67	.70	.72	.73	.75	.76	.77	.77
	75,000	30,000	.63	.68	.71	.73	.75	.76	.77	.78	.78
4000	40,000	16,000	.63	.65	.67	.68	.70	.71	.72	.73	.73
	50,000	20,000	.63	.65	.68	.69	.71	.72	.73	.74	.75
	60,000	24,000	.63	.66	.68	.70	.72	.73	.74	.75	.76
	75,000	30,000	.63	.67	.69	.71	.73	.74	.75	.76	.77
5000	40,000	16,000	.63	.64	.66	.67	.69	.70	.71	.71	.72
	50,000	20,000	.63	.65	.67	.68	.70	.71	.72	.73	.73
	60,000	24,000	.63	.65	.67	.69	.71	.72	.73	.74	.74
	75,000	30,000	.63	.66	.68	.70	.72	.73	.74	.75	.76

PLOTTING INTERACTION DIAGRAMS

The design procedures outlined in the foregoing paragraphs involve tedious trial and error methods. A more direct process for design may be devised by preparing interaction diagrams for specific cases of f'_c, f_s and type of column. The diagrams are more useful if the parameters are dimensionless, and values of $N/f'_c bt$ are plotted versus $Ne/f'_c bt^2$ for rectangular columns, or $N/f'_c D^2$ and $Ne/f'_c D^3$ are used for round columns.

The process involved in preparing interaction design diagrams will be illustrated in Problem 9.18.

EQUIVALENT CIRCULAR COLUMNS AND BIAXIAL BENDING

The ACI Code permits the design of a column as circular, and then construction of the column having any shape, the sides of which lie outside of the assumed circular section. An example of such a column is shown in Fig. 9-4.

There is often an advantage involved in using this provision. When biaxial bending occurs for a section which is not symmetrical about either axis, it is necessary to locate the principal axes and then obtain the sectional properties about those axes. This often involves detailed computations which can

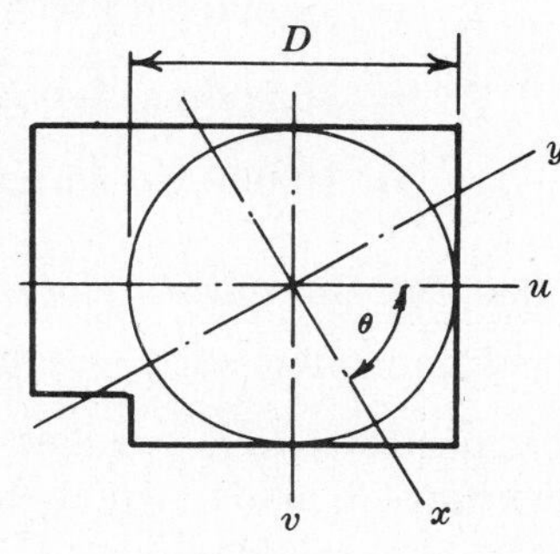

Fig. 9-4

be circumvented by using the equivalent circular section inscribed within the boundaries of the column. For a circular section, all axes are principal axes when they are centroidal axes. The problem is then reduced to one of bending about a single axis. It is often uneconomical to follow this procedure, and desirable to utilize the principal axes of the actual, unsymmetrical section. It is important, therefore, to discuss principal axes.

PRINCIPAL AXES

The principal axes of any section (with regard to the centroid) are those axes about which the maximum and minimum moments of inertia occur, and about which the *product of inertia* is zero. For *any set* of perpendicular centroidal axes u and v, the moments of inertia I_u and I_v and the product of inertia I_{uv} may be calculated. The principal centroidal axes may then be located at an angle θ therefrom, using the equation

$$\tan 2\theta = -2I_{uv}/(I_u - I_v) \tag{9.55}$$

and the principal moments of inertia may then be computed as

$$I_x = (I_u + I_v)/2 + (I_u - I_v)\cos 2\theta/2 - I_{uv}\sin 2\theta \tag{9.56}$$

and

$$I_y = (I_u + I_v)/2 - (I_u - I_v)\cos 2\theta/2 + I_{uv}\sin 2\theta \tag{9.57}$$

Unfortunately, the equations for balanced eccentricity e_{bx} and e_{by} are very approximate for the x and y axes in this instance, and one should proceed with caution before applying the equations for balanced eccentricity for cases other than those for which they were specifically devised.

MISCELLANEOUS COMBINATION COLUMNS

Columns consisting of a combination of structural steel sections (or cast-iron sections) and reinforced or plain concrete may be designed using the appropriate equations which follow. All of the equations refer to short columns which are not subjected to bending moments.

Composite Columns

A structural steel or cast-iron column thoroughly encased in concrete and reinforced with vertical bars and spirals can safely support an axial load equal to

$$P = 0.225A_g f'_c + f_s A_{st} + A_r f_r \tag{9.58}$$

in which A_g = gross area of concrete, not deducting the area of the steel bars or the core, in^2

$f_s = 0.4f_y \leqq 30{,}000$ psi

A_r = area of the metal core, in^2

f_r = 18,000 psi for ASTM A-36 steel

f_r = 16,000 psi for ASTM A-7 steel or

f_r = 10,000 psi for cast-iron

Limitations for Composite Columns

(a) The column must satisfy equation (9.58) at all points along its length.

(b) At connections and brackets attached to the metal core, the concrete shall carry all superimposed loads, which may not exceed $0.35f'_c A_g$.

(c) The metal core may not have a total area greater than $0.2A_g$.

(*d*) If a hollow pipe column is used, it must be filled with concrete.

(*e*) Properly designed spirals must be provided.

(*f*) All spirals must be at least 3″ clear of the metal core unless the core is an H-beam, in which case the minimum clearance is 2″.

(*g*) The metal core must be capable of supporting all of the dead load prior to the proper hardening of the concrete.

Steel Columns Encased in Concrete

For concrete encased structural steel columns, all metal (except rivet heads) must be covered by at least $2\frac{1}{2}''$ of concrete having strength at least equal to 2500 psi. The concrete must be reinforced with the equivalent of a mesh having wires of No. 10 AS & W gage or larger. The vertical wires may not be spaced farther apart than 8″ and the horizontal wires must not be farther apart than 4″.

The fabric shall be placed 1″ from the outer concrete surface. At laps, splices of at least 40 wire diameters must be used. The steel section must carry all loads imposed before the concrete has thoroughly hardened.

When all of the conditions stated above have been fulfilled, the finished section is assumed to be capable of supporting a load not exceeding

$$P = A_r f_r'(1 + A_g/100\,A_r) \qquad (9.59)$$

in which A_g = gross area of the concrete, not deducting the steel area

A_r = the area of the structural steel section

and

$$f_r' = 17{,}000 - 0.485(h/K_s)^2 \qquad (9.60)$$

in which h = effective length of the column as defined for conditions of restraint for the Euler load (see texts on Strength of Materials)

K_s = the radius of gyration of the steel section about its weakest axis, and shall be such that h/K_s does not exceed 120.

(Equation (*9.60*) applies only when $f_y \geqq 33.0$ ksi.)

Concrete Filled Pipe Columns

The total load on an axially loaded steel pipe column filled with concrete may not exceed

$$P = 0.25 f_c' A_c[1.0 - 0.000025(h/K_s)^2] + A_r f_r' \qquad (9.61)$$

for which all variables have been previously defined. The pipe must have a yield strength of at least 33,000 psi and a ratio h/K_s not greater than 120.

Solved Problems

9.1. A short spirally reinforced concrete column is 20″ square and is reinforced with 16 No. 9 bars. Using $f_c' = 3000$ psi and $f_y = 60{,}000$ psi, determine the maximum permissible axial load for the column.

The governing equation is

$$P = A_g(0.25 f_c' + f_s p_g)$$

where $A_g = (20)(20) = 400\text{ in}^2$, $f_s = (0.4)(60{,}000) = 24{,}000$ psi, $p_g = A_{st}/A_g = 16/400 = 0.04$. By substitution into the equation, obtain $P = 684{,}000$ lb.

9.2. A short round column is spirally reinforced. The diameter D of the section is 20″, $f'_c = 3000$ psi, $f_y = 50{,}000$ psi and 16 No. 9 bars are used. Determine the maximum permissible axial load for the column.

The governing equation is

$$P = A_g(0.25f'_c + f_s p_g)$$

where $A_g = \pi(20)^2/4 = 314\ \text{in}^2$, $f_s = (0.4)(50{,}000) = 20{,}000$ psi, $p_g = A_{st}/A_g = 16/314 = 0.051$. Substituting values, $P = 555{,}500$ lb.

9.3. A short tied column measures 15″ × 20″ and is reinforced with 20 No. 9 bars. Using $f'_c = 5000$ psi and $f_y = 75{,}000$ psi, determine the maximum permissible axial load.

The governing equation for tied columns is

$$P = A_g(0.2125f'_c + 0.85f_s p_g)$$

Here $A_g = (15)(20) = 300\ \text{in}^2$, $f_s = (0.4)(75{,}000) = 30{,}000$ psi, $p_g = A_{st}/A_g = 20/300 = 0.0667$. Substituting values, $P = 701{,}250$ lb.

9.4. Determine the balanced eccentricity e_{bx} and e_{by} for the section shown in Fig. 9-5. Use $f_y = 60.0$ ksi and $f'_c = 3.0$ ksi.

For symmetrical tied columns,

$$e_{bx} = [0.67(p_g m) + 0.17](t - d')$$

and

$$e_{by} = [0.67(p_g m) + 0.17](b - d')$$

Here $p_g = A_{st}/A_g = 12.48/(14)(24) = 0.0372$, $m = f_y/(0.85f'_c) = 23.53$, $t = 24$, $d' = 2$, $b = 14$. Then $e_{bx} = 16.64''$, $e_{by} = 9.08''$.

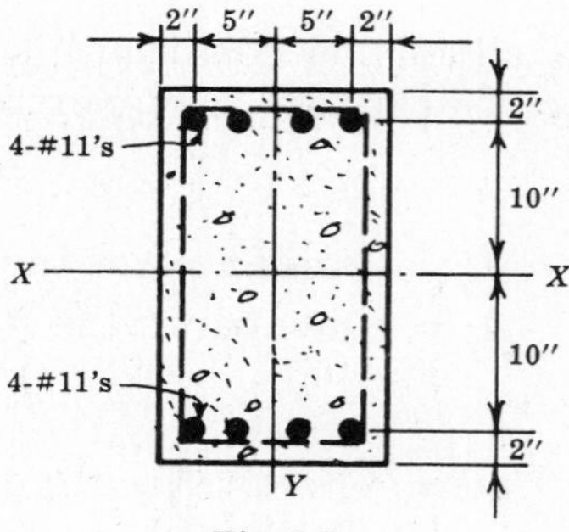

Fig. 9-5

9.5. Determine the balanced eccentricity for the spirally reinforced column shown in Fig. 9-6, for which $f'_c = 3.0$ ksi, $f_y = 60.0$ ksi and the reinforcement consists of 8 No. 11 bars.

For round spirally reinforced columns,

$$e_b = 0.43p_g m D_s + 0.14D$$

Here $p_g = A_{st}/A_g = 12.48/[(0.7854)(24)^2] = 0.0277$, $m = f_y/0.85f'_c = 23.53''$, $p_g m = 0.649''$, $D_s = 20$, $D = 24$. Then $e_b = 8.94''$.

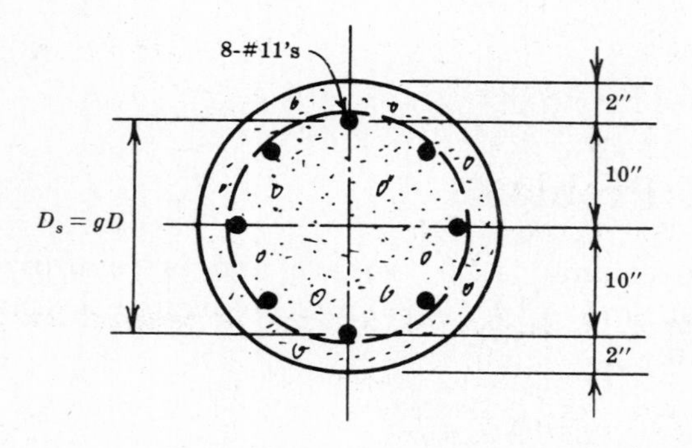

Fig. 9-6

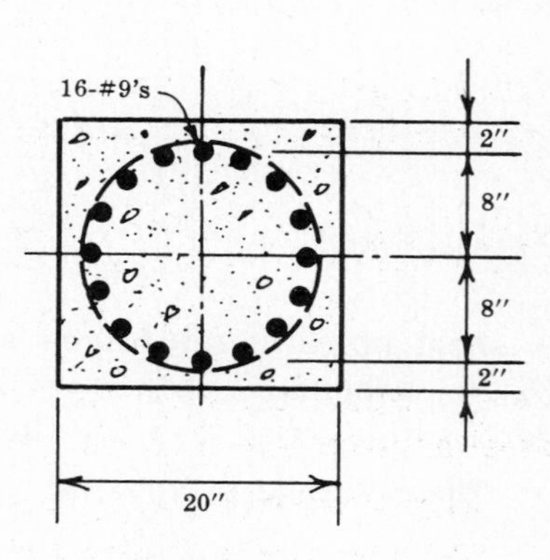

Fig. 9-7

9.6. Determine the balanced eccentricity about the x-axis for the spirally reinforced column shown in Fig. 9-7 above. Use $f'_c = 4.0$ ksi, $f_y = 60.0$ ksi and $A_{st} = 16.0$ in^2.

For square spirally reinforced columns,

$$e_{bx} = 0.43 p_g m D_s + 0.14t$$

where $p_g = A_{st}/A_g = 16/400 = 0.04$, $m = f_y/(0.85f'_c) = 17.65$, $D_s = 16$, $t = 20$. Thus $e_{bx} = 7.66''$.

9.7. Determine the balanced eccentricity e_{bx} for the unsymmetrical tied column shown in Fig. 9-8, for which $f'_c = 3.0$ ksi, $f_y = 60.0$ ksi, $A_s = 6.24$ in^2 and $A'_s = 4.0$ in^2.

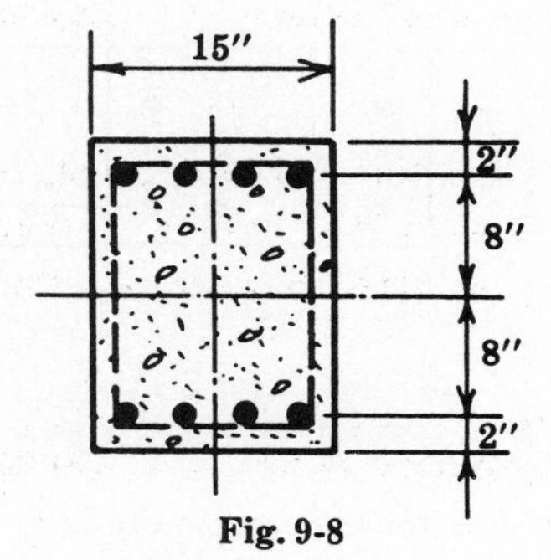

Fig. 9-8

The appropriate equation is

$$e_{bx} = \frac{p'm(t-2d') + 0.1(t-d')}{(p'-p)m + 0.6}$$

where $p' = 4.0/(15)(20) = 0.0133$, $p = 6.24/(15)(20) = 0.0208$, $m = f_y/(0.85f'_c) = 23.53''$, $t = 20$, $d' = 2$. Hence $e_{bx} = 16.1''$.

9.8. Prove equation (*9.21*) assuming that the steel area consists of a thin tubular ring having a total area equal to $(2n-1)A_{st}$. Refer to Fig. 9-9.

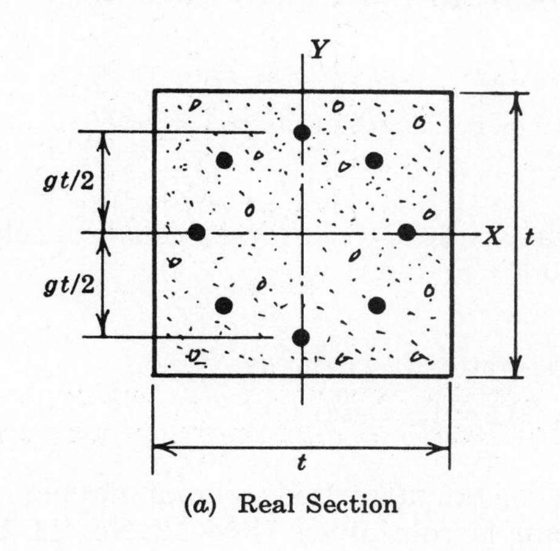

(*a*) Real Section

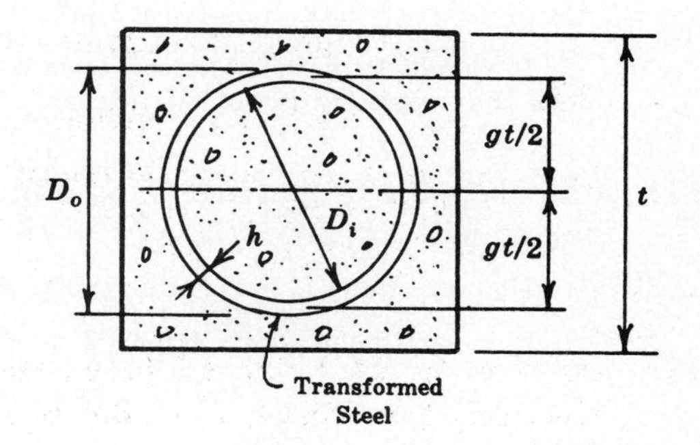

(*b*) Substitute Section

Fig. 9-9

For the square concrete section, $I_c = bt^3/12$.

For the steel, assuming the transformed area to be uniformly distributed as a thin ring of mean diameter gt,

$$I_s = \pi(D_o^4 - D_i^4)/64 = (\pi/64)(D_o + D_i)(D_o - D_i)(D_o^2 + D_i^2)$$

Substituting $h = (D_o - D_i)/2 = A_{st}(2n-1)/(\pi gt)$, $(D_o + D_i)/2 = gt$, $D_o = gt + h$, $D_i = gt - h$ and $D_o^2 - D_i^2 = 2(gt)^2$ (neglecting the term $2h^2$), we obtain

$$I_s = A_{st}(2n-1)(gt)^2/8$$

Finally, since $I = I_c + I_s$, and $b = t$ for a square section,

$$I = t^4/12 + A_{st}(2n-1)(gt)^2/8$$

9.9. Derive equation (*9.24*) for the moment of inertia of a circular section having the steel arranged in a circle.

From Problem 9.8, $I_s = A_{st}(2n-1)(gt)^2/8$. Here $D = t$.

The moment of inertia of a circle about any diameter is $I_c = \pi D^4/64$. Hence

$$I = I_c + I_s = \pi D^4/64 + A_{st}(2n-1)(gD)^2/8$$

9.10. Derive equation (*9.32*) for the moment of inertia I_x for a rectangular tied column having steel equally distributed along the four faces. Refer to Fig. 9-10.

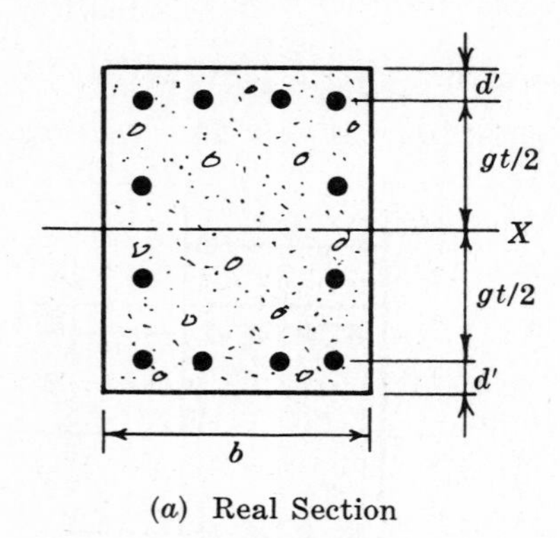

(*a*) Real Section

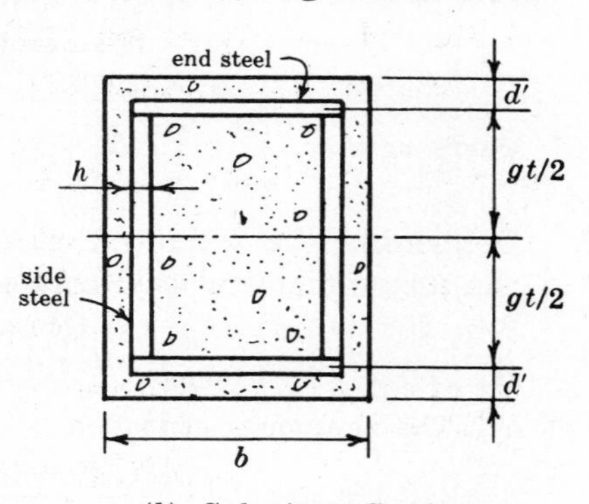

(*b*) Substitute Section

Fig. 9-10

For the concrete rectangle, $I_c = bt^3/12$, and for the end steel

$$I_{se} = (2A_{st}/4)(gt/2)^2(2n-1) = A_{st}(2n-1)(gt)^2/8$$

For the side steel (two thin rectangles), $I_{ss} = 2h(gt)^3/12$ where $h = A_{st}(2n-1)/4gt$,

$$I_{ss} = A_{st}(2n-1)(gt)^2/24$$

Since $I = I_c + I_s$,

$$I = bt^3/12 + A_{st}(2n-1)(gt)^2/8 + A_{st}(2n-1)(gt)^2/24 = bt^3/12 + A_{st}(2n-1)(gt)^2/6$$

It should be noted that the term $bt^3/12$ has been ignored for the end steel moment of inertia, since that quantity is not significant.

9.11. Calculate the properties I_x, A_g and S_x for a square, spirally reinforced concrete column for which $t = 20''$, $D_s = gt = 16''$, $A_{st} = 20.0$ in^2 and $n = 9$.

Use equations (*9.21*), (*9.22*) and (*9.23*).

$$I_x = t^4/12 + A_{st}(2n-1)(gt)^2/8 = (20)^4/12 + (20)(17)(16)^2/8 = 24{,}210 \text{ in}^4.$$

$$A_g = t^2 = (20)^2 = 400 \text{ in}^2 \quad \text{and} \quad S_x = I_x/(t/2) = 24{,}210/10 = 2421 \text{ in}^3.$$

9.12. Calculate the properties I_x, A_g and S_x for a rectangular tied column having steel equally distributed on four faces. The column is reinforced with 12 No. 11 bars, $b = 16''$, $t = 20''$, $d' = 2''$ and $n = 9$.

The governing equations are

$$I_x = bt^3/12 + (2n-1)(gt)^2/6, \qquad S_x = I_x/(t/2), \qquad A_g = bt$$

Since $gt = (t - 2d')$, we find $I_x = 24{,}245$ in^4, $S_x = 2424$ in^3, $A_g = 320$ in^2.

9.13. Using the ACI Code criteria for Region I, derive equation (*9.44*) for the minimum eccentricity e_a. Refer to Fig. 9-11.

The Code criteria for Region I is $0 \leqq e \leqq e_a$.

Using similar triangles in Fig. 9-11,

$$M_a/(N_m - N_a) = M_s/N_m$$

Substituting the identity $M_a = N_a e_a$ and rearranging,

$$N_a e_a = (M_s/N_m)(N_m - N_a)$$

Dividing both sides of the equation by N_a,

$$e_a = M_s(1/N_a - 1/N_m)$$

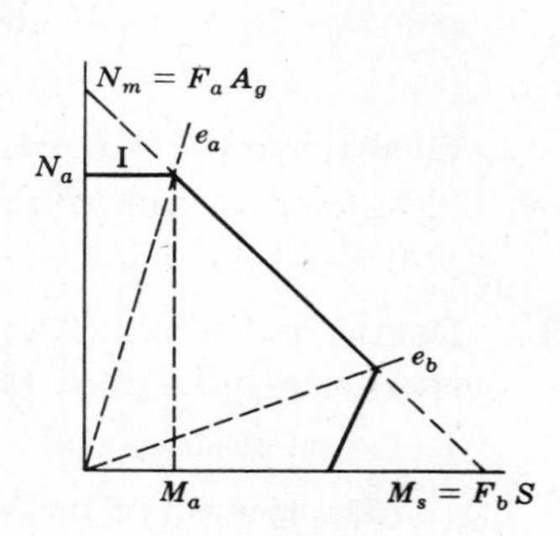

Fig. 9-11

9.14. Derive equation (*9.46*) for use in design in Region II. Use the ACI Code criteria. Refer to Fig. 9-12.

The Code criteria for this region is $e_a \leqq e \leqq e_b$.

Using similar triangles in Fig. 9-12 with M in ft-kips and M_s in inch-kips, obtain

$$N_m/M_s = (N_m - N)/(12M)$$

Since $M_s = F_b S$ and $N_m = F_a A_g$,

$$F_a A_g/(F_b S) = (F_a A_g/12M) - (N/12M)$$

If each term is multiplied by $12M$ and only the first term is multiplied by t/t,

$$(F_a/F_b)(A_g/S)(12M)(t/t) + N = F_a A_g$$

Now substitute the definitions $C = F_a/F_b$ and $D_1 = A_g t/S$ and multiply each term by $G = N_q/F_a A_g$ to obtain

$$G[N + CD_1(12M/t)] = F_a A_g N_q/F_a A_g \quad \text{or} \quad N_q = G[N + CD_1(12M/t)]$$

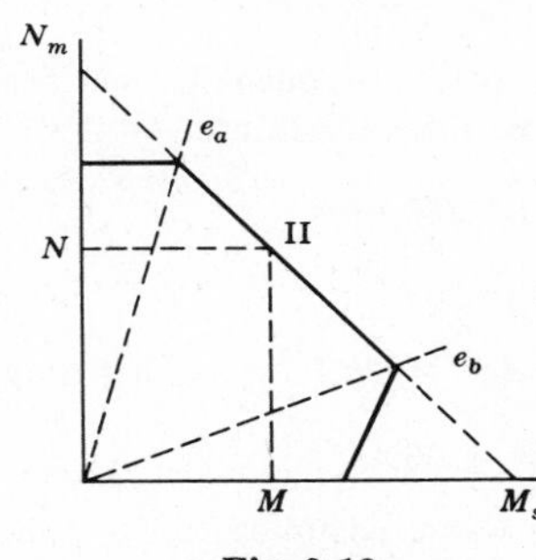

Fig. 9-12

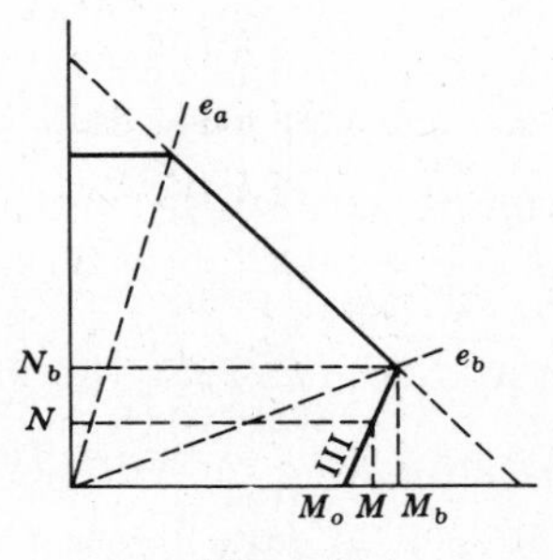

Fig. 9-13

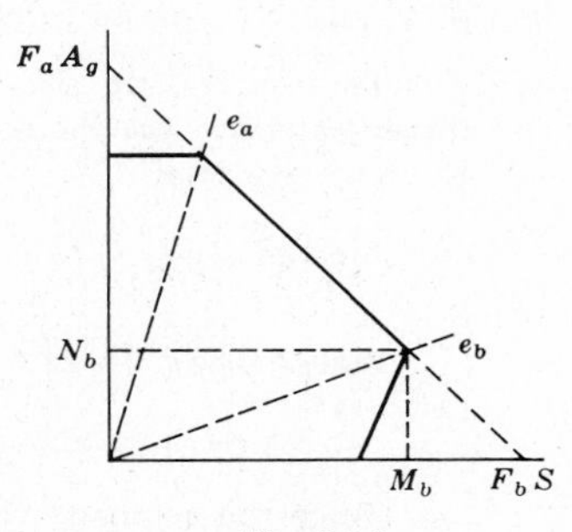

Fig. 9-14

9.15. Derive equation (*9.47*) for use in design in Region III. Use the ACI Code criteria. Refer to Fig. 9-13.

The Code criteria for Region III is $e > e_b$ and $N < N_b$.

Using similar triangles in Fig. 9-13 with M in ft-kips and M_o in inch-kips, observe that

$$(12M - M_o)/N = (M_b - M_o)/N_b$$

Multiply both sides of the equation by $(-N)$ and rearrange to obtain

$$M_o = 12M + (-M_b N/N_b) + (M_o N/N_b)$$

Multiply the last two terms by $12t/12t$ and get

$$M_o = 12M - (M_b/N_b t - M_o/N_b t)(12Nt/12)$$

Now multiply the numerator and denominator of the last term by $f'_c A_g$ and define $M'_b = M_b/(f'_c A_g t)$, $M'_o = M_o/(f'_c A_g t)$, $N'_b = N_b/(f'_c A_g)$. Then

$$M_o = 12M - [(M'_b - M'_o)/N'_b][12Nt/12]$$

Finally, define $D' = (M'_b - M'_o)/N'_b$ and substitute the term M_{oe} for M_o to differentiate between the modified value of M_o and the basic definition of M_o, using M_{oe} for the former quantity, to obtain

$$M_{oe} = 12(M - D'Nt/12)$$

9.16. Derive an equation for the balanced axial load N_b in terms of the section properties and the definitions established in Problem 14. Refer to Fig. 9-14.

From similar triangles in Fig. 9-14,

$$(F_a A_g - N_b)/M_b = F_a A_g/(F_b S)$$

Substitute $N_b e_b = M_b$ and multiply both sides of the relation by t to obtain

$$(t)(F_a A_g - N_b)/(N_b e_b) = (F_a/F_b)(A_g t/S)$$

Apply the definitions $C = F_a/F_b$ and $D_1 = A_g t/S$, and rearrange terms to get

$$N_b = F_a A_g t/(CD_1 e_b + t) \quad \text{or} \quad N_b = F_a A_g/[1 + (CD_1 e_b/t)]$$

9.17. Determine the required area of steel for a rectangular column having steel symmetrically placed parallel to two faces if $b = 15''$, $t = 24''$, $f'_c = 3.0$ ksi, $f_y = 50.0$ ksi, $f_s = 20.0$ ksi, $n = 9.2$ and $g = 0.8$. The axial load and moment are $N = 480$ kips, $M = 240$ ft-kips.

Compute $N/bt = 480/(15)(24) = 1.33$.

Observe in Table 9.1, that J varies from 0.20 to 0.19 for $g = 0.8$ and $f_s = 20.0$ ksi. Using the average value, $Jf'_c = (0.195)(3.0) = 0.585$. Since $Jf'_c = 0.585 < N/bt = 1.33$, *compression controls* and either Region I or Region II will be involved.

The actual eccentricity is $e = 12M/N = (12)(240)/480 = 6.0''$, and $e/t = 6.0/24.0 = 0.25$.

From the table note that e_a/t ranges from 0.13 to 0.10, and that the entire range for Region I is exceeded. Region II is involved.

The design is made using an equivalent axial load $N_q = G[N + CD_1(12M/t)]$.

For a *trial value* of $p_g = 0.06$ (or $100p_g = 6.0$), the tabular values are $CD_1 = 3.3$ and $G = 0.75$ in Tables 9.1 and 9.2. Thus $N_q = 0.75[480 + (3.3)(12)(240)/24] = 657$ kips.

The equivalent axial load is used with equation (*9.1*) in order to develop the design. The latter equation can be separated into two parts, that of the concrete effect and that of the steel. For the concrete,

$$0.2125 f'_c A_g = (0.2125)(3.0)(15.0)(24.0) = 230 \text{ kips}$$

and for the steel,

$$0.85 f_s A_{st} = 657 - 230 = 427 \text{ kips}$$

Since $0.85f_s = 17.0$ ksi, $A_{st} f_s = 17.0A_{st} = 427$ kips or $A_{st} = 25.1$ in². The required steel ratio is

$$p_g = A_{st}/A_g = 25.1/(15.0)(24.0) = 0.069$$

Since the required value of $100p_g$ is quite different than the value assumed (6.0), it is necessary to perform another trial series of calculations. After several trials the value $100p_g = 6.35$ is found to be satisfactory. Therefore select 18 No. 10 bars (9 each end face) for which $A_{st} = 22.86$ and $p_g = 0.0635$.

The section must now be checked to prove satisfaction of $f_a/F_a + f_b/F_b \leq 1.0$.

The necessary computations include evaluation of: $f_a = N/A_g = 1.33$; $F_a = 0.34(1 + p_g m)f'_c$ where $m = f_y/(0.85f'_c) = 19.61$, so $F_a = 0.34(3.0)[1 + (0.0635)(19.61)] = 2.29$ ksi; $F_b = 0.45f'_c = 1.35$ ksi; $f_b = Mc/I$ where $M = (240)(12) = 2880$ inch-kips and $I = (15)(24)^3/12 + [(2)(9.2) - 1][(22.86)(12)^2] = 74{,}590$ in⁴, so that $f_b = (2880)(12)/74{,}590 = 0.463$ ksi. Thus

$$f_a/F_a + f_b/F_b = 1.33/2.29 + 0.463/1.35 = 0.924$$

and the Code requirements are therefore satisfied.

9.18. Outline a procedure for preparing dimensionless *interaction diagrams* for use in design by the ACI Code equations. Consider bending about one axis only for symmetrical sections. Refer to Fig. 9-15.

(*a*) Locate point R_1 using

$R_1 = M_o/f'_c bt^2$ for tied columns,

$R_1 = M_o/f'_c t^3$ for square cols. with spirals,

$R_1 = M_o/f'_c D^3$ for round cols. with spirals.

For this condition of pure moment there is no axial load, so $N = 0$. Use the ACI equations for M_o.

(*b*) Locate point R_4 using $f_a = 0$ and $M_4/F_b S = 1$; then

$R_4 = F_b S/f'_c bt^2$ for tied columns,

$R_4 = F_b S/f'_c t^3$ for square cols. with spirals,

$R_4 = F_b S/f'_c D^3$ for round cols. with spirals.

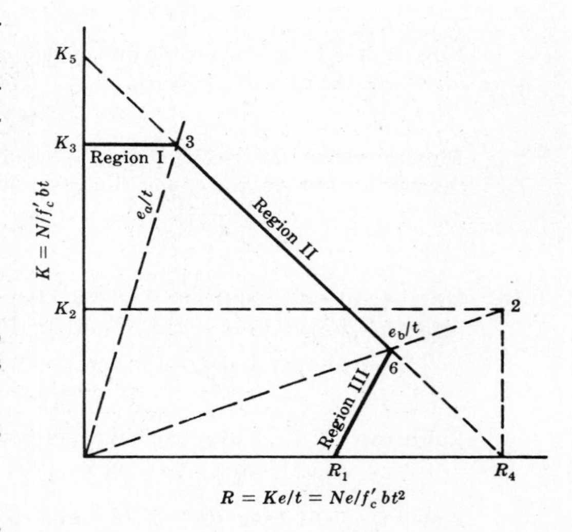

Fig. 9-15

(c) Locate point K_5 using $f_b = 0$ and $N_5/F_a A_g = 1$; then

$K_5 = F_a A_g / f'_c bt$ for tied columns,

$K_5 = F_a A_g / f'_c t^2$ for square columns with spirals,

$K_5 = F_a A_g / f'_c D^2$ for round columns with spirals.

(d) Draw a diagonal line from R_4 to K_5, thus locating a line along which Region II lies.

(e) Locate point K_3 using $M = 0$ with $K_3 = N_a / f'_c bt$ for tied columns or $K_3 = N_a / f'_c t^2$ for spiral columns. For N_a use P from equation (9.1) or (9.2).

(f) Draw a horizontal line from K_3 to intersect the diagonal line between R_4 and K_5. The point of intersection is noted as point 3. The horizontal line defines solutions in Region I, and point 3 establishes the boundary between Regions I and II.

(g) Determine e_b/t using the appropriate Code equation for e_b and divide by t to render dimensionless.

(h) Determine K_2 as being equal to $R_4/(e_b/t)$.

(i) Locate point 2 which has coordinates (R_4, K_2). Draw a diagonal line from the origin to point 2. This line will intersect the diagonal line (R_4, K_5) at point 6, which establishes the boundary of Regions II and III, since Region II is fully described by segment (3, 6).

(j) Draw line $(R_1, 6)$ to describe Region III.

The interaction diagram is completely represented by line segments $(R_1, 6)$, (3, 6) and $(K_3, 3)$. The other lines are no longer useful and should be erased.

The steel ratio p_t may be varied in increments, and a family of interaction diagrams may be plotted for any desired set of variables including f'_c, f_y and g. The resulting design aids are very valuable as time savers when designing by the working strength design method.

Fig. 9-16 illustrates a typical set of interaction diagrams for square columns with spirals, for $g = 0.8$, $f'_c = 3.0$ ksi and $f_s = 16{,}000$ psi.

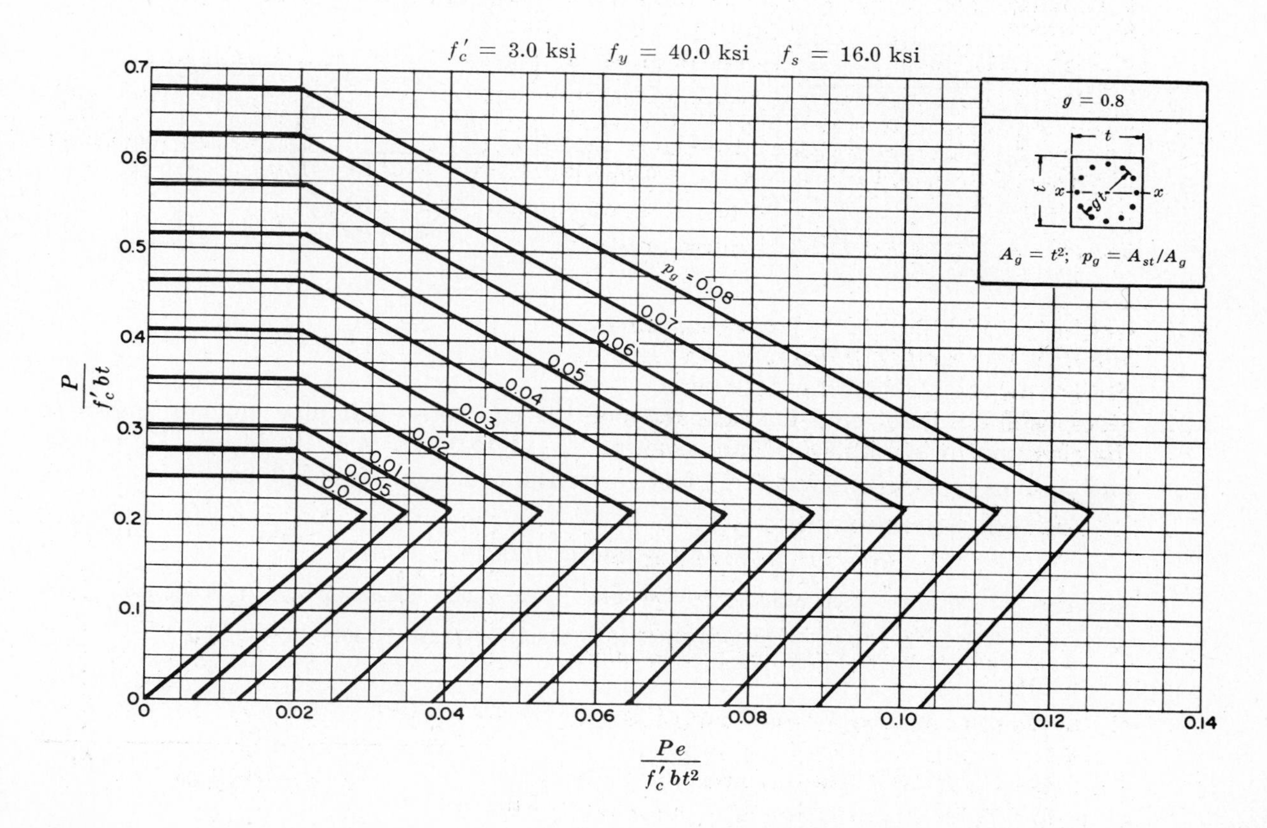

Fig. 9-16

9.19. Determine the safe axial load capacity of the combination section shown in Fig. 9-17. The metal core has area 6.0 in² and is ASTM A-36 steel. The area of the vertical reinforcing bars is 18.0 in² and $f_y = 60.0$ ksi. Use $f'_c = 3.0$ ksi.

Equation (9.58) applies,

$$P = 0.225 f'_c A_g + f_s A_{st} + A_r f_r$$

where $A_g = (20)(20) = 400\ \text{in}^2$, $f_s = 0.4 f_y = 24.0$ ksi, and $f_r = 18.0$ ksi for ASTM A-36 steel. Thus,

$$P = (0.225)(3)(400) + (24)(18) + (6)(18) = 810 \text{ kips}$$

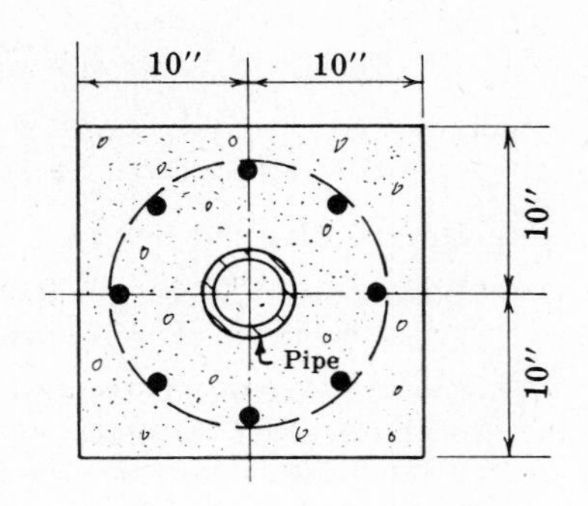

Fig. 9-17

9.20. Determine the safe axial load capacity for the column shown in Fig. 9-18. The steel column is a 14″ wide flange section weighing 53 lb/ft. The properties of the steel section are: $K_{sx} = 5.9''$, $K_{sy} = 1.92''$, $A_r = 15.59\ \text{in}^2$, $f_y = 40.0$ ksi. The unsupported height is 12 ft.

The pertinent equation is $P = A_r f'_r [1 + A_g/(100 A_r)]$ where $f'_r = 17{,}000 - 0.485(h/K_s)^2$. Using the least radius of gyration, $K_{sy} = 1.92''$, calculate

$$f'_r = 17{,}000 - 0.485[(12)(12)/1.92]^2 = 14{,}270 \text{ psi}$$

Hence $$P = (15.59)(14.27)[1 + (20)(20)/(100)(15.59)] = 279.5 \text{ kips}$$

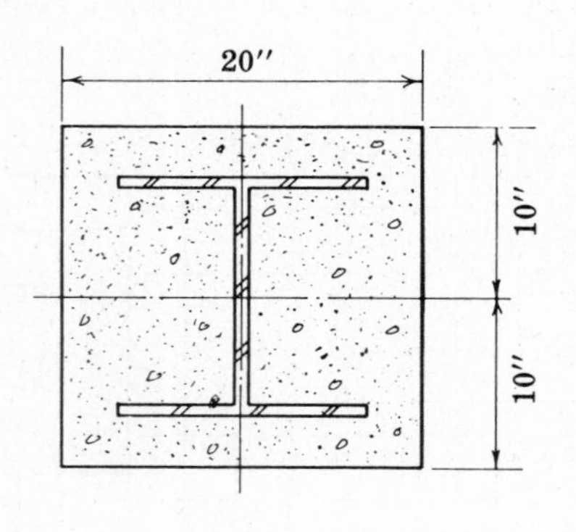

Fig. 9-18

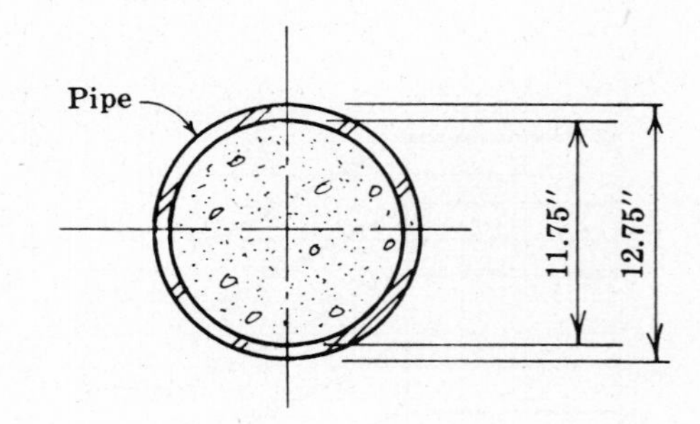

Fig. 9-19

9.21. Determine the safe axial load capacity of the concrete filled pipe column shown in Fig. 9-19. For the 12″ steel pipe section, $f_y = 40.0$ ksi, $A = 19.24\ \text{in}^2$, $K_s = 4.34$, and the unsupported height is 12 feet. For the concrete, $f'_c = 5.0$ ksi.

The applicable equation is

$$P = 0.25 f'_c [1 - 0.000025(h/K_s)^2] A_c + f'_r A_r$$

where $A_c = 108.5\ \text{in}^2$, $0.000025(h/K_s)^2 = 0.0275$, $f'_r = 17{,}000 - 0.485(h/K_s)^2 = 16{,}466$ psi. Then

$$P = (0.25)(5)[1 - (0.000025)(33.18)^2](108.4) + (19.24)(16.466) = 449 \text{ kips}$$

Supplementary Problems

9.22. A short, spirally reinforced concrete column is 20″ in diameter and contains 16 No. 10 bars. Using $f'_c = 3.0$ ksi and $f_y = 50.0$ ksi, determine the maximum permissible axial load.
Ans. $P = 642$ kips

9.23. A column has properties identical to those stated for Problem 9.21, except that the column is 20″ square. Calculate the permissible axial load capacity. *Ans.* $P = 706$ kips

9.24. Determine the balanced eccentricity e_{bx} for the column shown in Fig. 9-20 for $f'_c = 3.0$ ksi and $f_y = 50.0$ ksi. The vertical steel consists of 12 No. 9 bars. *Ans.* $e_{bx} = 9.91''$

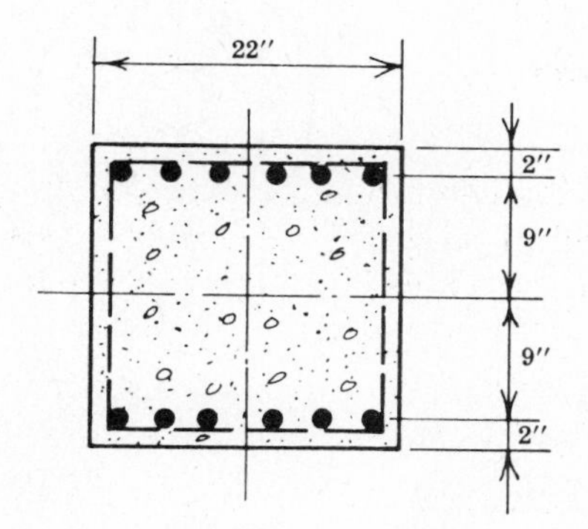

Fig. 9-20

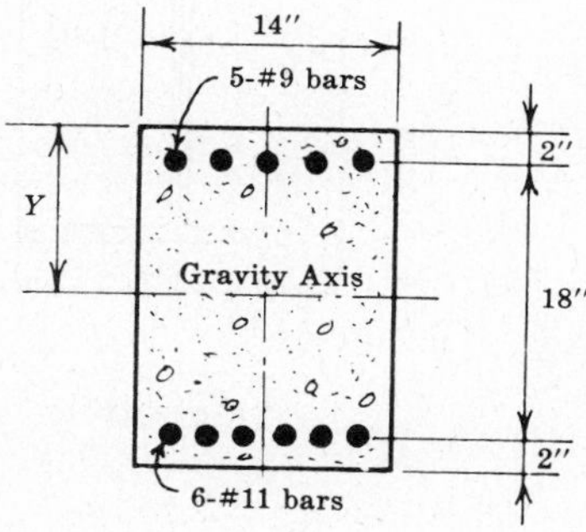

Fig. 9-21

9.25. Determine the moment of inertia for the gross transformed section of the column shown in Fig. 9-20. Use $n = 9.2$. *Ans.* $I_x = 36{,}400$ in^4

9.26. Determine the section moduli for the steel and concrete for the column shown in Fig. 9-20.
Ans. $S_s = 4050$ in^3, $S_c = 3310$ in^3

9.27. Locate the gravity axis for the gross transformed section of the column shown in Fig. 9-21 if $n = 9.2$. (*Hint.* Use $2n - 1$ for steel.) *Ans.* $y = 12.22''$ from the top

9.28. Calculate the moment of inertia about the gravity axis for the column of Problem 9.27.
Ans. $I_x = 31{,}360$ in^4

9.29. Determine the maximum permissible axial load for the section shown in Fig. 9-20 if $f'_c = 5.0$ ksi and $f_y = 60.0$ ksi. *Ans.* $P = 760$ kips

9.30. Determine the balanced load N_b for the column shown in Fig. 9-20. Use $f'_c = 5.0$ ksi, $f_y = 60.0$ ksi and $n = 9.2$. *Ans.* $N_b = 450$ kips

9.31. A column measures 16″ x 20″ and is reinforced with steel on two faces. Using $f'_c = 3.0$ ksi, $f_s = 20.0$ ksi, $d' = 2''$ and $n = 9.2$, select the required number of No. 9 bars if $M = 150$ ft-kips and $N = 250$ kips. (First prove that Region II governs.)
Ans. Use 12 No. 9 bars ($p_g = 0.037$)

9.32. Prove that $f_a/F_a + f_b/F_b$ does not exceed unity in Problem 9.31. *Ans.* Sum = 0.99

9.33. Design the steel for the section shown in Fig. 9-22 using $f'_c = 3.0$ ksi, $f_y = 50.0$ ksi and $n = 9.2$, $M = 100$ ft-kips, $N = 300$ kips. Prove that Region II governs. Use No. 10 bars.
Ans. Use 8 No. 10 bars. $f_a/F_a + f_b/F_b = 0.88$.

9.34. Solve Problem 9.31 using $M = 25$ ft-kips and $N = 550$ kips. Show that Region I governs.
Ans. $A_{st} = 20.35$ in^2 required

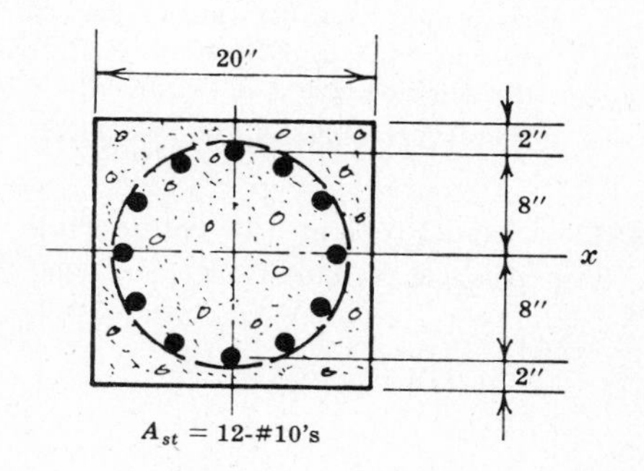

Fig. 9-22

9.35. Solve Problem 9.31 using $M = 150$ ft-kips and $N = 150$ kips.
Ans. Region III governs. $A_{st} = 9.60$ in^2 required.

For Problems 9.36 through 9.39, determine the safe capacity of each section shown. For all reinforcement, $f_y = 40.0$ ksi; and for the concrete, $f'_c = 5.0$ ksi. The unsupported height is 14 feet in all cases. For pipe sections consider ASTM A-7 steel.

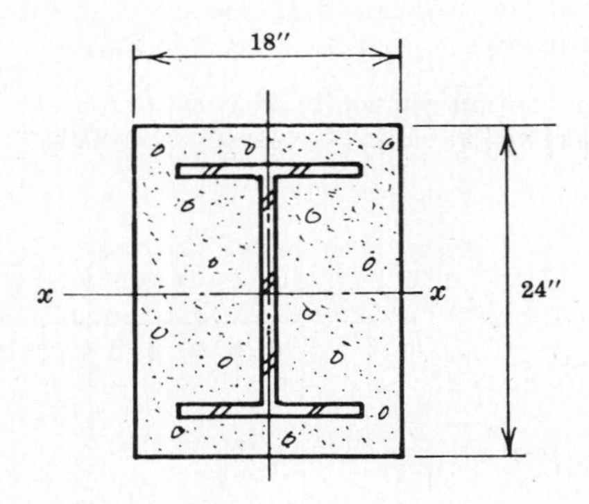

Fig. 9-23

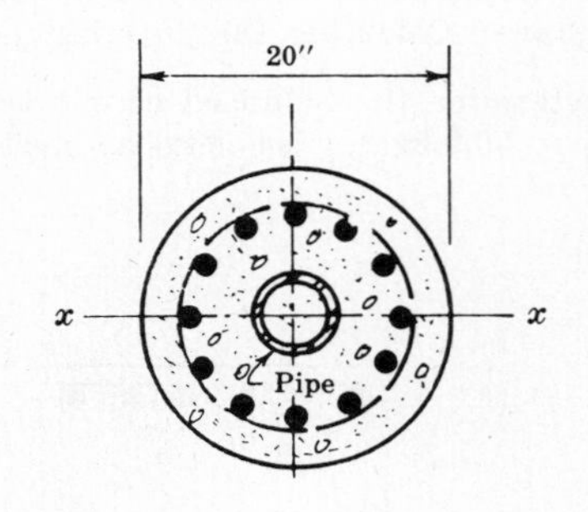

Fig. 9-24

9.36. $A_r = 55.86$ in^2, $K_s = 3.25''$
Ans. $P = 947$ kips

9.37. $A_{st} = 12.0$ in^2, $A_r = 21.3$ in^2, $K_s = 2.76''$
Ans. $P = 886$ kips

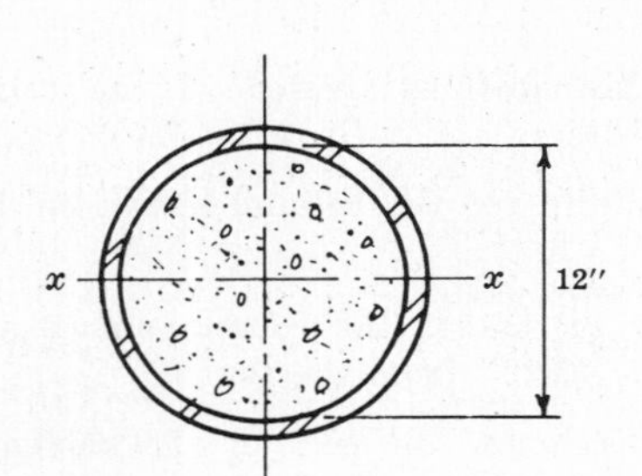

Fig. 9-25

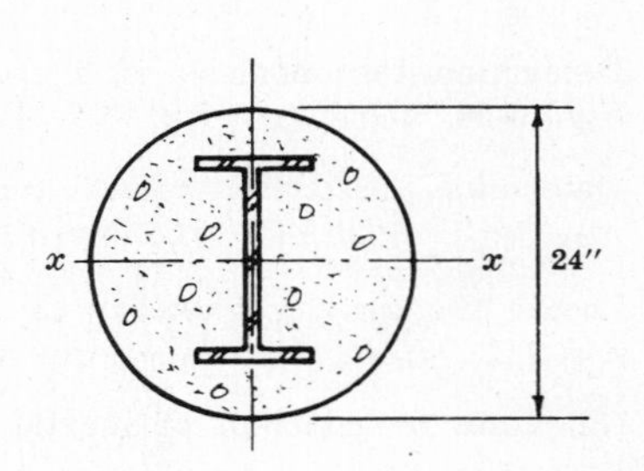

Fig. 9-26

9.38. $A_r = 19.24$ in^2, $K_s = 4.34''$
Ans. $P = 449$ kips

9.39. $A_r = 11.17$ in^2, $K_s = 1.49''$
Ans. $P = 170$ kips

9.40. Calculate the values of all key coordinates for plotting a dimensionless interaction diagram for design of square, spirally reinforced concrete columns by working stress design principles. Refer to Fig. 9-15 and Problem 9.18. Use $p_t = 0.03$, $f_y = 40.0$ ksi, $f'_c = 3.0$ ksi and $g = 0.8$. (Column dimensions are not required.)
Ans. $R_1 = 0.0384$, $R_4 = 0.1125$, $K_2 = 0.3726$, $K_3 = 0.4100$, $K_5 = 0.500$

9.41. Solve Problem 9.40 considering a circular column.
Ans. $R_1 = 0.0301$, $R_4 = 0.0736$, $K_2 = 0.244$, $K_3 = 0.322$, $K_5 = 0.3927$

Chapter 10

Short Columns — Ultimate Strength Design

NOTATION

The notation and definitions relating to this chapter are identical to those listed for Chapter 8. Additional definitions are provided in connection with the pertinent material.

DEVELOPMENT OF ULTIMATE STRENGTH PROCEDURES

For many years columns were designed using the straight-line stress distribution, considering concrete to be an elastic material. Numerous tests have indicated that reinforced concrete column behavior cannot be predicted using the classical elastic theory equation which states that Stress $= P/A \pm Mc/I$.

As a result of tests made throughout the world, experts have concluded that other methods should be developed for use in the design of reinforced concrete columns. The pioneering work in this field was performed by Charles Whitney, who devised semi-empirical equations for use in the design of reinforced concrete columns considering the ultimate strength of the materials. Those equations have been adopted in the 1963 ACI Code, and will be discussed in this chapter. In addition, the limitations of those equations will be discussed and more accurate methods will be introduced. The more accurate methods were approved by the American Concrete Institute and published in ACI Special Publication No. 7, *Ultimate Strength Design of Reinforced Concrete Columns,* by Everard and Cohen.

SPECIAL CONDITIONS FOR ULTIMATE STRENGTH DESIGN

Special conditions placed on the design of columns using the ultimate strength design method are listed in the paragraphs which follow. (See also Chapter 5.)

(1) The forces and moments due to dead load D, live load L, and wind load W, as calculated using elastic procedures for analysis, shall be increased to an ultimate value U using the most severe of the following conditions:

$$(a) \quad U = 1.5D + 1.8L \qquad (10.1)$$

$$(b) \quad U = 1.25(D + L + W) \qquad (10.2)$$

$$(c) \quad U = 0.9D + 1.1W \qquad (10.3)$$

Earthquake or other phenomena which produce conditions which can be converted to an *equivalent wind force* shall be included in the same manner as wind forces. Creep, elastic deformation, shrinkage and temperature shall be considered on the same basis as dead load.

(2) If the steel used has a yield strength greater than 60.0 ksi, the yield strength used in design shall be reduced to $0.85f_y$ or 60.0 ksi, whichever is greater, unless tension tests of the steel used show that the strain of the steel does not exceed 0.003 in/in at the specified yield stress.

(3) Yield strength f_y used in design may never exceed 75.0 ksi. *Tension reinforcement* may not be considered to be stressed to more than 60.0 ksi unless *full scale tests* indicate that crack widths due to *service load* at the extreme tension edge of the concrete will not exceed 0.015″ for interior members nor 0.01″ for exterior members. Members tested must be typical of those used in the structure, and not merely scaled models.

ASSUMPTIONS FOR ULTIMATE STRENGTH DESIGN

The following assumptions are taken directly from the ACI Code, with minor changes.

(1) Ultimate strength design of members for bending and axial load shall be based on the assumptions given herein, and on satisfaction of the applicable conditions of equilibrium and compatibility of strains. The simplified design equations are satisfactory for steel ratios less than 0.04 and values of f_y not exceeding 50.0 ksi.

(2) Strain in the concrete shall be assumed directly proportional to the distance from the neutral axis. Except in anchorage regions, strain in reinforcing bars shall be assumed equal to the strain in the concrete at the same position.

(3) The maximum strain at the extreme compression fiber at ultimate strength shall be assumed equal to 0.003.

(4) Stress in reinforcing bars below the yield strength f_y for the grade of steel used shall be taken as 29,000,000 psi times the steel strain. For strain greater than that corresponding to the design yield strength f_y, the reinforcement stress shall be considered independent of strain and equal to the design yield strength f_y.

(5) Tensile strength of the concrete shall be neglected in flexural calculations.

(6) At ultimate strength, concrete stress is not proportional to strain. The diagram of compressive concrete stress distribution may be assumed to be a rectangle, trapezoid, parabola, or any other shape which results in predictions of ultimate strength in reasonable agreement with the results of comprehensive tests.

(7) The requirements of (6) may be considered satisfied by the equivalent rectangular concrete stress distribution which is defined as follows: At ultimate strength, a concrete stress intensity of $0.85f'_c$ shall be assumed uniformly distributed over an equivalent compression zone bounded by the edges of the cross section and a straight line located parallel to the neutral axis at a distance $a = k_1 c$ from the fiber of maximum compressive strain. The distance c from the fiber of maximum strain to the neutral axis is measured in the direction perpendicular to that axis. The fraction k_1 shall be taken as 0.85 for strengths f'_c up to 4000 psi and shall be reduced continuously at a rate of 0.05 for each 1000 psi of strength in excess of 4000 psi.

SAFETY PROVISIONS

In addition to the *load factors* which are applied to the elastically calculated forces and moments, a *capacity reduction factor* ϕ must be applied to the ultimate strength of columns.

For tied columns $\phi = 0.7$, and for spiral columns $\phi = 0.75$.

ADDITIONAL DEFINITIONS

The *plastic centroid* of a section is the centroid of the resistance to load computed for the assumptions that the concrete is stressed uniformly to $0.85f'_c$ and the steel is stressed uniformly to f_y. For symmetrically reinforced members, the plastic centroid will correspond to the centroid of the cross section.

Balanced conditions exist when, at ultimate strength of a member, the tension reinforcement reaches its yield stress just as the concrete in compression reaches its assumed ultimate strain of 0.003.

LIMITATIONS

(1) All members subjected to a compression load shall be designed for the eccentricity e corresponding to the maximum moment which can accompany this loading condition, but not less than $0.05t$ for spirally reinforced columns or $0.10t$ for tied columns, about either principal axis.

(2) The maximum load capacities for members subject to axial load as determined by the requirements of this chapter apply only to *short members* and shall be reduced for the effects of length according to the requirements of Chapter 11.

(3) Members subjected to small compressive loads may be designed for the maximum moment $P_u e$, in accordance with the provisions of Chapter 5 and disregarding the axial load, but the resulting section shall have a capacity P_b greater than the applied compressive load.

SEMI-EMPIRICAL EQUATIONS

The ACI Code provides *semi-empirical design equations* for use in the design of rectangular tied columns, and round and square spiral columns. The equations were devised using test data in which the steel yield stress was usually less than 50.0 ksi, the dimensions of the scaled models were reasonably small and the steel ratio p_g was less than 0.04. Nevertheless, the Code permits steel stresses up to 75.0 ksi and steel ratios between 0.01 and 0.08. It follows from the test data that the equations provide reasonably accurate results when used in the range of the test data. Extrapolation beyond $f_y = 50.0$ ksi and $p_g = 0.04$ is very dangerous and often produces errors as much as 50 percent on the *unsafe side.*

Recognizing that deficiencies exist in the semi-empirical equations, the ACI Code requires that proof of yielding of the compression steel must be shown in order for a design to be acceptable. Solved problems will show that proof of yielding of the compression steel is insufficient to insure a satisfactory design under certain very practical conditions. Those problems are presented to caution the designer not to use the Code equations indiscriminately.

RECTANGULAR SECTIONS

When A'_s and A_s are parallel to the axis about which bending occurs the *approximate balanced ultimate load* P_b *and moment* M_b may be computed using equations *(10.4)* and *(10.5)* respectively, assuming $f'_s = f_y$ and $c = c_b$.

$$P_b \;=\; \phi[0.85f'_c bk_1c_b + A'_sf_y - A_sf_y] \tag{10.4}$$

$$M_b \;=\; P_b e_b \;=\; \phi[0.85f'_c bk_1c_b(d-d''-0.5k_1c_b) + A'_sf_y(d-d'-d'') + A_sf_yd''] \tag{10.5}$$

where $k_1 \;=\; 0.85$ for $f'_c \leqq 4.0$ ksi and $[0.85 - (0.05)(f'_c - 4.0)]$ for $f'_c > 4.0$ ksi,

$c_b \;=\; 87d/(87 + f_y)$ where f_y is in ksi.

The section is considered to be subjected to balanced failure when the yield strain in the tension steel occurs simultaneously with a maximum concrete strain of 0.003.

When P_u is less than P_b, *tension controls* the design; when P_u exceeds P_b, *compression controls* the design.

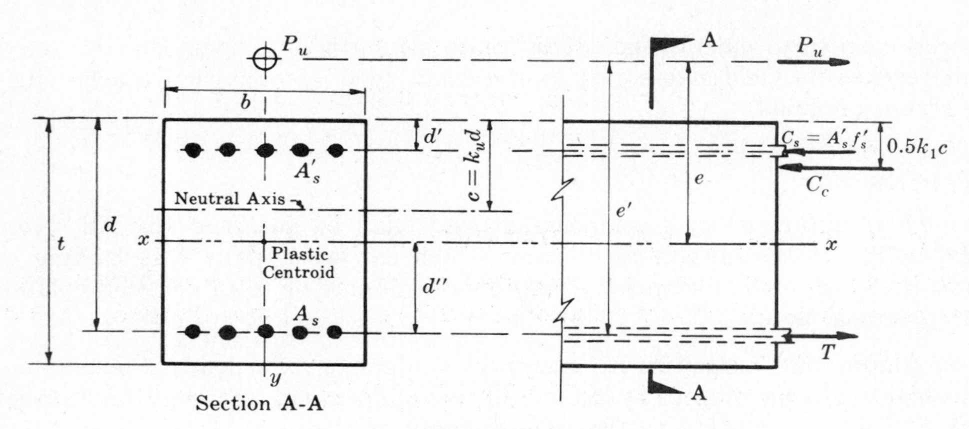

Fig. 10-1

For *tension failure,* when the section has reinforcement on one or two faces, each parallel to the axis of bending, and all of the reinforcement in one face is located at approximately the same distance from the axis of bending, the *approximate* ultimate load may be computed as

$$P_u = \phi\Big[0.85f'_c bd\,\{p'm' - pm + (1 - e'/d) + \sqrt{(1-e'/d)^2 + 2[(e'/d)(pm - p'm') + p'm'(1 - d'/d)]}\,\}\Big] \qquad (10.6)$$

in which $e' = e + d - t/2$.

For symmetrical reinforcement in two faces,

$$P_u = \phi\Big[0.85f'_c bd\,\{-p + 1 - e'/d + \sqrt{(1-e'/d)^2 + 2p[m'(1-d'/d) + e'/d]}\,\}\Big] \qquad (10.7)$$

With no compression reinforcement ($A'_s = 0$),

$$P_u = \phi\{0.85f'_c bd\,[-pm + 1 - e'/d + \sqrt{(1-e'/d)^2 + 2e'pm/d}\,]\} \qquad (10.8)$$

For *compression failure,* when the reinforcement is placed in two symmetrical single layers, the *approximate* ultimate load may be computed as

$$P_u = \phi\left[\frac{A'_s f_y}{e/(d-d') + 0.5} + \frac{btf'_c}{(3te/d^2) + 1.18}\right] \qquad (10.9)$$

For more general conditions of reinforcement, when *compression controls,* the ultimate load shall be assumed to decrease linearly from P_o to P_b as the moment is increased from zero to M_b, where

$$P_o = \phi[0.85f'_c(A_g - A_{st}) + A_{st}f_y] \qquad (10.10)$$

which is the theoretical capacity of an *axially loaded* column not subjected to moment.

The *approximate* value of P_u may be then computed using either equation (*10.11*) or (*10.12*), each of which satisfies the linearity requirement.

$$P_u = \frac{P_o}{1 + [(P_o/P_b) - 1]e/e_b} \qquad (10.11)$$

or

$$P_u = P_o - (P_o - P_b)M_u/M_b \qquad (10.12)$$

It is important to note that the designer is obligated to prove that the compression steel yields in the foregoing equations. This can be done only by assuming a neutral axis location, solving for P_u and M_u and establishing strain compatibility. The process is one of trial and error using the general equations for rectangular sections:

$$P_u = \phi[0.85f'_c bk_1 c + A'_s f'_s - A_s f_s] \qquad (10.13)$$

$$M_u = P_u e' = \phi[0.85f'_c bk_1 c(d - 0.5k_1 c) + A'_s f'_s (d - d')] \qquad (10.14)$$

where $f'_s = [87(c - d')/c] \leqq f_y$, $f_s = [87(d - c)/c] \leqq f_y$, $e' = e + d - t/2$.

CIRCULAR SECTIONS. STEEL ARRANGED IN A CIRCLE

The ultimate strength shall be computed on the basis of equilibrium and inelastic deformations, or the *approximate* equations (*10.15*) or (*10.16*) may be used. (See Fig. 10-2.)

When *tension controls,*

$$P_u = \phi\{0.85f'_c D^2[\sqrt{(0.85e/D - 0.38)^2 + p_t m D_s/2.5D} - (0.85e/D - 0.38)]\} \qquad (10.15)$$

When *compression controls,*

$$P_u = \phi\left[\frac{A_{st}f_y}{3e/D_s + 1} + \frac{A_g f'_c}{9.6De/(0.8D + 0.67D_s)^2 + 1.18}\right] \qquad (10.16)$$

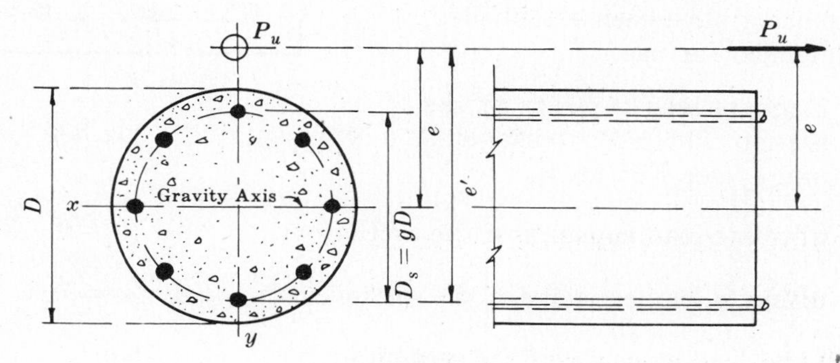

Fig. 10-2

SQUARE SECTIONS. STEEL ARRANGED IN A CIRCLE

The ultimate strength shall be computed on the basis of equilibrium and inelastic deformations, or the *approximate* equations (*10.17*) or (*10.18*) may be used. (See Fig. 10-3.)

When *tension controls,*

$$P_u = \phi\{0.85btf'_c[\sqrt{(e/t - 0.5)^2 + 0.67(D_s/t)p_t m} - (e/t - 0.5)]\} \qquad (10.17)$$

When *compression controls,*

$$P_u = \phi\left[\frac{A_{st}f_y}{3e/D_s + 1} + \frac{A_g f'_c}{12te/(t + 0.67D_s)^2 + 1.18}\right] \qquad (10.18)$$

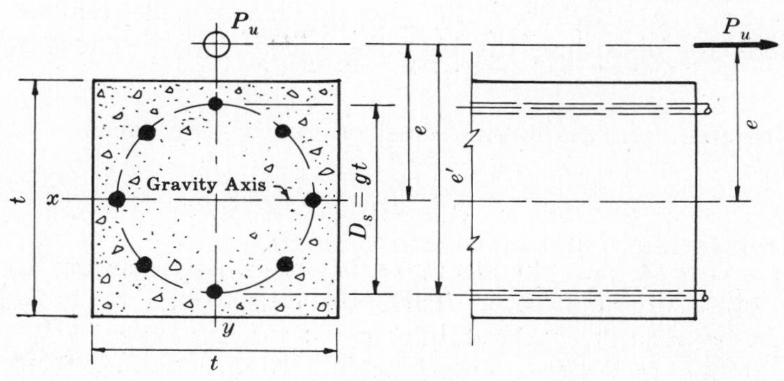

Fig. 10-3

GENERAL CASE

For any shape section, reinforced in any manner, the ultimate load shall be computed using the equations of equilibrium and strain compatibility with the general assumptions concerning stress distribution over the section.

BIAXIAL BENDING

The ACI Code does not contain empirical equations for design of sections subject to bending about two axes. However, experimental results indicate that a reciprocal type of interaction equation sometimes provides satisfactory results. Equation (*10.19*) represents such an expression. (The range of applicability of equation (*10.19*) has not been clearly established. Until such time that the limits of the equation have been established, it is not recommended for use.)

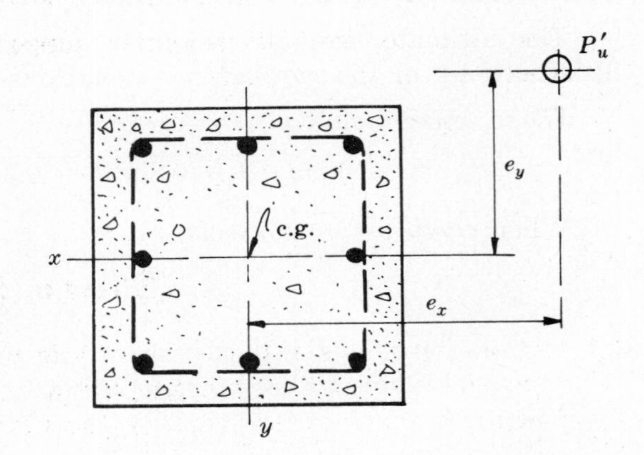

Fig. 10-4

Let P'_u = ultimate load capacity of the section with eccentricities e_x and e_y (see Fig. 10-4),

P'_x = ultimate load capacity of the section with e_x only ($e_y = 0$),

P'_y = ultimate load capacity of the section with e_y only ($e_x = 0$),

P'_o = axial load capacity of the section ($e_x = e_y = 0$). Then

$$(1/P'_u) = (1/P'_x) + (1/P'_y) - (1/P'_o) \qquad (10.19)$$

BALANCED ECCENTRICITY

The 1963 ACI Code does not contain specific empirical equations for use in obtaining the balanced eccentricity for ultimate strength design of columns. However, a method of obtaining the balanced eccentricity is implied inasmuch as equations for balanced moment M_b and balanced load P_b are given.

By definition, $M_b = P_b e_b$, so one can obtain the balanced eccentricity in this manner.

Since the calculations for M_b and P_b are somewhat lengthy and are approximate, the *ACI Code Commentary,* ACI SP-10 which was published in September 1965, provides approximate equations for obtaining the balanced eccentricity. Those equations are stated as follows:

For *rectangular tied columns* having steel on one or two faces,

$$e_b = (0.20 + 0.77p_t m)t \qquad (10.20)$$

and for spirally reinforced round or square columns

$$e_b = (0.24 + 0.39p_t m)t \qquad (10.21)$$

where $p_t = A_{st}/A_g$, $m = f_y/(0.85f'_c)$ and t is the overall dimension parallel to the direction of bending.

MORE ACCURATE METHODS

Recognizing the inaccuracies of the approximate equations given in the ACI Code, ACI Committee 340 has prepared computer programs which automatically account for strain compatibility and thus limits the steel stresses to the actual values. The computer programs have been used to prepare design charts and tables which have been published as ACI Special Publication 7, *Ultimate Strength Design of Reinforced Concrete Columns*, by Everard and Cohen. Complete derivations of equations appear therein.

An example of the design charts presented in that manual is provided in the Solved Problems section of this chapter.

Solved Problems

10.1. An analysis was made using elastic procedures. The axial loads and moments obtained therefrom are listed below. Determine the combinations of P_u and M_u required for use in design by the ultimate strength theory.

Dead Load:	$P = 150$ kips	$M = 75$ ft-kips
Live Load:	$P = 200$ kips	$M = 85$ ft-kips
Wind Load:	$P = 80$ kips	$M = 130$ ft-kips

The load factors must be applied to the forces and moments calculated using the elastic theory. Thus:

(*a*) For $U = 1.5D + 1.8L$,
$$\begin{cases} P_u = (1.5)(150) + (1.8)(200) = 585.0 \text{ kips} \\ M_u = (1.5)(75) + (1.8)(85) = 265.5 \text{ ft-kips} \end{cases}$$

(*b*) For $U = 1.25(D + L + W)$,
$$\begin{cases} P_u = 1.25(150 + 200 + 80) = 537.5 \text{ kips} \\ M_u = 1.25(75 + 85 + 130) = 362.5 \text{ ft-kips} \end{cases}$$

(*c*) For $U = 0.9D + 1.1W$,
$$\begin{cases} P_u = (0.9)(150) + (1.1)(80) = 223.0 \text{ kips} \\ M_u = (0.9)(75) + (1.1)(130) = 210.5 \text{ ft-kips} \end{cases}$$

All three of the combinations must be investigated in order to provide a satisfactory design.

10.2. The column cross-section shown in Fig. 10-5 contains steel having a yield stress of 50.0 ksi. Determine the actual stresses in the reinforcement, considering $0.85f'_c$ to be deducted from the steel in the compression zone. Use $f'_c = 3.0$ ksi.

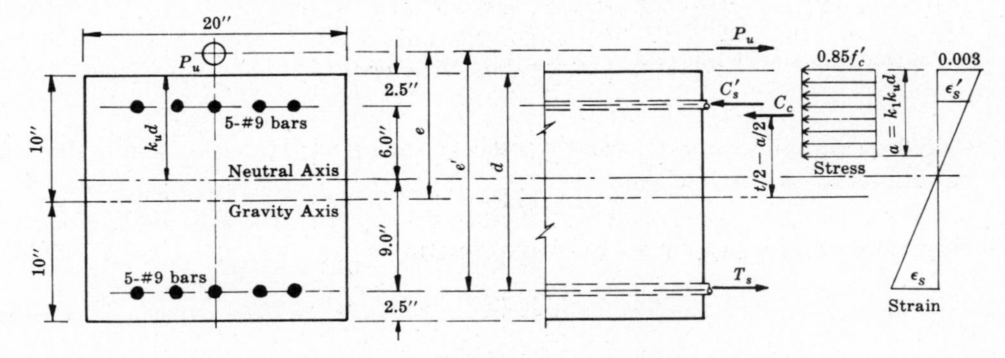

Fig. 10-5

Using similar triangles compute the unit stress in the compression steel based on strain compatibility and the assumption that the maximum strain in the concrete is 0.003 in/in. Thus

$$\epsilon_s' = \epsilon_u(k_u d - d')/(k_u d) = (0.003)(6.0)/8.5 = 0.00212 \text{ in/in}$$

and
$$f_s' = \epsilon_s' E_s = (0.00212)(29 \times 10^6) = 61.4 \text{ ksi}$$

Note that the calculated stress exceeds f_y, so limit the stress to f_y. The net effective compressive stress in the steel will be

$$(f_s' - 0.85 f_c') = 50.0 - (0.85)(3.0) = 47.45 \text{ ksi}$$

Similarly, the strain in the steel on the tension side will be

$$\epsilon_s = \epsilon_u(d - k_u d)/(k_u d) = (0.003)(9.0)/8.5 = 0.00318 \text{ in/in}$$

Therefore $f_s = \epsilon_s E_s = (0.00318)(29 \times 10^6) = 92.2$ ksi.

Note that the calculated stress exceeds f_y, so limit f_s to f_y. Thus, $f_s = 50.0$ ksi.

10.3. Determine the ultimate load P_u' that will satisfy statics in Fig. 10-5. Use the results of Problem 10.2. ($P_u = \phi P_u'$ where $\phi = 0.7$ for tied columns.)

Since statics requires that $\Sigma F_x = 0$, then $P_u' = C_c + C_s' - T_s$ where T_s is in the negative direction.

The forces are calculated as the stresses times the areas; thus

$$C_c = 0.85 f_c' k_1 k_u db = 0.85(3)(0.85)(8.5)(20) = 368.0 \text{ kips}$$

$$C_s' = A_s'(f_s' - 0.85 f_c') = (5.0)(47.45) = 237.3 \text{ kips}$$

$$T_s = A_s f_s = (5.0)(50.0) = 250.0 \text{ kips}$$

Hence

$$P_u' = 368.0 + 237.3 - 250.0 = 355.3 \text{ kips} \quad \text{and} \quad P_u = \phi P_u' = 0.7(355.3) = 248.7 \text{ kips}$$

10.4. Determine the eccentricity of P_u' required to establish the conditions of equilibrium shown in Fig. 10-5.

The moment of the forces may be summed about the gravity axis or the neutral axis or any axis parallel thereto. Using the neutral axis, and transferring to the plastic centroid,

$$M_c = C_c(k_u d - k_1 k_u d/2) = 368(8.5 - 3.61) = 1800 \text{ in-kips}$$

$$M_s' = C_s'(k_u d - d') = (237.3)(6.0) = 1424.0 \text{ in-kips}$$

$$M_s = T_s(d - d' - k_u d) = (250.0)(9.0) = 2250.0 \text{ in-kips}$$

The moment about the plastic centroid is

$$\Sigma M_{NA} + \Sigma F(t/2 - k_u d) = 6007.0 \text{ in-kips}$$

Then $\phi M_u' = M_u = (6007)(0.7) = 4204.9$ in-kips.

About the plastic centroid, $e = M_u/P_u = 16.9$ in.

10.5. Calculate the allowable P_u for the section shown in Fig. 10-5 using the approximate equations of the ACI Code.

First determine whether compression or tension controls. The balanced k_u or k_b must therefore be computed.

$$k_b = 87{,}000/(87{,}000 + f_y) = 0.635$$

The true k_u is $8.5/17.5 = 0.486$ which is less than k_b, so tension controls the design and

$$P_u = \phi\left[0.85 f_c' bd\,\{-p + 1 - e'/d + \sqrt{(1 - e'/d)^2 + 2p[m'(1 - d'/d) + e'/d]}\,\}\right]$$

in which $d = 17.5''$, $b = 20.0''$, $f'_c = 3.0$ ksi, $A_s = 5.0$ in^2, $d' = 2.5''$

$$\phi(0.85f'_c bd) = (0.7)(0.85)(3)(20.0)(17.5) = 625.0 \text{ kips}$$

$$p = A_s/bd = 5.0/[(20)(17.5)] = 0.0143$$

$$e' = e + \tfrac{1}{2}(t - 2d') = 24.4'', \quad e'/d = 24.4/17.5 = 1.40, \quad d'/d = 2.5/17.5 = 0.143$$

$$m = f_y/(0.85f'_c) = 50.0/2.55 = 19.61, \quad m' = m - 1 = 18.61.$$

Putting these values into the equation, we obtain $P_u = 247$ kips. This value does not differ greatly from the more correct value $P_u = 248$ kips obtained using statics and strain compatibility. In general, the empirical equations for rectangular tied columns will usually provide satisfactory results in practical cases.

10.6. Use statics and strain compatibility to determine the ultimate load P_u for the spiral column shown in Fig. 10-6. Pertinent data is as follows: $f'_c = 3.0$ ksi, $f_y = 60.0$ ksi, $b = t = 20''$. Total steel consists of 12 No. 9 bars, or $A_{st} = 12.0$ in^2; thus $p_t = 0.03$. Also, $g = 0.8$, $k_1 k_u t = 8.925''$ and the neutral axis lies $5.537''$ below the top of the section.

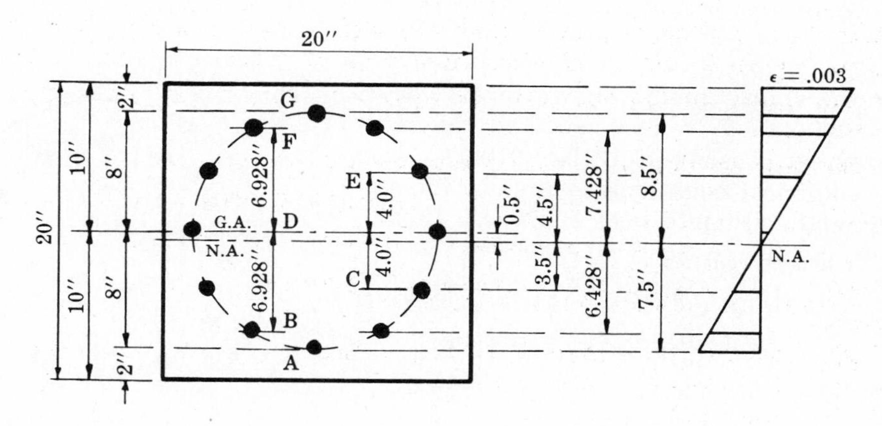

Fig. 10-6

Using strain compatibility and statics, the steel forces may be computed and tabulated as follows:

STEEL STRAINS, FORCES, AND MOMENTS

Row	A_s sq in.	$\epsilon_s \times 10^2$ in. per in.	f_s, ksi	$f_s - 0.85f'_c$, ksi	Force, kips	Y, in.	Moment, in-kips
G	1.0	0.243	60.00	57.45	57.45	8.000	460.0
F	2.0	0.212	60.0	57.45	114.90	6.925	796.0
E	2.0	0.129	37.40	34.85	69.70	4.000	279.0
D	2.0	0.014	4.15	—	8.30	0.000	0.0
C	2.0	−0.100	−29.00	—	−58.00	−4.000	232.0
B	2.0	−0.183	−53.00	—	−106.00	−6.925	734.0
A	1.0	−0.214	−60.00	—	−60.00	−8.000	480.0

$$\Sigma F_s = 26.35 \text{ kips} \qquad \Sigma M_s = 2981.0 \text{ in-kips}$$

The compressive force in the concrete and its moment about the gravity axis will be

$$C_c = 0.85 \times 3.0 \times 8.925 \times 20.0 = 455 \text{ kips} \quad \text{and} \quad M_c = 455 \times 5.537 = 2520 \text{ in-kips}$$

Hence

$$P_u = \phi(C_c + F_s) = 0.75(455.00 + 26.35) = 360 \text{ kips}$$

$$M_u = \phi(M_c + M_s) = 0.75(2520 + 2981) = 4125 \text{ in-kips}$$

Thus

$$e = M_u/P_u = 11.55 \text{ in.} \quad \text{and} \quad e/t = 11.55/20 = 0.5775$$

10.7. Use the results of Problem 10.6 and the ACI Code empirical equations to obtain the approximate ultimate load for the column (Fig. 10-6).

Refer to Problem 10.6 for the appropriate values and substitute into the empirical equations.

Check ACI Code equation for compression failure:

$$P_u \;=\; 0.75\left\{\frac{(12)(60)}{\dfrac{(3)(11.55)}{16}+1} + \frac{(400)(3)}{\dfrac{(12)(20)(11.55)}{[20+(0.67)(16)]^2}+1.18}\right\} \;=\; 389 \text{ kips}$$

Check ACI Code equation for tension failure:

$$p_t = 0.03,\quad m = 60.0/[(0.85)(3)] = 23.55,\quad p_t m = 0.705$$

$$0.85f'_c\,\phi t^2 \;=\; 0.85\times 3.0\times 0.75\times 400 \;=\; 765$$

$$P_u = 765\{\sqrt{[11.55/20-0.5]^2+[(0.67)(16)/(20)](0.705)} - [11.55/20-0.5]\} = 415 \text{ kips}$$

Compression governs, and $P_u = 389$ kips.

10.8. For the spiral column shown in Fig. 10-7, $f'_c = 3.0$ ksi, $f_y = 60.0$ ksi, and the reinforcement consists of 12 No. 9 bars. Use the empirical equations to predict the approximate ultimate load P_u if $e = 13.85''$.

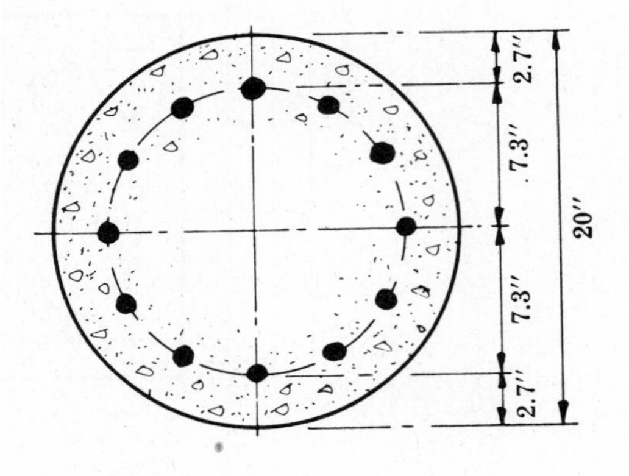

Fig. 10-7

If tension controls,

$$P_u = \phi\{0.85f'_c D^2[\sqrt{(0.85e/D-0.38)^2+p_t m D_s/2.5D} - (0.85e/D-0.38)]\} = 263 \text{ kips}$$

where $A_{st} = 12.0\text{ in}^2,\; A_g = \pi(20)^2/4 = 314.2\text{ in}^2$

$p_t = A_{st}/A_g = 0.0382$

$m = f_y/0.85f'_c = 60.0/2.55 = 23.53$

$D_s = gD = 20.0-(2)(2.7) = 14.6'',\; \phi = 0.75$

If compression controls,

$$P_u \;=\; \phi\left[\frac{A_{st}f_y}{3e/D_s+1} + \frac{A_g f'_c}{9.6De/(0.8D+0.67D_s)^2+1.18}\right] \;=\; 262 \text{ kips}$$

Note that both equations provide almost identical results. A balanced failure condition exists according to the empirical equations.

A complete check of statics and strain compatibility indicates that $k_u D = 10.0''$ and $P_u = 234$ kips. In this case, the empirical equations underestimate the strength of the section by 12.0%.

The proof of the foregoing statements is left as an exercise for the reader.

10.9. Fig. 10-8 illustrates a typical interaction design diagram for use in ultimate strength design of columns. Use the chart to design a column to withstand an ultimate moment of 2880 in-kips and an axial load of 720 kips. The chart is plotted for $f'_c = 3.0$ ksi, $f_y = 60.0$ ksi, $g = 0.8$.

Assume that a 20″ diameter column will be used. Calculate

$$e = M_u/P_u = 2880/720 = 4.0'',\quad e/D = 4.0/20.0 = 0.2,\quad K = P_u/f'_c D^2 = 720/[(3.0)(20)^2] = 0.60$$

Enter the chart with $K = 0.6$ and $e/D = 0.2$, and at the intersection of the two lines read $p_t = 0.05$. Then

$$A_{st} \;=\; p_t A_g \;=\; 0.05(\pi)(20)^2/4 \;=\; 15.71\text{ in}^2$$

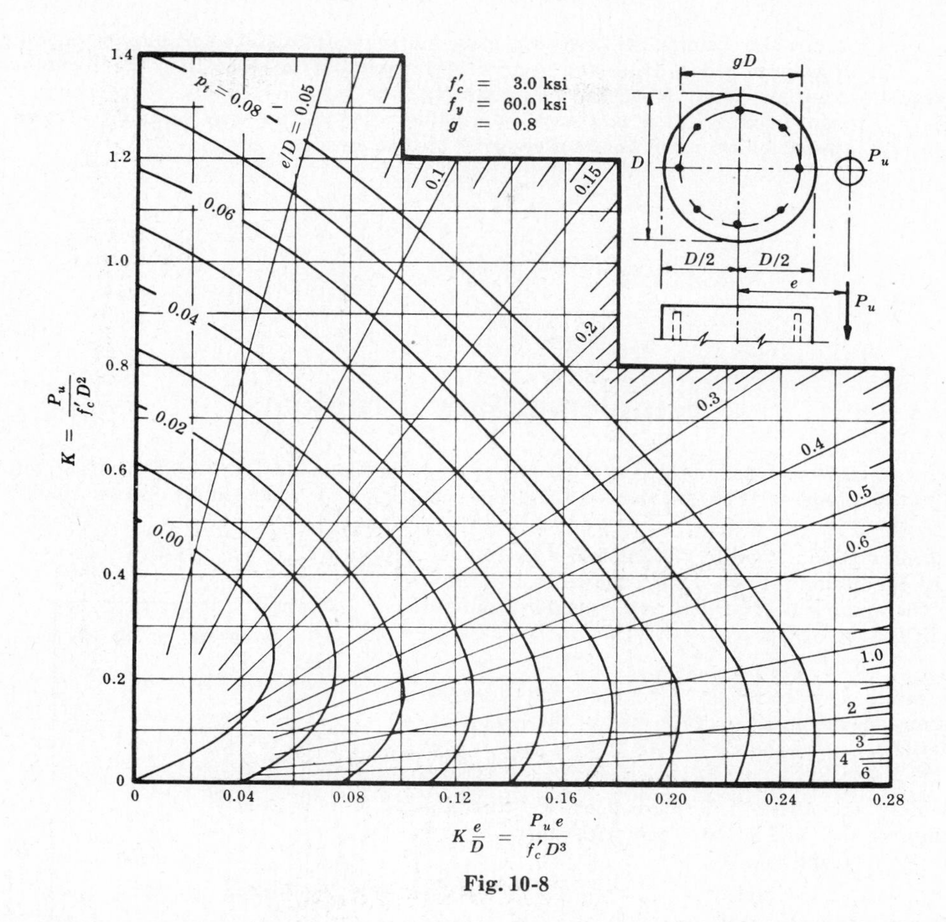

Fig. 10-8

Any size bars (greater than No. 5) may be used to provide the required steel area, but the clearances between bars must be maintained in accord with the detailed requirements of the Code. If the required amount of steel cannot be provided because of clearance requirements, it will be necessary to increase the size of the column and redesign the steel.

10.10. Use the section designed in Problem 10.9 to check the validity of the reciprocal equation (*10.19*) for biaxial bending. Consider bending about a line 45 degrees to the x axis, with $e = 5.656''$.

Since the design charts already contain ϕ, the primes are dropped and the equation becomes

$$1/P_u = 1/P_x + 1/P_y - 1/P_o$$

We may multiply all terms by $f'_c D^2$ to obtain a dimensionless form

$$1/K = 1/K_x + 1/K_y - 1/K_o$$

Since $e = 5.656''$ along a line 45° to the x axis, $e_x = e_y = 5.656/1.414 = 4.0''$ as in Problem 10.9.

The section is circular, so $K_x = K_y = 0.6$ as determined in Problem 10.9.

The term K_o refers to pure axial load and may be obtained as 1.065 at the intersection of the vertical axis $(Ke/D = 0)$ with $p_t = 0.05$. Thus

$$1/K = 1/0.6 + 1/0.6 - 1/1.065 \qquad \text{or} \qquad K = 0.418$$

Now, using $e/D = 5.656/20 = 0.2828$ and $p_t = 0.05$, return to the chart to find $K = 0.49$. This value is the correct solution, and indicates that the reciprocal formula provides a solution that is incorrect by 14.3% on the conservative side.

Since the circular column will provide more satisfactory results for biaxial bending than will a rectangular column when using the reciprocal formula, it is clear that the formula should not be applied to either type of section. Fortunately, one need not apply the formula for circular sections, since a direct solution is always obtainable. For rectangular columns, always use statics and strain compatibility when confronted with biaxial bending.

Supplementary Problems

Unless otherwise specified, use $f'_c = 3.0$ ksi, $f_y = 60.0$ ksi and $d' = 2.5''$ in all supplementary problems for this chapter. For Problems 10.11-10.19, use the ACI Code empirical equations.

10.11. Determine the location of the neutral axis for balanced conditions for the section shown in Fig. 10-9. Use the empirical equations and consider balanced conditions. *Ans.* $k_b d = 12.73''$

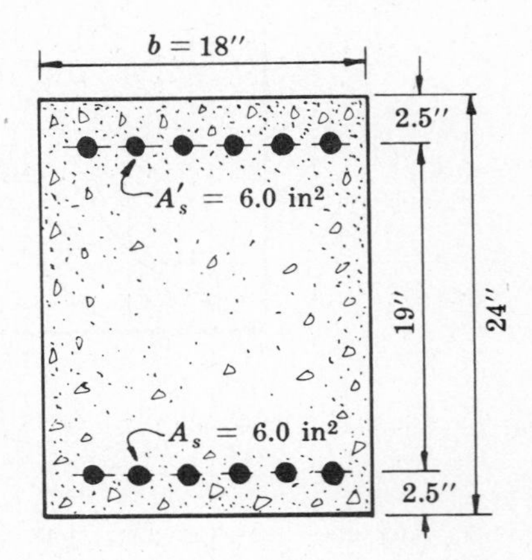

Fig. 10-9

10.12. Determine the balanced ultimate load for the column shown in Fig. 10-9 using the empirical equations. *Ans.* $P_b = 348$ kips

10.13. Calculate the theoretical axial load P_o that the column shown in Fig. 10-9 can withstand.
Ans. $P_o = 1252$ kips

10.14. Calculate the balanced eccentricity for the column shown in Fig. 10-9.
Ans. $e_b = 25.8''$ using M_b/P_b

10.15. Calculate P_u for the section shown in Fig. 10-9 if $e = 6.0''$. *Ans.* $P_u = 738$ kips

10.16. Find the ultimate load for the column shown in Fig. 10-10 if $e = 18.0''$. *Ans.* $P_u = 139.7$ kips

10.17. Find the ultimate load for the column shown in Fig. 10-10 if $e = 6.0''$. *Ans.* $P_u = 405.5$ kips

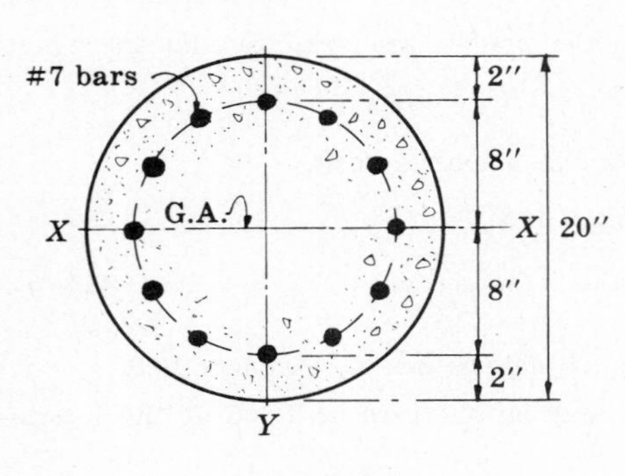

Fig. 10-10

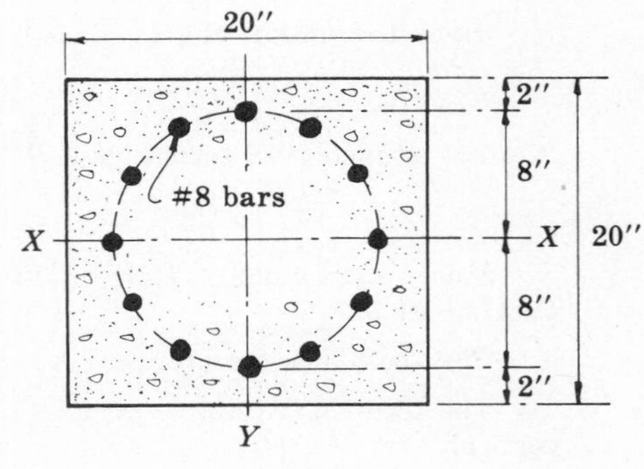

Fig. 10-11

10.18. Calculate the ultimate load for the column shown in Fig. 10-11 if $e = 20.0''$.
Ans. $P_u = 183.5$ kips

10.19. Calculate the ultimate load for the column shown in Fig. 10-11 if $e = 5.0''$.
Ans. $P_u = 587.0$ kips

10.20. Use the general method of statics and strain compatibility to determine P_u and M_u for the section shown in Fig. 10-12. The total steel area is 8.5 in^2 and $k_u t = 14.0''$.
Ans. $P_u = 537$ kips, $M_u = 3085$ in-kips

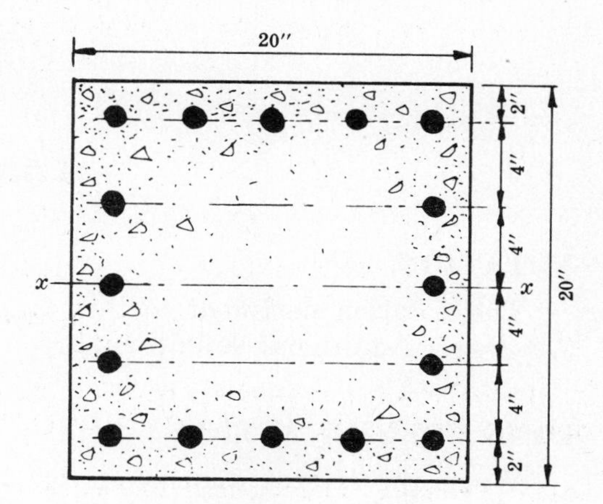

Fig. 10-12

10.21. Solve Problem 10.20 for $k_u t = 8.0''$. Use $A_{st} = 6.8$ in^2.
Ans. $P_u = 202$ kips, $M_u = 3189$ in-kips

10.22. Solve Problem 10.20 for $k_u t = 8.0''$ and $A_{st} = 8.5$ in^2.
Ans. $P_u = 191$ kips, $M_u = 3583$ in-kips

10.23. Solve Problem 10.20 for $A_{st} = 5.1$ in^2 and $k_u t = 16.0''$. *Ans.* $P_u = 578$ kips, $M_u = 2274$ in-kips

10.24. Calculate the balanced eccentricity for the column of Fig. 10-10 using equation (*10.21*).
Ans. $e_b = 9.0''$

10.25. Calculate the balanced eccentricity for the column of Fig. 10-11 using equation (*10.21*).
Ans. $e_b = 10.3''$

10.26. Calculate the balanced eccentricity for the column of Fig. 10-9 using equation (*10.20*).
Ans. $e_b = 16.85''$

10.27. Using Fig. 10-11 with No. 9 bars, $k_u t = 16.0''$, $f_c' = 3.0$ ksi and $f_y = 60.0$ ksi, calculate the true ultimate load and ultimate moment using statics and strain compatibility.
Ans. $P_u = 771.0$ kips, $M_u = 2940.0$ in-kips

10.28. Considering that the true eccentricity of P_u from Problem 10.27 is 3.82″, use the appropriate ACI Code empirical equation to obtain P_u. *Ans.* $P_u = 732.0$ kips (compression controls)

Chapter 11

Long Columns

NOTATION

The notation pertinent to this chapter is listed in Chapter 8, *General Provisions for Columns*. Additional definitions are provided as required with the related material.

LONG COLUMN ACTION

When the lateral dimensions of a compression member are small compared to the length, buckling may cause failure before the materials yield or fail. For individual members composed of perfectly elastic materials it is possible to obtain a theoretical ultimate load which can be verified in the testing laboratory.

Reinforced concrete members, being continuous and fundamentally inelastic, do not behave in the ideal manner predicted by the mathematical solution. Nevertheless, the *Euler equations* may be used for predicting the behavior of reinforced concrete compression members. The equations are of the form

$$P_{cr} = CE_r/(L/r)^2 \qquad (11.1)$$

where P_{cr} = the buckling failure load, kips

C = a constant, depending on the end conditions and loading conditions

E_r = a reduced modulus of elasticity of concrete, ksi (usually $E_c/3$)

L = unsupported length of the member, inches

r = least radius of gyration of the member, inches.

It is difficult to determine accurately the applicable constant C, and a reduced modulus E_r can only be estimated. The ACI Code requires that E_r may not exceed one-third the value of the modulus of elasticity for concrete. For E_c in psi (E_r is in ksi),

$$E_c = 33\,W^{1.5}\sqrt{f'_c} \qquad (11.2)$$

Because of the difficulty in accurately predicting buckling loads using the Euler equations, semi-empirical equations have been developed for use in slender column design. The *radius of gyration* is important in those equations, and can be calculated for any member as

$$r = \sqrt{I/A} \qquad (11.3)$$

where r = radius of gyration, inches

I = moment of inertia *about the axis of bending*, in^4

A = area of the cross-section.

The radius of gyration can be closely approximated for rectangular sections as $(0.3)t$, and for round sections as $(0.25)D$.

Inasmuch as the area A, moment of inertia I, and radius of gyration r are very important in long column design, it is useful to have a table which provides equations for these factors for commonly used sections. Table 11.1 provides such a design aid. The table is reproduced with the permission of the American Concrete Institute.

TABLE 11.1
PROPERTIES OF SECTIONS

Dash-and-dot lines are drawn through centers of gravity
A = area of section I = moment of inertia R = radius of gyration

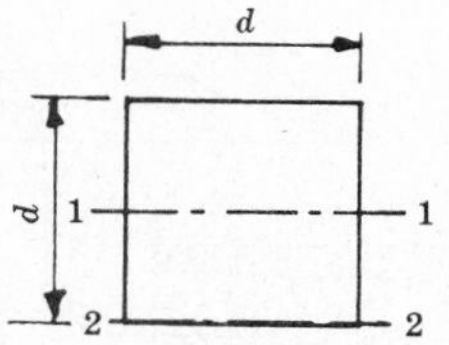 $A = d^2$ $I_1 = \dfrac{d^4}{12}$ $I_2 = \dfrac{d^4}{3}$ $R_1 = 0.2887d$ $R_2 = 0.5774d$	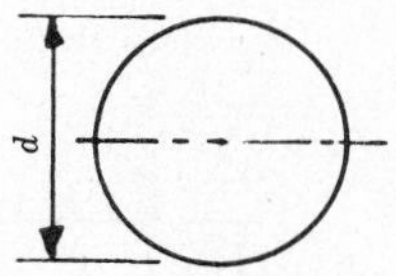 $A = \dfrac{\pi d^2}{4} = 0.7854d^2$ $I = \dfrac{\pi d^4}{64} = 0.0491d^4$ $R = \dfrac{d}{4}$
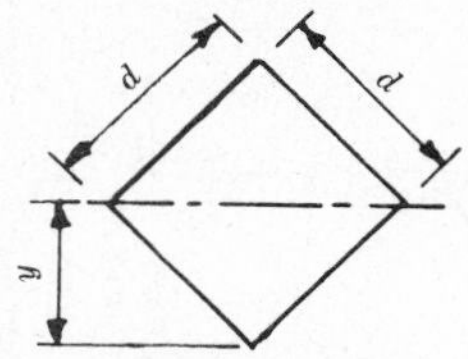 $A = d^2$ $y = 0.7071d$ $I = \dfrac{d^4}{12}$ $R = 0.2887d$	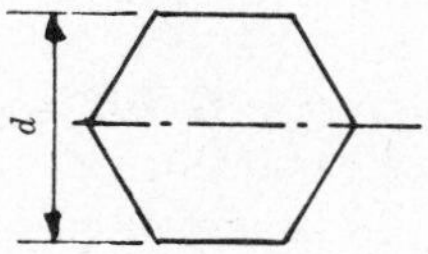 $A = 0.8660d^2$ $I = 0.060d^4$ $R = 0.264d$
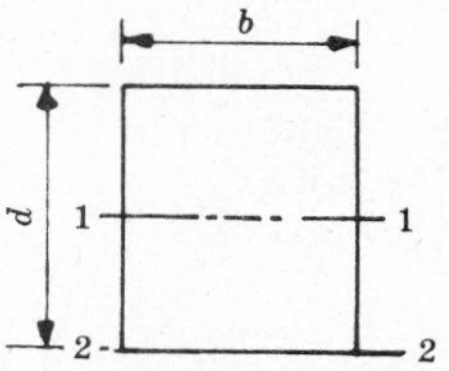 $A = bd$ $I_1 = \dfrac{bd^3}{12}$ $I_2 = \dfrac{bd^3}{3}$ $R_1 = 0.2887d$ $R_2 = 0.5774d$	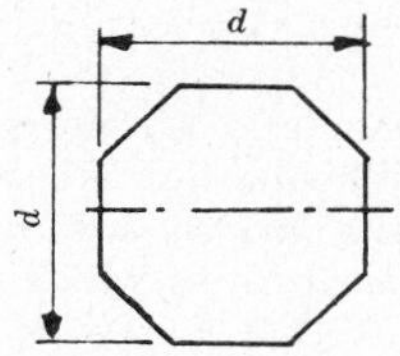 $A = 0.8284d^2$ $I = 0.055d^4$ $R = 0.257d$
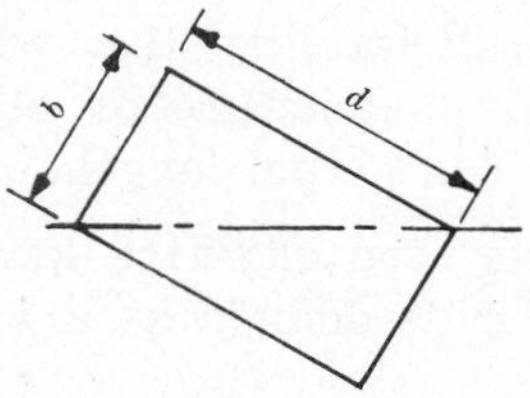 $A = bd$ $y = \dfrac{bd}{\sqrt{b^2+d^2}}$ $I = \dfrac{b^3d^3}{6(b^2+d^2)}$ $R = \dfrac{bd}{\sqrt{6(b^2+d^2)}}$	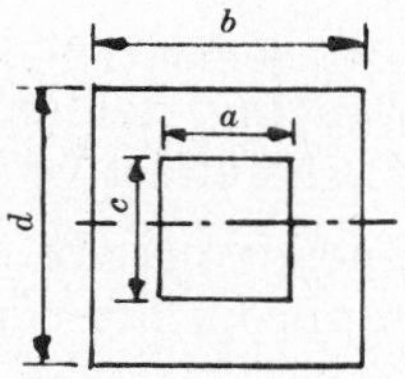 $A = bd - ac$ $I = \dfrac{bd^3 - ac^3}{12}$ $R = \sqrt{\dfrac{bd^3 - ac^3}{12(bd - ac)}}$
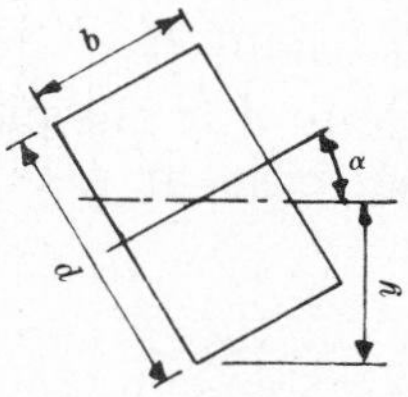 $A = bd$ $y = \dfrac{b \sin\alpha + d \cos\alpha}{2}$ $I = \dfrac{bd(b^2 \sin^2\alpha + d^2 \cos^2\alpha)}{12}$ $R = \sqrt{\dfrac{b^2 \sin^2\alpha + d^2 \cos^2\alpha}{12}}$	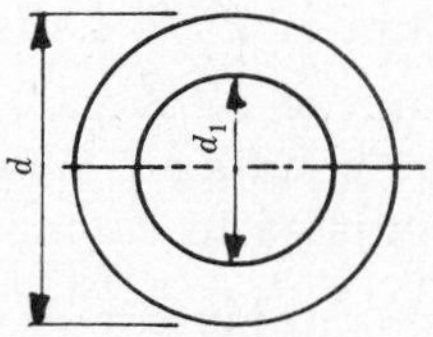 $A = \pi(d^2 - d_1^2)/4$ $= 0.7854(d^2 - d_1^2)$ $I = \pi(d^4 - d_1^4)/64$ $= 0.0491(d^4 - d_1^4)$ $R = \frac{1}{4}\sqrt{d^2 + d_1^2}$
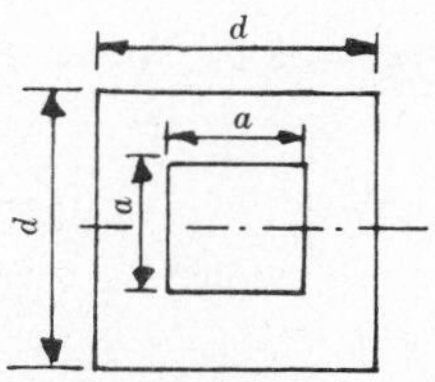 $A = d^2 - a^2$ $I = \dfrac{d^4 - a^4}{12}$ $R = \sqrt{\dfrac{d^2 + a^2}{12}}$	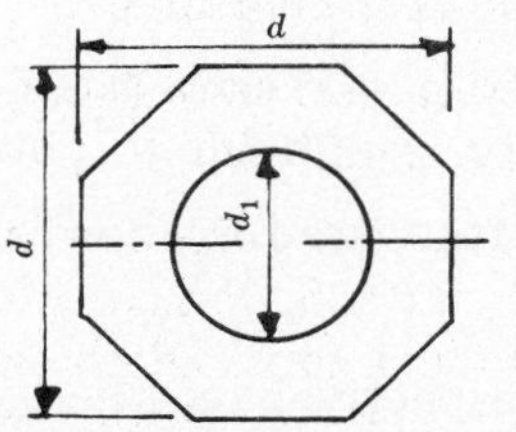 $A = 0.8284d^2 - 0.7854d_1^2$ $= 0.7854(1.055d^2 - d_1^2)$ $I = 0.055d^4 - 0.0491d_1^4$ $= 0.0491(1.12d^4 - d_1^4)$ $R = 0.257d - 0.25d_1$ $= 0.25(1.028d - d_1)$

HEIGHT OF COLUMNS

As the height of a column increases, the allowable load decreases because of the tendency of long columns to *buckle*. The height of a column is therefore an important factor in design. Fig. 11-1 indicates the heights that must be used in design for different situations.

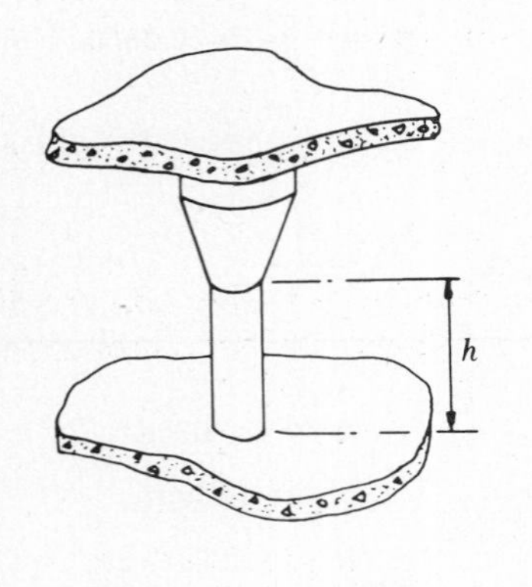

(*a*) Flat slab construction with column capital

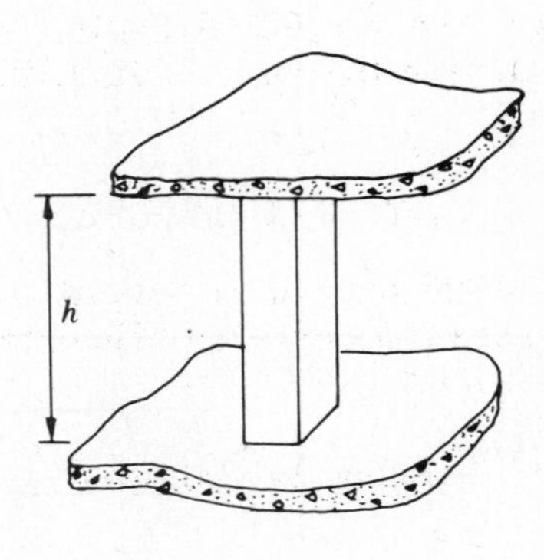

(*b*) Flat slab (or plate) construction without column capital

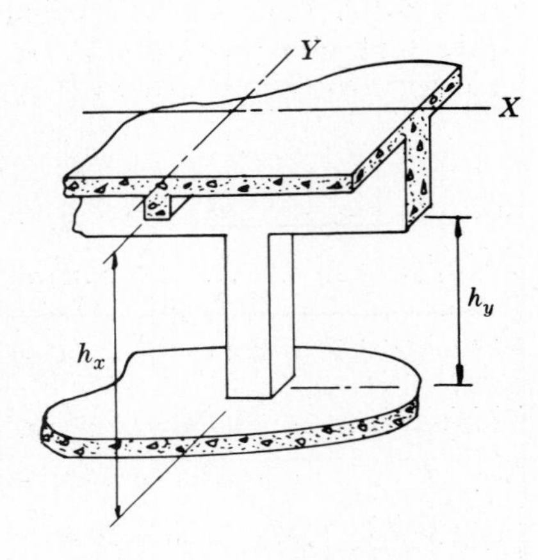

(*c*) Beam and girder construction

Note: In beam and girder construction, use the height in the plane of bending. When design is made for *minimum eccentricity*, use h which will provide the *largest* h/r *ratio*. (Minimum eccentricity and h/r are discussed in paragraphs which follow.)

Fig. 11-1

EFFECTIVE LENGTH

When the structure *does not* depend on the columns for lateral stability (i.e. when either shear walls, rigid bracing, fastening to a very stiff adjoining structure or other lateral support is provided) the effective length h' shall be equal to the actual length.

When the structure depends on the columns for lateral stability, the effective length h' shall be obtained using appropriate conditions which follow. If r'_J is defined as a *joint factor* for joint J, then

$$r'_J = \frac{\Sigma K \text{ columns}}{\Sigma K \text{ beams}} \tag{11.4}$$

where ΣK columns = sum of EI/h values for all columns which meet at joint J.

ΣK beams = sum of EI/L values for all beams which meet at joint J in the plane of bending. When E is constant as is the usual case, it may be omitted in calculating K.

The following conditions apply for r' values:

(1) Although r' is infinite for a *pinned end*, a column end is assumed pinned if r' exceeds 25.0.

(2) Although $r' = 0$ for a *perfectly fixed end*, it is practical to use $r' = 1.0$ for a fixed end. Complete fixity is difficult to obtain in practice.

The *average* r' is often used. This is designated as

$$r'_A = 0.5(r'_T + r'_B) \tag{11.5}$$

where T and B refer respectively to the top and bottom of the column.

CAPACITY REDUCTION FACTORS

For slender columns, the load capacity of the member is reduced by multiplying the *short column* capacity of the section by a reduction factor R. The factor R varies for different end conditions and is a function of r' when lateral bracing is not provided.

For design purposes it is convenient to *increase* the ideal short column requirements by **dividing** both the axial load and moment by R. The column is then designed as an equivalent short column to sustain the **amplified** loads. If the *design* axial load and moment are designated as P' and M' for the long column, then

$$P' = P/R \tag{11.6}$$

and

$$M' = M/R \tag{11.7}$$

where P and M are the calculated axial load and moment to be resisted by the section if it were a short column. The values of R must be calculated using the appropriate case of those which follow.

Compression Controls and Lateral Bracing Is Provided

Case 1. No sidesway, double curvature.

For $60 \leqq h/r \leqq 100$, use

$$R = [1.32 - (0.006)h/r] \leqq 1.0 \tag{11.8}$$

For $h/r < 60$, use $R = 1.0$.

For $h/r > 100$, use a comprehensive Euler analysis or any procedure using deflection calculations.

The column may also be fixed at one end ($r' = 1$) and partially fixed or restrained at the other end ($r' \leqq 25$).

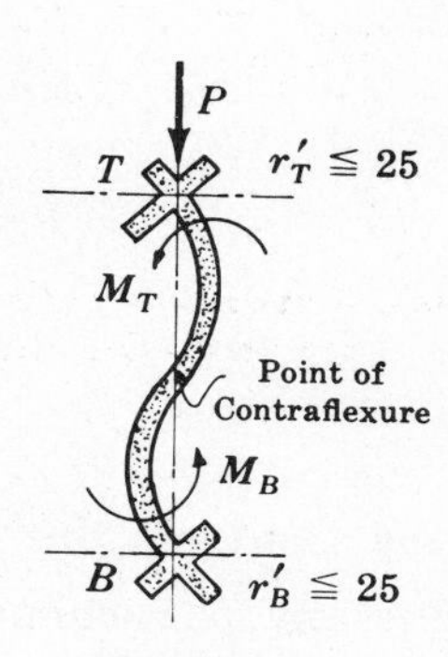

Fig. 11-2

Case 2. No sidesway, single curvature.

$$R = [1.07 - (0.008)h/r] \leqq 1.0 \tag{11.9}$$

The column ends may be partially fixed or restrained ($r' \leqq 25$) or pinned ($r' > 25$).

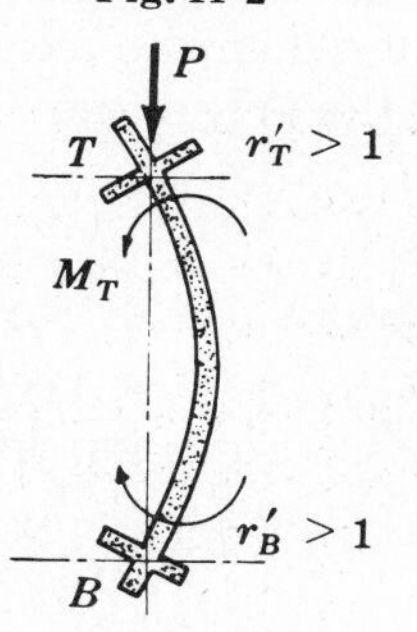

Fig. 11-3

Compression Controls and Structure Depends on Columns for Lateral Stability

(*When short term loads as wind, earthquake or blast cause critical condition, add 10 percent to R.*)

Case 3. Sidesway occurs, double curvature.

$$h' = h(0.78 + 0.22r'_A) \geqq h \tag{11.10}$$

$$R = [1.07 - (0.008)h'/r] \leqq 1.0 \tag{11.11}$$

where $r'_A = \frac{1}{2}(r'_T + r'_B)$.

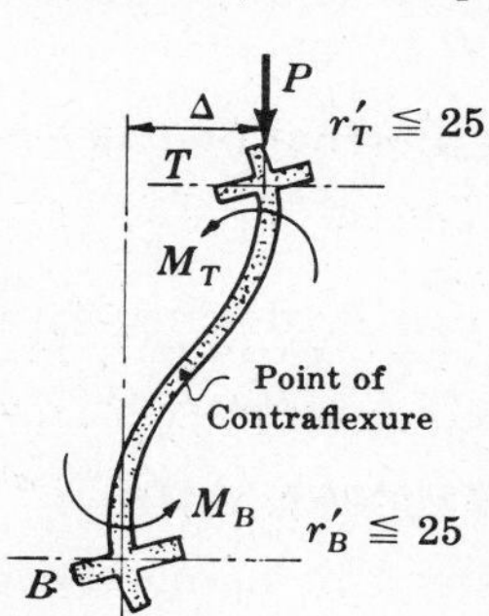

Fig. 11-4

Case 4. Sidesway occurs, single curvature.

$$h' = 2h(0.78 + 0.22r'_T) \geqq 2h \qquad (11.12)$$

$$R = [1.07 - (0.008)(h'/r)] \leqq 1.0 \qquad (11.13)$$

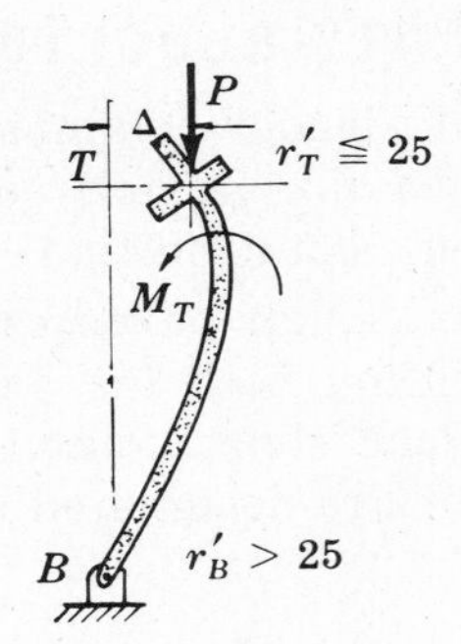

Fig. 11-5

Case 5. Sidesway occurs, cantilever column.

$$h' = 2h \qquad (11.14)$$

$$R = [1.07 - (0.008)(h'/r)] \leqq 1.0 \qquad (11.15)$$

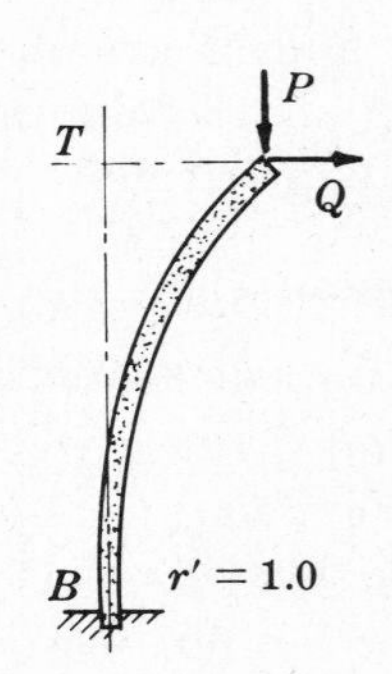

Fig. 11-6

Tension Controls the Design

For any particular condition shown in Cases 1 through 5, obtain R as though compression controls the design. Designate this value as balanced R, or R_b. Assuming that $R = 1.0$ when the axial load is zero, the value of R for the column under control of tension varies linearly from R_b at balanced load P_b, to R_c for the column at P_c, or

$$R_c = R_b + (1 - R_b)(1 - P_c/P_b) \geqq R_b \qquad (11.16)$$

A more useful form of this equation may be provided by defining $K_c = P_c/f'_c bt$ and $K_b = P_b/f'_c bt$ for rectangular columns and $K_c = P_c/f'_c D^2$ and $K_b = P_B/f'_c D^2$ for round columns. Equation 11.16 then becomes

$$R_c = R_b + (1 - R_b)(1 - K_c/K_b) \geqq R_b \qquad (11.17)$$

Design aids are available for obtaining K_c and K_b.

The solution of equation (*11.17*) involves successive trials since the design load P_c is a function of R_c, or

$$P_c = P/R_c \qquad (11.18)$$

where P is the *short column* axial load.

An alternate equation for (*11.16*) and (*11.17*) may be stated as

$$R_c = 1 - (1 - R_b)(e_b/e) \geqq R_b \qquad (11.19)$$

which can be derived from equation (*11.16*).

ADDITIONAL AIDS

Instead of using the value of the capacity reduction factor R for long column design, it is permissible (and necessary when h/r exceeds 100) to increase the design moment by the term Py, where P is the axial load and y is the deflection at the critical section.

When a *beam-column* is subjected to axial compression, moments and lateral loads, and is *bent in single curvature*, the total deflection will be

$$y = y_o/(1 - P/C'P_{cr}) \qquad (11.20)$$

where y_o = the deflection due to moments and lateral load

P = the applied axial load

P_{cr} = the Euler buckling load given by equation (*11.1*)

C' = a constant depending on the end conditions and the relative end moments.

Values of C' may be obtained from texts concerned with the *Theory of Elastic Stability*. Values of y_o may be obtained from standard texts on *Strength of Materials*.

COLUMNS GOVERNED BY MINIMUM ECCENTRICITY

If the computed eccentricities at *both ends* of the column are less than $(0.05)t$ or $(0.05)D$ for spiral columns or $(0.1)t$ for tied columns, use the appropriate condition of Cases 1 through 5 with regard to end restraint.

If the computations indicate that there is no eccentricity at one or both ends of the column, use the equation

$$R = 1.07 - (0.008)h/r \leqq 1.0 \qquad (11.21)$$

CONTROLLING CONDITIONS

When the eccentricity e is equal to or less than the balanced eccentricity e_b, compression controls the design. When e is greater than e_b, tension controls the design.

The ACI Code provides empirical expressions for e_b for *working stress design*, and empirical equations have been devised for e_b for *ultimate strength design*. Although very approximate, those empirical equations are as sound as the empirical equations for the capacity reduction factors.

BALANCED ECCENTRICITY, WORKING STRESS DESIGN

The ACI Code specifies empirical equations for balanced eccentricity about a principal axis for three types of columns:

For symmetrical spiral columns $$e_{bx} = 0.43p_g mD_s + 0.14t \qquad (11.22)$$

For symmetrical tied columns $$e_{bx} = (0.67p_g m + 0.17)d \qquad (11.23)$$

For unsymmetrical tied columns $$e_{bx} = \frac{p'm(d-d') + 0.1d}{(p'-p)m + 0.6} \qquad (11.24)$$

Expressions for e_{by} are obtained by substituting b for t, and $b - d'$ for d. Note that d' may differ in the x and y directions.

BALANCED ECCENTRICITY, ULTIMATE STRENGTH DESIGN

Although the ACI Code does not specify equations for balanced eccentricity for ultimate strength design, approximate expressions were derived by J. G. MacGregor for this condition, and may be used. The equations are not contained in the 1963 ACI Code, but their use has been sanctioned in the Commentary to the 1963 ACI Code, published September 1965.

For *symmetrically reinforced rectangular sections* with ties or spirals,

$$e_b = (0.20 + 0.77p_t m)t \tag{11.25}$$

and for *round and square spirally reinforced sections,*

$$e_b = (0.24 + 0.39p_t m)t \tag{11.26}$$

where

$$p_t = (A'_s + A_s)/A_g = A_{st}/A_g \tag{11.27}$$

MORE ACCURATE VALUES OF e_b

The equations stated for e_b for use in ultimate strength design are very approximate. More accurate tabulations have been obtained by electronic computer analysis and are presented in ACI-SP-7. Table 11.2 is a sample table. It provides values of $K_b = P_b/f'_c bt$, e_b/t, and $K_m = P_m/f'_c bt$ where P_m is the axial load corresponding to minimum eccentricity.

Fig. 11-7 shows a typical design curve taken from ACI-SP-7 which contains curves for short columns and also values of e_b/t for use in long column design when tension controls.

TABLE 11.2

SQUARE COLUMNS WITH SPIRALS; REINFORCEMENT ARRANGED IN A CIRCLE

f'_c ksi	f_y ksi	$p_t m$	$g = 0.6$			$g = 0.7$			$g = 0.8$			$g = 0.9$			K_o
			K_b	e_b/t	K_m*	K_b	e_b/t	K_m*	K_b	e_b/t	K_m*	K_b	e_b/t	K_m*	
		0.0	0.2191	0.3281	0.5737	0.2328	0.3174	0.5737	0.2464	0.3066	0.5737	0.2601	0.2959	0.5737	0.6375
		0.1	0.2058	0.3855	0.6139	0.2244	0.3718	0.6163	0.2424	0.3604	0.6183	0.2599	0.3504	0.6200	0.6976
		0.2	0.1925	0.4508	0.6540	0.2160	0.4305	0.6588	0.2384	0.4159	0.6628	0.2596	0.4050	0.6665	0.7577
		0.3	0.1792	0.5258	0.6942	0.2076	0.4939	0.7015	0.2344	0.4734	0.7077	0.2593	0.4598	0.7128	0.8179
		0.4	0.1659	0.6128	0.7345	0.1991	0.5627	0.7447	0.2303	0.5329	0.7525	0.2590	0.5146	0.7596	0.8780
5.0	75.0	0.5	0.1526	0.7149	0.7751	0.1907	0.6375	0.7876	0.2263	0.5945	0.7976	0.2587	0.5696	0.8063	0.9381
		0.6	0.1393	0.8366	0.8159	0.1823	0.7193	0.8308	0.2223	0.6583	0.8425	0.2584	0.6247	0.8530	0.9983
		0.7	0.1260	0.9838	0.8566	0.1739	0.8089	0.8744	0.2183	0.7245	0.8879	0.2582	0.6800	0.8996	1.0584
		0.8	0.1128	1.1658	0.8974	0.1655	0.9076	0.9181	0.2142	0.7932	0.9333	0.2579	0.7353	0.9467	1.1185
		0.9	0.0995	1.3964	0.9385	0.1571	1.0169	0.9616	0.2102	0.8645	0.9788	0.2576	0.7908	0.9967	1.1787
		1.0	0.0862	1.6980	0.9797	0.1487	1.1385	1.0055	0.2062	0.9386	1.0244	0.2573	0.8464	1.0473	1.2388
		0.0	0.2054	0.3388	0.5737	0.2182	0.3288	0.5737	0.2310	0.3187	0.5737	0.2439	0.3086	0.5737	0.6375
		0.1	0.1920	0.4011	0.6145	0.2096	0.3875	0.6167	0.2268	0.3763	0.6185	0.2434	0.3668	0.6203	0.6969
		0.2	0.1786	0.4727	0.6553	0.2011	0.4512	0.6596	0.2226	0.4361	0.6632	0.2429	0.4252	0.6666	0.7563
		0.3	0.1652	0.5559	0.6962	0.1925	0.5206	0.7029	0.2183	0.4983	0.7084	0.2423	0.4838	0.7132	0.8157
		0.4	0.1517	0.6539	0.7373	0.1839	0.5964	0.7462	0.2141	0.5628	0.7533	0.2418	0.5427	0.7599	0.8751
6.0	75.0	0.5	0.1383	0.7708	0.7786	0.1753	0.6797	0.7899	0.2099	0.6300	0.7986	0.2413	0.6018	0.8065	0.9345
		0.6	0.1249	0.9127	0.8197	0.1668	0.7715	0.8335	0.2056	0.7000	0.8440	0.2408	0.6612	0.8535	0.9939
		0.7	0.1115	1.0888	0.8610	0.1582	0.8733	0.8771	0.2014	0.7729	0.8895	0.2403	0.7208	0.9004	1.0534
		0.8	0.0981	1.3130	0.9025	0.1496	0.9867	0.9210	0.1972	0.8489	0.9350	0.2398	0.7808	0.9474	1.1128
		0.9	0.0847	1.6080	0.9440	0.1411	1.1140	0.9648	0.1929	0.9282	0.9806	0.2392	0.8409	0.9942	1.1722
		1.0	0.0713	2.0139	0.9857	0.1325	1.2376	1.0089	0.1887	1.0112	1.0265	0.2387	0.9013	1.0413	1.2316

*$e/t = 0.05$

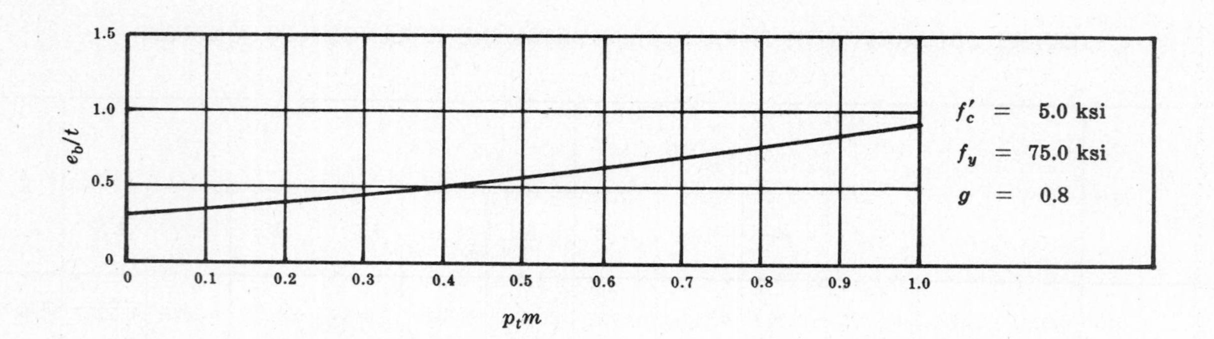

e_b/t
$p_t m$
f'_c = 5.0 ksi
f_y = 75.0 ksi
g = 0.8

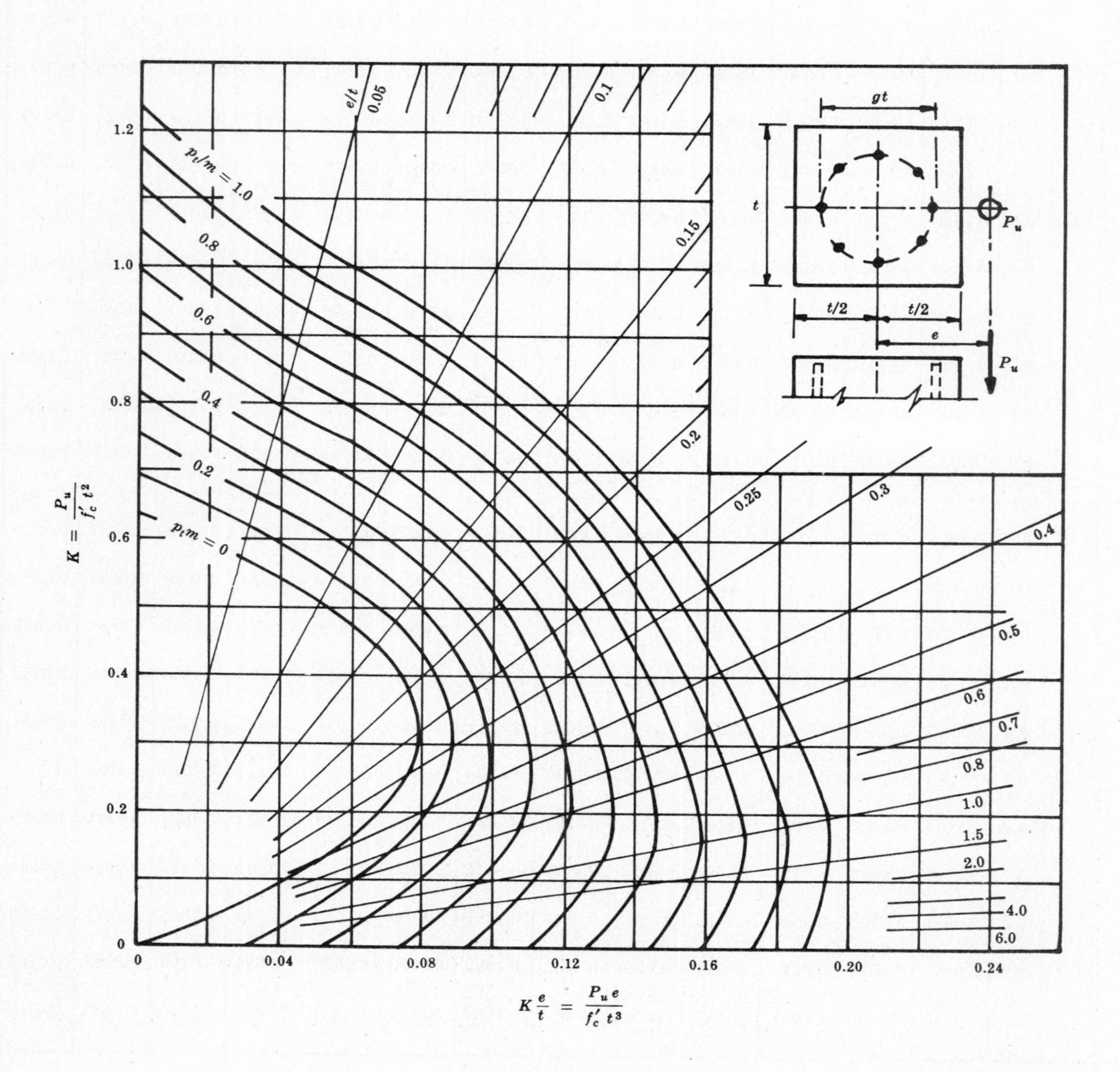

$K = \frac{P_u}{f'_c t^2}$
$K\frac{e}{t} = \frac{P_u e}{f'_c t^3}$
e/t
$p_t/m = 1.0$
$p_t m = 0$
gt
t
t/2
e
P_u

Fig. 11-7

Solved Problems

11.1. Using equation (*11.2*), provide a list of values of E_c for use with ordinary weight concrete for which $W = 145$ lb/ft³. Use values of f'_c equal to 3, 4, 5 and 6 ksi.

The applicable equation is $E_c = 33\,W^{1.5}\sqrt{f'_c}$.

By direct substitution and evaluation, obtain for

$$f'_c = 3{,}000 \text{ psi} \qquad \sqrt{f'_c} = 54.77 \text{ psi} \qquad E_c = 3.15 \times 10^6 \text{ psi}$$

$$f'_c = 4{,}000 \text{ psi} \qquad \sqrt{f'_c} = 63.25 \text{ psi} \qquad E_c = 3.64 \times 10^6 \text{ psi}$$

$$f'_c = 5{,}000 \text{ psi} \qquad \sqrt{f'_c} = 70.71 \text{ psi} \qquad E_c = 4.07 \times 10^6 \text{ psi}$$

$$f'_c = 6{,}000 \text{ psi} \qquad \sqrt{f'_c} = 77.46 \text{ psi} \qquad E_c = 4.45 \times 10^6 \text{ psi}$$

11.2. Calculate the radius of gyration about the x axis for (*a*) a 20″ square column and (*b*) a 20″ round column. Use the ACI Code approximate values.

$$(a)\quad r = 0.3t = 0.3(20) = 6.0'' \qquad (b)\quad r = 0.25D = 0.25(20) = 5.0''$$

11.3. A 20″ square column has an unsupported height of 20′. Calculate the stiffness factor K for the column.

$K = I/12h = (13{,}333)/[(12)(20)] = 55.5$ in³ where $I = bt^3/12 = (20)(20)^3/12 = 13{,}333$ in⁴.

11.4. A 20″ diameter section has an unsupported height of 20″. Calculate its stiffness factor K.

$K = I/12h = (0.0491)(20)^4/[(12)(20)] = 32.74$ in³ where $I = 0.0491D^4$.

11.5. Use Table 11.1 to calculate the stiffness factor K for an octagonal column having a depth $d = 20''$ and an unsupported height of 20′.

From the Table, $I = 0.055D^4$. Then $K = I/12h = (0.055)(20)^4/[(12)(20)] = 36.67$ in³.

11.6. Fig. 11-8 shows a joint into which beams and columns frame. Use the information shown on the figure to calculate the stiffness factors for the members and the r' factor for the joint.

For all members, $K = I/12h = bt^3/[(12)(12)h]$. Then

for BE, $K = (12)(18)^3/[(12)(12)(10)] = 48.6$ in³

for DE, $K = 48.6$ in³

for CE, $K = (12)(24)^3/[(12)(12)(20)] = 57.6$ in³

Thus $\Sigma K_{\text{beams}} = 57.6$ in³, $\Sigma K_{\text{columns}} = 48.6 + 48.6 = 97.2$ in³, and $r'_E = 97.2/57.6 = 1.69$.

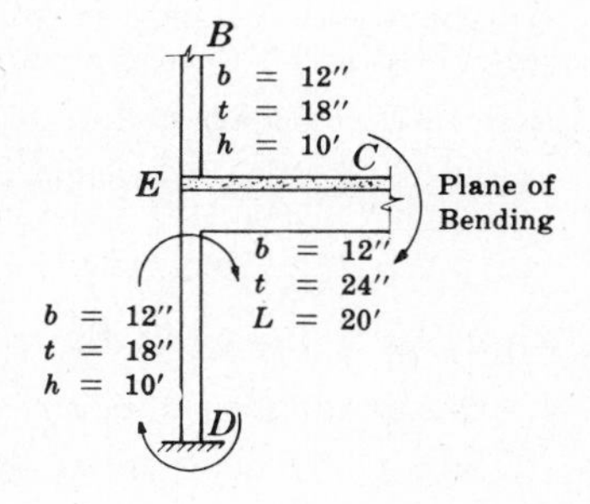

Fig. 11-8

11.7. If column DE of Fig. 11-8 is free to sway so that a relative horizontal deflection occurs between D and E, calculate the capacity reduction factor considering that compression controls the design and double curvature exists. Use the results of Problem 11.6.

Since Case 3 governs, use the average r' factor for the two ends. The base is fixed, thus $r'_D = 1.0$ is used. Then $r' = \frac{1}{2}(1.0 + 1.69) = 1.1345$.

The effective height is

$$h' = h(0.78 + 0.22r') \geqq h \quad \text{or} \quad h' = (10)(12)[0.78 + (0.22)(1.1345)] = 123.6''$$

Finally, using $r = 0.3t = 0.3(18) = 5.4''$,

$$R = 1.07 - (0.008)h'/r = 1.07 - 0.008(123.6/5.4) = 0.89$$

11.8. The column shown in Fig. 11-9 is bent in double curvature and is braced against sidesway. Determine the amplification factor for long column action if compression controls the design.

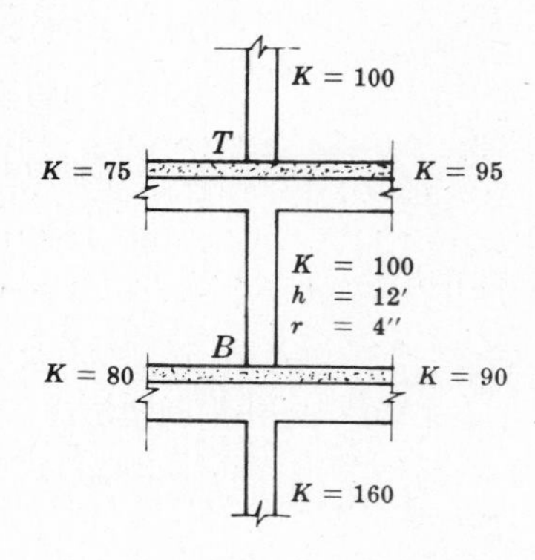

Fig. 11-9

The values r' at the top and bottom respectively are calculated using the equation

$$r' = (\Sigma K_{col})/(\Sigma K_{beams})$$

At the top,

$$r'_T = (100 + 100)/(75 + 95) = 1.175$$

and at the bottom,

$$r'_B = (100 + 160)/(80 + 90) = 1.53$$

The values of r' are close to 1.0, so definite restraint exists and Case 1 is involved. Then $h/r = (12)(12)/4 = 36 < 60$, and $R = 1.0$.

11.9. The column shown in Fig. 11-10 is subjected to axial load and bending and is braced against sidesway. Compression controls the design and $P_u = 400$ kips. Calculate the amplified load due to long column action.

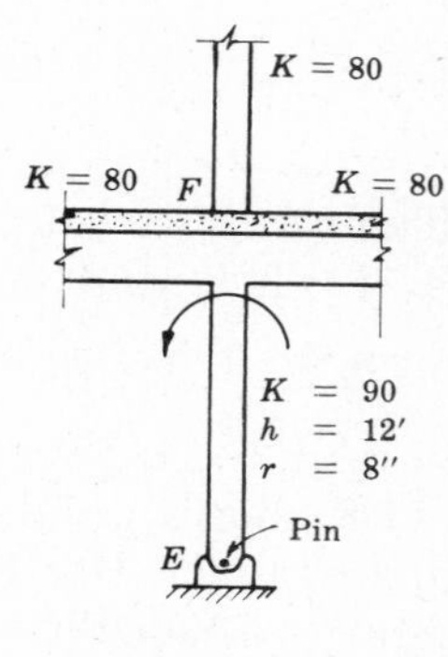

Fig. 11-10

At F,

$$r' = (\Sigma K_{col})/(\Sigma K_{beams}) = (80 + 90)/(80 + 80) = 1.06$$

Since the column is pinned at the base and *restrained* at F, Case 2 governs. Thus

$$R = 1.07 - 0.008h/r = 1.07 - (0.008)(12)(12)/8 = 0.926$$

and $P' = P_u/R = 400/0.926 = 432$ kips.

11.10. Find the value of R for the column of Problem 11.8 if h is 24′.

Note that Case 1 governs in Problem 11.8.

Since $h/r = (24)(12)/4 = 72$ and $60 < h/r < 100$, ACI equations may be used. Hence

$$R = 1.32 - (0.006)h/r = 1.32 - (0.006)(72) = 0.888$$

11.11. If the column of Problem 11.8 is subjected to sidesway, calculate the capacity factor R if compression controls the design.

From Problem 11.8, $r'_T = 1.175$ and $r'_B = 1.53$. Double curvature exists, so Case 3 governs.

The average value of r' is $r'_A = \frac{1}{2}(1.175 + 1.530) = 1.353$, and $h' = h(0.78 + 0.22r'_A) = 1.08h$. Hence

$$R = 1.07 - (0.008)(1.08)(12)(12)/4 = 0.76$$

11.12. Determine the design load P' for the column of Problem 11.9 if the column is subjected to sway.

Case 4 is involved. The modified height is

$$h' = 2h(0.78 + 0.22r_T') = (2)(12)[0.78 + (0.22)(1.06)] = 24.31'$$

Then $$R = 1.07 - (0.008)h'/r = 1.07 - (0.008)(26)(12)/8 = 0.778$$

and $P' = 400/0.778 = 514$ kips.

11.13. The cantilever column shown in Fig. 11-11 is subjected to an axial load and is free to sway. Determine the design axial load if $P_u = 400$ kips and the design is controlled by compression.

Case 5 is involved, and $h' = 2h = 32'$.

The radius of gyration is $0.3t$ or $6.0''$. Then

$$R = 1.07 - (0.008)(32)(12)/6 = 0.558$$

and $P' = 400/0.558 = 717$ kips.

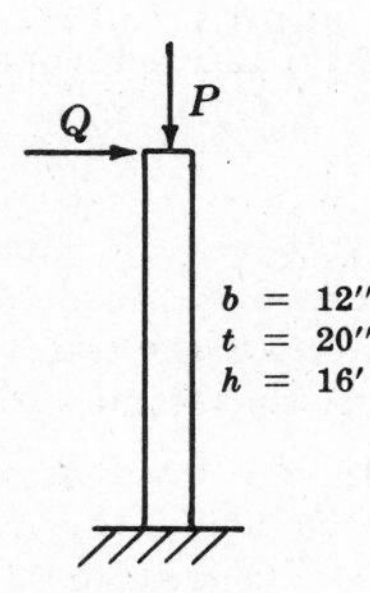

Fig. 11-11

11.14. Use equation (*11.20*) to determine the design conditions for the column of Fig. 11-11 if $Q = 2$ kips, $P = 200$ kips. Since the beam-column is cantilevered, use $C' = 1.0$ and $C = 1/4$ in the equation. The modulus of elasticity of the concrete is 3×10^6 psi.

For the cantilever beam subjected to a lateral load only at its end, the end deflection is

$$y_o = Qh^3/(3EI) = (2000)(16)^3(1728)/[(3)(3 \times 10^6/3)(8000)] = 0.59''$$

where $I = bt^3/12 = (12)(20)^3/12 = 8000$ in^4, and using an effective modulus of elasticity of $E/3$ in accord with the Code.

The deflection due to the beam-column action is

$$y = y_o[1 - (P/P_{cr})] = (0.59)/[1 - (200/535)] = 0.942''$$

where $P_{cr} = \pi^2 EI/(4L^2) = 535{,}000$ lb $= 535$ kips. Then $Py = (200)(0.942/12) = 15.7$ ft-kips.

The total design moment is therefore $M' = (16)(2) + 15.7 = 47.7$ ft-kips.

Thus the design should be accomplished for $P_u = 200$ kips and $M_u' = 47.7$ ft-kips.

11.15. The beam-column shown in Fig. 11-12 is subjected to an axial load of 400 kips and not free to sway. If $M = 100$ ft-kips, determine the design value of M' using the comprehensive analysis if $L = 20'$, $C' = 1.0$, $E_r = 1.0 \times 10^6$ psi, $I = 10{,}000$ in^4.

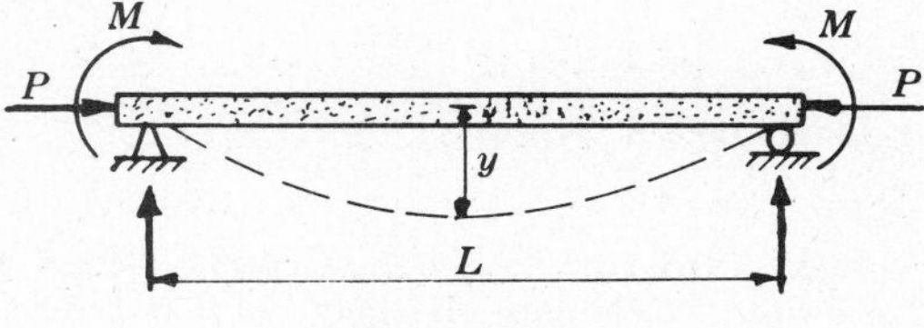

Fig. 11-12

The basic deflection due to M alone is

$$y_o = ML^2/8EI = (100)(20)^2(1728)/[(8)(10^3)(10{,}000)] = 0.864''$$

Since $P_{cr} = \pi^2 EI/L^2 = 1710$ kips,

$$y = y_o/(1 - P/P_{cr}) = 0.864/(1 - 400/1710) = 1.125''$$

and $$M_u' = 100.0 + (400)(1.125/12) = 137.5 \text{ ft-kips}$$

11.16. **The value of R is 0.926 for compression controlling in Problem 11.9. Use equation (*11.19*) to compute R_c if tension controls. $P = 400$ kips, the section is a symmetrical tied column, $d' = 2.5''$, $f'_c = 3.0$ ksi, $f_y = 40.0$ ksi, $t = 20''$, $p_g = 0.03$. Use working stress design. Use $e = 12.58''$.**

By working stress design equations,

$$e_{bx} = (0.67p_g m + 0.17)d = [(0.67)(0.03)(15.7) + 0.17]17.5 = 8.5''$$

where $d = t - d' = 20.0 - 2.5 = 17.5''$, $m = f_y/(0.85f'_c) = 15.7$, $p_g = 0.03$.

At balanced conditions R is taken as that value which applies when compression governs, or $R_b = 0.926$ here. Hence

$$R_c = [1 - (1 - R_b)(e_b/e)] = 0.95$$

Note that R increases when tension governs. Since the method is merely an approximation at best, it is reasonable to use the average of R_b and unity, which in this case would be 0.963. However, the latter value, although reasonable, does not conform to the Code. For a preliminary design, the latter value would be satisfactory, and it is easy to obtain.

In the case of design by ultimate strength procedures, the balanced eccentricity may be obtained using the definition $M_b = P_b e_b$, or the approximate equations (*11.25*) and (*11.26*) may be used. The latter equations are given in the 1965 Commentary to the Code.

Supplementary Problems

11.17. Calculate the moment of inertia of the hollow octagonal concrete section shown in Fig. 11-13 (exclude reinforcement) about the x axis. *Ans.* $I = 17{,}231$ in^4

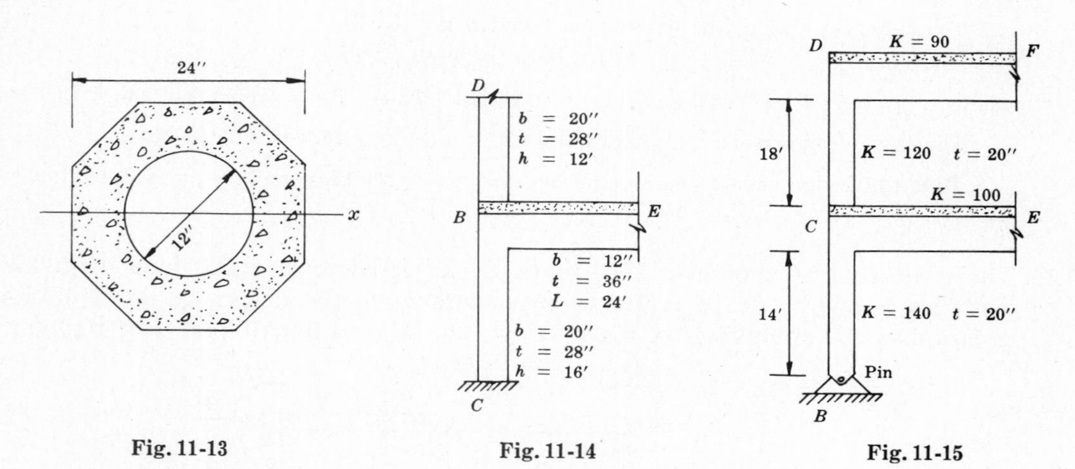

Fig. 11-13 Fig. 11-14 Fig. 11-15

11.18. In Fig. 11-14 the columns are free to sway. Column BC is bent in double curvature, and compression controls the design. Calculate the long column reduction factor. *Ans.* $R = 0.852$

11.19. In Fig. 11-15 column BC is not free to sway and is bent in single curvature, and compression controls the design. Calculate the long column reduction factor. The columns are 20" square. *Ans.* $R = 0.846$

11.20. Column DC in Fig. 11-15 is restrained against sway. Compression controls the design and the section is bent in single curvature. Wind load is the dominant factor. Calculate the long column reduction factor. *Ans.* $R = 0.7815$

11.21. Solve Problem 11.20 if column DC is bent in double curvature and free to sway. *Ans.* $R = 0.79$

11.22. Use the deflection method to determine the design moment for the beam-column of Fig. 11-16. $E_c = 4 \times 10^3$ ksi and $I = 12{,}000$ in^4. (*Hint.* y_o at center $= QL^3/48EI$ due to Q only; $P_{cr} = \pi^2 E_r I/L^2$.) *Ans.* $M = 122.2$ ft-kips

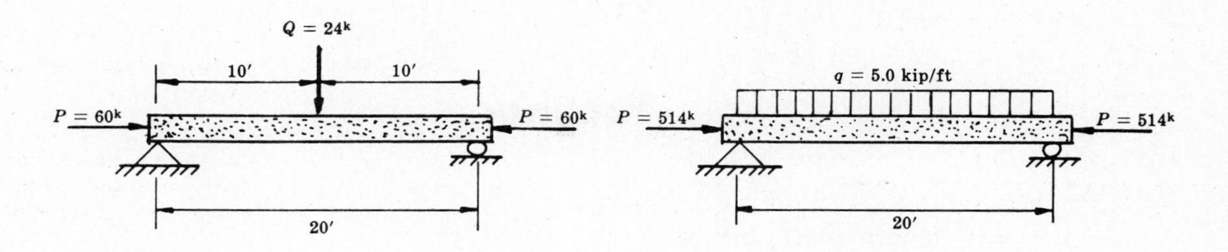

Fig. 11-16 **Fig. 11-17**

11.23. Use the deflection method to determine the design moment for the beam-column of Fig. 11-17. $E_c = 3 \times 10^3$ ksi and $I = 10{,}000$ in^4. (*Hint.* y_o at center $= 5qL^4/384E_r I$, $P_{cr} = \pi^2 E_r I/L^2$).
Ans. $M = 60.2$ ft-kips

Chapter 12

Footings

NOTATION

a = column width, inches

a = a coefficient for design of steel, WSD

ϕa_u = a coefficient for design of steel, USD

A_F = area of footing, ft^2

A_s = steel area, in^2

A_{sB} = steel area in band width B, in^2

A_{st} = total steel area, in^2

A_p = pile area, ft^2

b = width of a beam, inches

b_o = perimeter of a shear section, inches

B = footing width, feet

c = distance between centers of columns, feet

c = distance from neutral axis to a stressed point

d = effective depth of a footing, inches

d = distance between pile centers, feet

D = the diameter of a reinforcing bar, inches

e = eccentricity of load, feet

f'_c = ultimate strength of concrete, psi

f_s = steel stress, psi

f_y = yield stress for steel, psi

F = any force, kips

h = distance from footing base to horizontal force, feet

H = horizontal force, kips

I = moment of inertia of footing base, ft^4

L = length of footing base, feet

M = bending moment, ft-kips

n = modular ratio E_s/E_c

N = number of piles

p = steel ratio

p_t = total steel ratio

P = any load, kips

P = pile load, kips

P' = change in pile load due to displacement from critical section

q = soil pressure, psf

R = a coefficient for concrete design, WSD, psi; also, Resultant force, kips

ϕR_u = a coefficient for concrete design, USD, psi

u = bond stress, psi

v = shear stress, psi

V = shear force, kips

W = column load, kips

W_F = footing weight, kips

ϕ = capacity reduction factor

GENERAL NOTES

Reinforced concrete foundations or *footings* are utilized to support columns and walls composed of a variety of materials, including concrete, steel, masonry and timber.

Spread footings are designed to distribute large loads over a large area of soil near the ground surface in order to reduce the intensity of the force per unit area so that the soil will safely support the structure.

Pile footings are designed to deliver large loads to individual piles. The piles transfer the forces to lower levels by means of *skin-friction* between the soil and the pile surface and *point-bearing* of the pile on a dense soil strata at its base.

Both spread footings and pile footings may be classified into sub-groups such as isolated footings, multiple column footings, wall footings and mat footings.

Isolated footings support the load of a single column. The foundation for a structure may be composed of many isolated footings and, in addition, other types of footings.

Multiple column footings support two or more columns, acting as a beam or slab resting on the soil or piles.

Wall footings usually support continuous concrete or masonry walls around the perimeter of a building. Interior partition walls may also rest on continuous wall footings.

A special application of wall footings exists for retaining walls, which are discussed in Chapter 15.

Raft footings may support many columns and walls, acting as a continuous slab to distribute the loads over a large area.

Special types of footings are also used for particular purposes. *Cantilever footings* may be utilized advantageously near property lines or other structures.

Fig. 12-1 illustrates the types of footings which are used most often in general practice. These may be used with or without piles.

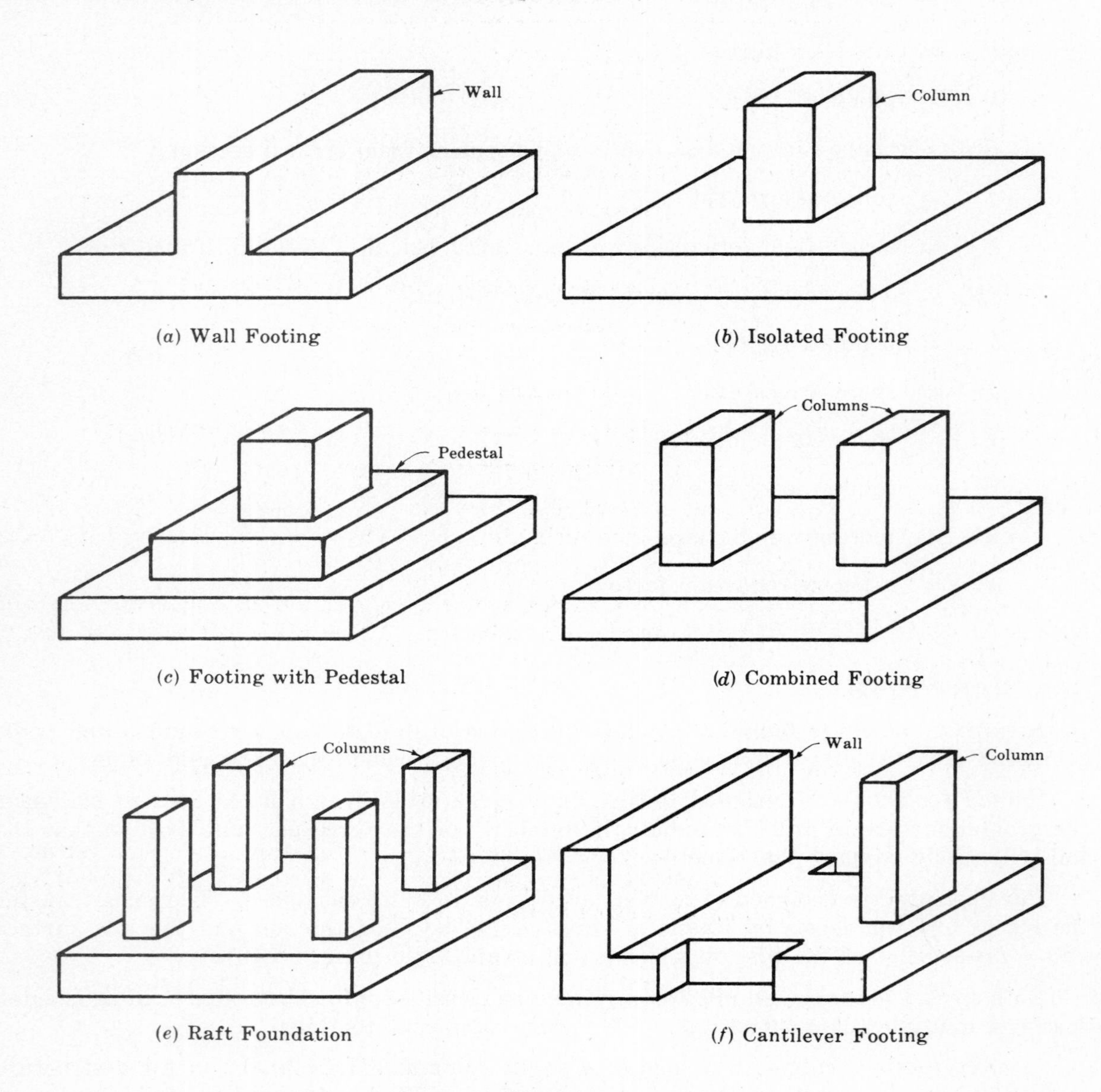

(*a*) Wall Footing

(*b*) Isolated Footing

(*c*) Footing with Pedestal

(*d*) Combined Footing

(*e*) Raft Foundation

(*f*) Cantilever Footing

Fig. 12-1

ALLOWABLE LOADS ON SOIL OR PILES

It is always desirable to determine safe bearing values for soil or allowable pile loads for the particular site upon which a structure is to be erected. Foundation engineers or soil testing experts should always be consulted in connection with the design of foundations, since it is usually more costly to rectify errors in judgment concerning foundations than to seek the advice of an authority before proceeding with the design of foundations.

In the absence of more authoritative information, most building codes permit the use of approximate values of safe soil bearing load. Typical values are listed in *tons per square foot* in Table 12.1.

TABLE 12.1

ALLOWABLE SOIL PRESSURES, tons/ft²

Soil	tons/ft²
Soft clay, medium density	1.5
Medium stiff clay	2.5
Sand, fine, loose	2
Sand, coarse, loose; compact fine sand; and loose sand-gravel mixture	3
Gravel, loose; and compact coarse sand	4
Sand-gravel mixture, compact	6
Hardpan and exceptionally compacted or partially cemented gravels or sands	10
Sedimentary rocks such as hard shales, sandstones, limestones, silt stones, in sound condition	15
Foliated rocks such as schist or slate in sound condition	40
Massive, bed rock, such as granite, diorite, gneiss, trap rock, in sound condition	100

It should be noted that the values listed refer to *working loads and not to ultimate loads.* Soil testing laboratories often report the ultimate values, allowing the structural engineer to select an appropriate safety factor.

It is not possible to state general values which should be used for designing pile footings. Pile tests should be initiated, or a soils expert consulted before proceeding to design pile footings.

DISTRIBUTION OF LOADS TO SOIL

The ACI Code permits the use of uniform soil pressure beneath spread footings so that

$$q = P/A_F \tag{12.1}$$

The soil pressure stated in equation (*12.1*) refers to footings for which the resultant column load is applied at the centroid of the base of the footing. If eccentricity of load exists, the pressure of the soil will vary uniformly, and one of two cases will exist. These are illustrated in Fig. 12-2.

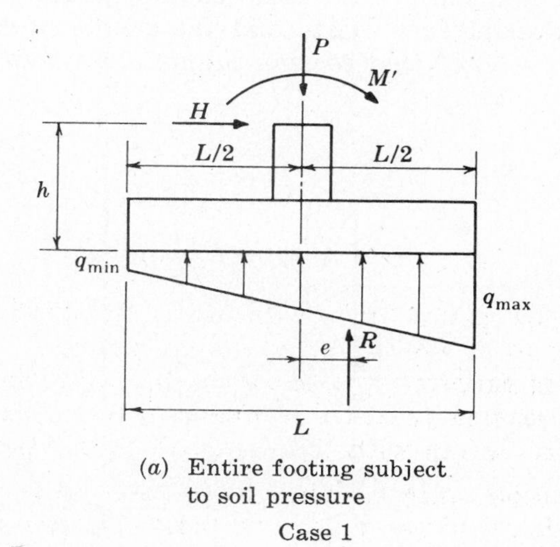

(*a*) Entire footing subject to soil pressure

Case 1

P
M′
H
L/2
L/2
h
q_{max}
e
R
2a
a
L

(*b*) Portion of footing not subject to soil pressure

Case 2

Fig. 12-2

Case 1 of Fig. 12-2 occurs for moderate values of M' or H, and Case 2 occurs for large values of M' or H. The resultant R consists of the applied load P plus the weight of the footing.

If the effects of M' and the moment due to H are combined as M, the maximum and minimum soil pressure can be obtained as follows, considering a constant width of footing B:

For Case 1,

$$q_{max} = R/BL + 6M/BL^2 \tag{12.2}$$

$$q_{min} = R/BL - 6M/BL^2 \tag{12.3}$$

and for Case 2,

$$q_{max} = 4R/[3B(L-2e)] \tag{12.4}$$

$$q_{min} = 0 \tag{12.5}$$

For both cases, the resultant R is located by determining the eccentricity of load,

$$e = (M' + Hh)/R \tag{12.6}$$

Case 1 will always exist if R lies within the *middle third* of the footing or if

$$e \leqq L/6 \tag{12.7}$$

otherwise, Case 2 will apply. If Case 2 applies,

$$a = L/2 - e \tag{12.8}$$

The foregoing equations for eccentrically loaded footings apply to bending about one axis only and are derived using the equation

$$q = R/A_F \pm Mc/I \tag{12.9}$$

in which A_F = base area of the footing, $M = Re$, $c = L/2$, $I = BL^3/12$, B = width of footing.

When bending occurs about both the x and y axes, and *the entire footing is subjected to pressure,*

$$q_{max} = R/A_F + M_x c_y/I_x + M_y c_x/I_y \tag{12.10}$$

$$q_{min} = R/A_F - M_x c_y/I_x - M_y c_x/I_y \tag{12.11}$$

The values of e_x and e_y are obtained using equation (*12.6*) first about the x-axis and then about the y-axis. If the resulting point of application of e falls outside of the *kern* of the section (shown shaded in Fig. 12-3), a special case exists and the points of zero pressure must be determined by trial. It should be noted that *tension cannot exist between the soil and the footing.*

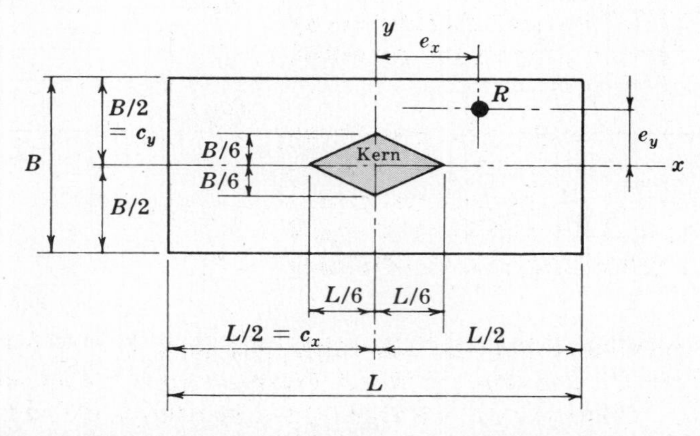

Fig. 12-3. Plan View of Base of Footing with Resultant Outside of Kern

FORCES ON PILES

Vertical Forces

When the resultant R is applied at the *centroid of a pile group* of N piles, each pile receives an identical load, so the pile load P will be

$$P = R/N \qquad (12.12)$$

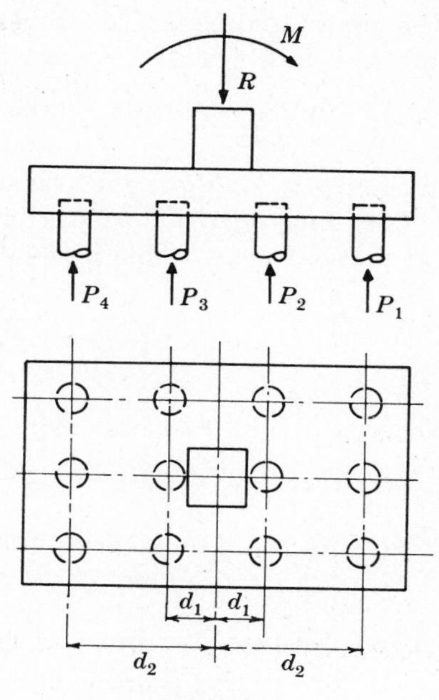

Fig. 12-4

When bending and axial load occur on pile footings, the approach is as follows: Let A_p be the area of each pile, so the average stress q_i in any pile i located at a distance d_i from the pile group centroid will be

$$q_i = R/NA_p \pm Md_i/I$$

where $I = \Sigma(A_p d^2)$; then if each pile has an identical area,

$$q_i = R/NA_p \pm Md_i/(A_p \Sigma d^2)$$

Multiplying both sides of the above equation by A_p and noting that qA_P is the *force* in a pile, the equation for any pile i becomes

$$P_i = R/N \pm Md_i/\Sigma d^2 \qquad (12.13)$$

Piles can assume tension loads if sufficient *negative skin-friction* can be developed, so only one case need be considered for bending about one principal axis. The *positive sign* applies to those piles on the side of the axis which is being rotated downward toward the piles.

When bending occurs about both principal axes, the load in pile i will be

$$P_i = R/N \pm M_x d_{ix}/\Sigma d_x^2 \pm M_y d_{iy}/\Sigma d_y^2 \qquad (12.14)$$

In all cases, the x and y principal axes must be directed through the centroid of the *pile group*.

Horizontal Forces

When horizontal loads are applied to spread footings, those loads must be transmitted to the soil by *friction* of concrete on soil or by *passive pressure* of the *side* of the footing (or a key) against the soil. Piles must resist the horizontal forces by shear stress in the piles and passive soil pressure, or batter piles (see Fig. 12-5).

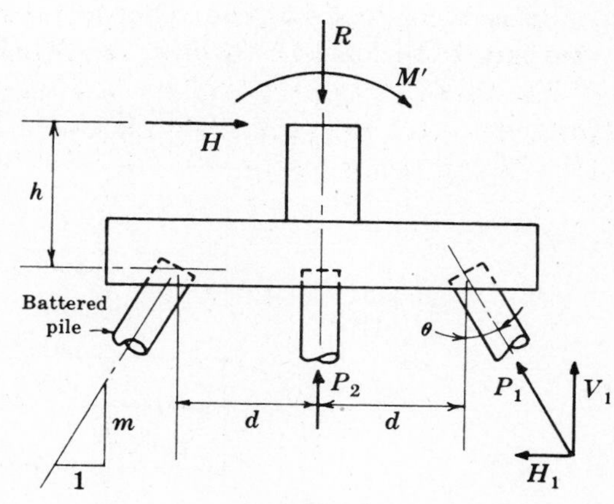

Fig. 12-5

Battered piles are not as effective in resisting vertical loads as are vertical piles. If θ is the batter angle, the efficiency of the battered pile is proportional to $\cos\theta$, and the values of N and I must be adjusted to reflect the loss in efficiency due to the batter.

The vertical component V_1 of the allowable pile load is $P\cos\theta$, and the horizontal component H_1 is $P\sin\theta$. The sum of all of the horizontal components of the pile loads must resist the applied horizontal force H, just as the sum of all the vertical components of the pile loads must resist the applied vertical force W.

WALL FOOTINGS

The equations presented previously for isolated column footings may be used for continuous wall footings if the following factors are utilized:

(1) For spread footings, the width B of the footing is taken as one foot along the length of the wall.

(2) For *pile footings*, one transverse row of piles is considered for uniform pile patterns, or, one repetition of the pile pattern (or one *panel*) is considered for non-uniform pile patterns. This is illustrated in Fig. 12-6.

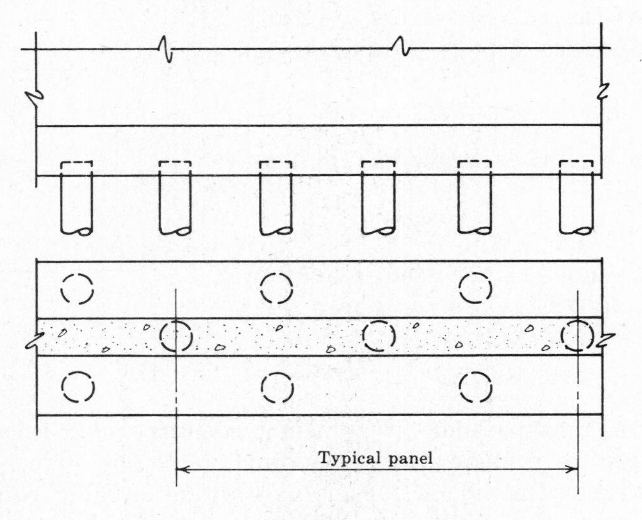

Fig. 12-6

MULTIPLE COLUMN FOOTINGS

The ACI Code Committee has not chosen to state specific requirements for the design of multiple column footings. Traditionally, such footings have been designed as *inverted beams*, using beam design criteria.

Whether spread footings or pile footings are used, it is necessary to locate the center of gravity of loads at the centroid of the base area of the footing or pile group, in order to spread the load uniformly. *Note.* For pile footings, $R = PN$.

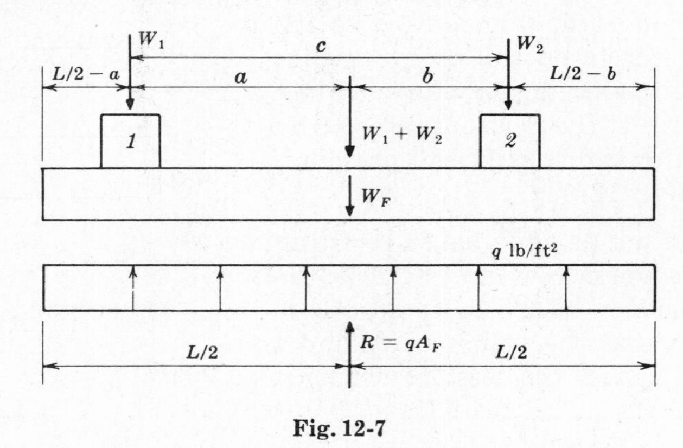

Fig. 12-7

The center of gravity of the total load is determined by summing moments about column *1*, from which

$$a/b = W_2/W_1 \qquad (12.15)$$

Since the distance c between columns is known and $b = c - a$, it follows that

$$a/c = W_2/(W_1 + W_2) \tag{12.16}$$

Whenever possible, the footing width is made constant so that the area of the footing can be obtained in a simple manner.

It is often impossible to use perfect rectangles for combined footings. Clearances with other footings or adjacent structures may prohibit this simplification. An odd shape may be necessary in order to place the center of gravity of the loads at the centroid of the footing base or at the centroid of the pile group. Fig. 12-8 illustrates some of the shapes which are often used.

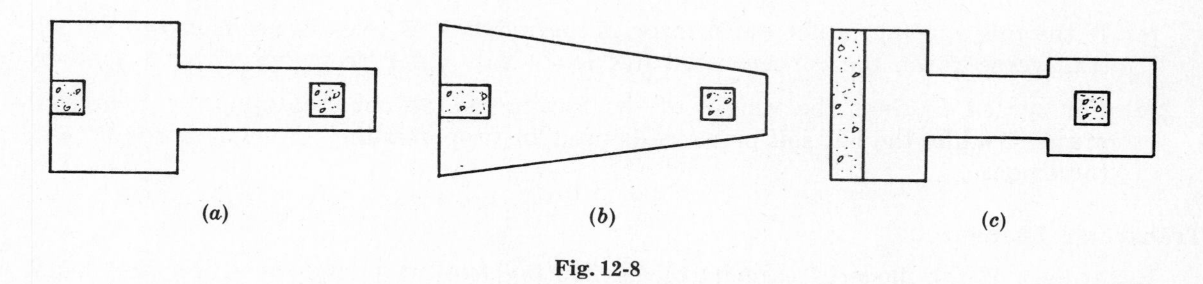

Fig. 12-8

When multiple column footings are utilized, the footing is treated as a beam on two supports in the long direction.

Transverse beams are designed to support the *longitudinal beam* and to transmit the soil pressure or pile reactions to the columns. Typical *qualitative shear and moment diagrams* for the longitudinal beam are illustrated in Fig. 12-9.

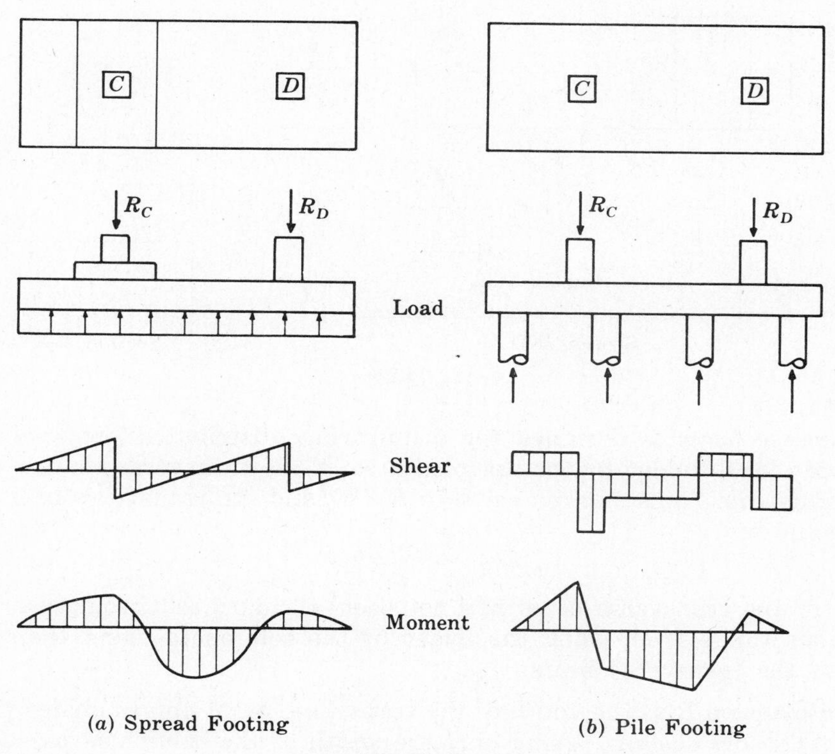

Fig. 12-9

Several comments must be made regarding Fig. 12-9.

(*a*) The column loads are actually distributed in some manner over the top of the footing, but the loads are usually considered as *line loads* in the transverse direction at the center of the columns. These loads may also be considered to be uniformly distributed over an area equal to the width of the column in the long direction and the total short width of the footing. In either case the solution is the same, since critical sections do not occur under the column.

(*b*) The entire weight of a pile cap is considered by some designers to be supported by the soil. Others assume that these loads are delivered to the piles. Either assumption has merit, although the latter is more conservative.

(*c*) If the pile spacing is not uniform or if the number of piles is not constant for all transverse rows, the concentrated live loads will differ from one row to another.

(*d*) For spread footings the weight of the footing is used in obtaining the gross soil pressure while the net soil pressure is used in proportioning the concrete and reinforcement.

Transverse Beams

Regardless of the mode of support of the footing (soil or piles), the transverse beam is required to transfer the loads to the column, and *each* pedestal must be designed to distribute the total load of the column (it supports) to the footing.

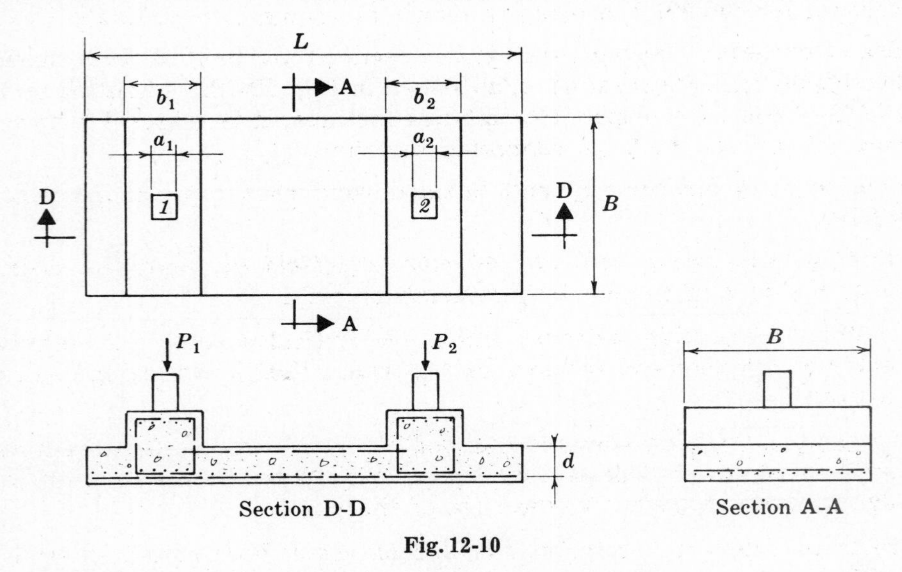

Fig. 12-10

Each *transverse beam* is designed for a uniformly distributed pressure equal to the respective column load divided by the assumed area of the transverse beam. For example, the beam under column *1* has an area equal to $B \times b_1$, and the transverse beam pressure p_T for design is equal to

$$p_T = P_1/Bb_1 \tag{12.17}$$

The width of the transverse beam has not been standardized in engineering practice. Recommendations range from using the width of the column to using the footing width as the width of the transverse beam.

Using the dimension B as the width of the transverse beam approximates the conditions used in isolated footing design. Using only the width of the column is based on ordinary beam and slab design.

A reasonable width for the transverse beam can be selected considering the usual mode of failure of footings as shown in Fig. 12-11.

Failure due to shear occurs approximately along 45 degree lines when footings are tested to destruction. This suggests that the width of the transverse beam should be taken as $b = a + 2d$, but not more than B.

As will be shown later, the critical section for shear in slabs and footings occurs at a distance $d/2$ from the face of a support. For this reason, it is quite reasonable to use a transverse beam width equal to $a + 2(d/2)$ or

$$b = a + d \qquad (12.18)$$

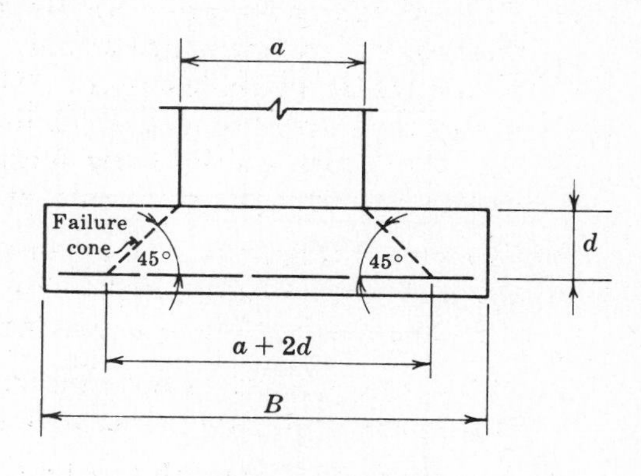

Fig. 12-11

When necessary in order to resist bending, shear or bond stresses, the transverse beam may be deepened to form a *pedestal.*

CRITICAL SECTIONS FOR BENDING AND FLEXURAL BOND STRESS

For *isolated footings* the ACI Code stipulates that the critical section for bending and flexural bond shall be taken as a section across the full width of the footing *at its top surface* (to account for tapered footings), at planes as shown in Fig. 12-12.

In the case of *combined footings,* the critical sections are used as for isolated footings with respect to bending and flexural bond, as shown in Fig. 12-12. Although no stipulation is contained in the Code with respect to combined footings, it is reasonable to regard such footings as several isolated footings connected together.

The critical sections for bending and flexural bond as shown on the figure may be described as follows:

(*a*) For footings supporting *concrete columns, pedestals or walls,* the critical section occurs at the *face* of the column, pedestal, or wall.

(*b*) For footings supporting *masonry walls,* the critical section lies midway between the center of the wall and the face of the wall. This is due to the flexibility of a masonry wall.

(*c*) For footings supporting columns founded on *metal base plates* which are in turn supported on a pier or pedestal, the critical section occurs at the midpoint between the edge of the base plate and the edge of the pedestal.

(*d*) For footings supporting columns founded on *metal base plates* without pedestals, the critical section occurs at the midpoint between the column face and the edge of the metal base plate.

In all of the above cases, the total shear force V_B for flexural bond is that force exerted by the soil pressure on area *1-2-3-4* as shown in Fig. 12-12. The design bending moment is the moment of V_B about the critical section.

When the footings are *stepped,* or at changes in depth or quantity of reinforcement, the section should be checked for bending and flexural bond.

Reinforcement

(*a*) In *two-way footings,* the reinforcement should be designed to resist bending and bond stresses at the critical section, and should be uniformly distributed across the footing. (The Code does not specify minimum A_s for footings.)

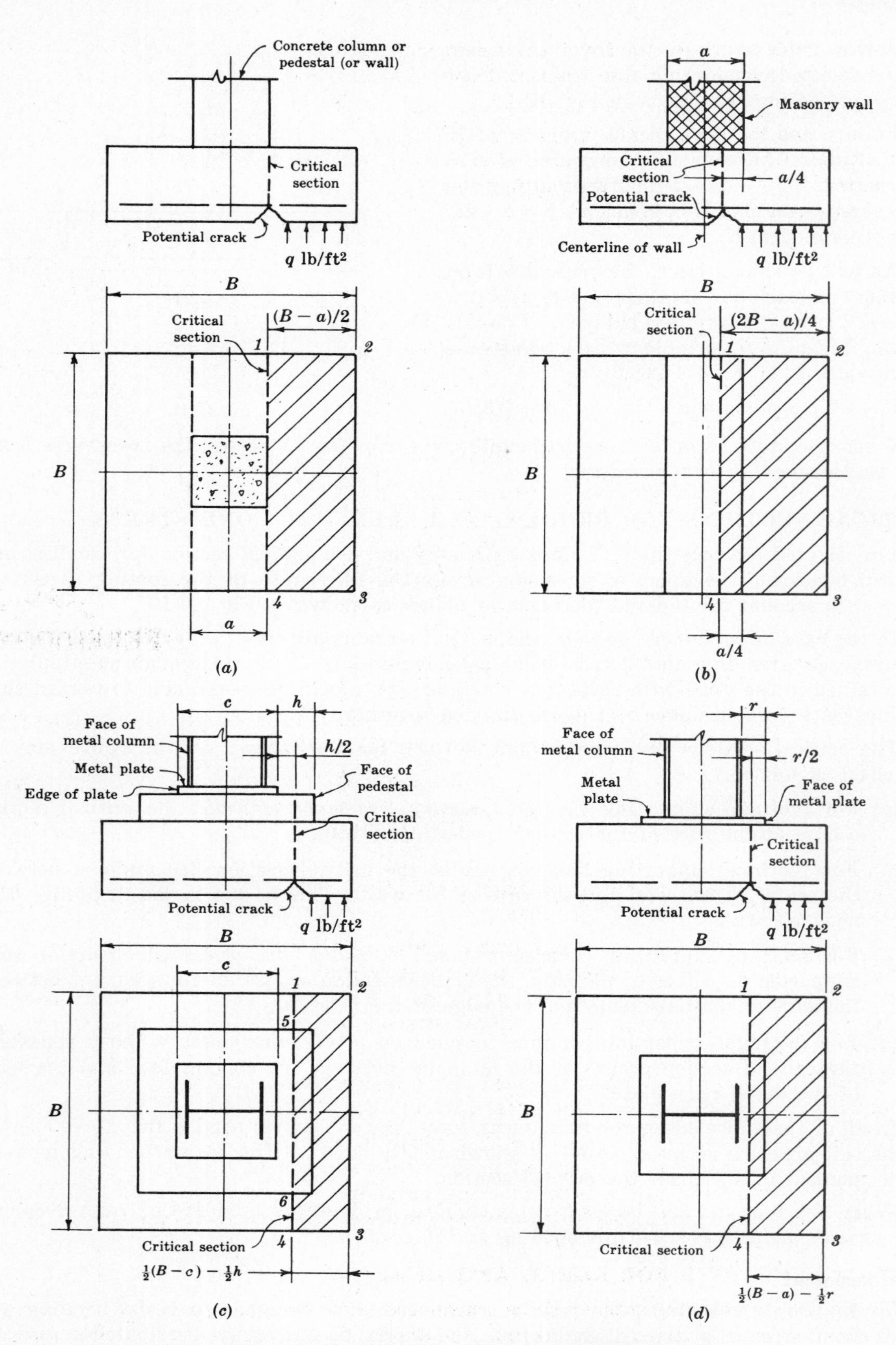

Fig. 12-12. Critical Sections for Bending and Flexural Bond

(*b*) In *square two-way footings,* the steel should be identical in both directions. (Two layers are provided, one in each direction.)

(*c*) In *rectangular two-way footings,* the steel must be computed separately in each direction, and spaced as follows: (1) The *long bars* should be equally distributed across the footing over the *short dimension B,* as shown in Fig. 12-13. (2) The *short bars* must be distributed in *three band widths* over the long dimension *L,* as shown in Fig. 12-13.

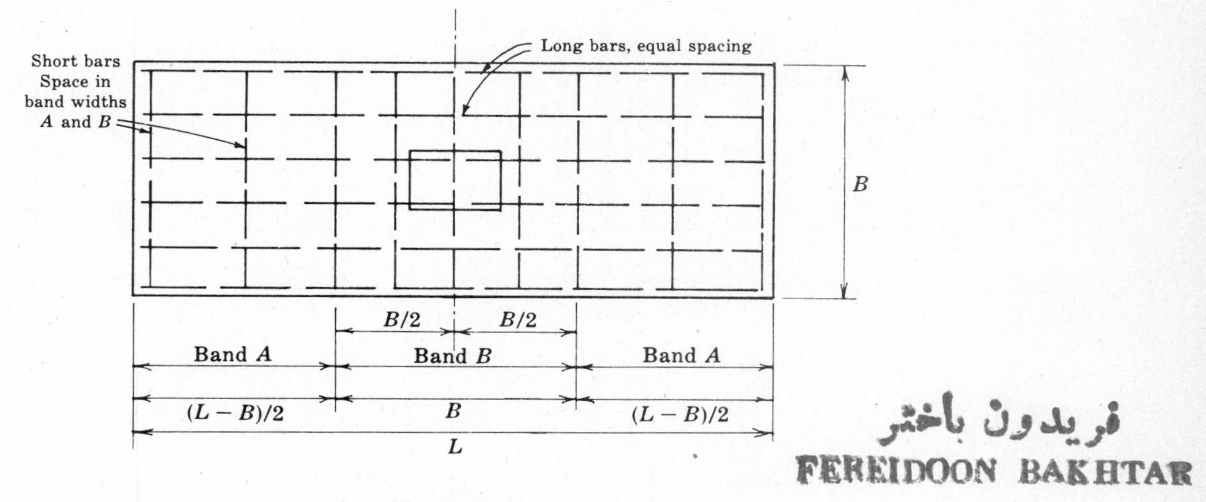

Fig. 12-13

The *area of steel* A_{sB} in band width B is obtained from equation (*12.19*), and the steel A_{sA} in band width A is obtained from equation (*12.20*):

$$A_{sB} = 2A_{st}/(S+1) \tag{12.19}$$

and

$$A_{sA} = (A_{st} - A_{sB})/2 \tag{12.20}$$

where

$$S = L/B \tag{12.21}$$

and A_{st} is the *total* area of short bars required. Evidently, $A_{st} = A_{sB}$ for a square footing, since $S = 1$ for that condition.

The spacing of the short bars stated above refers *only* to isolated footings, supporting a single column. The short bars are spaced uniformly across the width of the *transverse beams* in multiple column footings. The short bars are designed separately under each column to resist the flexural stresses and flexural bond stresses in each transverse beam.

Some quantity of steel is always placed in the segments between the transverse beams, and outside of the transverse beams toward the ends of the footing.

The steel provided *should* be *at least* that which is specified for temperature reinforcement, but the steel should be designed in a manner similar to that for isolated square footings or continuous concrete wall footings, assuming an *imaginary wall face* to exist along the faces of the columns. (*Authors' recommendation*: min. $p = 200/f_y$.)

CRITICAL SECTIONS FOR SHEAR AND DIAGONAL TENSION

Two separate requirements must be considered with respect to shearing stresses on *single column* footings and *multiple column footings,* whether square, rectangular or any other shape. The conditions are shown in Fig. 12-14 below.

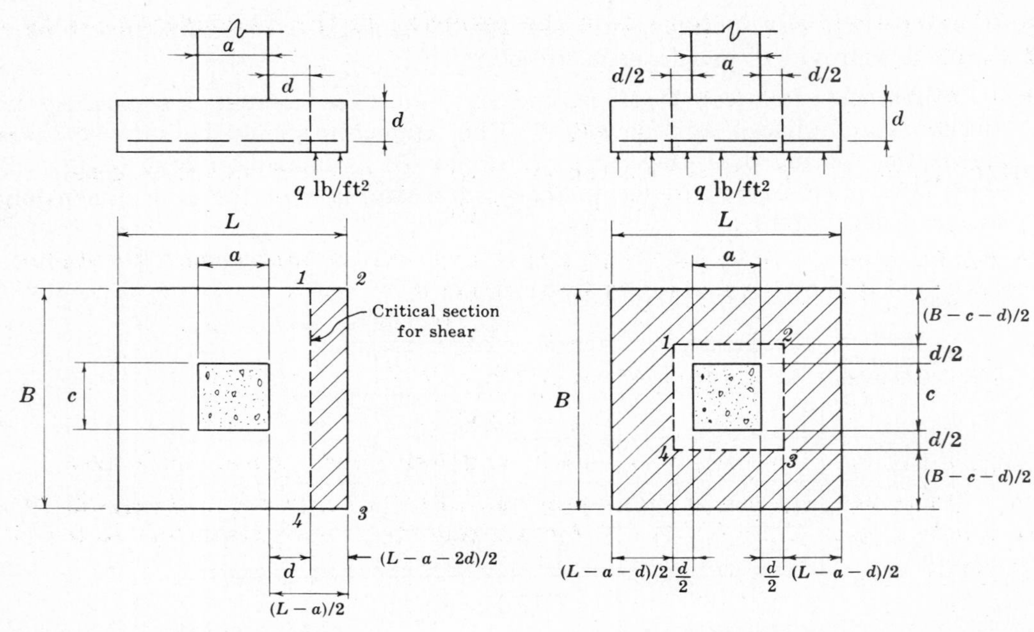

Critical section along line *1, 4*.
(*a*)

Critical section along perimeter *1, 2, 3, 4*.
(*b*)

Fig. 12-14

The first condition, shown in Fig. 12-14(*a*), assumes that the section is a wide beam, and all of the provisions for shear in beams apply in this case. The shear force (V or V_u) on Section *1-4* consists of all of the net soil pressure or pile reactions applied on area *1-2-3-4* as shown in Fig. 12-14(*a*). The width b of the section resisting shear extends the full width of the footing. In this case, $b = B$. Nominal shear stress is computed as

$$v = V/bd \tag{12.22}$$

for working stress design and

$$v_u = V_u/bd \tag{12.23}$$

for ultimate strength design.

When using the simplified forms, the allowable stress is

$$v_c = 1.1\sqrt{f'_c} \tag{12.24}$$

for working stress design and

$$v_{cu} = 2\phi\sqrt{f'_c} \tag{12.25}$$

for ultimate strength design.

The *more detailed analysis* may be used and web reinforcement may be provided as described for shear in beams.

The second condition considers perimeter shear or *punching shear* on the critical Section *1-2-3-4* as shown in Fig. 12-14(*b*). The shear force V or V_u consists of all of the net forces on the footing due to soil pressure or pile reactions which exist *outside* of Section *1-2-3-4*. Nominal shear stress is calculated as

$$v = V/b_o d \tag{12.26}$$

for working stress design and

$$v_u = V_u/b_o d \tag{12.27}$$

for ultimate strength design.

The effective perimeter b_o consists of the perimeter of the rectangle *1-2-3-4* as shown on Fig. 12-14(*b*).

The allowable shear stress v_c is

$$v_c = 2\sqrt{f'_c} \tag{12.28}$$

for working stress design unless web reinforcement is provided, in which case

$$v_{max} = 3\sqrt{f'_c} \tag{12.29}$$

for the combination of concrete shear stress and web reinforcement assistance. For ultimate strength design the maximum shear stress is

$$v_{cu} = 4\phi\sqrt{f'_c} \tag{12.30}$$

unless web reinforcement is provided, in which case

$$v_{cu} = 6\phi\sqrt{f'_c} \tag{12.31}$$

for the combination of concrete shear stress and web reinforcement assistance.

Note. When web reinforcement is provided to satisfy perimeter shear, the *stress in the web reinforcement is limited* to $f_s/2$ for working stress design and to $f_y/2$ for ultimate strength design.

ADDITIONAL REQUIREMENTS

(1) Piles are considered as contributing reactions concentrated at the pile center, except in the case of a pile near the critical section. A pile having its center located *at least 6″ outside of the critical section* delivers a *full pile load* to the critical section. If the center of the pile is located *6″ inside of the critical section,* it delivers no load to the critical section. Referring to Fig. 12-15, an equation may be derived for these conditions as

$$0 \leqq P' = P(X+6)/12 \leqq P \quad (12.32)$$

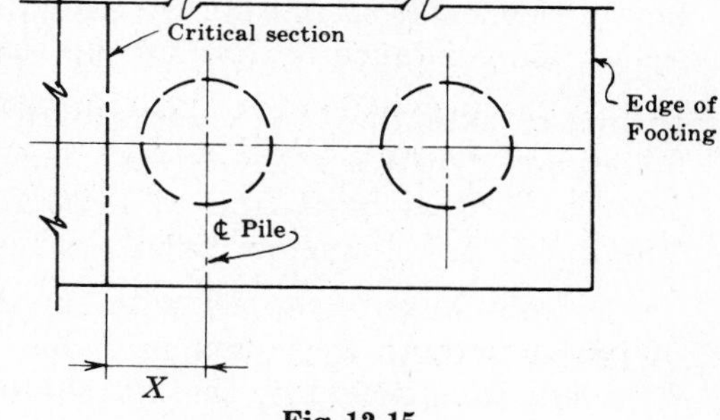

Fig. 12-15

where P' = The proportion of pile load P delivered to the critical section.

X = The distance of the pile center from the critical section in inches. X is positive when within the critical section, and negative when outside the critical section. When $X \geqq 6$, $P' = P$; and when $X = 0$, $P' = P/2$. When $-X \geqq -6$, $P' =$ zero. (Based on an assumed pile top diameter of 12″.)

(2) Minimum clearances for footings on piles are shown on Fig. 12-16. When reinforcement is not used (plain concrete footings), min. $d = 14″$. Embedment of piles should

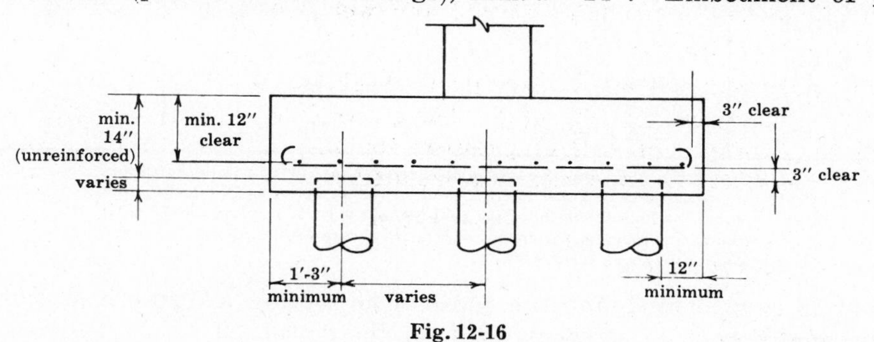

Fig. 12-16

be 4″ to 6″ for timber piles, and 6″ to 12″ for concrete or metal piles. For heavily loaded piles, the larger values apply.

(3) Minimum clearances for footings on soil (spread footings) are shown on Fig. 12-17.

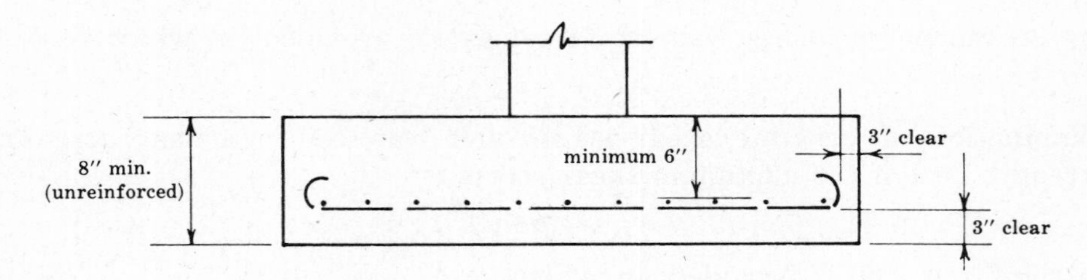

Fig. 12-17

(4) For *round columns* (or pedestals) an equivalent concentric square section having the same area as the round section shall be used for determining the critical sections. Although not specified in the Code, this is also practical for use with other symmetrical shapes, i.e. octagonal, hexagonal, cross, etc.

TRANSFER OF STRESS AT THE BASE OF A COLUMN

(1) The stresses in the longitudinal steel in any column shall be transferred into its supporting pedestal or footing by extending the vertical bars into the pedestal or footing a sufficient distance to develop the full bar capacity by bond.

(2) In lieu of extending the column bars into the pedestal or footing, the use of *dowels* is a common practice. The dowels must have a cross-sectional area at least equal to the column vertical bars, and at least four dowel bars must be used for each member. The dowel bar diameter may not exceed that of the column bars by more than $\frac{1}{8}$″.

Dowels must be lapped with the column bars in the column for distances in terms of bar diameters as shown in Table 12.2. The table refers to $f'_c \geqq 3000$ psi. When f'_c is less than 3000 psi, the lap shall be increased by one-third of the tabular values.

TABLE 12.2

LAP FOR DEFORMED COLUMN BARS AND DOWELS

(Number of bar diameters for lapping)			
f_y, psi	Lap of Bars in Terms of Diameters		Min. Lap inches
	Tension Bars	Compression Bars	
30,000	24	20	12
40,000	24	20	12
50,000	30	20	12
60,000	36	24	12
75,000	Not Allowed	30	12

Note: For plain bars the lapped splice shall be twice the values shown for deformed bars in Table 12.2. Ties or spirals must be used to enclose the splice for its full length, or the splice length must be increased by 20 percent. For columns, the spirals or ties are always continued down into the footing to space the dowels.

The dowels must extend into the pedestal or footing a distance at least sufficient to develop the *full working stress in bond* for the dowel.

ALLOWABLE BEARING STRESSES ON THE TOP SURFACES OF REINFORCED CONCRETE PEDESTALS AND FOOTINGS

The compression stress in the concrete at the base of a column shall be assumed to be distributed to the pedestal or footing as *bearing stress.*

In considering the bearing stress on the pedestal, two conditions must be checked:

(1) Bearing on the *full area* of the top of the pedestal or footing, and

(2) Bearing on *one-third* (or less than one-third) of the top of the pedestal or footing.

Both conditions (1) and (2) apply also to transfer of stress from the pedestal base to the footing top.

For condition (1), the bearing stress on the *full area* of the pedestal or footing may not exceed $0.25f_c'$ for working stress design, nor $0.475f_c'$ for ultimate strength design.

Condition (2) permits larger stresses than condition (1), but is subject to certain restrictions, as shown in Fig. 12-18. The *loaded area* is equal to $A_L = ca$, and the *unloaded area* is equal to $A_u = LB - A_L$.

Restrictions for Fig. 12-18:

(1) the smaller of b_1 or b_2 must be $\geqq c/4$.

(2) the smaller of L_1 or L_2 must be $\geqq a/4$.

If the restrictions are met, the allowable bearing stress on area $BL/3$ may be $0.375f_c'$ when using the working stress design method and $0.7225f_c'$ when using the ultimate strength method. Interpolation may be used between the values for full area and area $BL/3$.

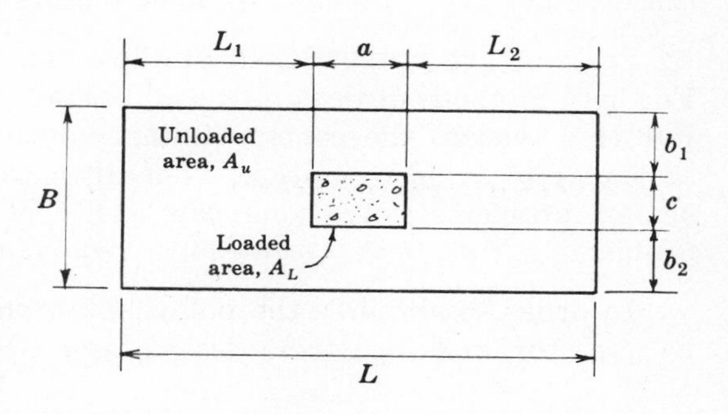

Fig. 12-18

BEARING FOR SLOPED OR STEPPED REINFORCED CONCRETE FOOTINGS

For sloped or stepped footings (as shown in Fig. 12-19) the supporting area for bearing may be taken as

(1) the top horizontal surface of the footing (area *9, 10, 11, 12*) or

(2) the largest base (area *5, 6, 7, 8*) of the frustrum of a cone wholly within the footing, formed by sides sloping from the actual loaded area and extending outward on a slope of 1 vertical to 2 horizontal.

Under certain circumstances, Case (1) will provide the larger area, and under other circumstances Case (2) will provide the larger area.

The area selected shall then be used as the unloaded area as shown in Fig. 12-18. The restrictions shown in that figure are fully applicable here.

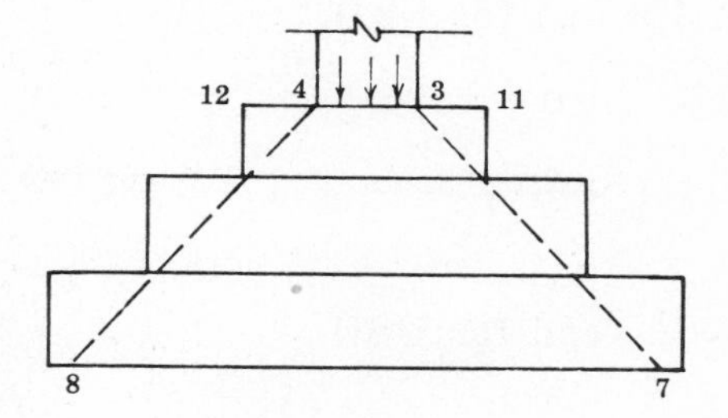

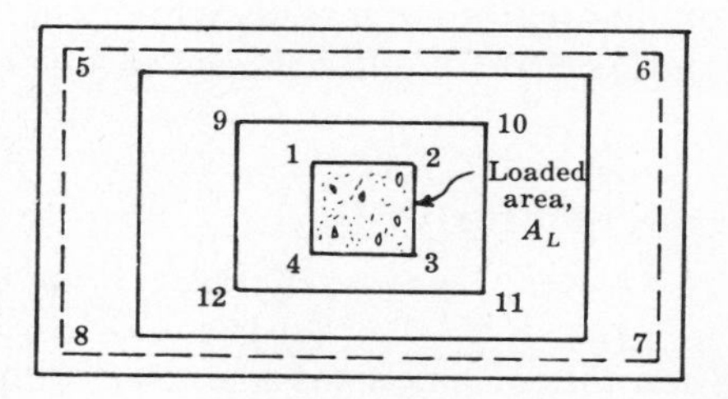

Fig. 12-19

PLAIN CONCRETE PEDESTALS AND FOOTINGS

The allowable compressive stress on the gross area of a concentrically loaded pedestal shall not exceed $0.25f'_c$ under service loads (working stress design). When this stress is exceeded, reinforcement shall be provided, considering the pedestal as though it is a column.

The depth and width of a pedestal or footing of plain concrete shall be such that the tension in the concrete shall not exceed $1.6\sqrt{f'_c}$ for working stress design nor $3.2\sqrt{f'_c}$ for ultimate strength design.

STRUCTURAL DESIGN OF FOOTINGS

The effective depth required and the steel reinforcement may be determined using either the working stress design method or the ultimate strength design method. For both procedures the methods of obtaining the basic elastic shears and moments are identical. The difference lies in the use of safety factors in using one method rather than the other. For the working stress method, the shears and moments are used as calculated, whereas those values are increased by load factors for the ultimate strength design method.

Sections are proportioned by the same methods as used for beams or slabs, in general. For both methods of design, the equations for proportioning can be placed in similar forms. For this reason, the corresponding equations for both methods will be discussed simultaneously to avoid repetition. The ultimate strength method will be used primarily in the design problems, since one can easily convert the problems to working stress design problems. *Note that the authors recommend* min. $p = 200/f_y$.

In order to simplify the notation for the many equations involved, the term WSD will be used for working stress design, and USD for ultimate strength design.

FLEXURAL DESIGN

The required effective depth to resist compressive stresses is

$$d = \sqrt{M/Rb} \tag{12.33}$$

for WSD; and for USD,

$$d = \sqrt{M_u/\phi R_u b} \tag{12.34}$$

The required area of reinforcement is

$$A_s = M/ad \tag{12.35}$$

for WSD; and for USD,

$$A_s = M_u/\phi a_u d \tag{12.36}$$

For flexural members in which the tension reinforcement is parallel to the compression face, the *flexural bond stresses* are calculated using

$$u = V_B/(\Sigma o\, jd) \tag{12.37}$$

for WSD; and for USD,

$$u_u = V_{Bu}/(\Sigma o\, \phi j_u d) \tag{12.38}$$

When mixed bar sizes are used, substitute $4A_s/D$ for Σo, using the largest bar diameter of the group for D.

The maximum permissible values of u and u_u are listed in Table 12.3 for *tension bars.* Top bars which do not have at least 12″ of concrete below the bars are treated as *other bars.* In the table, D is the nominal diameter of the reinforcing bar.

TABLE 12.3

MAXIMUM BOND STRESS, u or u_u, psi

	Tension bars conforming to ASTM A-305		Tension bars conforming to ASTM A-408	
	WSD	USD	WSD	USD
Top Bars	$3.4\sqrt{f'_c}/D \leqq 350$	$6.7\sqrt{f'_c}/D \leqq 560$	$2.1\sqrt{f'_c}/D$	$4.2\sqrt{f'_c}/D$
Other Bars	$4.8\sqrt{f'_c}/D \leqq 500$	$9.5\sqrt{f'_c}/D \leqq 800$	$3.0\sqrt{f'_c}/D$	$6.0\sqrt{f'_c}/D$

For compression bars of any grade or type of steel, for WSD $u \leqq 6.5\sqrt{f'_c} \leqq 400$ psi; and for USD, $u_u \leqq 13.0\sqrt{f'_c} \leqq 800$ psi. (Deformed bars).

For *plain bars,* the allowable bond stresses shall be one-half of those values permitted for bars conforming to ASTM A-305, but shall not exceed 160 psi for WSD, nor 250 psi for USD.

It should be noted that flexural bond stress need not be considered in compression steel nor in cases where *tension anchorage bond* is less than 0.8 times the permissible anchorage bond stresses.

Anchorage bond (or development bond) stress is calculated by dividing the bar forces by Σo times the embedded length L_b, or

$$u_a = A_s f_s/(\Sigma o\, L_b) \qquad (12.39)$$

for WSD; and for USD,

$$u_{ua} = M_u/(\Sigma o\, L_b \phi j_u d) \qquad (12.40)$$

L_b may be furnished by *hooks* for tension bars, when necessary. The permissible bond stresses for anchorage bond are identical to those stated in Table 12.3.

The values of a, j, R, ϕa_u, ϕj_u and ϕR_u may be obtained from Tables 4.2 and 5.2, Pages 71 and 103. For convenience, the equations for calculating those variables are restated.

For *working stress design* (WSD):

$$a = f_s j/12{,}000 \qquad (12.41)$$

$$j = 1 - \tfrac{1}{3}[\sqrt{2np + (np)^2} - np] \qquad (12.42)$$

or

$$j = 1 - 1/3(1 + f_s/nf_c) \qquad (12.43)$$

and

$$R = f_s pj = \tfrac{1}{2} f_c kj \qquad (12.44)$$

For *ultimate strength design* (USD):

$$\phi a_u = \phi f_y(1 - 0.59pf_y/f'_c)/12{,}000 \qquad (12.45)$$

$$\phi j_u = \phi a_u(12{,}000)/f_y = \phi(1 - 0.59pf_y/f'_c) \qquad (12.46)$$

$$\phi R_u = \phi pf_y(1 - 0.59pf_y/f'_c) \qquad (12.47)$$

Solved Problems

12.1. Derive an equation for the required effective depth for flexure of a square spread footing which supports a square isolated column as shown in Fig. 12-20. The maximum soil pressure is q lb/ft². Use the working strength design method.

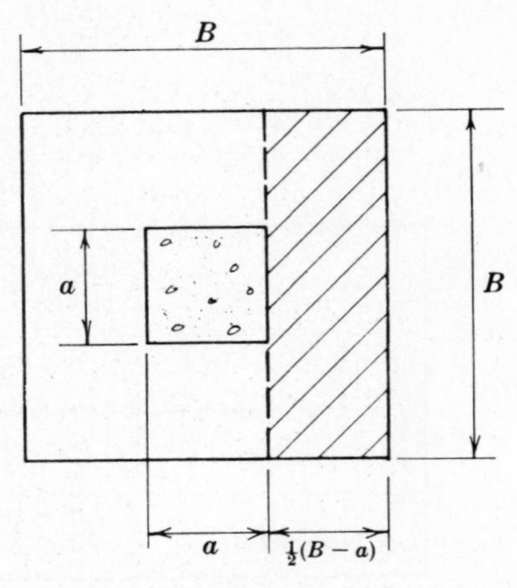

Fig. 12-20

The total shear force V_B on the critical section is $V_B = qB(B-a)/2$, and the moment of that force about the critical section is $M = qB(B-a)^2/8$.

The required depth for flexure is

$$d = \sqrt{M/RB} = \sqrt{qB(B-a)^2/8RB}$$

or

$$d = \tfrac{1}{2}(B-a)\sqrt{q/2R} \qquad (12.48)$$

which applies to working strength design directly. The equation may be modified for use with ultimate strength design by replacing q with q_u and R with ϕR_u.

12.2. Derive an equation for the required effective depth of the footing in Problem 12.1 to resist shear as a measure of diagonal tension. Refer to Fig. 12-21.

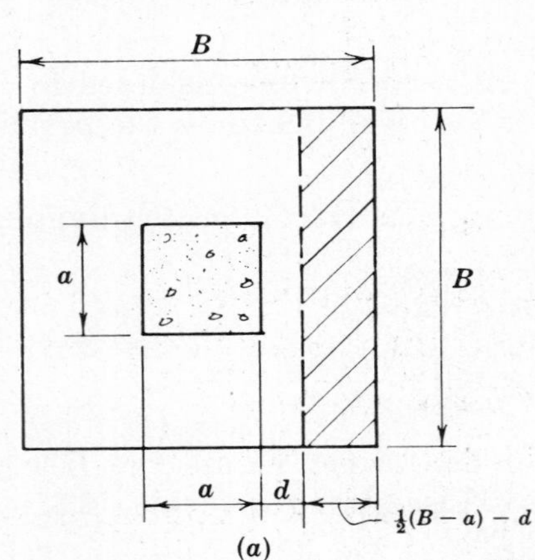

(a)

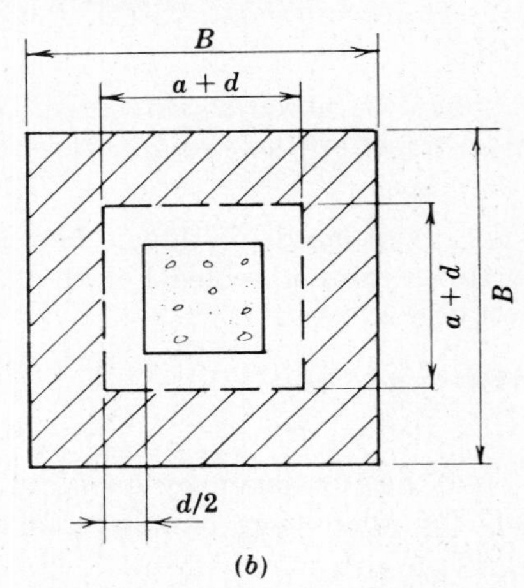

(b)

Fig. 12-21

For slab shear as shown in Fig. 12-21(a), $V = (B-a-2d)Bq/2$.

The unit shear is $v = V/bd = V/Bd$, since here $b = B$; then $d = (B-a-2d)q/2v$.

The allowable shear stress is $\alpha\sqrt{f'_c}$, where α is 1.1 for WSD and 2ϕ for USD; thus

$$(B-a-2d)(q/2d) = \alpha\sqrt{f'_c} \qquad (12.49)$$

or

$$\tfrac{1}{2}(B/d - a/d - 2) = \alpha\sqrt{f'_c}/q \qquad (12.50)$$

which can be used effectively for developing design aids since it is dimensionless.

For perimeter shear as shown in Fig. 12-21(b), $V = [B^2-(a+d)^2]q$; and $v = V/b_o d$ where $b_o = 4(a+d)$. Thus

$$\frac{[B^2-(a+d)^2]q}{4d(a+d)} = \gamma\sqrt{f'_c} \qquad (12.51)$$

where γ is equal to 2 for WSD and 4ϕ for USD. This equation becomes

$$\frac{(B/d)^2-(a/d+1)^2}{a/d+1} = 4\gamma\sqrt{f'_c}/q \qquad (12.52)$$

which may be readily utilized to produce design aids since it is dimensionless.

12.3. An 18″ square column is required to support a 200 kip dead load force and a 145 kip live load force. The allowable soil pressure is 4.0 ksf. Using $f'_c = 3.0$ ksi and $f_y = 40.0$ ksi, design the footing using ultimate strength design. The column contains 10 No. 9 vertical bars.

Since allowable soil pressure is stated rather than ultimate soil pressure, the footing area is calculated using working loads. Thus considering the footing to weigh *approximately* 7.5% of the total applied load, $W_F = (0.075)(200+145) = 25.9$ kips (use 26 kips).

The total load is therefore $W = 200+145+26 = 371$ kips, and the required footing area is $A_F = W/q = 371/4 = 92.75\ \text{ft}^2$. Thus $B = \sqrt{92.75} = 9.63'$. Try a footing 9′-8″ square for which $A_F = 93.5\ \text{ft}^2$.

The *ultimate applied load* will be $P_u = 1.5W_D + 1.8W_L = (1.5)(200)+(1.8)(145) = 561$ kips. The net ultimate soil pressure will then be: net $q_u = P_u/A_F = 561/93.5 = 6.0\ \text{kips/ft}^2$.

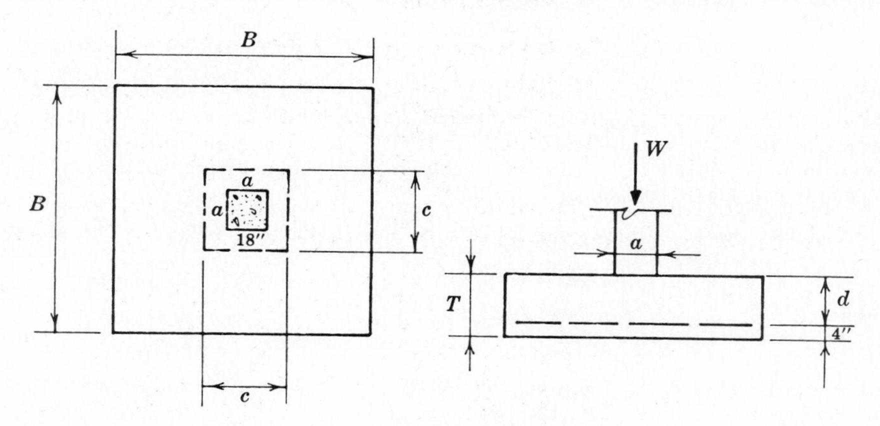

Fig. 12-22

Refer to Fig. 12-22 and check perimeter shear. Assume $d = 19''$. Thus

$$c = a + d = 18'' + 19'' = 37'' = 3.08' \quad \text{and} \quad b_o = (4)(37) = 148''$$

The total shear force outside the critical section will be $V_o = 6.0[93.5-(3.08)^2] = 504$ kips. Hence

$$v_{ou} = V_o/(b_o d) = 504/[(148)(19)] = 0.179\ \text{ksi} = 179\ \text{psi}$$

The allowed perimeter shear is $v_{cu} = 4\phi\sqrt{f'_c} = 186$ psi. The section therefore satisfies the conditions for perimeter shear.

Refer to Fig. 12-23 and check beam shear. The critical section is at D-D′, and the total shear force outside of the section is $V_u = (6.0)(30/12)(9.67) = 145$ kips.

The unit shear stress is

$$v_u = V_u/Bd = 145{,}000/[(116)(19)] = 65.8\ \text{psi}$$

The allowable beam shear is $v_{cu} = 2\phi\sqrt{f'_c} = 93.5$ psi. The section is satisfactory for beam shear.

Refer to Fig. 12-23 and design the section for flexure. The critical section is at E-E′. The ultimate moment about E-E′ is $M_u = (6.0)(49)^2/[(2)(12)^2] = 50.0$ ft-kips/ft. Now

$$\phi R_u = 12{,}000 M_u/bd^2$$

$$= (50)(12{,}000)/[(12)(19)^2] = 139\ \text{psi/ft}$$

Using Table 5.2, Page 103, for $\phi R_u = 139$ note that $\phi a_u = 2.906$ and $\phi j_u = 0.872$. Thus

$$A_s = M_u/\phi a_u d = 0.905\ \text{in}^2/\text{ft}$$

Using the entire width now, the total area of steel required is

$$A_{st} = (0.905)(9.67) = 8.77\ \text{in}^2$$

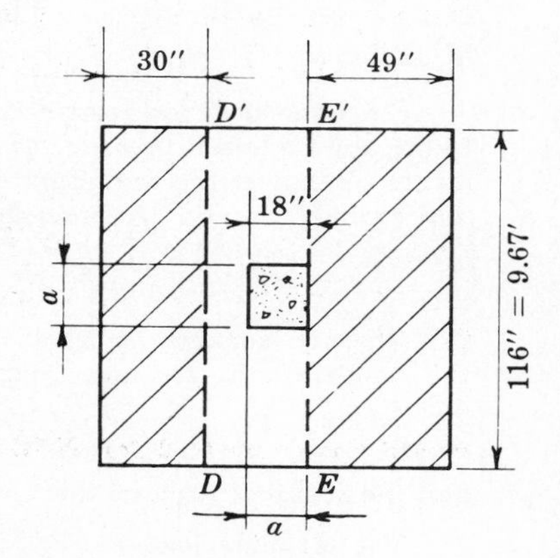

Fig. 12-23

Considering 20 No. 6 bars each way, $A_{st} = 8.8 \text{ in}^2$ and $\Sigma o = 47.2''$. The bar diameter D is $0.75''$.

For flexural bond stress, the critical section is also at E-E'. The total force is

$$V_{Bu} = (6.0)(49/12) = 24.5 \text{ kips/ft}$$

and the bond stress is

$$u_u = V_{Bu}/(\Sigma o \, \phi j_u d) = (24{,}500)(9.67)/[(47.2)(0.872)(19)] = 303 \text{ psi/ft}$$

The allowable bond stress is

$$u = 9.5\sqrt{f_c'}/D = 9.5\sqrt{3000}/0.75 = 694 \text{ psi} < 800$$

The steel and the section satisfy bond requirements.

The development length for the column dowels must now be checked. For the column bars, $D = 1.128''$, $A_s = 1.0 \text{ in}^2/\text{bar}$ and $\Sigma o = 3.54''/\text{bar}$. The allowable development stress is $9.5\sqrt{f_c'}/D = 461$ psi so the development length is

$$L_d = A_s f_y/(\Sigma o \, u) = (1.0)(40{,}000)/[(3.54)(461)] = 24.5''$$

Since the effective depth of the footing is 19″, the full development length of the column dowels cannot be provided. Therefore additional dowels must be used. By proportion, the allowable force for 10 bars is $(19/24.5)A_{st}f_y = 310{,}000$ lb. The total capacity of the 10 No. 9 bars is $(10)(40{,}000) = 400{,}000$ lb. Additional dowels must be provided to account for a force equal to $400{,}000 - 310{,}000 = 90{,}000$ lb.

Consider No. 5 bars for the additional dowels, so $D = 0.625''$. The allowable bond stress is

$$u = 9.5\sqrt{f_c'}/D = 833 \text{ psi} > 800; \quad \text{use 800 psi}$$

Thus

$$\Sigma o = 90{,}000/[(19)(800)] = 5.9''$$

For symmetry, use 4 No. 5 dowels for which $\Sigma o = 7.9''$. The dowels should be extended 19″ into the footing and also 19″ into the column.

The footing weight is actually 27.0 kips compared to the 26.0 kips assumed.

12.4. Design a footing to support a continuous concrete wall as shown in Fig. 12-24. Allowable soil pressure q is 4.75 ksf. The wall dead load reaction is 12 kips/ft and the live load is 6 kips/ft. Use ultimate strength design with $f_c' = 3.0$ ksi, $f_y = 40.0$ ksi. The vertical steel in the wall consists of No. 5 bars at 8″ center to center on both faces. The wall width is 12″.

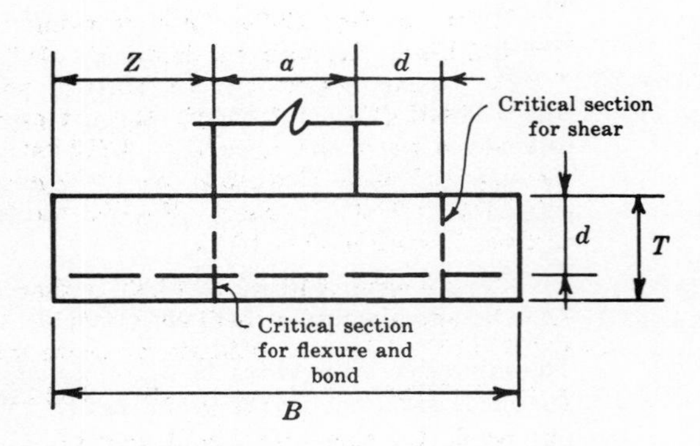

Fig. 12-24

Since the loads are relatively large compared to the allowable soil pressure, the footing will be large. Further, the minimum depth conditions will probably apply. Assume therefore that

$$T = 6'' + \text{bar diameter} + 3'' \text{ cover} = \text{approximately } 10''$$

The *net* working load soil pressure will be equal to the maximum allowable soil pressure less the weight of the footing. Since the footing weighs 12.5 psf per inch thickness,

$$\text{net } q = 4.75 - (10)(12.5/1000) = 4.625 \text{ ksf}$$

and the required area of the footing is $A_F = 18.0/4.625 = 3.89 \text{ ft}^2/\text{ft}$ (say $B = 4.0'$).

The ultimate load is $P_u = 1.5(12) + 1.8(6) = 28.8$ kips/ft. Hence the net ultimate soil pressure is $q_u = P_u/A_F = 28.8/4 = 7.2$ ksf.

The ultimate shear force is $V_u = (7.2)(24 - 6 - 6.5)/12 = 6.9$ kips/ft, and the unit shear stress is $v_u = V_u/bd = 6900/[(12)(6.5)] = 88.5$ psi. Since the allowable ultimate shear stress is $v_{cu} = 2\phi\sqrt{f'_c} = 93.5$ psi, the section satisfactorily resists the shear stress.

For flexural bond, $V_{Bu} = (7.2)(48 - 12)/[(2)(12)] = 10.8$ kips/ft; and for flexural stress, $M_u = (10.8)(48 - 12)(1/2)/[(2)(12)] = 8.1$ ft-kips/ft. Then $\phi R_u = (8.1)(12{,}000)/[(12)(6.5)^2] = 191.7$ psi.

From Table 5.2, Page 103, obtain $\phi a_u = 2.86$ and $\phi j_u = 0.86$; so

$$A_s = M_u/[\phi a_u d] = 8.1/[(2.86)(6.5)] = 0.436 \text{ in}^2/\text{ft}$$

Considering No. 4 bars, u_u is found to be 800 psi, and

$$\Sigma o = 10{,}800/[(800)(0.86)(6.5)] = 2.42 \text{ in/ft}$$

Thus use No. 4 bars at $5\frac{1}{2}''$ center to center in the short direction.

The authors suggest using as the minimum steel in the longitudinal direction: $A_s = 0.002bd = (0.002)(12)(6.5) = 0.156$ in²/ft, which can be obtained by providing No. 3 bars at 8″ center to center.

The wall dowels are No. 5 bars for which $A_s = 0.31$ in²/bar and $\Sigma o = 2.0''$/bar. The maximum allowable bond stress is 800 psi. Thus, the embedment length of the dowels must be

$$L_d = (0.31)(40{,}000)/[(800)(2)] = 7.75''$$

which is greater than the effective depth which is 6.5″.

Additional dowels could be provided, or the footing can be made deeper beneath the wall. The latter is more economical since a continuous *lug* can be provided under the wall as shown in Fig. 12-25. This necessitates merely excavating for the lug and providing a negligible quantity of concrete. The labor involved in placing additional dowels would be more costly than providing a lug to obtain the development length of the dowels.

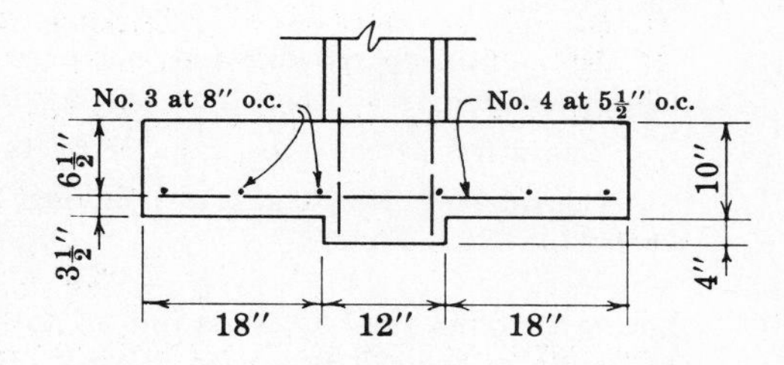

Fig. 12-25

12.5. Use the data given for Problem 12.3 and design a footing considering that its width may not exceed 7′.

Refer to Fig. 12-26. For working loads, $P = 345$ kips. Assuming a depth $d = 19''$ and $T = 23''$, the weight of the footing will be $W_F = (23/12)(150) = 288$ psf; the net pressure will be net $q = 4.0 - 0.288 = 3.712$ ksf; and the required area of footing $A_F = P/(\text{net } q) = 345/3.712 = 92.9$ ft². Since $B = 7'$, the length will be $L = 92.9/7.0 = 13.3'$.

Try a length of 13′-4″ or 13.33′ so that $A_F = 93.3$ ft² and dimension c will be $c = 18'' + 19'' = 37'' = 3.08'$. The net ultimate pressure will be

$$q_u = [(1.5)(200) + (1.8)(145)]/93.3 = 6.02 \text{ kips/ft}^2$$

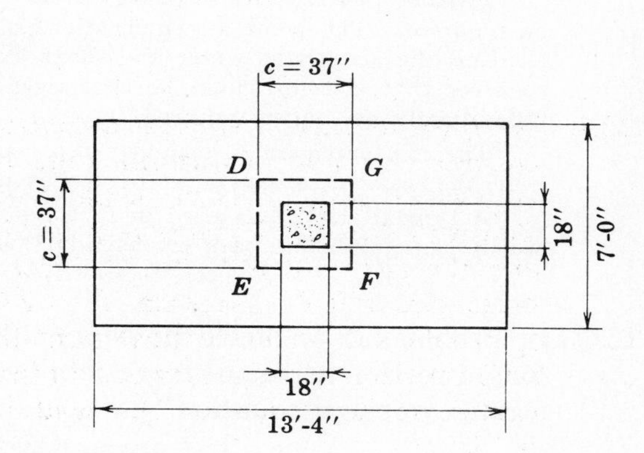

Fig. 12-26

The perimeter shear force will then be

$$V_o = (6.02)\{93.3 - [(37)(37)/144]\} = 504.5 \text{ kips}$$

and the unit shear stress will be

$$v_{ou} = V_o/b_o d = 504{,}500/[(4)(37)(19)] = 179.4 \text{ psi}$$

The allowable unit shear stress around the perimeter is $v_{cu} = 4\phi\sqrt{f'_c} = 187$ psi. The section is satisfactory for punching shear.

Beam shear must now be checked (see Fig. 12-27). This will be more critical in the long direction than in the short direction. The total shear force will be

$$V_u = (52/12)(7)(6.02) = 182.6 \text{ kips}$$

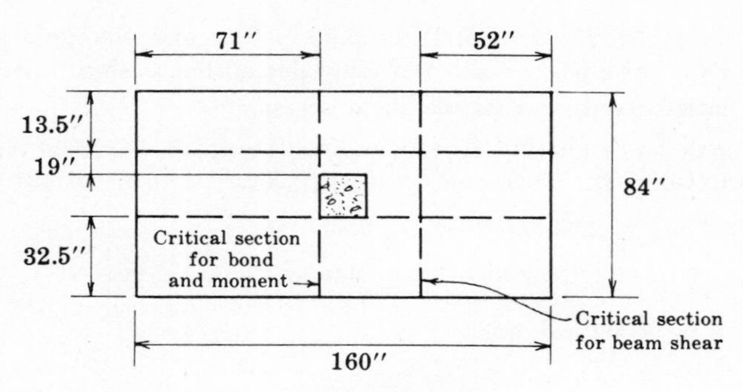

Fig. 12-27

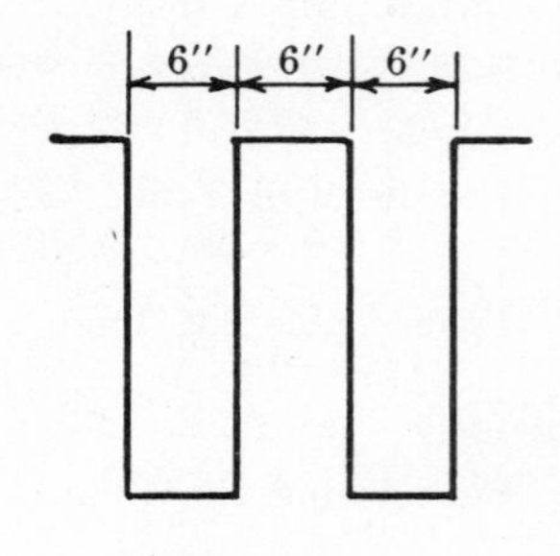

Fig. 12-28

The unit shear stress will be

$$v_u = V_u/bd = 182{,}600/[(84)(19)] = 114.4 \text{ psi}$$

The allowable unit shear stress is $v_{cu} = 2\phi\sqrt{f'_c} = 93.5$ psi, so the section is overstressed.

It is usually considered more desirable to increase the effective depth to resist shear in footings than to provide web reinforcement. However, web reinforcement will be designed here to illustrate the process.

If the stirrup spacing does not exceed $d/2$, the total shear stress may be as much as $6\phi\sqrt{f'_c}$ or 280 psi, which is greater than the actual shear stress, 114.4 psi.

The stirrups must be capable of taking the excess stress $v'_u = 114.4 - 93.5 = 20.9$ psi.

Considering No. 3 stirrups having 2 vertical legs per foot laterally, the area provided will be 0.22 in²/ft. Thus,

$$s = A_v(\tfrac{1}{2}f_y)\phi/v'_u b = (0.22)(20{,}000)(0.85)/[(20.9)(12)] = 14.9''$$

noting that the steel stress is limited to $\frac{1}{2}f_y$. The bars may be spaced at approximately 14″ center to center to satisfy this requirement. However, since $d/2 = 19/2 = 9.5''$, the stirrups may not be placed farther than 9.5″ apart in the area where such are required.

Farther out from the critical section the shear stress decreases rapidly, so stirrups will not be required. The point at which stirrups may theoretically be discontinued is obtained simply by locating the section at which the shear stress allowed is not exceeded. However, the ACI Code requires that stirrups must be continued for a distance d beyond the point at which they are theoretically no longer required.

The reinforcement is designed in a manner identical to that illustrated in Problem 12.3. However, the transverse bars must be spaced according to the criteria indicated in Fig. 12-13.

A typical stirrup as used in footings is illustrated in Fig. 12-28. In order to provide a more sturdy *cage*, alternate bars are usually inverted.

12.6. In Problem 12.5, check the beam shear in the long direction using the more detailed ultimate strength design equation

$$v_{cu} = \phi(1.9\sqrt{f'_c} + 2500p_w Vd/M)$$

Use $A_s = 14.4$ in².

Refer to Fig. 12-29. Recall from Problem 12.5 that $V_u = 182.6$ kips/ft, $v_u = 114.4$ psi, $d = 19''$.

At the critical section, $M_u = 394$ ft-kips/ft and $Vd/M = (182.6)(19)/[(394)(12)] = 0.733$. Then

$$2500p_w Vd/M = [(2500)(14.4)/(84)(19)](0.733) = 16.5 \text{ psi}$$

and $1.9\sqrt{f'_c} = 104.0$ psi. Hence

$$v_{cu} = (0.85)(104.0 + 16.5) = 102.5 \text{ psi}$$

Stirrups are also required using the detailed analysis. In this case the detailed analysis does not provide any advantage over the approximate analysis.

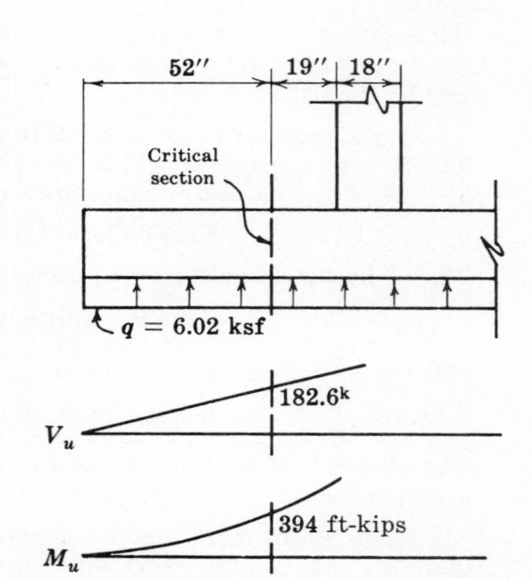

Fig. 12-29

12.7. Design a pile footing to support an 18″ square column subjected to a live load reaction of 170 kips and a dead load reaction of 150 kips at working conditions. The testing laboratory recommends an ultimate pile load of 70 kips per pile, and a working pile load of 40 kips per pile. The vertical steel in the column consists of 12 No. 7 bars. Use $f'_c = 3.0$ ksi, $f_y = 40.0$ ksi, and 12″ diameter piles.

Since the footing weight will be about 3 kips/pile, the net working load per pile is $40.0 - 3.0 = 37.0$ kips/pile. The number of piles required is $N = W/P = 320/37 = 8.65$ or 9 piles. Use a pile pattern as shown in Fig. 12-30.

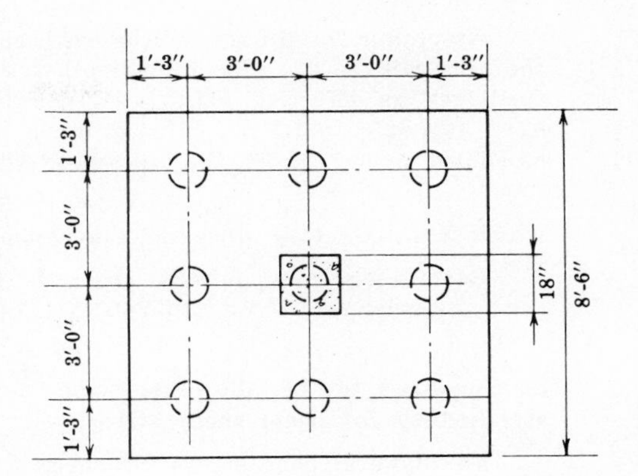

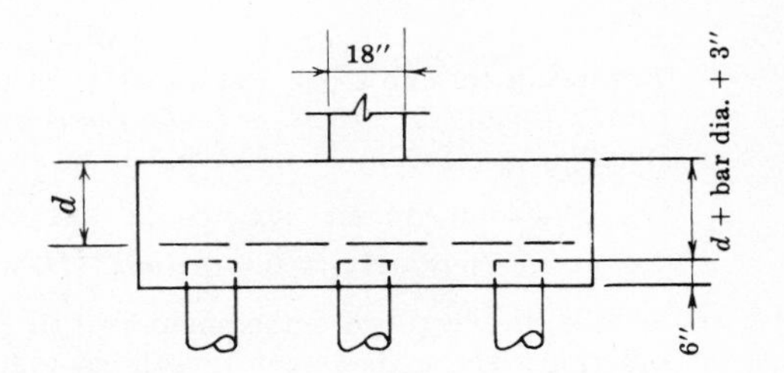

Fig. 12-30

The net ultimate load is used to design the footing; thus $W_u = (1.5)(150) + (1.8)(170) = 531$ kips, and the load per pile is $P_u = 531/9 = 59.0$ kips/pile, which is less than the maximum ultimate load, 70 kips/pile.

Punching shear around a single pile often governs the footing depth determination, except in cases in which the loads are small. In this case, it will be shown that beam shear governs.

Referring to Fig. 12-31, we calculate the punching shear stress.

After several trials, assume $d = 19.5''$. The shear perimeter is $b_o = \pi(12 + d) = 99.0''$, and the unit shear stress is

$$v_{ou} = P_u/b_o d = 59{,}000/(99)(19.5) = 30.6 \text{ psi}$$

The allowable shear stress is $v_{cu} = 4\phi\sqrt{f'_c} = 187$ psi, so the depth is satisfactory for pile punching shear.

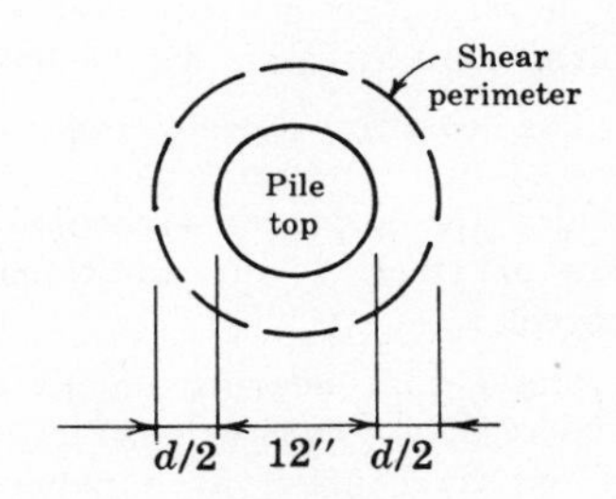

Fig. 12-31

Perimeter shear around the column must now be checked. Refer to Fig. 12-32.

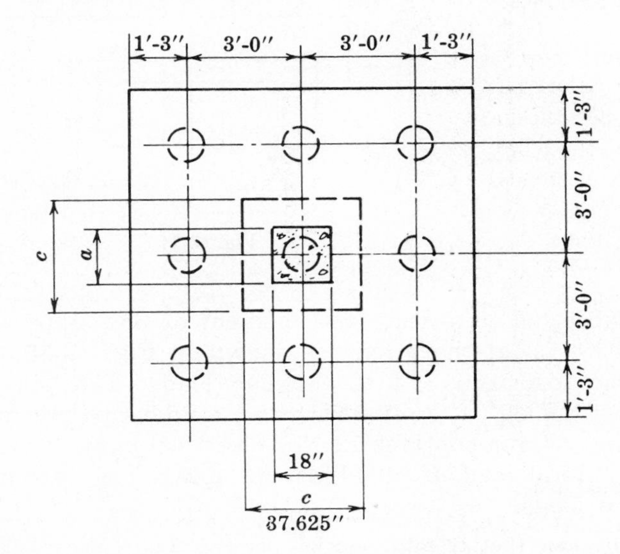

Fig. 12-32

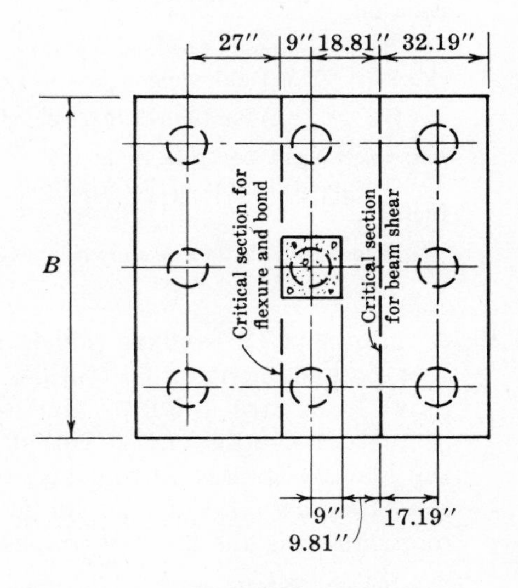

Fig. 12-33

Assuming No. 6 bars will be used, clearance above the pile butts will be 3″ and embedment of the piles will be 6″. The total depth required will be 28.75″. For practical reasons use 29″; this furnishes an effective depth $d = 19.625''$. Thus $c = a + d = 18.0 + 19.625 = 37.625''$ and $b_o = 4(37.625) = 150.0''$. Hence, since $V_{ou} = 472$ kips on 8 piles outside of the critical section as shown on Fig. 12-32, the unit shear stress will be

$$v_{ou} = V_{ou}/b_o d = 472{,}000/(150)(19.625) = 160.4 \text{ psi}$$

which is satisfactory since the allowable shear stress is 187 psi.

Beam shear must now be checked. Refer to Fig. 12-33 above. Three piles exist beyond the critical section, so $V_u = (3)(59.0) = 177.0$ kips. Since $b = B = 8'\text{-}6'' = 102''$,

$$v_u = V_u/bd = 177{,}000/(102)(19.625) = 88.4 \text{ psi}$$

as compared to the allowable beam shear stress $v_{cu} = 2\phi\sqrt{f_c'} = 93.5$ psi, so the section is satisfactory for beam shear stress.

For flexural bond stress the shear force is also 177.0 kips and the bending moment about the face of the column is $M_u = (177.0)(27/12) = 398.3$ ft-kips. Then

$$\phi R_u = 12{,}000 M_u/bd^2 = 12.17 \text{ psi}$$

Referring to Table 5.2, Page 103, note that p is less than 0.002. *The authors recommend providing* min. $p = 200/f_y$. (*Not a Code requirement.*) Thus $p = 200/40{,}000 = 0.005$ and $A_s = pbd = (0.005)(102)(19.625) = 10.0$ in² or 17 No. 7 bars each way.

The embedment length to satisfy flexural bond is $L_d = A_s f_y/(u\Sigma o) = 14.9''$, where $\Sigma o = 2.7''/\text{bar}$, $A_s = 0.6$ in²/bar, $D = 0.875$, and $u = 9.5\sqrt{f_c'}/D = (9.5)(54.8)/0.875 = 595$ psi.

If the required embedment length is less than 0.8 times the embedment length provided, flexural bond stress is not calculated. To exercise this provision the available embedment length must not be less than $14.9/0.8 = 18.63''$. Since the bar extension beyond the critical section exceeds 18.67″, flexural bond stress is not considered.

The column dowels require an embedment length of 14.9″ for No. 7 bars, and the footing depth provides for a larger embedment above the two layers of steel. Thus additional dowels are not required.

The actual bearing stress on the footing is $531{,}000/(18)^2 = 1638$ psi directly beneath the column on the bearing area, whereas the allowable bearing stress is $f_{bu} = (1.9)(0.375 f_c') = 2138$ psi, so the section does not require a load distributing pedestal.

The assumed footing weight must now be checked. The total weight is

$$W_F = (8.5)(8.5)(29)(12.5)/1000 = 26.2 \text{ kips}$$

and the weight per pile is $26.2/9 = 2.91$ kips/pile. The assumed weight of 3.0 kips/pile is most satisfactory.

The final details are shown in Fig. 12-34.

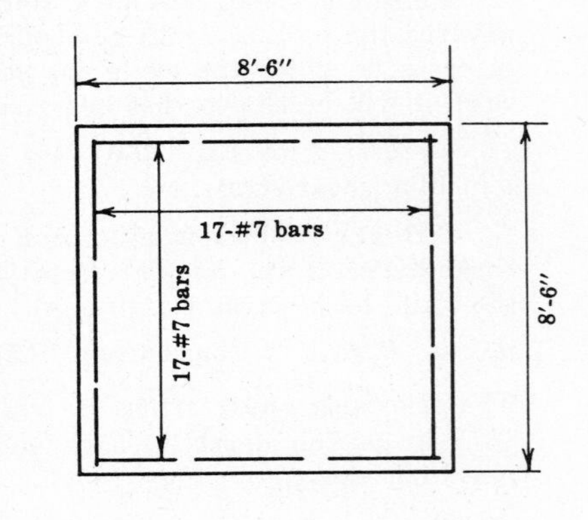

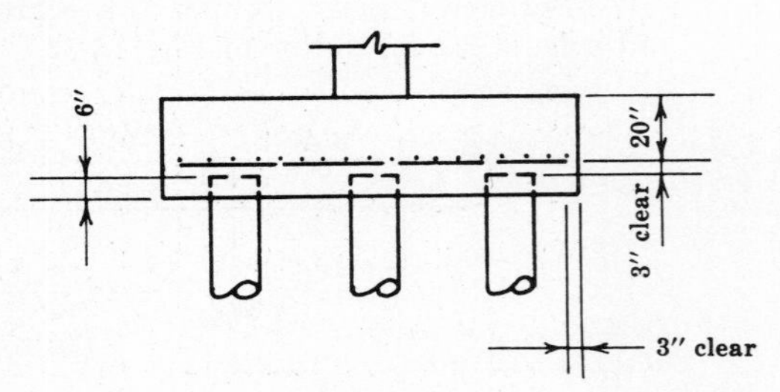

Fig. 12-34

12.8. A column is 18″ square and is subjected to a dead load moment of 85 ft-kips and a live load moment of 33 ft-kips, both about the y axis as shown in Fig. 12-35. The axial loads are 143 kips for dead load and 59 kips for live load. The pile butt diameters average 12″. Use $f_c' = 3.0$ ksi, $f_y = 40.0$ ksi, and a 25.0 kips/pile working load to design the footing. The column contains 12 No. 7 vertical bars. Consider that the *test piles* failed under a total load of 50.0 kips per pile. The forces and moments are due to superimposed loads.

Several sets of trial calculations indicate that 12 piles will be required, and a pile pattern as shown in Fig. 12-35 is selected. The total working load is 202 kips, and $M = 118$ ft-kips.

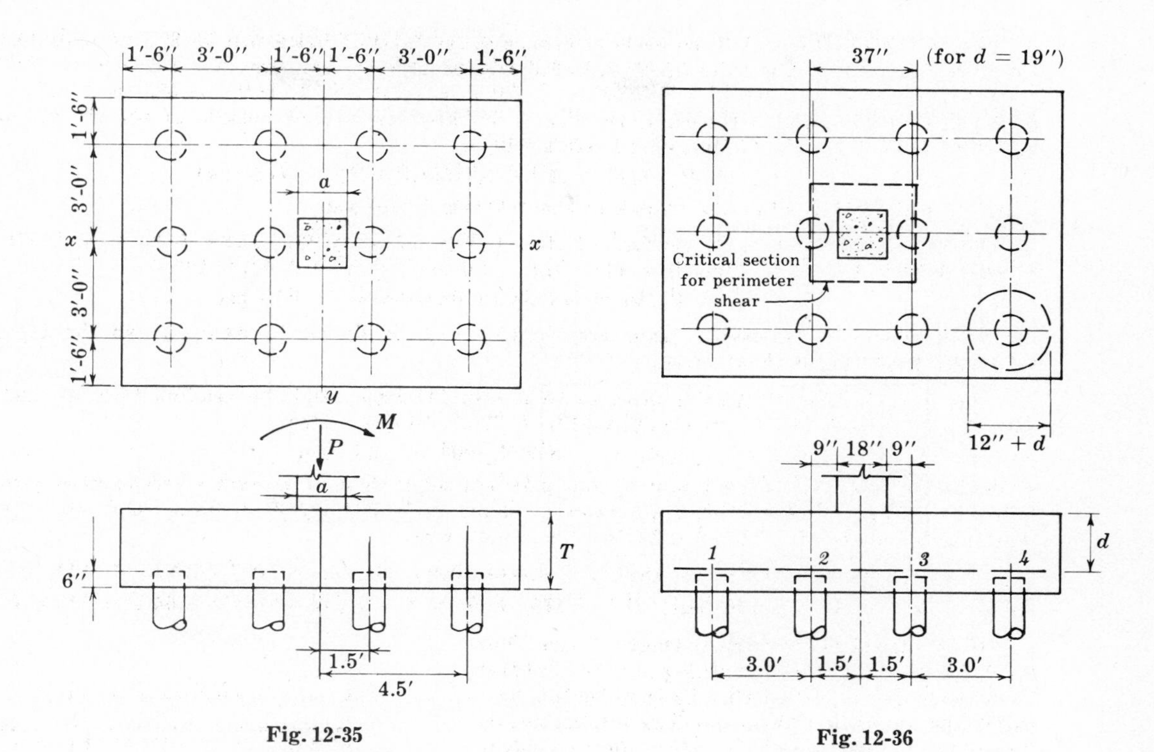

Fig. 12-35

Fig. 12-36

Since the footing is subjected to bending, the *equivalent moment of inertia* of the piles about the centroid of the footing must be calculated:

$$\Sigma d_y^2 \;=\; (2)(3)(1.5)^2 + (2)(3)(4.5)^2 \;=\; 135 \text{ ft}^2$$

The load on any pile is given by

$$P_i \;=\; Md_i/\Sigma d_y^2 \;\pm\; R/N \qquad (a)$$

Then $\qquad \max. P_i \;=\; (118)(4.5)/135 \;+\; 202/12 \;=\; 20.76 \text{ kips}$

This allows $25.0 - 20.76 = 4.24$ kips/pile for the weight of the footing.

The ultimate load on each pile can be obtained by substituting the appropriate distance d into equation (a). Then, referring to Fig. 12-36,

for row *1*, $P_u = 20.5$ kips/pile $\qquad$ for row *3*, $P_u = 28.9$ kips/pile

for row *2*, $P_u = 24.7$ kips/pile $\qquad$ for row *4*, $P_u = 33.1$ kips/pile

based on the ultimate load and moment. Since the largest value of P_u plus 1.5 times the footing weight per pile will not exceed the test load or 50.0 kips, pile failure is not anticipated.

Punching shear around a single pile is governed by the most heavily loaded pile, as shown in Fig. 12-36.

After several trials an approximate depth d is obtained as 19". The shear perimeter around the pile is $b_o = \pi(d + 12) = 97.4''$, and the maximum shear stress is

$$v_{ou} \;=\; V_u/b_o d \;=\; 33{,}100/(97.4)(19) \;=\; 17.9 \text{ psi}$$

The allowable punching shear is $v_{cu} = 4\phi\sqrt{f'_c} = 187$ psi, so the section is satisfactory for punching shear.

For slab shear note that rows *2* and *3* are $\frac{1}{2}''$ inside of the critical section. Therefore only partial effects of those piles are considered. The governing equation for the modified pile load is $P' = P(6 - X)/12 = 0.46P$, where X is $\frac{1}{2}''$, as shown in Fig. 12-37 below. The slab shear is

$$V_{ou} \;=\; 3(20.5) + 2(24.7) + 0.46(24.7) + 0.46(28.9) + 2(28.9) + 3(33.1) \;=\; 292.65 \text{ kips}$$

and the shear perimeter is $b_o = 4(37) = 148''$. Then the unit shear stress is

$$v_{ou} = V_{ou}/b_o d = 292{,}650/(148)(19) = 104 \text{ psi}$$

which is less than the previously established allowable value, 187 psi.

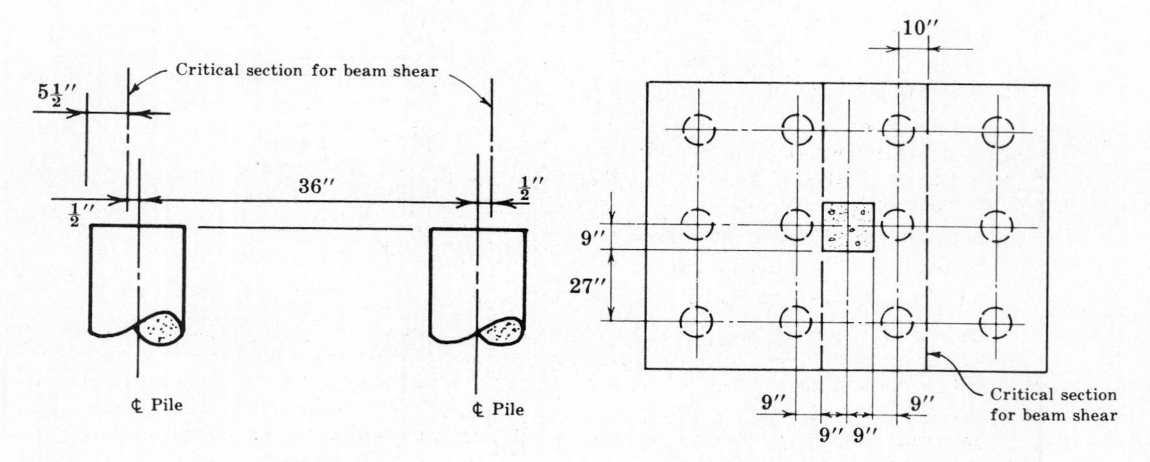

Fig. 12-37

Fig. 12-38

Beam shear must now be checked. The most stressed piles have loads $P_u = 33.1$ kips. From Fig. 12-38, for shear in the short direction, $V_u = 3(33.1) = 99.3$ kips; hence the unit shear stress is $v_u = 99{,}300/(108)(19) = 48.4$ ksi. For the long side, the shear force is $V_u = 20.5 + 24.7 + 28.9 + 33.1 = 107.2$ kips, and the unit shear stress is $v_u = 107{,}200/(144)(19) = 39.2$ psi. The allowable shear stress for beam action is $v_{cu} = 2\phi\sqrt{f_c'} = 93.5$ psi, so the section is adequate to resist shear.

The critical section for moment and bond occurs at the column face on the side of the heavy pile loads. Hence $V_{Bu} = 3(28.9) + 3(33.1) = 186.0$ kips, $M_u = (86.7)(1.5) + (99.3)(4.5) = 577$ ft-kips, and

$$\phi R_u = 12{,}000 M_u/bd^2 = 178.0 \text{ psi}$$

From Table 5.2, Page 103, note that $\phi a_u = 2.875$, $\phi j_u = 0.863$ and $p = 0.00518$. Thus $A_s = M_u/\phi a_u d = 577/(2.875)(19) = 10.6 \text{ in}^2$, or alternatively, $A_s = pbd = 0.00518(108)(19) = 10.6 \text{ in}^2$.

A very large embedment length is available, so flexural bond will not be critical.

In the transverse direction the same process is followed, except that a decision must be made concerning the pile loads to be used since each pile is loaded differently. For practical reasons, the largest load is usually used, and the lesser loads are considered to be equal to this load. Although this might produce an excess of steel over the more lightly loaded piles, a safeguard against reversals of load is developed in this manner.

A check will show that an average moment over the most heavily loaded pile is 24.8 ft-kips/ft and that $A_s = 5.35 \text{ in}^2$ total in the transverse direction satisfies moment requirements. However, since that area is less than $200/f_y$, *the authors recommend using* $(1.33)(5.35) = 7.12 \text{ in}^2$.

Axial load bearing on the pile cap must also be checked, using the average load R/N and ignoring the effect of moment. Thus, under the column, $f_b = 321{,}000/324 = 990$ psi compared to the allowable stress $f_{bu} = 1.9(0.375)(3000) = 2140$ psi.

Using a total depth of 30″, the weight of the footing per pile is 4.05 kips compared to the value 4.24 kips/pile permitted.

12.9. Fig. 12-39 shows a multiple column footing which must be designed to support two columns, A and B. For column A, $P_{DL} = 110$ kips and $P_{LL} = 80$ kips. For column B, $P_{DL} = 190$ kips and $P_{LL} = 120$ kips. The columns are oriented near the property lines as shown in the figure. The *ultimate* soil pressure is 6.5 ksf. Establish the dimensions of the footing and draw the shear and bending moment diagrams for the long direction.

For column A,

$$P_u = 1.5(110) + 1.8(80) = 309 \text{ kips}$$

For column B,

$$P_u = 1.5(190) + 1.8(120) = 501 \text{ kips}$$

The total load is 810 kips.

Assuming the soil pressure due to 1.5 times the footing weight will be approximately 0.525 ksf (after several trials), the net *ultimate* soil pressure is

$$\text{net } q_u = 6.5 - 0.525 = 5.975 \text{ ksf}$$

and the required footing area is

$$A_F = 810/5.975 = 135.5 \text{ ft}^2$$

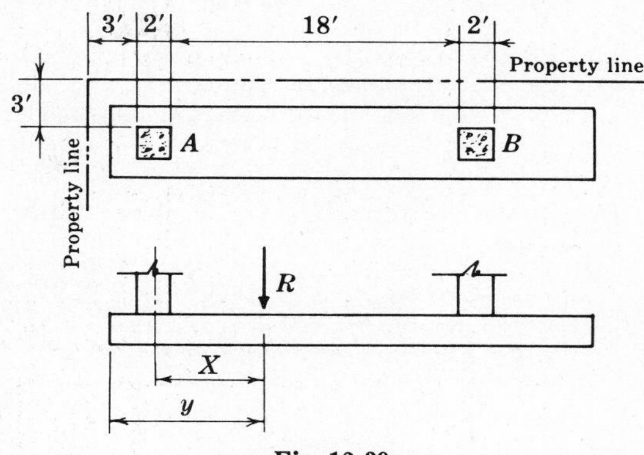

Fig. 12-39

The center of gravity of the loads is located at a distance $X = (501)(20)/810 = 12.37'$ from the center of column A. Considering the overhang at A to be 1.63′, $y = 12.37 + 1 + 1.63 = 15.0' = L/2$. Hence the length of the footing must be 30′ in order that the centroid of the footing and the center of gravity of the loads will coincide.

Since the length is 30′ and the area is 135.5 ft², the width of the footing must be 4.5′, and the net soil pressure is 6.0 ksf.

The shear and bending moment diagrams for the footing design are shown in Fig. 12-40. Although it is usually considered desirable to make footings as nearly square as possible, it is often necessary to build long narrow footings of the type developed in this problem.

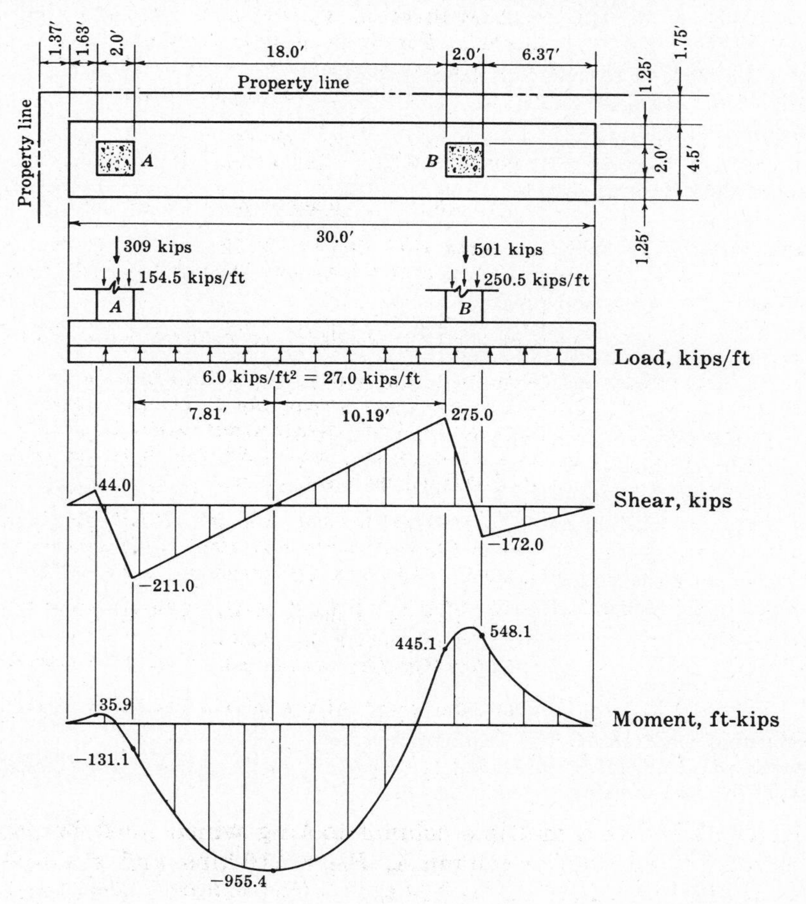

Fig. 12-40

Supplementary Problems

12.10. For the footing shown in Fig. 12-41, $P_{DL} = 180$ kips, $P_{LL} = 120$ kips, $f'_c = 3.0$ ksi, $f_y = 40.0$ ksi and $f_s = 20.0$ ksi. Determine the required effective depth. Use working strength design and net $q = 3.7$ ksf. *Ans.* $d = 18.1''$

12.11. Design the reinforcement for Problem 12.10. (Fig. 12-41) *Ans.* $A_s = 9.5$ in²

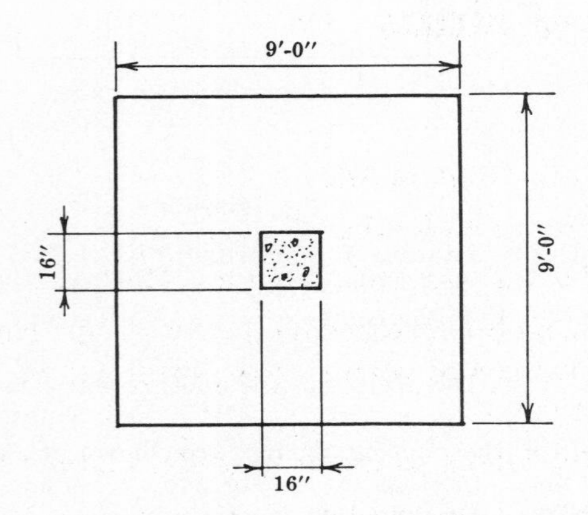

Fig. 12-41

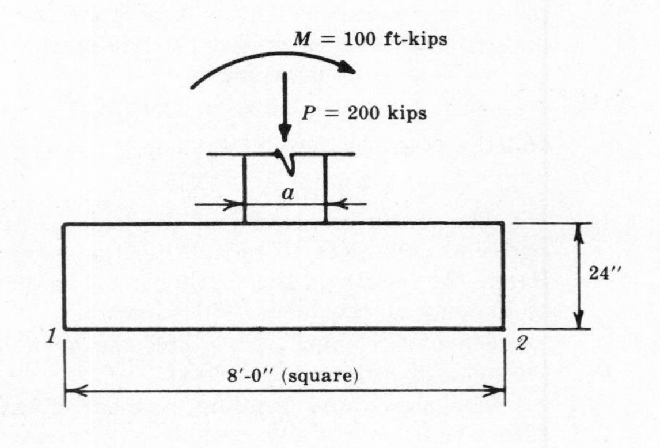

Fig. 12-42

12.12. Solve Problem 12.10 using ultimate strength design and net $q_u = 6.0$ ksf.
Ans. $d = 17.6''$

12.13. Design the reinforcement for Problem 12.12. (Fig. 12-41)
Ans. $A_s = 9.5$ in² ($200/f_y$ governs)

12.14. Determine the gross soil pressure at points *1* and *2* for the footing shown in Fig. 12-42. Include the weight of the footing.
Ans. $q_1 = 2.25$ ksf, $q_2 = 4.59$ ksf

12.15. Determine the maximum and minimum net pile loads for the footing shown in Fig. 12-43 (not including footing weight).
Ans. $P_{max} = 36.11$ kips, $P_{min} = 30.55$ kips

12.16. If $M = 0$, $f'_c = 3.0$ ksi and $f_y = 40.0$ ksi, determine the required effective depth for the footing shown in Fig. 12-43. The pile butts are 12″ in diameter. Use the net pile load and working strength design. *Ans.* $d = 17.3''$

12.17. Determine the area of steel required for the footing of Problem 12-16. Use $d = 17.3''$, $j = 0.875$, and working strength design.
Ans. $A_s = 8.9$ in²

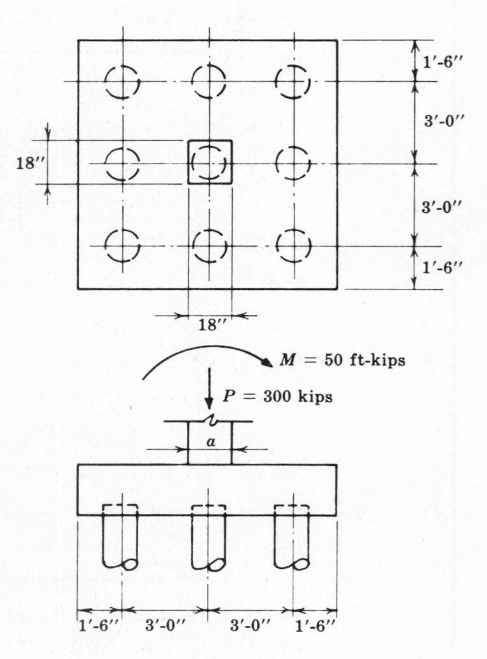

Fig. 12-43

Chapter 13

Two-way Slabs

NOTATION

A = length of clear span in short direction, Method 3

B = length of clear span in long direction, Method 3

B = bending moment coefficient for one-way construction, Method 1

C = factor modifying bending moments prescribed for one-way construction for use in proportioning the slabs and beams in the direction of L of slabs supported on four sides, Method 1

C = moment coefficient for two-way slabs as given in table, Method 2

C = moment coefficients for two-way slabs as given in tables. Coefficients have identifying indexes, such as $C_{A\ neg}$, $C_{B\ neg}$, $C_{A\ DL}$, $C_{B\ DL}$, $C_{A\ LL}$, $C_{B\ LL}$, Method 3.

C_s = ratio of the shear at any section of a slab strip distant xL from the support to the total load W on the strip in direction of L, Method 1

C_b = ratio of the shear at any section of a beam distant xL from the support to the total load W on the beam in the direction of L, Method 1

g = ratio of span between lines of inflection to L in the direction of span L, when span L only is loaded, Method 1

g_1 = ratio of span between lines of inflection to L_1 in the direction of span L_1, when span L_1 only is loaded, Method 1

L = length of clear span, Method 1

L_1 = length of clear span in the direction normal to L, Method 1

m = ratio of short span to long span for two-way slabs, Method 2

r = gL/g_1L_1, Method 1

S = length of short span for two-way slabs. The span shall be considered as the center-to-center distance between supports or the clear span plus twice the thickness of slab, whichever value is the smaller, Method 2

w = total uniform load per sq ft

W = total uniform load between opposite supports on slab strip of any width or total slab load on beam when considered as one-way construction, Method 1

w = uniform load per sq ft. For negative moments and shears, w is the total dead load plus live load. For positive moments, w is to be separated into dead and live loads, Method 3

w_A, w_B = percentages of load w in A and B directions. These shall be used for computations of shear and for loadings on supports, Method 3

x = ratio of distance from support to any section of slab or beam, to span L or L_1, Method 1

INTRODUCTION

The 1963 ACI Building Code defines two-way slabs with supports on four sides: "This construction, reinforced in two directions, includes solid reinforced concrete slabs; concrete joists with fillers of hollow concrete units or clay tile, with or without concrete top slabs; and concrete joists with top slabs placed monolithically with the joists. The slab shall be supported by walls or beams on all sides..."

Fig. 13-1 shows a view of a two-way solid slab supported on all sides by concrete beams.

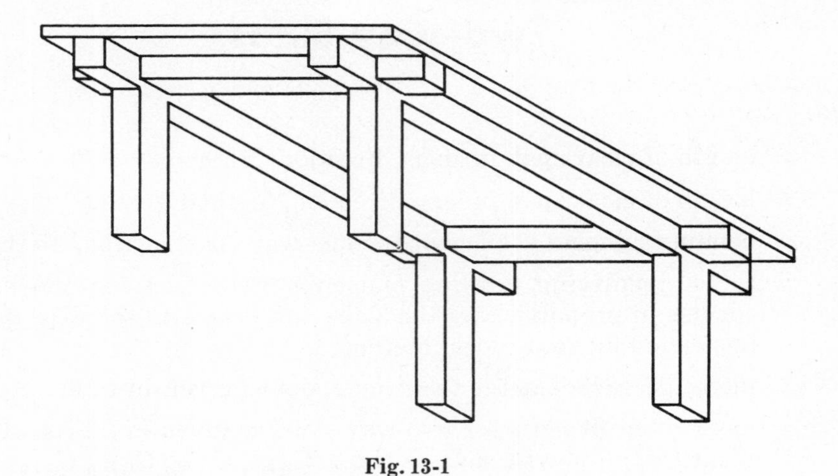

Fig. 13-1

The 1963 ACI Building Code also states: "The slab shall be designed by approved methods which shall take into account the effect of continuity and fixity at supports, the ratio of length to width of slab and the effect of two-way action."

Hence in two-way slabs the reinforcing steel is usually placed so that flexural resistance will be provided in two directions.

Slabs reinforced in two directions but not supported on four sides will be discussed in Chapter 14. This type slab supported on columns only at the corners of the panel is termed a flat slab or flat plate.

ANALYSIS OF TWO-WAY SLABS

Two-way slabs are extremely complex and statically indeterminate. Many attempts, analytical and empirical, have been made to determine the division of the moments and shears between the two spans and the distribution of these along the principal axes of the slab.

The slab shown in Fig. 13-2 is subjected to a uniform vertical load. If the slab is supported on all sides by non-yielding supports, the deflection at e of a central strip ab must be equal to the deflection at e of the other central strip cd. The deflection of a simple beam uniformly loaded is $5wl^4/384EI$. If the strips have equal thickness, the deflection of strip ab is $kw_{ab}S^4$ and the deflection of strip cd is $kw_{cd}L^4$, where w_{ab} and w_{cd} are the portions of the load carried by strips ab and cd respectively. If the total load

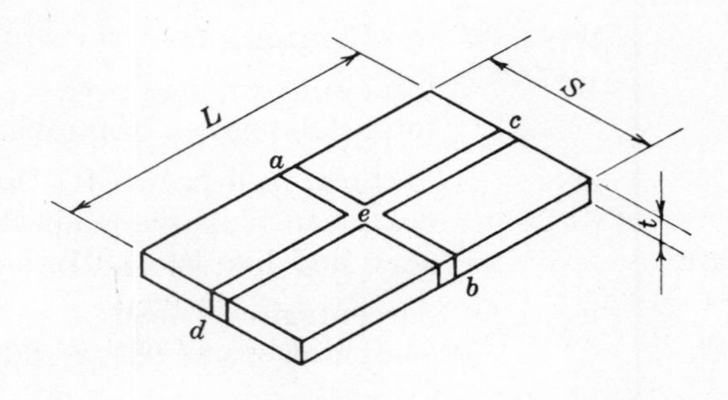

Fig. 13-2

$w = w_{ab} + w_{cd}$, equating deflections of each strip at point e, $w_{ab} = wL^4/(L^4 + S^4)$ and $w_{cd} = wS^4/(L^4 + S^4)$.

If $L = 1.5S$, $w_{ab} = 0.835w$ and $w_{cd} = 0.165w$. If $L = 2.0S$, $w_{ab} = 0.941w$ and $w_{cd} = 0.059w$. Thus the shorter span must resist the major portion of the load. If L/S is greater than 2, the portion of the load carried by the long span becomes less significant and the slab is often proportioned as a one-way slab.

Many elastic analyses of two-way slabs have been proposed. All have shortcomings. They neglect Poisson's ratio, torsion, changes in stiffness, ultimate capacity, edge restraint, variation of moments and shears along span, and others. As an example: if strip ab in Fig. 13-2 is not a central strip and is moved toward one end of the panel, point e is no longer at the midpoint of strip cd. Hence the constant in the expression for the deflection of cd is not 5/384 and the results of equating deflections would not be the same as obtained before.

As strip ab approaches the edge of the panel, the deflection at e approaches zero. Hence the load sustained by strip ab approaches zero while that sustained by strip cd remains the same. Empirical test results and analytical investigations by Dr. H. M. Westergaard, Joseph Di Stasio, M. P. van Buren and others have demonstrated within reasonable limits the distribution of the bending moments in two-way slabs.

The 1963 ACI Building Code presents three methods of design of two-way slabs. These are based on other than an elastic analysis like that presented herein. Rather than determining the exact distribution of bending moments, the slab is divided into column and middle strips as shown in Fig. 13-3 and the value of the moment is assumed constant across the full width of a strip.

The analysis of two-way slab systems is further complicated by the variation in shear along the panel edge. The methods of analysis proposed by the 1963 ACI Building Code contain different requirements concerning the load distribution to the supporting beams. These requirements will be discussed in the solved problems.

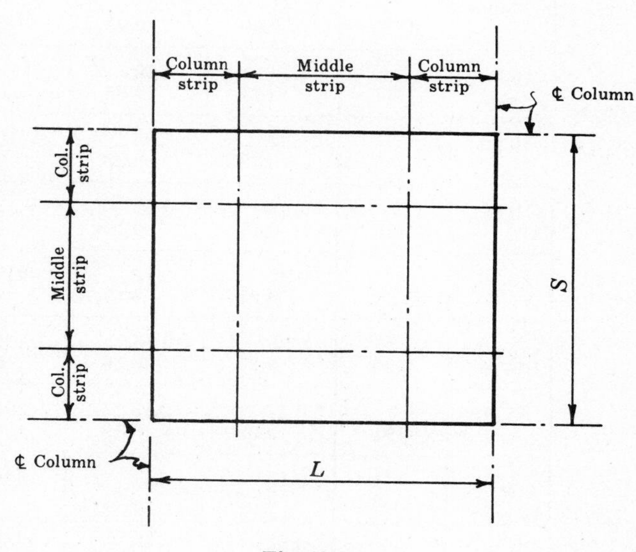

Fig. 13-3

The 1963 ACI Building Code requires that the minimum thickness of two-way slabs shall not be less than $3\frac{1}{2}$ inches nor less than the perimeter of the slab divided by 180.

When a two-way slab is subjected to a vertical load, the slab deflects forming a dish-like surface and the corners tend to lift off of the supports. This action may cause distress in the vicinity of the corners. Hence the 1963 ACI Building Code requires special reinforcement in these regions. This reinforcement should be provided for a distance from each corner not less than 1/5 of the longest span. The steel area should be equal to that of the maximum positive moment slab steel and be placed in the top and bottom. The top bars must be placed parallel to the diagonal of the panel and the bottom bars are placed perpendicular to the diagonal or parallel to both sides of the slab.

TABLES OF DESIGN COEFFICIENTS

Tables 13.1 and 13.2 contain coefficients for slab moments and shears and beam moments and shears respectively for what is termed Method 1. Tables 13.1 and 13.2 are reproduced with permission of the ACI.

TABLE 13.1

METHOD 1 — SLABS

Upper figure Lower figure		C_s C_{s1}					C C_1
r	$r_1 = \frac{1}{r}$	Values of x					
		0.0	0.1	0.2	0.3	0.4	
0.00	∞	0.50 0.00	0.40 0.00	0.30 0.00	0.20 0.00	0.10 0.00	1.00 0.00
0.50	2.00	0.44 0.06	0.36 0.03	0.27 0.02	0.18 0.00	0.09 0.00	0.89 0.06
0.55	1.82	0.43 0.07	0.33 0.04	0.23 0.02	0.15 0.01	0.07 0.00	0.79 0.08
0.60	1.67	0.41 0.09	0.30 0.05	0.20 0.03	0.12 0.01	0.05 0.00	0.70 0.10
0.65	1.54	0.39 0.11	0.28 0.06	0.18 0.03	0.10 0.01	0.04 0.00	0.64 0.13
0.70	1.43	0.37 0.13	0.26 0.08	0.16 0.04	0.09 0.01	0.03 0.00	0.58 0.15
0.80	1.25	0.33 0.17	0.22 0.10	0.13 0.06	0.07 0.02	0.02 0.00	0.48 0.21
0.90	1.11	0.29 0.21	0.19 0.13	0.11 0.07	0.05 0.03	0.01 0.01	0.40 0.27
1.00	1.00	0.25 0.25	0.16 0.16	0.09 0.09	0.04 0.04	0.01 0.01	0.33 0.33
1.10	0.91	0.21 0.29	0.13 0.19	0.07 0.11	0.03 0.05	0.01 0.01	0.28 0.39
1.20	0.83	0.18 0.32	0.11 0.21	0.06 0.13	0.02 0.06	0.00 0.02	0.23 0.45
1.30	0.77	0.16 0.34	0.10 0.23	0.05 0.14	0.02 0.07	0.00 0.03	0.19 0.51
1.40	0.71	0.13 0.37	0.08 0.25	0.04 0.16	0.02 0.09	0.00 0.03	0.16 0.57
1.50	0.67	0.11 0.39	0.07 0.27	0.04 0.17	0.01 0.10	0.00 0.04	0.14 0.61
1.60	0.63	0.10 0.40	0.06 0.29	0.03 0.19	0.01 0.11	0.00 0.05	0.12 0.66
1.80	0.55	0.07 0.43	0.04 0.33	0.02 0.23	0.01 0.15	0.00 0.07	0.08 0.79
2.00	0.50	0.06 0.44	0.03 0.36	0.02 0.27	0.00 0.18	0.00 0.09	0.06 0.89
∞	0.00	0.00 0.50	0.00 0.40	0.00 0.30	0.00 0.20	0.00 0.10	0.00 1.00

TABLE 13.2

METHOD 1 — BEAMS

Upper figure: C_b Lower figure: C_{b1}		Values of x					$1-C$ $1-C_1$
r	$r_1 = \frac{1}{r}$	0.0	0.1	0.2	0.3	0.4	
0.00	∞	0.00 0.50	0.00 0.40	0.00 0.30	0.00 0.20	0.00 0.10	0.00 1.00
0.50	2.00	0.06 0.44	0.04 0.37	0.03 0.28	0.02 0.20	0.01 0.10	0.11 0.94
0.55	1.82	0.07 0.43	0.07 0.36	0.07 0.28	0.05 0.19	0.03 0.10	0.21 0.92
0.60	1.67	0.09 0.41	0.10 0.35	0.10 0.27	0.08 0.19	0.05 0.10	0.30 0.90
0.65	1.54	0.11 0.39	0.12 0.34	0.12 0.27	0.10 0.19	0.06 0.10	0.36 0.87
0.70	1.43	0.13 0.37	0.14 0.32	0.14 0.26	0.11 0.19	0.07 0.10	0.42 0.85
0.80	1.25	0.17 0.33	0.18 0.30	0.17 0.24	0.13 0.18	0.08 0.10	0.52 0.79
0.90	1.11	0.21 0.29	0.21 0.27	0.19 0.23	0.15 0.17	0.09 0.09	0.60 0.73
1.00	1.00	0.25 0.25	0.24 0.24	0.21 0.21	0.16 0.16	0.09 0.09	0.67 0.67
1.10	0.91	0.29 0.21	0.27 0.21	0.23 0.19	0.17 0.15	0.09 0.09	0.72 0.61
1.20	0.83	0.32 0.18	0.29 0.19	0.24 0.17	0.18 0.14	0.10 0.08	0.77 0.55
1.30	0.77	0.34 0.16	0.30 0.17	0.25 0.16	0.18 0.13	0.10 0.07	0.81 0.49
1.40	0.71	0.37 0.13	0.32 0.15	0.26 0.14	0.18 0.11	0.10 0.07	0.84 0.43
1.50	0.67	0.39 0.11	0.33 0.13	0.26 0.13	0.19 0.10	0.10 0.06	0.86 0.39
1.60	0.63	0.40 0.10	0.34 0.11	0.27 0.11	0.19 0.09	0.10 0.05	0.88 0.34
1.80	0.55	0.43 0.07	0.36 0.07	0.28 0.07	0.19 0.05	0.10 0.03	0.92 0.21
2.00	0.50	0.44 0.06	0.37 0.04	0.28 0.03	0.20 0.02	0.10 0.01	0.94 0.11
∞	0.00	0.50 0.00	0.40 0.00	0.30 0.00	0.20 0.00	0.10 0.00	1.00 0.00

Table 13.3 contains slab moment coefficients for Method 2. Table 13.3 is reproduced with permission of the ACI.

TABLE 13.3

METHOD 2 — MOMENT COEFFICIENTS

Moments	Short span						Long span, all values of m
	Values of m						
	1.0	0.9	0.8	0.7	0.6	0.5 and less	
Case 1—Interior panels							
Negative moment at—							
Continuous edge	0.033	0.040	0.048	0.055	0.063	0.083	0.033
Discontinuous edge	—	—	—	—	—	—	—
Positive moment at midspan	0.025	0.030	0.036	0.041	0.047	0.062	0.025
Case 2—One edge discontinuous							
Negative moment at—							
Continuous edge	0.041	0.048	0.055	0.062	0.069	0.085	0.041
Discontinuous edge	0.021	0.024	0.027	0.031	0.035	0.042	0.021
Positive moment at midspan	0.031	0.036	0.041	0.047	0.052	0.064	0.031
Case 3—Two edges discontinuous							
Negative moment at—							
Continuous edge	0.049	0.057	0.064	0.071	0.078	0.090	0.049
Discontinuous edge	0.025	0.028	0.032	0.036	0.039	0.045	0.025
Positive moment at midspan	0.037	0.043	0.048	0.054	0.059	0.068	0.037
Case 4—Three edges discontinuous							
Negative moment at—							
Continuous edge	0.058	0.066	0.074	0.082	0.090	0.098	0.058
Discontinuous edge	0.029	0.033	0.037	0.041	0.045	0.049	0.029
Positive moment at midspan	0.044	0.050	0.056	0.062	0.068	0.074	0.044
Case 5—Four edges discontinuous							
Negative moment at—							
Continuous edge	—	—	—	—	—	—	—
Discontinuous edge	0.033	0.038	0.043	0.047	0.053	0.055	0.033
Positive moment at midspan	0.050	0.057	0.064	0.072	0.080	0.083	0.050

Tables 13.4, 13.5, 13.6 and 13.7 contain coefficients for slab moments and beam shears for Method 3. These tables are reproduced with permission of the ACI.

TABLE 13.4

METHOD 3 — COEFFICIENTS FOR NEGATIVE MOMENTS IN SLABS*

$$\left.\begin{aligned} M_{A\ neg} &= C_{A\ neg} \times w \times A^2 \\ M_{B\ neg} &= C_{B\ neg} \times w \times B^2 \end{aligned}\right\} \quad \text{where } w = \text{total uniform dead plus live load}$$

Ratio $m = \frac{A}{B}$		Case 1 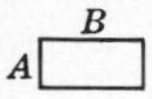	Case 2 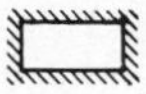	Case 3 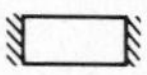	Case 4 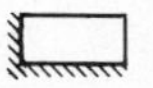	Case 5 	Case 6 	Case 7 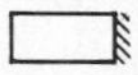	Case 8 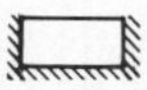	Case 9 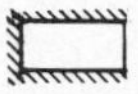
1.00	$C_{A\ neg}$		0.045		0.050	0.075	0.071		0.033	0.061
	$C_{B\ neg}$		0.045	0.076	0.050			0.071	0.061	0.033
0.95	$C_{A\ neg}$		0.050		0.055	0.079	0.075		0.038	0.065
	$C_{B\ neg}$		0.041	0.072	0.045			0.067	0.056	0.029
0.90	$C_{A\ neg}$		0.055		0.060	0.080	0.079		0.043	0.068
	$C_{B\ neg}$		0.037	0.070	0.040			0.062	0.052	0.025
0.85	$C_{A\ neg}$		0.060		0.066	0.082	0.083		0.049	0.072
	$C_{B\ neg}$		0.031	0.065	0.034			0.057	0.046	0.021
0.80	$C_{A\ neg}$		0.065		0.071	0.083	0.086		0.055	0.075
	$C_{B\ neg}$		0.027	0.061	0.029			0.051	0.041	0.017
0.75	$C_{A\ neg}$		0.069		0.076	0.085	0.088		0.061	0.078
	$C_{B\ neg}$		0.022	0.056	0.024			0.044	0.036	0.014
0.70	$C_{A\ neg}$		0.074		0.081	0.086	0.091		0.068	0.081
	$C_{B\ neg}$		0.017	0.050	0.019			0.038	0.029	0.011
0.65	$C_{A\ neg}$		0.077		0.085	0.087	0.093		0.074	0.083
	$C_{B\ neg}$		0.014	0.043	0.015			0.031	0.024	0.008
0.60	$C_{A\ neg}$		0.081		0.089	0.088	0.095		0.080	0.085
	$C_{B\ neg}$		0.010	0.035	0.011			0.024	0.018	0.006
0.55	$C_{A\ neg}$		0.084		0.092	0.089	0.096		0.085	0.086
	$C_{B\ neg}$		0.007	0.028	0.008			0.019	0.014	0.005
0.50	$C_{A\ neg}$		0.086		0.094	0.090	0.097		0.089	0.088
	$C_{B\ neg}$		0.006	0.022	0.006			0.014	0.010	0.003

*A cross-hatched edge indicates that the slab continues across or is fixed at the support; an unmarked edge indicates a support at which torsional resistance is negligible.

TABLE 13.5

METHOD 3 — COEFFICIENTS FOR DEAD LOAD POSITIVE MOMENTS IN SLABS*

$$\left.\begin{aligned} M_{A\ \text{pos DL}} &= C_{A\ \text{DL}} \times w \times A^2 \\ M_{B\ \text{pos DL}} &= C_{B\ \text{DL}} \times w \times B^2 \end{aligned}\right\} \text{ where } w = \text{total uniform dead load}$$

Ratio $m = \frac{A}{B}$		Case 1	Case 2	Case 3	Case 4	Case 5	Case 6	Case 7	Case 8	Case 9
1.00	$C_{A\ DL}$	0.036	0.018	0.018	0.027	0.027	0.033	0.027	0.020	0.023
	$C_{B\ DL}$	0.036	0.018	0.027	0.027	0.018	0.027	0.033	0.023	0.020
0.95	$C_{A\ DL}$	0.040	0.020	0.021	0.030	0.028	0.036	0.031	0.022	0.024
	$C_{B\ DL}$	0.033	0.016	0.025	0.024	0.015	0.024	0.031	0.021	0.017
0.90	$C_{A\ DL}$	0.045	0.022	0.025	0.033	0.029	0.039	0.035	0.025	0.026
	$C_{B\ DL}$	0.029	0.014	0.024	0.022	0.013	0.021	0.028	0.019	0.015
0.85	$C_{A\ DL}$	0.050	0.024	0.029	0.036	0.031	0.042	0.040	0.029	0.028
	$C_{B\ DL}$	0.026	0.012	0.022	0.019	0.011	0.017	0.025	0.017	0.013
0.80	$C_{A\ DL}$	0.056	0.026	0.034	0.039	0.032	0.045	0.045	0.032	0.029
	$C_{B\ DL}$	0.023	0.011	0.020	0.016	0.009	0.015	0.022	0.015	0.010
0.75	$C_{A\ DL}$	0.061	0.028	0.040	0.043	0.033	0.048	0.051	0.036	0.031
	$C_{B\ DL}$	0.019	0.009	0.018	0.013	0.007	0.012	0.020	0.013	0.007
0.70	$C_{A\ DL}$	0.068	0.030	0.046	0.046	0.035	0.051	0.058	0.040	0.033
	$C_{B\ DL}$	0.016	0.007	0.016	0.011	0.005	0.009	0.017	0.011	0.006
0.65	$C_{A\ DL}$	0.074	0.032	0.054	0.050	0.036	0.054	0.065	0.044	0.034
	$C_{B\ DL}$	0.013	0.006	0.014	0.009	0.004	0.007	0.014	0.009	0.005
0.60	$C_{A\ DL}$	0.081	0.034	0.062	0.053	0.037	0.056	0.073	0.048	0.036
	$C_{B\ DL}$	0.010	0.004	0.011	0.007	0.003	0.006	0.012	0.007	0.004
0.55	$C_{A\ DL}$	0.088	0.035	0.071	0.056	0.038	0.058	0.081	0.052	0.037
	$C_{B\ DL}$	0.008	0.003	0.009	0.005	0.002	0.004	0.009	0.005	0.003
0.50	$C_{A\ DL}$	0.095	0.037	0.080	0.059	0.039	0.061	0.089	0.056	0.038
	$C_{B\ DL}$	0.006	0.002	0.007	0.004	0.001	0.003	0.007	0.004	0.002

*A cross-hatched edge indicates that the slab continues across or is fixed at the support; an unmarked edge indicates a support at which torsional resistance is negligible.

TABLE 13.6

METHOD 3 — COEFFICIENTS FOR LIVE LOAD POSITIVE MOMENTS IN SLABS*

$$\left.\begin{aligned} M_{A\ \text{pos LL}} &= C_{A\ \text{LL}} \times w \times A^2 \\ M_{B\ \text{pos LL}} &= C_{B\ \text{LL}} \times w \times B^2 \end{aligned}\right\} \quad \text{where } w = \text{total uniform live load}$$

Ratio $m = \frac{A}{B}$		Case 1	Case 2	Case 3	Case 4	Case 5	Case 6	Case 7	Case 8	Case 9
1.00	$C_{A\ LL}$	0.036	0.027	0.027	0.032	0.032	0.035	0.032	0.028	0.030
	$C_{B\ LL}$	0.036	0.027	0.032	0.032	0.027	0.032	0.035	0.030	0.028
0.95	$C_{A\ LL}$	0.040	0.030	0.031	0.035	0.034	0.038	0.036	0.031	0.032
	$C_{B\ LL}$	0.033	0.025	0.029	0.029	0.024	0.029	0.032	0.027	0.025
0.90	$C_{A\ LL}$	0.045	0.034	0.035	0.039	0.037	0.042	0.040	0.035	0.036
	$C_{B\ LL}$	0.029	0.022	0.027	0.026	0.021	0.025	0.029	0.024	0.022
0.85	$C_{A\ LL}$	0.050	0.037	0.040	0.043	0.041	0.046	0.045	0.040	0.039
	$C_{B\ LL}$	0.026	0.019	0.024	0.023	0.019	0.022	0.026	0.022	0.020
0.80	$C_{A\ LL}$	0.056	0.041	0.045	0.048	0.044	0.051	0.051	0.044	0.042
	$C_{B\ LL}$	0.023	0.017	0.022	0.020	0.016	0.019	0.023	0.019	0.017
0.75	$C_{A\ LL}$	0.061	0.045	0.051	0.052	0.047	0.055	0.056	0.049	0.046
	$C_{B\ LL}$	0.019	0.014	0.019	0.016	0.013	0.016	0.020	0.016	0.013
0.70	$C_{A\ LL}$	0.068	0.049	0.057	0.057	0.051	0.060	0.063	0.054	0.050
	$C_{B\ LL}$	0.016	0.012	0.016	0.014	0.011	0.013	0.017	0.014	0.011
0.65	$C_{A\ LL}$	0.074	0.053	0.064	0.062	0.055	0.064	0.070	0.059	0.054
	$C_{B\ LL}$	0.013	0.010	0.014	0.011	0.009	0.010	0.014	0.011	0.009
0.60	$C_{A\ LL}$	0.081	0.058	0.071	0.067	0.059	0.068	0.077	0.065	0.059
	$C_{B\ LL}$	0.010	0.007	0.011	0.009	0.007	0.008	0.011	0.009	0.007
0.55	$C_{A\ LL}$	0.088	0.062	0.080	0.072	0.063	0.073	0.085	0.070	0.063
	$C_{B\ LL}$	0.008	0.006	0.009	0.007	0.005	0.006	0.009	0.007	0.006
0.50	$C_{A\ LL}$	0.095	0.066	0.088	0.077	0.067	0.078	0.092	0.076	0.067
	$C_{B\ LL}$	0.006	0.004	0.007	0.005	0.004	0.005	0.007	0.005	0.004

*A cross-hatched edge indicates that the slab continues across or is fixed at the support; an unmarked edge indicates a support at which torsional resistance is negligible.

TABLE 13.7

METHOD 3 — RATIO OF LOAD W IN A AND B DIRECTIONS FOR SHEAR IN SLAB AND LOAD ON SUPPORTS*

Ratio $m = \frac{A}{B}$		Case 1	Case 2	Case 3	Case 4	Case 5	Case 6	Case 7	Case 8	Case 9
1.00	W_A	0.50	0.50	0.17	0.50	0.83	0.71	0.29	0.33	0.67
	W_B	0.50	0.50	0.83	0.50	0.17	0.29	0.71	0.67	0.33
0.95	W_A	0.55	0.55	0.20	0.55	0.86	0.75	0.33	0.38	0.71
	W_B	0.45	0.45	0.80	0.45	0.14	0.25	0.67	0.62	0.29
0.90	W_A	0.60	0.60	0.23	0.60	0.88	0.79	0.38	0.43	0.75
	W_B	0.40	0.40	0.77	0.40	0.12	0.21	0.62	0.57	0.25
0.85	W_A	0.66	0.66	0.28	0.66	0.90	0.83	0.43	0.49	0.79
	W_B	0.34	0.34	0.72	0.34	0.10	0.17	0.57	0.51	0.21
0.80	W_A	0.71	0.71	0.33	0.71	0.92	0.86	0.49	0.55	0.83
	W_B	0.29	0.29	0.67	0.29	0.08	0.14	0.51	0.45	0.17
0.75	W_A	0.76	0.76	0.39	0.76	0.94	0.88	0.56	0.61	0.86
	W_B	0.24	0.24	0.61	0.24	0.06	0.12	0.44	0.39	0.14
0.70	W_A	0.81	0.81	0.45	0.81	0.95	0.91	0.62	0.68	0.89
	W_B	0.19	0.19	0.55	0.19	0.05	0.09	0.38	0.32	0.11
0.65	W_A	0.85	0.85	0.53	0.85	0.96	0.93	0.69	0.74	0.92
	W_B	0.15	0.15	0.47	0.15	0.04	0.07	0.31	0.26	0.08
0.60	W_A	0.89	0.89	0.61	0.89	0.97	0.95	0.76	0.80	0.94
	W_B	0.11	0.11	0.39	0.11	0.03	0.05	0.24	0.20	0.06
0.55	W_A	0.92	0.92	0.69	0.92	0.98	0.96	0.81	0.85	0.95
	W_B	0.08	0.08	0.31	0.08	0.02	0.04	0.19	0.15	0.05
0.50	W_A	0.94	0.94	0.76	0.94	0.99	0.97	0.86	0.89	0.97
	W_B	0.06	0.06	0.24	0.06	0.01	0.03	0.14	0.11	0.03

*A cross-hatched edge indicates that the slab continues across or is fixed at the support; an unmarked edge indicates a support at which torsional resistance is negligible.

The use of Tables 13.1 through 13.7 will be demonstrated in the solved problems.

The solved problems have been selected to demonstrate the specific requirements in this chapter. Usually the detailed proportioning and determination of stresses will not be discussed.

Solved Problems

13.1. Determine the central strip bending moments for a simply supported square two-way slab. Assume a concentrated vertical load at the center of the panel. Use the "crossed sticks" method shown in Fig. 13-2.

The deflection of a simply supported beam with a concentrated load at the midpoint is $PL^3/48EI$. The deflection at e of strips ab and cd are equal. The load required to deflect ab is P_{ab} and the load required to deflect cd is P_{cd}, and $P = P_{ab} + P_{cd}$. Equating deflections of the two strips,

$$P_{ab}\,S^3/48EI \;=\; P_{cd}\,L^3/48EI$$

If $L = S$, then $P_{ab}\,L^3 = P_{cd}\,L^3$ and

$$P_{ab} \;=\; (P - P_{ab})L^3/L^3 \;=\; P/2$$

Likewise, $P_{cd} = P/2$.

The maximum simple beam bending moment for a concentrated load at the midpoint is $M = PL/4$. Hence $M_{ab} = M_{cd} = PL/8$.

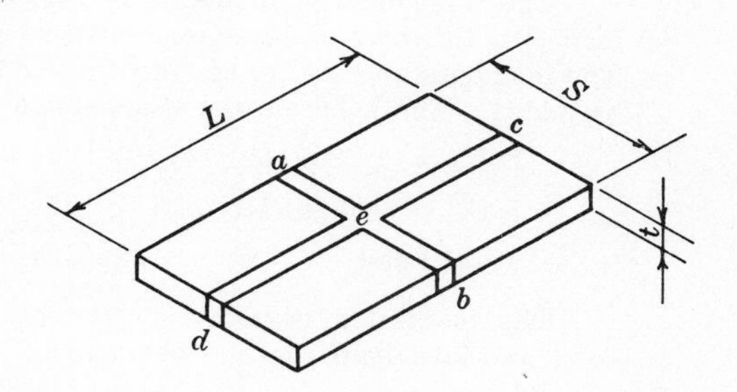

Fig. 13-4

13.2. Repeat Problem 13.1 and assume $L = 2.0S$.

As before, $P_{ab}\,S^3/48EI = P_{cd}\,L^3/48EI$ and

$$P_{ab} \;=\; (P - P_{ab})L^3/S^3 \;=\; 8(P - P_{ab})S^3/S^3 \;=\; 8P/9$$

Likewise, $P_{cd} = P/9$.

The maximum bending moments for the strips are

$$M_{ab} \;=\; (8P/9)(S/4) \;=\; 2PS/9, \qquad M_{cd} \;=\; (P/9)(L/4) \;=\; PL/36 \;=\; PS/18$$

13.3. An interior two-way slab panel 25′-0″ by 25′-0″ must carry a live load of 300 psf. The slab is $8\frac{1}{2}''$ thick and is supported on masonry walls. Determine the principal bending moments and shears in the slab.

The 1963 ACI Building Code requires that in the design of two-way slabs according to Method 2 the supports must be built monolithically with the slab. Hence Method 1 or 3 must be used. Method 1 will be used in this solution.

For a continuous slab, $g = g_1 = 0.76$. If $L = L_1$, then $r = gL/gL = 1.00$. From Table 13.1 for $r = 1.00$, $C = C_1 = 0.33$. For a continuous one-way slab $B = 1/16$ and $1/11$ for positive and negative moments respectively.

The bending moments in the slab are $M = CBWL$ and $M_1 = C_1BW_1L_1$, where W is the total uniform load between opposite supports on the slab strip.

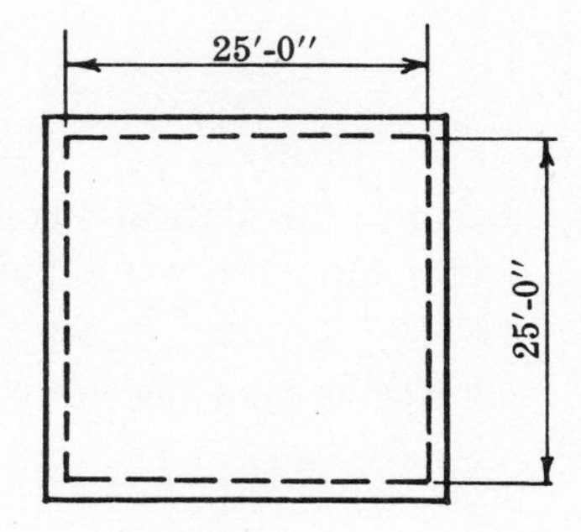

Fig. 13-5

The loads are

D.L. – $8\frac{1}{2}''$ slab	=	106 psf
L.L.	=	300 psf
Total Load	=	406 psf

Hence $W = 25.0(406) = 10{,}100$ lb/ft.

Because the panel is square, $M = M_1 = CBWL$. The maximum moments are

$$+M \;=\; \frac{0.33(10{,}100)(25.0)}{1000(16)} \;=\; 5.22, \qquad -M \;=\; \frac{0.33(10{,}100)(25.0)}{1000(11)} \;=\; 7.60 \text{ ft-kips/ft}$$

The shear in the slab is $V = C_sW$ and $V_1 = C_{s1}W_1$, where the coefficients C_s and C_{s1} are taken from Table 13.1. At $x = 0$ for $r = 1.0$, $C_s = C_{s1} = 0.25$. Hence at the support $V = V_1 = 0.25(10{,}100)/1000 = 2.53$ kips/ft.

13.4. Assume that the slab in Problem 13.3 is supported on concrete beams 18″ wide by 36″ deep. Determine the principal beam shears and bending moments. Use Method 1.

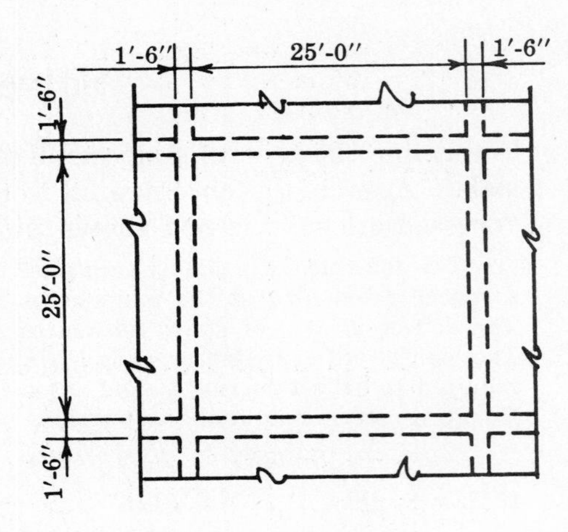

Fig. 13-6

The bending moments in the beams due to the slab load are $M = (1-C)BWL$ and $M_1 = (1-C_1)BW_1L_1$.

The total load on the beam is part of the slab load plus the dead and live load on the beam itself. The moments and shears of both these effects must be added. The loads on the beam itself are

$$\text{D.L.} = 18(36)(150)/144 = 674 \text{ lb/ft}$$
$$\text{L.L.} = 300(18)/12 = 450 \text{ lb/ft}$$
$$\text{Total Load} = 1124 \text{ lb/ft}$$

The maximum moments in the beam due to dead and live loads on the beam are

$$+M = \frac{wl^2}{16} = \frac{1124(25.0)^2}{16(1000)} = 44.0, \qquad -M = \frac{wl^2}{11} = \frac{1124(25.0)^2}{11(1000)} = 63.9 \text{ ft-kips}$$

The maximum moments in the beam due to slab loads are

$$+M = (1-C)BWL = \frac{0.67(10{,}100)(25.0)^2}{16(1000)} = 264 \text{ ft-kips}$$

$$-M = (1-C)BWL = \frac{0.67(10{,}100)(25.0)^2}{11(1000)} = 385 \text{ ft-kips}$$

Summing the total beam moments,

$$+M = 264 + 44.0 = 308 \text{ ft-kips at midspan}$$
$$-M = 385 + 63.9 = 449 \text{ ft-kips at support}$$

The shears in the beams due to the slab load are $V = C_bW$ and $V_1 = C_{b1}W_1$. The maximum shear at the face of the support due to the dead and live load on the beam is approximately $V = wl/2 = 14.0$ kips. The shear at the face of the support or at $x = 0$ due to the slab loads is $V = V_1 = C_bW = 0.25(10{,}100)(25.0)/1000 = 63.1$ kips. C_b is obtained from Table 13.2 for $r = 1.0$. Summing the total shears,

$$\text{Shear due to slab load} = 63.1 \text{ kips}$$
$$\text{Shear due to beam load} = 14.0 \text{ kips}$$
$$\text{Total maximum shear} = 77.1 \text{ kips}$$

13.5. Assume the slab in Problem 13.3 is supported on monolithically cast concrete beams. Determine the principal bending moments and shears in the slab and beams. Use Method 2.

(*a*) The loads on the slab are

$$\text{D.L.} - 8\tfrac{1}{2}'' \text{ slab} = 106 \text{ psf}$$
$$\text{L.L.} = 300 \text{ psf}$$
$$\text{Total Load} = 406 \text{ psf}$$

(*b*) Method 2 defines the span length as the distance center-to-center of supports or the clear span plus twice the slab thickness, whichever is smaller. Assuming 18″ wide supporting beams, $25.0 + 1.5 = 26.5$ ft and $25.0 + 2(8.5)/12 = 26.4$ ft. Hence $S = 26.4$ ft. Because the panel is square, $m = 1.0$.

From Table 13.3 the moment coefficients for a panel with four edges continuous are 0.033 and 0.025 for negative and positive moments respectively.

The bending moments in the middle strip are $M = CwS^2$ or

$$+M = \frac{0.025(406)(26.4)^2}{1000} = 7.1, \qquad -M = \frac{0.033(406)(26.4)^2}{1000} = 9.3 \text{ ft-kips/ft}$$

The bending moments in the column strips are 2/3 the above values.

(c) The shear in the slab at the supports is computed assuming that the load on the supporting beams is that within an area bounded by the 45 degree lines from the panel corners and the centerline of the long span. See Fig. 13-7. (See Problems 2.17 and 2.18.)

The maximum shear in the slab is $V = 406(25.0)/(2)(1000) = 5.06$ kips/ft. The average shear in the slab is one-half this value or 2.53 kips/ft.

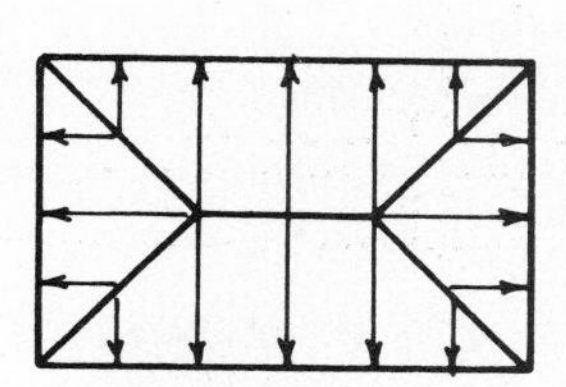
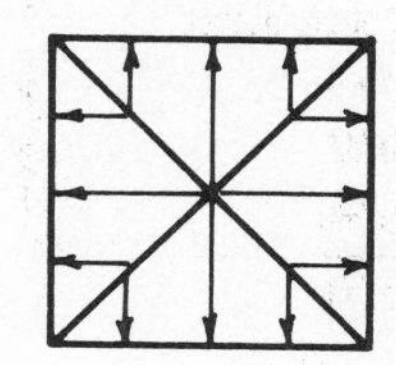

Fig. 13-7

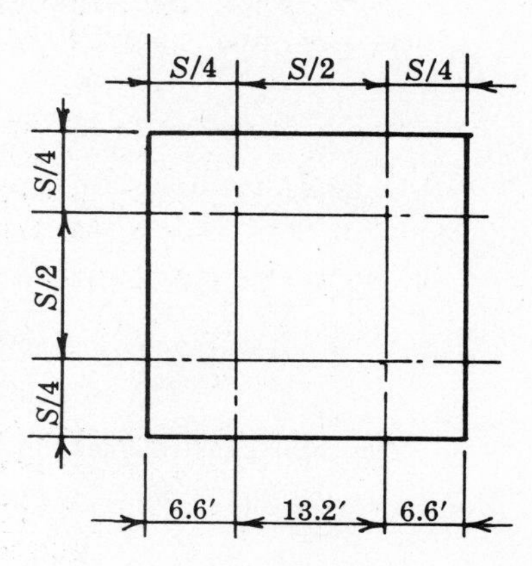

Fig. 13-8

(d) When $2S > L$, the middle strip is one-half panel wide symmetrical about the slab centerline. See Fig. 13-8.

The total panel moment in the slab is approximately

$$+M = 7.1(13.2) + 0.67(7.1)(13.2) = 156 \text{ ft-kips}$$
$$-M = 9.3(13.2) + 0.67(9.3)(13.2) = 205 \text{ ft-kips}$$
$$\text{Total panel moment in slab} = 361 \text{ ft-kips}$$

(e) The equivalent uniform load on the beams used to determine moments may be approximated by $wS/3$ for each panel supported. Hence the moments in the supporting beams due to slab loads are

$$+M = \left(\frac{2wS}{3}\right)\frac{S^2}{16} = \frac{2(406)(26.4)^3}{3(16)(1000)} = 312, \qquad -M = \left(\frac{2wS}{3}\right)\frac{S^2}{11} = 452 \text{ ft-kips}$$

(f) The total moment in panel slab and beams in one direction is

$$\text{Slab moments, positive and negative} = 361 \text{ ft-kips}$$
$$\text{Beam moments, positive and negative} = 764 \text{ ft-kips}$$
$$\text{Total panel moment} = 1125 \text{ ft-kips}$$

From Problem 13.3(c) and 13.4(b), the total panel moment due to slab loads obtained by Method 1 is

$$\text{Slab moment} = 25.0(5.22 + 7.60) = 320 \text{ ft-kips}$$
$$\text{Beam moment} = 264 + 385 = 649 \text{ ft-kips}$$
$$\text{Total panel moment} = 969 \text{ ft-kips}$$

The difference in the above answers is due to the difference in effective span lengths used in Methods 1 and 2. In Problems 13.3 and 13.4, the span length was 25.0 feet. Therefore $(969)(26.4/25.0)^3 = 1140$, which is approximately equal to 1125.

13.6. If an interior two-way slab has a width to length ratio 0.7, compare the total panel moments determined by Methods 1 and 2. Assume $L < L_1$.

(a) For Method 1, $g = g_1 = 0.76$ and $r = gL/g_1L_1 = 0.7$. From Table 13.1, $C = 0.58$ and $C_1 = 0.15$. If the coefficients for positive and negative bending moments are assumed to be 1/16 and 1/11 respectively, the slab bending moments in the long direction are

$$+M_1 = C_1BW_1L_1 = 0.15wLL_1^2/16 = 0.00938wLL_1^2$$
$$-M_1 = C_1BW_1L_1 = 0.15wLL_1^2/11 = 0.0136wLL_1^2$$

Likewise, the slab bending moments in the short direction are

$$+M = CBWL = 0.58wL^2L_1/16 = 0.0363wL^2L_1$$
$$-M = CBWL = 0.58wL^2L_1/11 = 0.0527wL^2L_1$$

With the same coefficients 1/11 and 1/16 the bending moments in the supporting beams in the long direction are

$$+M_1 = (1-C_1)BW_1L_1 = 0.85wLL_1^2/16 = 0.0531wLL_1^2$$
$$-M_1 = (1-C_1)BW_1L_1 = 0.85wLL_1^2/11 = 0.0773wLL_1^2$$

Likewise, the beam bending moments in the short direction are

$$+M = (1-C)BWL = 0.42wL^2L_1/16 = 0.0263wL^2L_1$$
$$-M = (1-C)BWL = 0.42wL^2L_1/11 = 0.0382wL^2L_1$$

The total panel moment in the long direction is

$$M_1 = (0.00938 + 0.0136 + 0.0531 + 0.0773)wLL_1^2 = 0.1534wLL_1^2$$

The total panel moment in the short direction is

$$M = (0.0363 + 0.0527 + 0.0263 + 0.0382)wL^2L_1 = 0.1535wL^2L_1$$

(*b*) For Method 2 the coefficients for bending moments are taken from Table 13.3. $C = 0.055$ and 0.041 for negative and positive moments respectively in the short directions and $C = 0.033$ and 0.025 for negative and positive moments respectively in the long direction. The slab bending moments in the long direction are

$$+M_1 = CwL^3/2 + CwL^3/3 = 5(0.025)wL^3/6 = 0.0208wL^3$$
$$-M_1 = CwL^3/2 + CwL^3/3 = 5(0.033)wL^3/6 = 0.0275wL^3$$

Likewise, the slab bending moments in the short direction are

$$+M = CwL^2L_1/2 + CwL^2L_1/3 = 5(0.041)wL^2L_1/6 = 0.0342wL^2L_1$$
$$-M = CwL^2L_1/2 + CwL^2L_1/3 = 5(0.055)wL^2L_1/6 = 0.0458wL^2L_1$$

The equivalent loadings on the supporting beams are $2(0.33)wL$ and $2(wL/3)(3-m^2/2) = 2(0.418)wL$ for the short and long spans respectively. Assuming moment coefficients of 1/16 and 1/11, the bending moments in the supporting beam in the long direction are

$$+M_1 = 2(0.418)wLL_1^2/16 = 0.0522wLL_1^2$$
$$-M_1 = 2(0.418)wLL_1^2/11 = 0.0760wLL_1^2$$

Likewise, the beam bending moments in the short direction are

$$+M = 2(0.33)wL^3/16 = 0.0412wL^3$$
$$-M = 2(0.33)wL^3/11 = 0.0600wL^3$$

The total panel moment in the long direction is

$$M_1 = (0.0208 + 0.0275)wL^3 + (0.0552 + 0.0760)wLL_1^2 = 0.1549wLL_1^2$$

The total panel moment in the short direction is

$$M = (0.0342 + 0.0458)wL^2L_1 + (0.0412 + 0.0600)wL^3 = 0.1508wL^2L_1$$

(*c*) Methods 1 and 2 yield practically the same results. This close correlation will be true for all two-way slabs of usual length to width ratios. Exterior and corner panels do not check as near as the above. It should be noted that the same effective span lengths were used in both methods.

13.7. Determine the principal slab and beam bending moments and shears in the two-way slab in Problem 13.3 by use of Method 3.

(*a*) From Problem 13.3, D.L. = 106 psf, L.L. = 300 psf, $A = B = 25.0$ ft and $m = 1.0$.

The bending moment in the middle strip of the slab is $M = CwA^2$ or CwB^2. For Case 2 in Table 13.4, $C_{A\,\text{neg}} = C_{B\,\text{neg}} = 0.045$. For w = total uniform load = 406 psf, $M_{A\,\text{neg}} = M_{B\,\text{neg}} = CwA^2 = 0.045(406)(25.0)^2/1000 = 11.4$ ft-kips/ft. For Case 2 in Tables 13.5 and 13.6, $C_{A\,\text{DL}} = C_{B\,\text{DL}} = 0.018$ and $C_{A\,\text{LL}} = C_{B\,\text{LL}} = 0.027$. Hence

$$M_{A\,\text{pos}} = M_{B\,\text{pos}} = C_{A\,\text{DL}}\, w_{\text{DL}} A^2 + C_{A\,\text{LL}}\, w_{\text{LL}} A^2$$

$$= 0.018(106)(25.0)^2/1000 + 0.027(300)(25.0)^2/1000 = 6.25 \text{ ft-kips/ft}$$

The average bending moments in the column strips are 2/3 of the above values.

(*b*) For Case 2 in Table 13.7, the ratio of total panel load to load resisted in A and B directions is 0.50. Hence,

$$V_A = V_B = \frac{0.50wAB}{2A} = \frac{0.50(406)(25.0)^2}{2(25.0)(1000)} = 2.53 \text{ kips/ft}$$

(*c*) The load on the supporting beams is determined by the shear in the slab. However, the total load on the short span beam must not be less than that of an area bounded by the 45 degree lines from the panel corners. The equivalent uniform load on the beams used to determine moments is $wA/3$ for each panel supported. If the coefficients 1/11 and 1/16 are assumed, the beam moments due to slab loads are

$$+M_A = +M_B = \left(\frac{2wA}{3}\right)\frac{A^2}{16} = \frac{2(406)(25)^3}{3(16)(1000)} = 264 \text{ ft-kips}$$

$$-M_A = -M_B = \left(\frac{2wA}{3}\right)\frac{A^2}{11} = \frac{2(406)(25)^3}{3(11)(1000)} = 385 \text{ ft-kips}$$

(*d*) The maximum beam shears at the supports are

$$V = \frac{2(2.53)A}{2} = \frac{2(2.53)(25.0)}{2(1000)} = 63.1 \text{ kips}$$

13.8. An apartment building is designed using 20′-0″ by 20′-0″ two-way slabs. The live load is 40 psf, the partition load is 20 psf and the floor finish is 5 psf. Design by working stress design techniques a typical exterior, interior and corner panel. Assume $f'_c = 4000$ psi, $f_s = 20{,}000$ psi and $w = 150$ pcf. The columns are 12″ by 12″ and the supporting beams are 12″ wide.

(*a*) The slab is cast monolithically with the supporting beams. Hence Method 2 will be used.

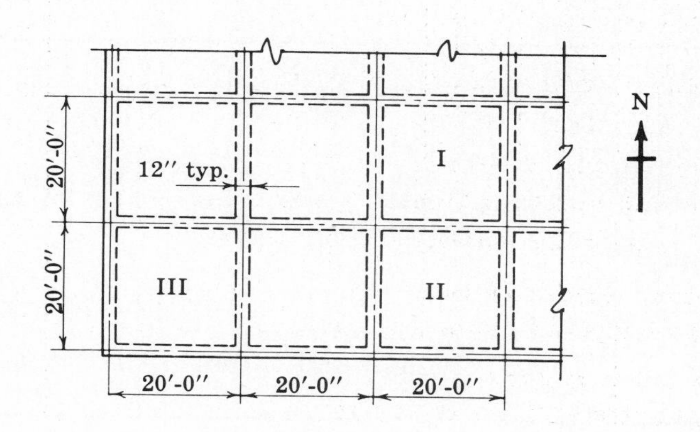

Fig. 13-9

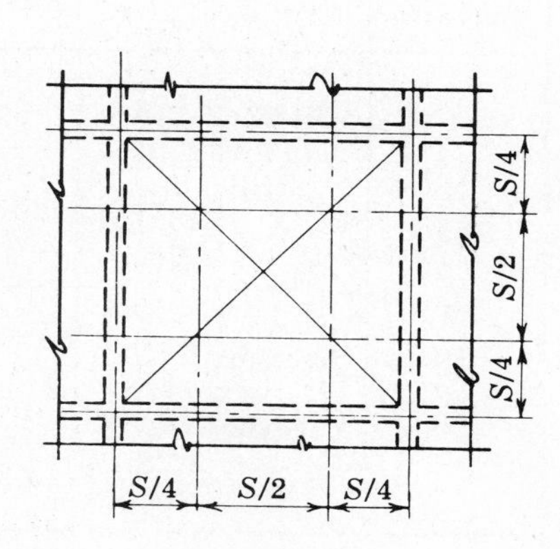

Fig. 13-10

(*b*) The 1963 ACI Building Code requires that the slab thickness t must be greater than $3\frac{1}{2}$″ and the slab perimeter divided by 180. The perimeter is approximately 76 feet. Hence 76/180 = 0.422 ft = 5.06″. Try a slab thickness of 5.5″. The loads on the slab are

D.L. — Slab	=	69 psf
Floor finish	=	5 psf
Partitions	=	20 psf
Total D.L.	=	94 psf
L.L.	=	40 psf
Total Load	=	134 psf

(c) By Method 2 the panel is divided into column and middle strips as shown in Fig. 13-10 above.

The 1963 ACI Building Code specifies the length of the short span S shall be taken as center-to-center of supports or the clear span plus two times the slab thickness, whichever is the smaller. The clear span is 19′-0″. Hence $19'0'' + 2(5.5'') = 19'\text{-}11'' < 20'\text{-}0''$ or $S = 19.92$ ft.

(d) Before computing the slab moments and shears, it is advisable to check the approximate slab thickness required for flexural and shearing stresses.

In Table 13.3 the maximum moment coefficient for $m = 1.0$ is $C = 0.049$ for the corner panel. Hence the maximum moment is $M = CwS^2 = 0.049(134)(19.92)^2/1000 = 2.61$ ft-kips/ft. The maximum shear in the slab is approximately $V = wS/2 = 134(19.92)/2(1000) = 1.33$ kips/ft.

From Table 4.2, a slab with a 4″ effective depth can resist a moment of 5.19 ft-kips/ft > 2.61.

The unit shear stress $v = V/bd = 1.33(1000)/12(4) = 27.7$ psi. From Table 4.1, the allowable shear is 70 psi > 27.7. This check is somewhat conservative because the actual shear stress at a distance d from the face of the support would be less than the above.

In the above, the value of d was assumed as the average depth to steel. See Fig. 13-11.

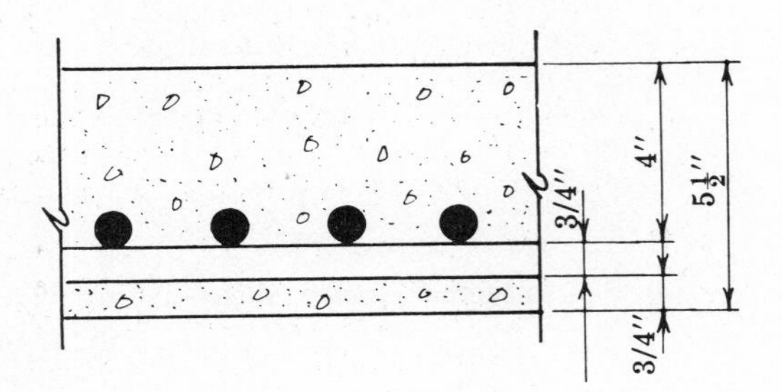

Fig. 13-11

(e) Perhaps the most efficient way to execute the slab design is by use of a table. The following table is self-explanatory and is complete.

TWO-WAY SLAB DESIGN

Direction of Span	Strip		Panel I		Panel II			Panel III		
			wS^2 = 53.2 ft-kips/ft		wS^2 = 53.2 ft-kips/ft			wS^2 = 53.2 ft-kips/ft		
			$+M$	$-M$	$-M$	$+M$	$-M$	$-M$	$+M$	$-M$
NORTH-SOUTH	Middle Strip	C	0.025	0.033	0.021	0.031	0.041	0.025	0.037	0.049
		M	1.33	1.76	1.12	1.65	2.18	1.33	1.97	2.61
		d	4.0	4.0	4.0	4.0	4.0	4.0	4.0	4.0
		A_s/ft	0.23	0.31	0.19	0.29	0.38	0.23	0.34	0.45
		A_s/Strip	2.29	3.09	1.89	2.89	3.78	2.29	3.39	4.48
		Reinf.	9-#5	10-#5	9-#5	10-#5	13-#5	9-#5	11-#5	15-#5
	Column Strip	C								
		M								
		d								
		A_s/ft								
		A_s/Strip	0.76	1.03	0.63	0.96	1.26	0.76	1.13	1.49
		Reinf.	5-#4	6-#4	5-#4	5-#4	7-#4	5-#4	6-#4	8-#4
EAST-WEST	Middle Strip	C								
		M								
		d								
		A_s/ft								
		A_s/Strip								
		Reinf.	9-#5	10-#5	9-#5	10-#5	13-#5	9-#5	11-#5	15-#5
	Column Strip	C								
		M								
		d								
		A_s/ft								
		A_s/Strip								
		Reinf.	5-#4	6-#4	5-#4	5-#4	7-#4	5-#4	6-#4	8-#4

The top and bottom portions of the table are the same because the panels are square.

The 1963 ACI Building Code specifies that the minimum reinforcement in two-way slabs is $0.0020bt = 0.0020(12)(5.5) = 0.132$ in²/ft. For a middle strip, minimum $A_s = 0.132(19.92)/2 = 1.31$ in². For a column strip, minimum $A_s = 0.132(19.92)/4 = 0.62$ in².

The maximum bar spacing is $3t = 3(5.5) = 16.5''$; and $10'\text{-}0''/1'\text{-}4.5'' = 8$ spaces or 9 bars.

(*f*) The reinforcement in the column strips is two-thirds of the corresponding reinforcement in the middle strips. If trussed bars are used, it is sometimes convenient to select the same bar size in the table for positive and negative moment regions. This practice coupled with the maximum spacing requirement often necessitates the furnishing of an excess of reinforcement. As an example, 8-#5's would be sufficient for the positive steel requirement in Panel I. However, 9 bars are furnished in order to comply with the maximum spacing of $3t$.

The column strip steel spacing may be varied from the spacing in the middle strip to a value three times this at the edge of the panel. In this example, such a practice cannot be used because of the $3t$ maximum.

(*g*) In the table, the average depth to the reinforcing steel was used for d. Some designers prefer to use the minimum depth in order to be conservative. Other designers prefer to use the two different depths and require specific placement sequences of the reinforcing steel.

(*h*) The shear is maximum and the area of negative reinforcement is minimum at the midpoint of the exterior long side. The shear is approximately $V = wS/2 = 134(19.92)/2(1000) = 1.33$ kips/ft. The perimeter of 9-#5's is $9(2.0)/10 = 1.8$ in/ft. Hence the bond stress is $u = V/\Sigma ojd = 1.33(1000)/(1.8)(0.86)(4.0) = 215$ psi. The allowable bond is $u_{\text{All}} = 4.8\sqrt{f'_c}/D = 4.8\sqrt{4000}/0.625 = 487 \text{ psi} > 215$.

The 1963 ACI Building Code specifies that flexural bond stresses need not be considered if anchorage bond stresses are less than 0.8 of that permitted. The required anchorage length is $l = f_s A_s / 0.8\Sigma ou$. For the #4 and #5 bars this length is 6.4″ and 8.1″ respectively if $f_s = 20{,}000$ psi. Because of the complexity of the nature of shear and bond stresses in two-way slabs, the anchorage bond check may assure a better design than a flexural bond check.

(*j*) In continuous beams and slabs, the longitudinal reinforcement is frequently bent up or trussed. See Fig. 13-12. Theoretically, the bars should be bent up or down at points where they are no longer needed to resist flexural stresses. These points would be points of inflection. Due to the practice of maximizing positive and negative moments in a span by alternate loading of the live load, the points of inflection are located at different positions for the various loading conditions.

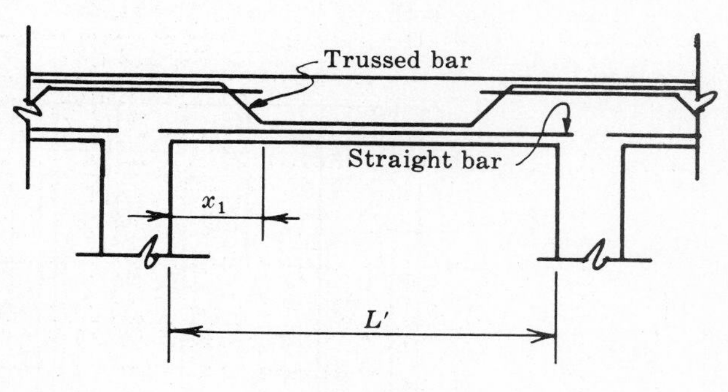

Fig. 13-12

If positive moment coefficients of 1/14 and 1/16 and negative moment coefficients of 1/10 and 1/11 are used, the various moment diagrams and points of inflection are as shown in Fig. 13-13 and 13-14.

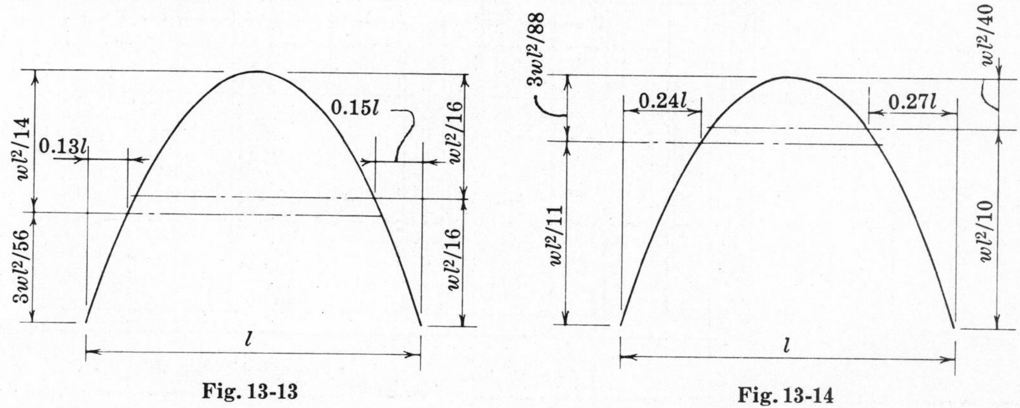

Fig. 13-13 **Fig. 13-14**

The 1963 ACI Building Code requires that in continuous beams at least 1/4 of the positive reinforcement must be extended into the support at least 6″. If in Fig. 13-12 $x_1 = L'/4$ for an interior span and $x_1 = L'/7$ for the exterior end of an exterior span, the bottom bar extension will provide adequate positive reinforcement for the various loading conditions.

The 1963 ACI Building Code requires that at least 1/3 of the negative reinforcement must be extended past the extreme interior point of inflection not less than $L'/16$ or d, whichever is greater. In addition, the anchorage bond requirements must be met. Two general rules have been used to provide adequate negative reinforcement for the various loading conditions: (1) Extend all bars to $L'/4$, or (2) Extend half of the bars to $L'/3$ and remainder to $L'/6$.

The general rules for bending bars are illustrated in Fig. 13-15. It should be noted that these bendup and cutoff points are based on moment coefficients and are approximate only, and they should be verified before adoption as a standard practice in designing a structure.

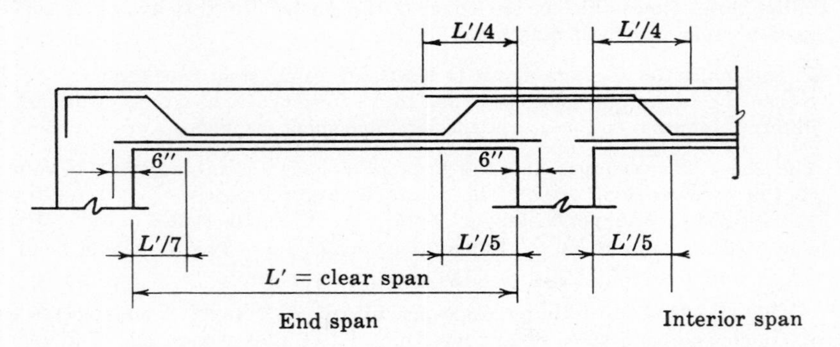

Fig. 13-15

(*k*) Fig. 13-16 is part plan showing the reinforcement for the three panels designed. The reinforcement in both directions is the same.

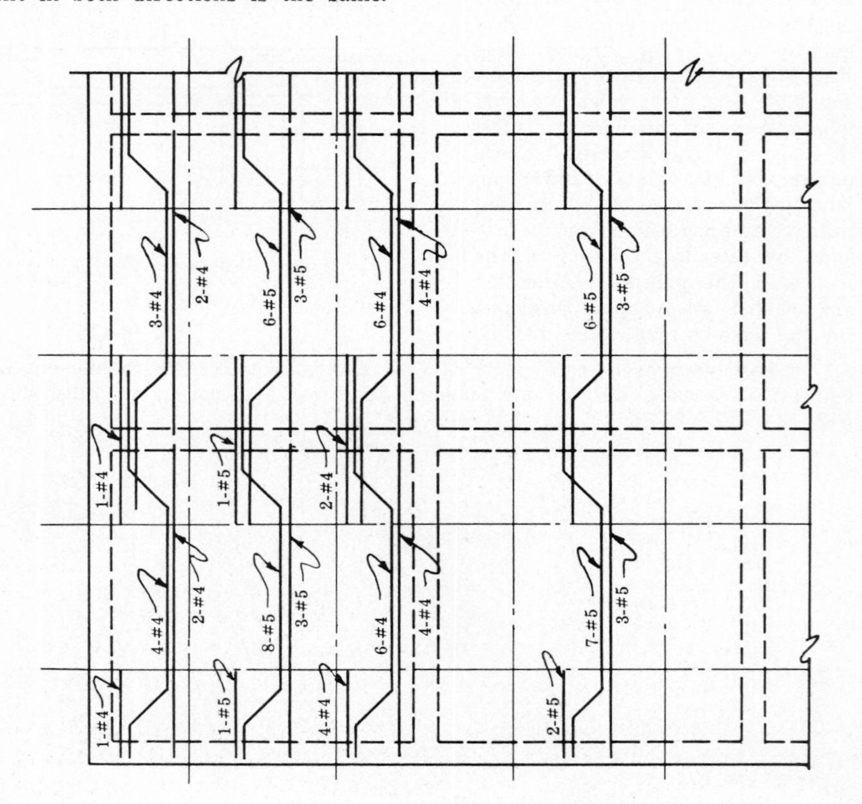

Fig. 13-16

(*l*) The load on the supporting beams is the slab load within the area shown in Fig. 13-7. If the clear span of 19′-0″ is used, the total slab load on an interior beam is $W = 2(19.0)^2(134)/4(1000) = 24.2$ kips. In addition there is dead and live load on the beam itself. If the total depth of the beam is assumed to be 19″, the dead and live loads on the beam are $W' = 19.0(5+20+40+237) = 5730$ lb. $= 5.73$ kips.

The shear at the support is approximately $V = (24.2+5.73)/2 = 15.0$ kips.

For computing moments an equivalent uniform load may be used, $w' = 2wS/3 = 2(134)(19.92)/3(1000) = 1.78$ kips/ft. The total equivalent uniform load acting on the beam is approximately $w'' = 1.78+0.30 = 2.08$ kips/ft. Using the moment coefficients 1/16 and 1/11, the beam moments are approximately

$$+M = 2.08(19.0)^2/16 = 46.9, \qquad -M = 2.08(19.0)^2/11 = 68.2 \text{ ft-kips}$$

(*m*) If $b = 12''$ and $d = 17.5''$, the unit shear stress in the beam at a distance d from the face of the support is approximately

$$v = \frac{V}{bd} = \frac{15.0 - 1.46(0.30) - (1.46)^2(0.134)/2}{12(17.5)} = 69 \text{ psi}$$

The allowable shear is $v_{All} = 1.1\sqrt{f'_c} = 70$ psi > 69. Hence no web reinforcement is required.

(*n*) For $f_c = 1800$ psi and $f_s = 20{,}000$ psi, $R = 324$. The resisting moment of the beams is $M = Rbd^2 = 324(12)(17.5)^2/12{,}000 = 99.0$ ft-kips > 68.2.

(*p*) The selection of the reinforcing size and details would be similar to that illustrated elsewhere and will not be done here. The design of the other supporting beams would be the same as above except that the spandrel beams carry less load and the moment and shear coefficients for end spans will not be the same as those in (*l*).

Supplementary Problems

13.9. Using the "crossed sticks" method, develop a curve showing the ratio of bending moments in each direction as a function of the ratio of the width to length of a two-way slab.

13.10. Repeat Problem 13.3 but assume USD. *Ans.* $+M = 9.0$, $-M = 13.05$ ft-kips/ft, $V = 4.35$ kips/ft

13.11. Repeat Problem 13.6 but assume the width to length is 0.85.
Ans. Total panel moment $= 0.15wL^2L_1$ or $0.15wLL_1^2$

13.12. Design a typical interior two-way slab panel that is 25′-0″ by 25′-0″. Assume $f'_c = 3000$ psi, $f_s = 20{,}000$ psi, superimposed D.L. = 10 psf, L.L. = 75 psf. The supporting beams are 14″ wide. Use WSD. *Partial Ans.* $t = 6''$

13.13. Repeat Problem 13.12 using USD and $f_y = 50{,}000$ psi.

13.14. Using the data given in Problem 13.12, design a corner panel.

13.15. An interior two-way slab panel is 20′-0″ by 20′-0″. Determine by Method 3 the slab bending moments and shears if $t = 7''$ and the live load 150 psf.
Partial Ans. Maximum negative moment $= 4.76$ ft-kips/ft

Chapter 14

Flat Slabs

NOTATION

A = distance in the direction of span from center of support to the intersection of the centerline of the slab thickness with the extreme 45-deg diagonal line lying wholly within the concrete section of slab and column or other support, including drop panel, capital and bracket

b_o = periphery of critical section for shear

c = effective support size

d = distance from extreme compression fiber to centroid of tension reinforcement

f'_c = compressive strength of concrete

h = distance from top of slab to bottom of capital

H = story height in feet of the column or support of a flat slab center to center of slabs

K = ratio of moment of inertia of column provided to I_c

L = span length of a flat slab panel center to center of supports

M_o = numerical sum of assumed positive and average negative moments at the critical design sections of a flat slab panel

R_n = factor for increasing negative moment

R_p = factor for increasing positive moment

t = thickness in inches of slab at center of panel

t_1 = thickness in inches of slab without drop panels, or through drop panel, if any

t_2 = thickness in inches of slab with drop panels at points beyond the drop panel

w' = uniformly distributed unit dead and live load

W = total dead and live load on panel

W_D = total dead load on panel

W_L = total live load on panel, uniformly distributed

INTRODUCTION

A concrete flat slab is a particular type of two-way slab. A flat slab is usually point-supported on columns rather than line-supported on perimeter beams like two-way slabs discussed in Chapter 13. A flat slab that has a constant thickness and is supported on prismatic columns is termed a flat plate. The 1963 ACI Building Code defines a flat slab as: "A concrete slab reinforced in two or more directions, generally without beams or girders to transfer the loads to supporting members..."

Many times the tops of the columns supporting a flat slab are enlarged by a cone forming a column capital. Also, portions of the slab which are usually symmetrical about the columns may be increased in thickness by lowering the soffit and forming a drop panel. Seldom does a column capital have an included angle greater than 90 degrees, because design specifications restrict the structural effect of capitals to this angle.

Fig. 14-1 is a part plan of a flat slab system showing the various components discussed above.

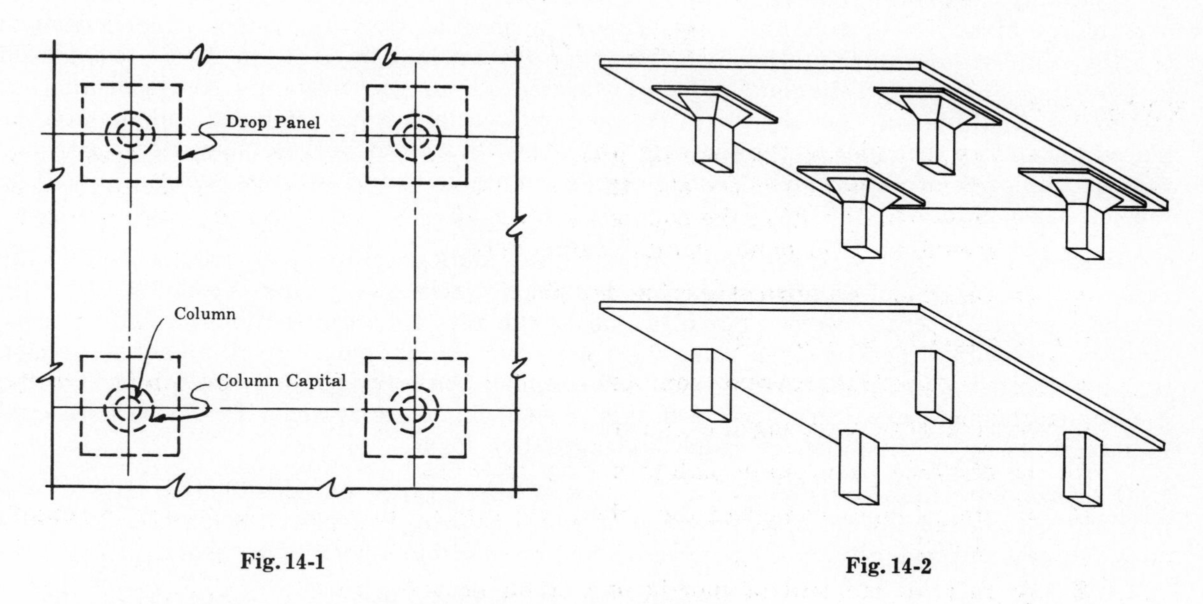

Fig. 14-1

Fig. 14-2

Fig. 14-2 illustrates flat slab and flat plate systems.

Because of the ever increasing use of flat slab and flat plate construction, the reinforced concrete structural designer should become proficient in their design. Flat slab construction with large drop panels and column capitals is particularly well suited for large industrial type structures such as manufacturing plants, warehouses, or parking garages where there are high live load to dead load ratios. However the use of column capitals and drop panels is sometimes aesthetically undesirable. Flat plates are best suited for relatively short spans and light live loads. Flat plate construction is a most prominent type of floor system used in the building of high-rise apartments and hotels.

There may be many advantages in flat slab and flat plate construction. They include:

(1) Lesser construction depth required for floor systems and subsequent reduction in story height.

(2) Reduction of dead loads and foundation loads due to decrease in overall weight and height of structure.

(3) Simpler formwork required for construction.

(4) Improved fire resistance as compared to other types of floor systems.

(5) Easier to illuminate due to absence of beams.

(6) Easier to install sprinkler systems and other piping and utilities due to absence of beams.

(7) Flexibility of column location due to the fact that columns need not be in exactly the same line.

(8) Easier to paint or apply acoustical treatment to under side.

(9) Capability of resisting concentrated loads.

Normally the reinforcement in flat slabs is placed parallel or perpendicular to the major axes of the structure or column lines. This is termed a two-way system. Some designs use the reinforcement orientation in a two-way system and in addition have steel placed at 45 degrees or along a diagonal. This is termed a four-way system. And still another method is the three-way system where the column locations form triangles and the steel is placed parallel to the sides of the triangle. This last system is very seldom used. Because the vast majority of all flat slab construction is adaptable to the two-way system, it will be the one used here.

It should be apparent that when drop panels and/or column capitals are used, the resistance to shear and flexural stress in the slab is increased. This is effected not only by the increase in cross section but also due to the fact that the sections at which these stresses are critical are removed some distance from the support. As the critical section becomes farther and farther away from the support, the total applied shear and, usually, the applied moment are decreased. For this reason, flat plate systems should be used only in structures of relatively short spans and light live loads.

Because flat slabs are not as rigid as some other types of construction due to the flexibility of the column to slab connection, the use of drop panels and column capitals not only decrease the magnitude of the stresses in the slab but also add to overall rigidity of the structure.

ANALYSIS OF FLAT SLABS

A flat slab is a very complex structure and it is extremely difficult to determine the location and magnitude of the stresses in the slab. There are two generally accepted techniques of analysis of flat slabs. They are termed the elastic analysis and the empirical method. The 1963 ACI Building Code contains specific rules for proportioning flat slabs by either of these two methods if certain prescribed requirements are met. If the structure does not comply with these requirements then the design must be accomplished by some other approach such as a yield-line theory.

In order to develop the statics of a flat slab, consider the slab shown in Fig. 14-3. The panel is a square interior panel surrounded by similar interior panels and it is loaded with a uniform load of w lb/ft^2. It is supported on circular column capitals that have a diameter c.

If a free body is taken as one-half of the panel as shown in Fig. 14-4 below, the boundaries of the element are free from shear. This is due to symmetry of the panel and because the panels are symmetrical about the column lines.

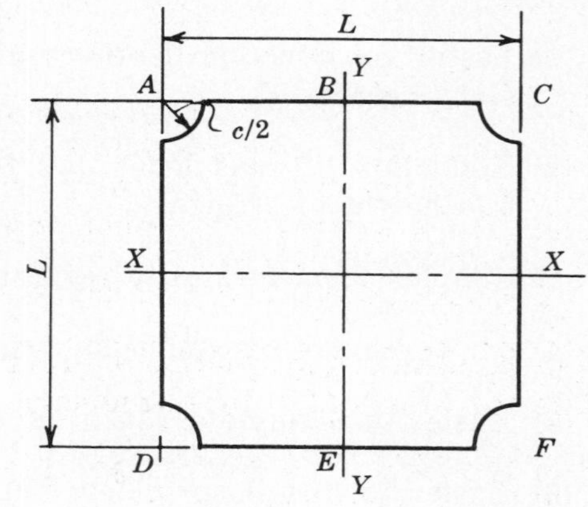

Fig. 14-3

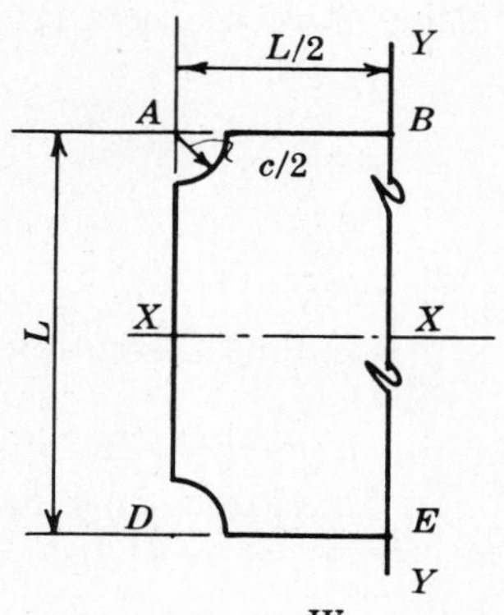

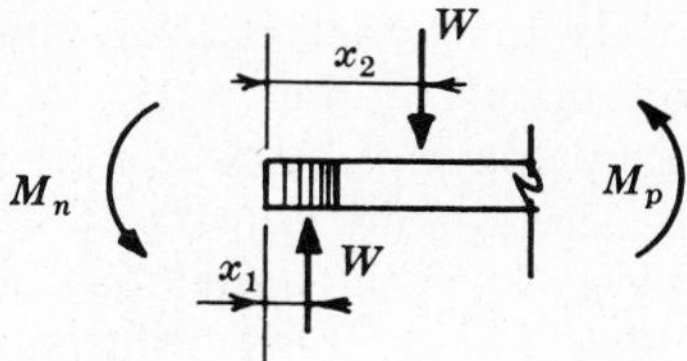

Fig. 14-4

If there is no shear on the straight lines AB, BE, ED and DA, then the vertical shear that must resist the downward load W must act along the curved surfaces at A and D. Assuming no torsion along the curved surfaces, summation of moments of the loads about line AD gives

$$dM = Wx_2 - Wx_1 \qquad (a)$$

And x_2 is determined by summing moments of the areas about AD:

$$x_2 = \frac{(L^2/2)(L/4) - (\pi c^2/8)(2c/3\pi)}{(L^2/2 - \pi c^2/8)} = \frac{3L^3 - 2c^3}{3(4L^2 - \pi c^2)} \qquad (b)$$

If it is assumed that the shear along the column capital is uniform, then the reaction acts at a distance from the centerline of the column, or $x_1 = c/\pi$. If the uniform load is w, then

$$W = wA = (w/8)(4L^2 - \pi c^2) \qquad (c)$$

From (a) and (c),

$$dM = (w/8)(4L^2 - \pi c^2)(x_2 - x_1)$$

The only other forces other than shear acting on the half-panel are the positive and negative moments, M_p and M_n. If $M_o = M_p + M_n = dM$, then $M_o = (w/8)(4L^2 - \pi c^2)(x_2 - x_1)$ where

$$x_2 - x_1 = \frac{3L^3 - 2c^3}{3(4L^2 - \pi c^2)} - \frac{c}{\pi} = \frac{3\pi L^3 - 12cL^2 + \pi c^3}{3\pi(4L^2 - \pi c^2)}$$

Hence

$$M_o = \frac{w}{8}(4L^2 - \pi c^2)\left[\frac{3\pi L^3 - 12cL^2 + \pi c^3}{3\pi(4L^2 - \pi c^2)}\right] = \frac{w}{24\pi}(3\pi L^3 - 12cL^2 + \pi c^3)$$

from which

$$M_o = \frac{wL^3}{8}\left(1 - \frac{4c}{\pi L} + \frac{c^3}{3L^3}\right) = \frac{WL}{8}\left(1 - \frac{4c}{\pi L} + \frac{c^3}{3L^3}\right) \qquad (d)$$

A very close approximation of (d) is

$$M_o = \frac{WL}{8}\left(1 - \frac{2c}{3L}\right)^2 \qquad (14.1)$$

Equation (*14.1*) is the expression for the sum of the positive and average negative moments in a flat slab. This expression was derived without consideration of the effect of Poisson's ratio, stress readjustments due to creep, and torsional moments around the support. The panel was assumed to be square and no attempt was made to determine the distribution of the moments along the edges of the panel.

In flat slabs as in two-way slabs supported on four sides, the static analysis yields results which are more conservative than results from test data. This fact coupled with the more theoretical analyses done largely by H. M. Westergaard demonstrate that the coefficient in equation (*14.1*) should be 0.09. Hence

$$M_o = 0.09\,WL\left(1 - \frac{2c}{3L}\right)^2 \qquad (14.2)$$

This is the form of the equation used for many years. The 1963 ACI Building Code further modifies equation (*14.2*) by introducing another factor F, so that

$$M_o = 0.09\,WLF\left(1 - \frac{2c}{3L}\right)^2 \qquad (14.3)$$

where $F = 1.15 - c/L \geqq 1$. Equation (*14.3*) will be the expression used here.

If a panel is rectangular rather than square so that $L_x \neq L_y$, then

$$M_{ox} = 0.09\,WL_xF\left(1 - \frac{2c}{3L_x}\right)^2 \qquad (14.4)$$

and

$$M_{oy} = 0.09\,WL_yF\left(1 - \frac{2c}{3L_y}\right)^2 \qquad (14.5)$$

In previous discussions of two-way slabs supported on four sides, it was shown that in a rectangular panel the short span strips were responsible for resisting the bulk of the bending moment. By study of equations (*14.4*) and (*14.5*), it is seen that the contrary is true in flat slabs. In a flat slab, the slab must support itself from column to column rather than from beam to beam.

As previously stated, the above analysis assumed a uniform distribution of both positive and negative moments. The moments are not constant along the panel width but the variations can be determined approximately by analytical and empirical analyses. Referring to Fig. 14-5, the panel may be divided into strips which are termed column strips and middle strips. The column strips have a width of $L/2$ and are centered along the column lines. The middle strips have a width of $L/2$ and are centered along the midline of the panel.

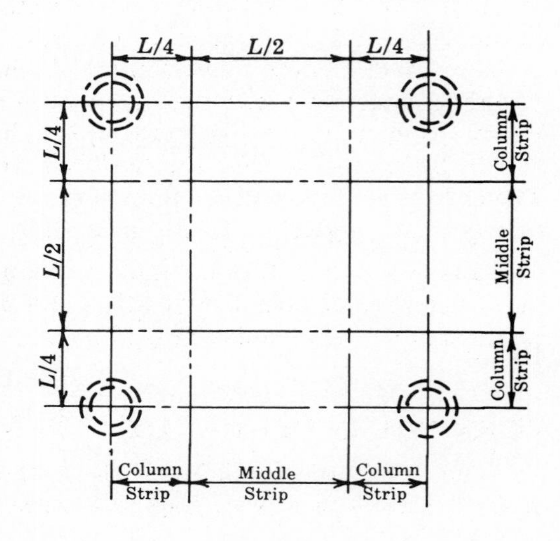

Fig. 14-5

In the design of flat slabs, the sum of the positive and average of the negative moments, M_o, is divided between the negative and positive moment regions. This is accomplished by utilizing coefficients when using the empirical method and by the structural analysis when using the elastic design technique. Then these moments must be apportioned to the column and middle strips. The column strips, like the supporting beams for two-way slabs, are subjected to heavier loads than the middle strips.

The 1963 ACI Building Code contains coefficients which specify what portion of the positive and negative moments each strip must resist. These coefficients will not be the same for the empirical method and the elastic analysis.

DESIGN OF FLAT SLABS

Because the design of flat slabs as discussed here is practically limited to the following of the rules or specifications of the 1963 ACI Building Code, the discussion of the design can best be accomplished through solved problems. However, there are several topics that should be discussed first.

Previous editions of the ACI Building Code did not recognize specifically the proportioning of flat slabs using ultimate strength design techniques. However, now ultimate strength may be used provided that in equation (*14.3*) the coefficient 0.09 is changed to 0.10 so that

$$M_o = 0.10\,WLF\left(1 - \frac{2c}{3L}\right)^2 \qquad (14.6)$$

Also, minimum slab thickness requirements must be met. These thicknesses are shown in Table 14.1.

TABLE 14.1
MINIMUM SLAB THICKNESSES FOR USE WITH ULTIMATE STRENGTH DESIGN

f_y	With Drop Panels	Without Drop Panels
40,000	$L/40$ or 4 inch	$L/36$ or 5 inch
50,000	$L/36$ or 4 inch	$L/33$ or 5 inch
60,000	$L/33$ or 4 inch	$L/30$ or 5 inch

SHEAR IN FLAT SLABS

The 1963 ACI Building Code specifies that for shear calculations two checks should be made. First, a check must be made at a critical section at a distance $d/2$ around the column, column capital, or drop panel which is along a periphery designated as b_o. Second, the shear must be checked at a distance d from the support considering the slab as a wide beam. Seldom, if ever, will the second check be critical. If shear reinforcement is required, it will be spaced so that the first row shall be not farther than $d/2$ from the face of the support.

As discussed in Chapter 6, bars, rods and wires shall not be used as shear reinforcement in slabs less than 10″ thick, due to difficulties in developing sufficient anchorage. However, this requirement does not preclude the use of fabricated shear-heads using steel beams or channels or other similar assemblies.

OPENINGS IN FLAT SLABS

The 1963 ACI Building Code contains specific requirements concerning openings in slabs. Referring to Fig. 14-6, there are four different types of holes illustrated.

Hole A is a hole common to two middle strips. Holes B and C are common to two column strips. Hole D is common to one middle strip and one column strip. The requirements concerning these various holes in flat slabs as discussed here will be those specified in the 1963 ACI Building Code.

It will be found that most holes penetrating flat slabs have a more adverse effect on shearing stresses than flexural stresses. Consequently hole A is located in a position that least affects the structural integrity of the slab. It is in an area of relatively low shear. There is no limitation placed on the size of the hole if the net section remaining is capable of resisting the applied moments. This is true for both the empirical and elastic designs.

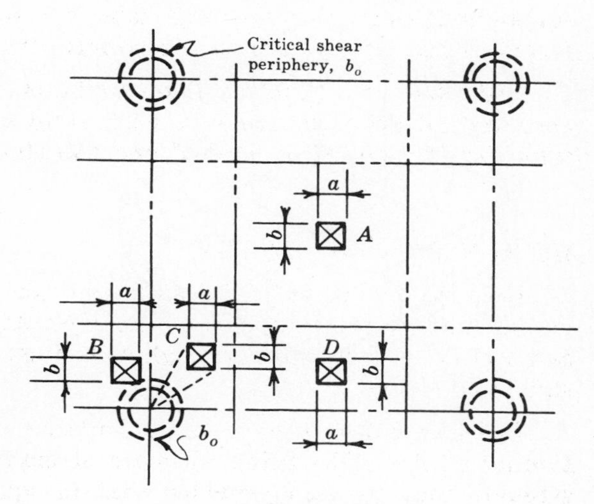

Fig. 14-6

When holes in flat slabs are common to two column strips, the critical shear periphery, b_o, must be reduced. That portion of b_o which is interrupted by an opening, hole B, and that portion of b_o which is contained within radial projections of the opening to the centroid of the reaction, hole C, shall not be considered as effective in resisting shear.

There is no limitation placed on the size of this type hole when this analysis is based on the elastic design if the moment and shear requirements can be met by the remaining section. If the slab is designed by the empirical method, the maximum width of the strip interrupted is limited to $L/16$ and the interrupted reinforcing steel must be replaced by an equal amount on all sides of the hole. In the empirical design also the net section must be capable of resisting the shear and flexural stresses.

There is no limitation in the elastic analysis placed on the size of hole D if the remaining section can withstand the loads. When using the empirical method, hole D may interrupt only one-fourth of the reinforcing steel (not concrete cross-section) in either the column or the middle strip. The interrupted reinforcing must be replaced by an equal amount on all sides of the opening. The net section must resist the applied loads.

If an opening is placed in a flat slab so that it is less than ten times the slab thickness from a concentrated load or reaction, the critical shear periphery b_o must be reduced as for holes B and C. That portion of b_o that is contained in the angle subtended by the opening with the centroid of the reaction or concentrated load as the apex must not be considered capable of resisting shear.

TRANSFER OF MOMENT AT COLUMNS

As previously mentioned, the connection between columns and flat slabs is a relatively weak one and gives rise to high concentrations of flexural and shear stresses. This is particularly true for flat plate systems. When there is an unbalanced moment in the slab at a support, it must be transferred to the column in a relatively limited width. If the slab shown in Fig. 14-7 must transfer an unbalanced moment of dM_1 or dM_2 into the column, the interchange is made by direct flexural moments, torsional moments, and direct

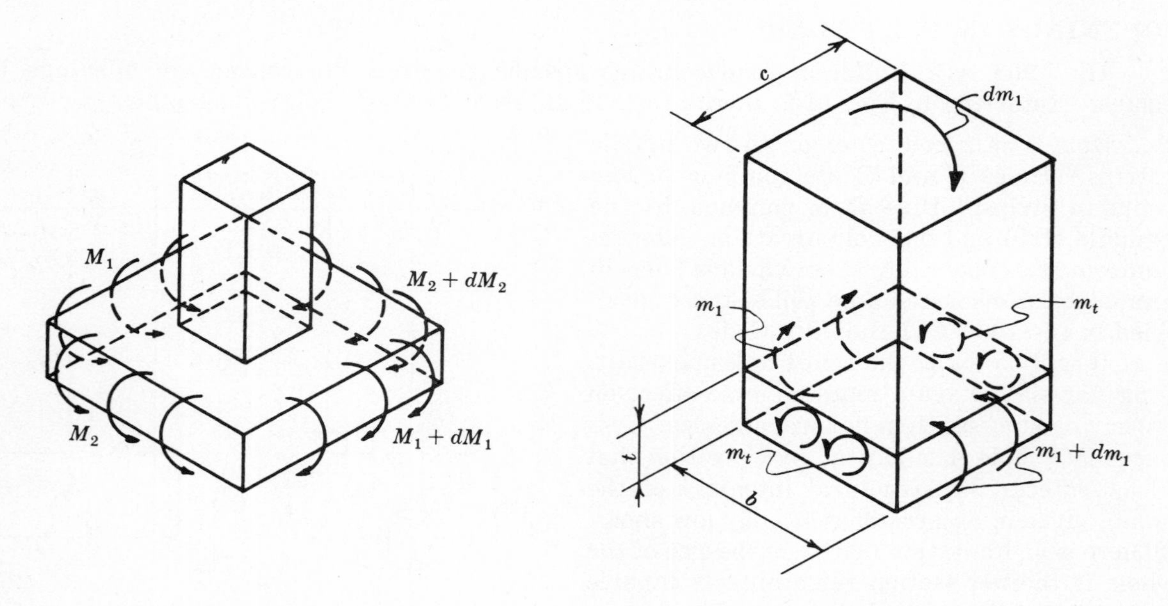

Fig. 14-7

Fig. 14-8

shear. Taking a free-body of a portion of the column in Fig. 14-8 and considering moments in one direction only, the flexural moments are m_1 and $m_1 + dm_1$ and the torsional moments are m_t. Applying the basic laws of mechanics, the flexural stresses due to dm_1 and the shear stresses due to m_t are readily determined. However, due to the extreme complexity of the problem, it is very difficult to determine what portion of dM_1 in Fig. 14-7 is resisted by flexural moments dm_1 and torsional moments m_t in Fig. 14-8. The total effect on shearing stresses of all the forces acting on the column are illustrated in Fig. 14-9.

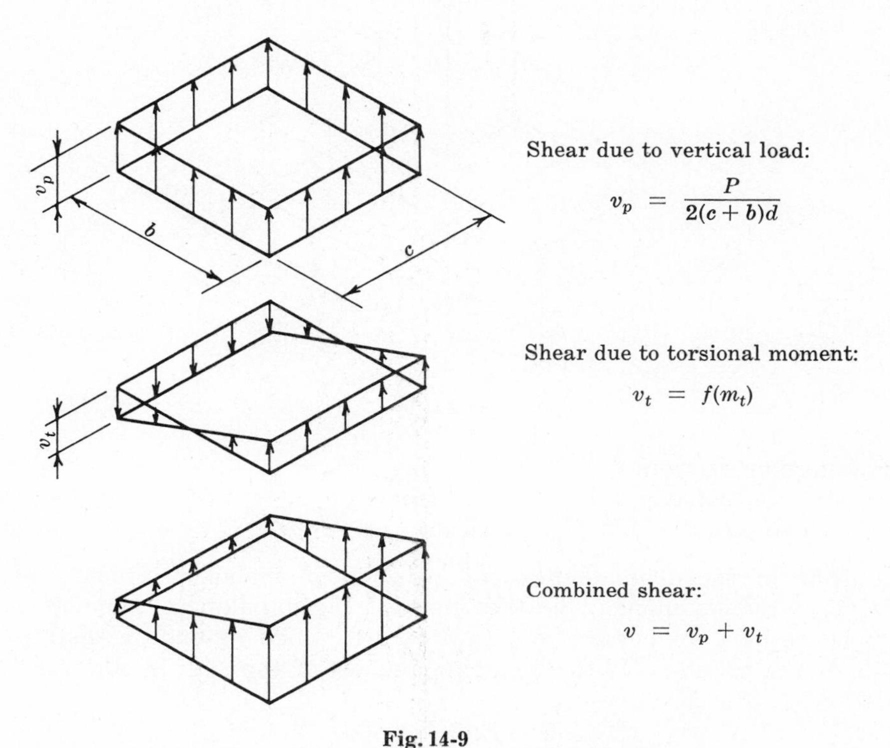

Fig. 14-9

An empirical and an analytical analysis of the division of the moments have been proposed. The report of the Joint American Concrete Institute-American Society of Civil Engineers Committee 326, Shear and Diagonal Tension, published in January, February and March, 1962, proposes an analysis based on test data. Whereas, Joseph DiStasio, Sr., and M. P. Van Buren in the September 1960 issue of The Journal of the American Concrete Institute propose an analysis based on mathematics. Both of these methods will be discussed and illustrated.

In Fig. 14-8, the section for the free body was cut at the face of the column. If the section is at some distance from the column, then the dimensions b and c are not those of the column and the calculated stresses are different. The Report of the ACI-ASCE Committee 326 recommends that the section that is critical for shear stress shall be at a distance of $d/2$ from the face of the column. Results of computations based on this section agree closely with test data. The section of Fig. 14-8 would now be as that shown in Fig. 14-10 below. The Report further recommends that the shear stress be computed by

$$v = v_p + v_t = \frac{V}{A_c} + \frac{kM}{J_c}\left(\frac{c}{2}\right) \qquad (14.7)$$

where M = moment at critical section, V = shear at critical section, $A_c = 2(b+c)d$, J_c = polar moment of inertia of section at critical section, c = length of side of critical section perpendicular to torsional axis, and k = ratio of moment transferred by torsion to total moment transferred at section. The Report recommends that $k = 0.2$ based on the limited test data.

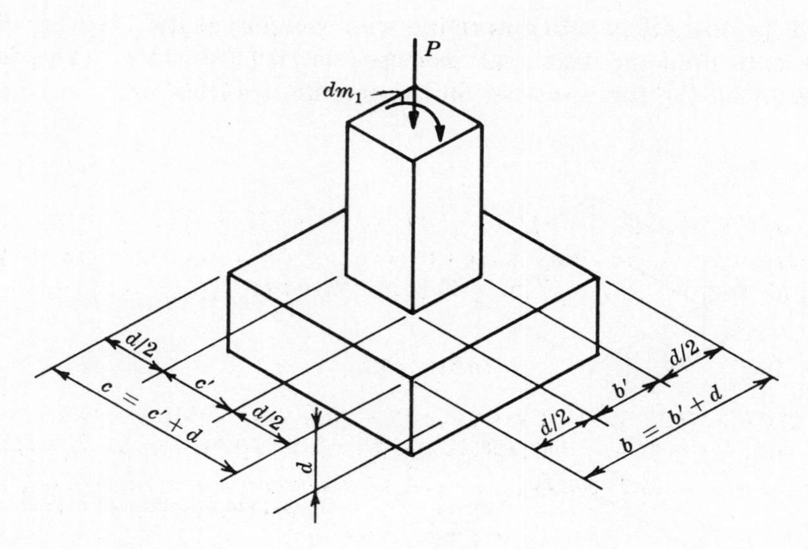

Fig. 14-10

For a rectangular section,

$$J_c = \frac{dc^3}{6} + \frac{cd^3}{6} + 2bd\left(\frac{c}{2}\right)^2 \tag{14.8}$$

In recognition of the ultimate flexural capacity of a slab, DiStasio and Van Buren require that the torsional moment be that part of the unbalanced moment at the section that can not be resisted in flexure by a slab width which extends a distance of $t - 1.5''$ on either side of the column. If the flexural capacity of the slab is $M_r = Rbd^2 = A_s f_s jd$, then in equation (14.7),

$$k = \frac{M - Rbd^2}{M} = \frac{M - A_s f_s jd}{M} \tag{14.9}$$

It should be noted that the critical section is assumed now to be as shown in Fig. 14-11.

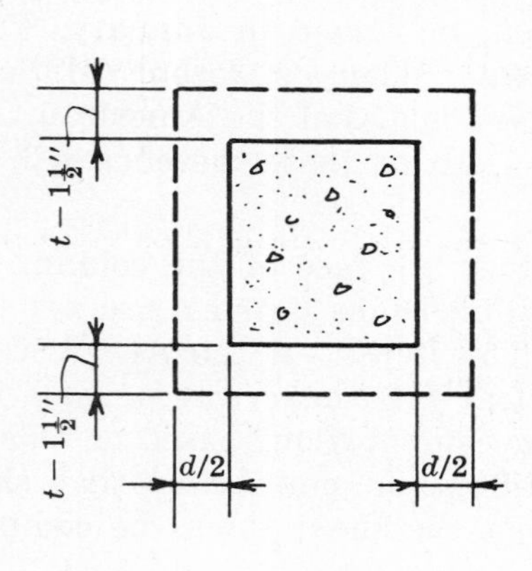

Fig. 14-11

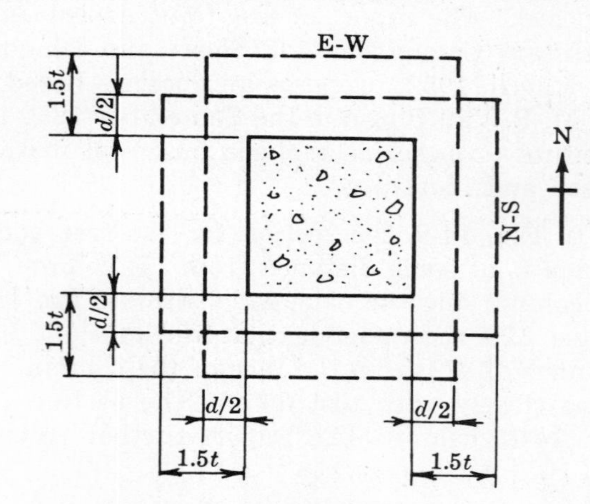

Fig. 14-12

The four boundary lines of the section shown in Fig. 14-11 are not equidistant from the face of the column. This means that if there are differential moments in two perpendicular planes, there are two different critical sections that are common at four points only. This is not true for the section shown in Fig. 14-10 where both critical sections would be identical.

The 1963 ACI Building Code specifies that a slab width between lines that are a distance of $1.5t$ on either side of the column may be considered effective in transferring moments between columns and flat slabs. This is shown in Fig. 14-12 above.

ELASTIC ANALYSIS

Table 14.2 contains the values of the 1963 ACI Building Code for flat slabs designed by the elastic analysis. The table gives the percent of total moment that must be resisted by the column and middle strips. Table 14.2 is reproduced with permission of the ACI.

TABLE 14.2

DISTRIBUTION BETWEEN COLUMN STRIPS AND MIDDLE STRIPS IN PERCENT OF TOTAL MOMENTS AT CRITICAL SECTIONS OF A PANEL

<table>
<tr><th colspan="2" rowspan="3">Strip</th><th colspan="4">Moment section</th></tr>
<tr><th rowspan="2">Negative moment at interior support</th><th rowspan="2">Positive moment</th><th colspan="2">Negative moment at exterior support</th></tr>
<tr><th>Slab supported on columns and on beams of total depth equal to the slab thickness*</th><th>Slab supported on reinforced concrete bearing wall or columns with beams of total depth equal or greater than 3 times the slab thickness*</th></tr>
<tr><td colspan="2">Column strip</td><td>76</td><td>60</td><td>80</td><td>60</td></tr>
<tr><td colspan="2">Middle strip</td><td>24</td><td>40</td><td>20</td><td>40</td></tr>
<tr><td rowspan="2">Half column strip adjacent and parallel to marginal beam or wall</td><td>Total depth of beam equal to slab thickness*</td><td>38</td><td>30</td><td>40</td><td>30</td></tr>
<tr><td>Total depth of beam or wall equal to or greater than 3 times slab thickness*</td><td>19</td><td>15</td><td>20</td><td>15</td></tr>
</table>

*Interpolate for intermediate ratios of beam depth to slab thickness.

Note: The total dead and live reaction of a panel adjacent to a marginal beam or wall may be divided between the beam or wall and the parallel half column strip in proportion to their stiffness, but the moment provided in the slab shall not be less than that given in Table.

EMPIRICAL METHOD

Table 14.3 contains values of the 1963 ACI Building Code for flat slabs designed by the empirical method. The table gives the percent of M_o that must be resisted by the column and middle strips. Table 14.3 is reproduced with permission of the ACI.

TABLE 14.3
MOMENTS IN FLAT SLAB PANELS IN PERCENTAGES OF M_o

Strip	Column head	Side support type	End support type	Exterior panel: Exterior negative moment	Exterior panel: Positive moment	Exterior panel: Interior negative moment	Interior panel: Positive moment	Interior panel: Negative moment
Column strip	With drop		A	44				
			B	36	24	56	20	50
			C	6	36	72		
	Without drop		A	40				
			B	32	28	50	22	46
			C	6	40	66		
Middle strip	With drop		A	10				
			B	20	20	17*	15	15*
			C	6	26	22*		
	Without drop		A	10				
			B	20	20	18*	16	16*
			C	6	28	24*		
Half column strip adjacent to marginal beam or wall	With drop	1	A	22				
			B	18	12	28	10	25
			C	3	18	36		
		2	A	17				
			B	14	9	21	8	19
			C	3	14	27		
		3	A	11				
			B	9	6	14	5	13
			C	3	9	18		
	Without drop	1	A	20				
			B	16	14	25	11	23
			C	3	20	33		
		2	A	15				
			B	12	11	19	9	18
			C	3	15	25		
		3	A	10				
			B	8	7	13	6	12
			C	3	10	17		

Percentage of panel load to be carried by marginal beam or wall in addition to loads directly superimposed thereon	Type of support listed above: Side support parallel to strip	Type of support listed above: Side or end edge condition of slabs of depth t	Type of support listed above: End support at right angles to strip
0	1	Columns with no beams	A
20	2	Columns with beams of total depth $1\frac{1}{4}t$	A
40	3	Columns with beams of total depth $3t$ or more	B
		Reinforced concrete bearing walls integral with slab	B
		Masonry or other walls providing negligible restraint	C

*Increase negative moments 30 percent of tabulated values when middle strip is continuous across support of Type B or C. No other values need be increased.

Note: For intermediate proportions of total beam depth to slab thicknesses, values for loads and moments may be obtained by interpolation.

The use of Tables 14.2 and 14.3 will be demonstrated in the solved problems.

DESIGN PROCEDURE

Before a design of a flat slab can be accomplished, several assumptions must be made. After the concrete and steel design stresses and live loads are established, the most important assumption or estimate is the slab thickness. The slab thickness influences the dead load, required column stiffness, drop panel size, etc. Hence the design involves a relatively complex series of trial and error solutions. An experienced structural designer may be able to eliminate the necessity of several trial designs by an ability to make the proper initial assumptions. However, it may be worthwhile to establish a design procedure which could be followed until the designer has the capacity to establish his own.

A suggested design procedure is as follows:

(1) Determine f_c', f_s, and load criteria.

(2) Select design technique, elastic analysis or empirical method, and check if requirements of method are met.

(3) Select method of proportioning, working stress or ultimate strength design.

(4) Estimate slab thickness based on minimum requirements and experience. Working stress and ultimate strength design procedures may have different requirements.

(5) Estimate drop panel and column capital size if used.

(6) Determine total loads.

(7) If not determined previously, estimate column size and check stiffness requirements.

(8) When designing by the empirical method, check minimum slab thickness by formulas.

(9) Determine shear stresses at critical section b_o.

(10) Determine shear stresses considering slab as wide beam.

(11) Consider effect of any openings on shear stresses.

(12) If shear stresses are excessive, increase concrete strength, slab thickness, column size, or use shear-head reinforcement and repeat (4) through (11).

(13) When designing by elastic analysis, determine moments, shears and thrust.

(14) If necessary, recompute shear stresses.

(15) When designing by the empirical method, determine M_o.

(16) Distribute moments to positive and negative moment regions in column and middle strips.

(17) Determine flexural stresses or capacity of slab.

(18) Select reinforcement to resist flexure. Check placement.

(19) When using the empirical method, modify positive and negative moments if column moment of inertia is less than required.

(20) Investigate effect of transfer of bending moment at column-slab connection.

(21) Determine minimum reinforcement requirements.

(22) Check bond and anchorage of reinforcement.

(23) Check spacing of column strip negative moment steel at support.

(24) If necessary, determine deflections.

As in other chapters of this book, the solved problems that follow are selected to demonstrate the specific requirements concerning information discussed in this chapter. The detailed proportioning and determination of such factors as flexural stresses, shear stresses, bond requirements, etc., will be shown but not discussed unless the item is specific to this chapter.

Solved Problems

14.1. A typical interior flat plate 17′-0″ by 18′-0″ is proportioned by working stress design techniques. Using the empirical method, determine the shears and moments at the critical sections if the total live load is 50 psf and the normal weight slab is 8″ thick. The columns are 14″ by 16″.

(*a*) The total uniform loads are

D.L.	= 8(150)/12	=	100 psf
L.L.		=	50 psf
Total Load		=	150 psf

The gross panel load is $W = 150(17.0)(18.0) = 45{,}900$ lb.

(*b*) The critical shear periphery around the column is located at a distance $d/2$ from the face of the column. Therefore if $d = 6.5''$, the critical section is as shown in Fig. 14-13.

The net shear on the section is $V = 45{,}900 - 150(20.5)(22.5)/144 = 45{,}400$ lb.

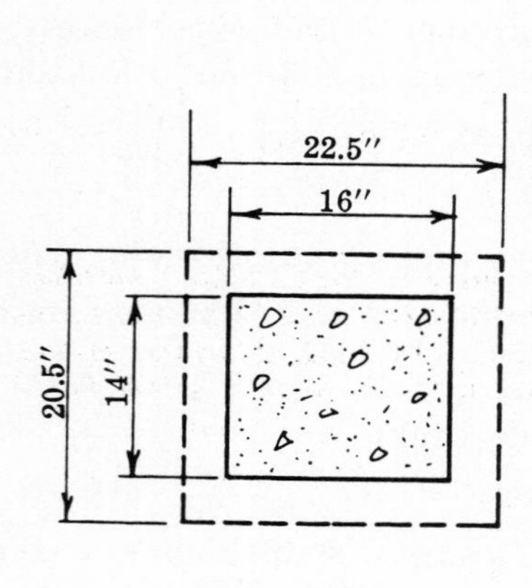

Fig. 14-13

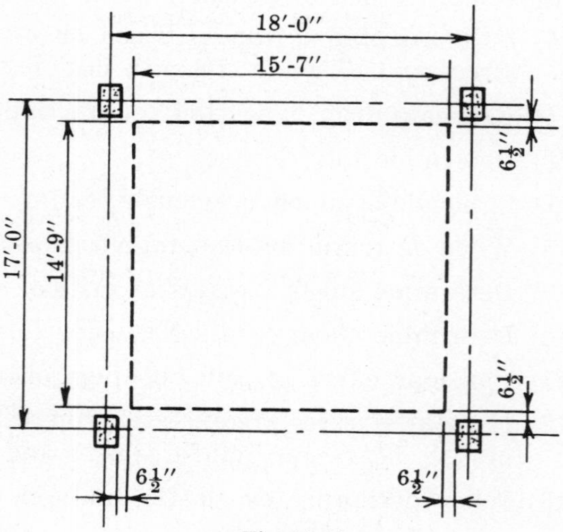

Fig. 14-14

(*c*) If the slab is considered as a wide beam, the critical section for shear is at a distance d from the face of the column. See Fig. 14-14.

The net shear on the section is $V = (15.58)(14.75)(150) = 34{,}500$ lb. This calculation is based on two-way action. The intent of the 1963 ACI Code is not obvious and some designers might assume one-way action. If the latter assumption is used, $V = 2(34{,}500) = 69{,}000$ lb. It will be shown later that this value is seldom, if ever, critical.

(*d*) The sum of the average of the negative moments and the positive moment, M_o, is given by equation (*14.3*). The 1963 ACI Building Code specifies that for columns without capitals the value of c shall be equal to the column dimension in the direction considered. For the 18′-0″ span, $c = 1.33$, $F = 1.08$, and

$$M_o = 0.09(45.9)(18.0)(1.08)\left[1 - \frac{2(1.33)}{3(18)}\right]^2 = 72.6 \text{ ft-kips}$$

For the 17′-0″ span, $c = 1.17$, $F = 1.08$, and

$$M_o = 0.09(45.9)(17.0)(1.08)\left[1 - \frac{2(1.17)}{3(17)}\right]^2 = 69.0 \text{ ft-kips}$$

(*e*) Table 14.3 contains the percentages of M_o that must be resisted by the column and middle strips. From the table, for an interior panel without drop panels,

$$\begin{aligned} \text{Column strip positive moment} &= +M_c = 0.22M_o \\ \text{Column strip negative moment} &= -M_c = 0.46M_o \\ \text{Middle strip positive moment} &= +M_m = 0.16M_o \\ \text{Middle strip negative moment} &= -M_m = 0.16M_o \end{aligned}$$

For the 18′-0″ span,

$$+M_c = 0.22(72.6) = 16.0 \text{ ft-kips} \qquad +M_m = 0.16(72.6) = 11.6 \text{ ft-kips}$$

$$-M_c = 0.46(72.6) = 33.4 \text{ ft-kips} \qquad -M_m = 0.16(72.6) = 11.6 \text{ ft-kips}$$

and for the 17′-0″ span

$$+M_c = 0.22(69.0) = 15.2 \text{ ft-kips} \qquad +M_m = 0.16(69.0) = 11.0 \text{ ft-kips}$$

$$-M_c = 0.46(69.0) = 31.8 \text{ ft-kips} \qquad -M_m = 0.16(69.0) = 11.0 \text{ ft-kips}$$

14.2. A typical interior flat slab 17′-0″ × 18′-0″ is proportioned by ultimate strength design procedures. Using the empirical method, determine the shears and moments at the critical sections if the total live load is 50 psf and the normal weight slab is 6″ thick. The column capital is 4′-0″ × 4′-0″, the drop panel is 7′-0″ × 7′-0″. The drop panel thickness is $8\frac{1}{2}$″ and the columns are 14″ × 14″.

(*a*) The total uniform loads are

D.L. 6(150)/12	=	75 psf
(Drop Panel)	=	5 psf
L.L.	=	50 psf
Total Load	=	130 psf

$$\frac{\text{D.L.}}{\text{T.L.}} = \frac{80}{130} = 0.615, \qquad \frac{\text{L.L.}}{\text{T.L.}} = \frac{50}{130} = 0.385$$

Because there are no wind load effects, the critical load factor equation is $U = 1.5D + 1.8L = 1.5(0.615)(D+L) + 1.8(0.385)(D+L) = 1.614(D+L)$. When the moments, shears, and thrusts of a structure are computed on the basis of total load, this form of the load factor equation is somewhat handier. The ultimate panel load is $W_u = 1.614(130)(17.0)(18.0) = 64{,}200$ lb.

(*b*) The critical shear periphery is around the column and the drop panel at a distance of $d/2$. If the effective depths of the slab and the drop panel are $4\frac{1}{2}$″ and 7″ respectively, the critical sections are as shown in Fig. 14-15.

The net shear acting on the section through the drop panel is

$$V_u = 64{,}200 - 1.614(130)(4.58)^2 = 59{,}800 \text{ lb}$$

The net shear acting on the section through the slab is

$$V_u = 64{,}200 - 1.614(130)(7.38)^2 = 52{,}800 \text{ lb}$$

If the slab is considered as a wide beam, the critical section for shear is at a distance d from the edge of the column capital. The net shear on the section around the panel is

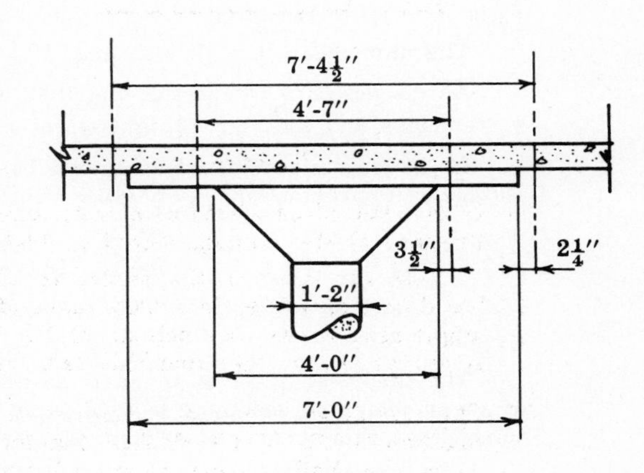

Fig. 14-15

$$V_u = 1.614(130)(12.83)(11.83) = 31{,}900 \text{ lb}$$

(*c*) The sum of the average of the negative moments and positive moment, M_o, is given by equation (*14.6*). The 1963 ACI Building Code specifies that for columns with capitals the value of c shall be the diameter of the 90° vertex right circular cone included in the capital. Hence for $c = 4.0'$ and for the 18′-0″ span,

$$M_o = 0.10(64.2)(18.0)(1.0)\left[1 - \frac{2(4.0)}{3(18.0)}\right]^2 = 83.9 \text{ ft-kips}$$

For the 17′-0″ span,

$$M_o = 0.10(64.2)(17.0)(1.0)\left[1 - \frac{2(4.0)}{3(17.0)}\right]^2 = 77.7 \text{ ft-kips}$$

(*d*) Table 14.3 contains the percentages of M_o that must be resisted by the column and middle strips. From the table, for an interior panel with drop panels,

$$\begin{aligned} \text{Column strip positive moment} &= +M_c = 0.20M_o \\ \text{Column strip negative moment} &= -M_c = 0.50M_o \\ \text{Middle strip positive moment} &= +M_m = 0.15M_o \\ \text{Middle strip negative moment} &= -M_m = 0.15M_o \end{aligned}$$

For the 18′-0″ span,

$$+M_{uc} = 0.20(83.9) = 16.8 \text{ ft-kips} \qquad +M_{um} = 0.15(83.9) = 12.6 \text{ ft-kips}$$
$$-M_{uc} = 0.50(83.9) = 42.0 \text{ ft-kips} \qquad -M_{um} = 0.15(83.9) = 12.6 \text{ ft-kips}$$

and for the 17′-0″ span,

$$+M_{uc} = 0.20(77.7) = 15.5 \text{ ft-kips} \qquad +M_{um} = 0.15(77.7) = 11.7 \text{ ft-kips}$$
$$-M_{uc} = 0.50(77.7) = 38.8 \text{ ft-kips} \qquad -M_{um} = 0.15(77.7) = 11.7 \text{ ft-kips}$$

14.3. An 18′-0″ × 18′-0″ flat plate must resist a total live load of 60 psf and is supported on 16″ round columns. Determine the working stress design depth required to satisfy shear requirements of the 1963 ACI Building Code. Assume the concrete is lightweight structural concrete with a strength of 3000 psi. The splitting strength ratio, F_{sp}, is 5.5 and the unit weight is 110 pcf.

Because the dead load must be determined, assume a depth of 9″. Then

$$\begin{aligned} \text{D.L.} = 9(110)/12 &= 82.5 \text{ psf} \\ \text{L.L.} &= 60 \text{ psf} \\ \text{Total Load} &= 142.5 \text{ psf.} \quad \text{Say 143 psf.} \end{aligned}$$

The total panel load is $W = 143(18.0)^2 = 46{,}300$ lb.

The critical shear periphery around the column is at a distance $d/2$. Therefore the length of b_o is $\pi(16+d)$. Assuming $d = t - 1.5'' = 7.5''$, $b_o = \pi(16+7.5) = 73.8''$.

The unit shear $v = V/b_o d$ and $V = W - \pi(23.5/12)^2(143)/4 = 45{,}900$ lb. Then $v = 83$ psi.

As discussed in Chapter 6, the 1963 ACI Building Code specifies that the allowable shear at b_o for slabs and footings of lightweight concrete shall be $v_c = 0.3F_{sp}\sqrt{f_c'} = 90.7 \text{ psi} > 83$.

If the slab is considered as a wide beam, the shear along one side at a distance of d from the column is approximately $V = (18.0 - 16/12 - 15.0/12)^2(143)/4 = 8500$ lb. The unit shear is

$$v = \frac{V}{bd} = \frac{8500}{(18.0 - 16/12 - 15/12)(12)(7.5)} = 6.1 \text{ psi}$$

As discussed in Chapter 6, the allowable shear is $v_c = 0.17F_{sp}\sqrt{f_c'} = 51.4 \text{ psi} > 6.1$. See Problem 14.1.

The thickness of 9″ is adequate to resist the shear stress. It should be noted that the value of shear computed assuming a wide beam is very low in magnitude. It has been previously mentioned that this shear will seldom, if ever, be critical.

14.4. Solve Problem 14.3 using ultimate strength design procedures.

Assume a depth of slab of $8\frac{1}{2}''$; then

$$\begin{aligned} \text{D.L.} = 8.5(110)/12 &= 78 \text{ psf} \\ \text{L.L.} &= 60 \text{ psf} \\ \text{Total Load} &= 138 \text{ psf} \end{aligned} \qquad \frac{\text{D.L.}}{\text{T.L.}} = \frac{78}{138} = 0.565, \quad \frac{\text{L.L.}}{\text{T.L.}} = \frac{60}{138} = 0.435$$

Because there are no wind load effects, the critical load factor equation is $U = 1.5D + 1.8L = 1.5(0.565)(D+L) + 1.8(0.435)(D+L) = 1.63(D+L)$. $W_u = 1.63(138)(18.0)^2 = 72{,}900$ lb.

The critical shear periphery has a length $b_o = \pi(16+d)$. Assuming $d = 7''$, $b_o = 72.2''$.

The unit shear $v_u = V_u/b_o d$ and $V_u = W_u - \pi(23.0/12)^2(1.63)(138)/4 = 72{,}200$ lb. Then $v_u = 143$ psi.

As discussed in Chapter 6, the allowable shear is $v_u = 0.6\phi F_{sp}\sqrt{f'_c}$, where $\phi = 0.85$ for shear. Then $v_u = 154$ psi > 143.

Assuming the slab acts as a wide beam, the shear along one side of the panel at a distance d from the column is approximately $V_u = (18.0 - 16/12 - 14/12)^2(225)/4 = 13{,}500$ lb. The unit shear is

$$v_u = \frac{V_u}{bd} = \frac{13{,}500}{(18.0 - 16/12 - 14/12)(12)(7.0)} = 10.4 \text{ psi}$$

The allowable shear is $v_c = 0.3\phi F_{sp}\sqrt{f'_c} = 72.1$ psi > 10.4. See Problem 14.1.

The assumed thickness of slab of $8\frac{1}{2}''$ is adequate. Again the shear stress as computed assuming a wide beam is very low in magnitude.

14.5. A 7″ flat slab having an effective depth of $5\frac{1}{2}''$ is supported on a 14″ × 16″ interior column that resists an unbalanced moment of 10.0 ft-kips in each principal direction. The vertical reaction at the column is 40.0 kips. If the reinforcing steel in the top of the slab is #4's at 12″, determine the shear stresses at the critical sections near the column. Assume working stress design.

This solution will be based on the report of ACI-ASCE Committee 326. The critical section is located at a distance $d/2$ from the face of the column.

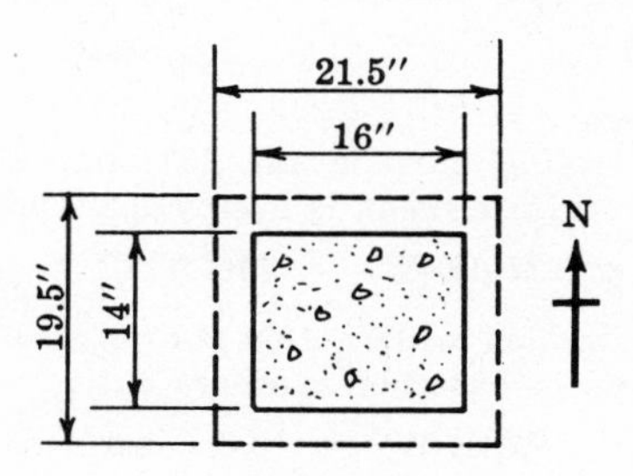

Fig. 14-16

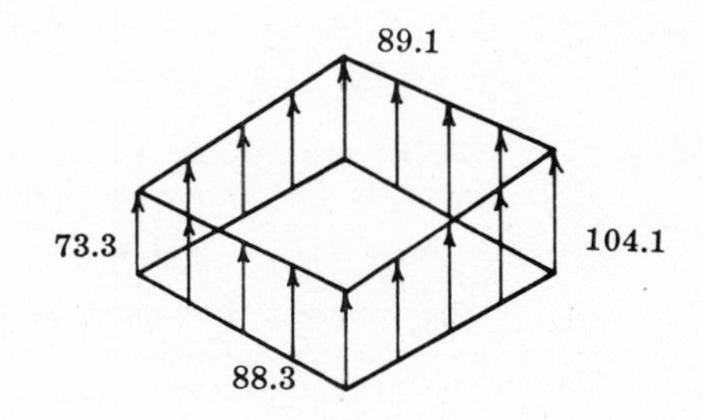

Fig. 14-17

The shear stress due to the vertical reaction is

$$v_p = V/A_c = 40{,}000/(2)(19.5 + 21.5)(5.5) = 88.7 \text{ psi}$$

The shear stress due to the torsion transmitted as a portion of the differential moment is

$$v_t = \frac{kM}{J_c}\left(\frac{c}{2}\right)$$

The Committee 326 recommendation is that the value of k be taken as 0.2 until additional information is available.

For the N-S direction, $c = 19.5$, $b = 21.5$, and

$$J_c = 5.5(19.5)^3/6 + 19.5(5.5)^3/6 + 2(21.5)(5.5)(19.5/2)^2 = 29{,}800 \text{ in}^4$$

For the E-W direction, $c = 21.5$, $b = 19.5$, and

$$J_c = 5.5(21.5)^3/6 + 21.5(5.5)^3/6 + 2(19.5)(5.5)(21.5/2)^2 = 34{,}500 \text{ in}^4$$

The shear stress due to torsion in the N-S direction is

$$v_t(\text{N-S}) = \frac{0.2(10.0)(12{,}000)(19.5)}{29{,}800(2)} = 7.86 \text{ psi}$$

The shear stress due to torsion in the E-W direction is

$$v_t(\text{E-W}) = \frac{0.2(10.0)(12{,}000)(21.5)}{34{,}500(2)} = 7.49 \text{ psi}$$

Combining all the factors as computed above in equation (*14.7*), the shear stresses at the corners of critical section are

$$v = \frac{V}{A_c} \pm \frac{kM_1}{J_{c1}}\left(\frac{c_1}{2}\right) \pm \frac{kM_2}{J_{c2}}\left(\frac{c_2}{2}\right) = 88.7 \pm 7.9 \pm 7.5 = 104.1,\ 89.1,\ 88.3, \text{ and } 73.3 \text{ psi}$$

See Fig. 14-17.

14.6. Repeat Problem 14.5 assuming the critical section has a width between lines which are at a distance of $1.5t$ on either side of the column. Assume $f'_c = 3000$ psi and $f_s = 20{,}000$ psi.

(*a*) The solution will be based on the use of equation (*14.9*). The critical sections are as shown in Fig. 14-18.

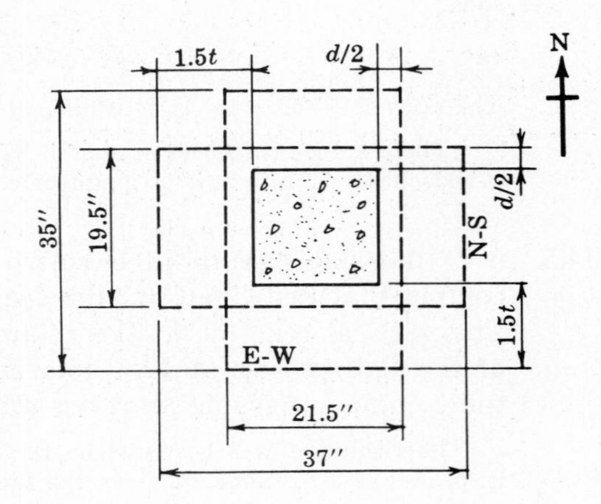

Fig. 14-18

(*b*) The shear stress due to the vertical reaction at the two different sections is

$$v_p(\text{N-S}) = \frac{V}{A_c} = \frac{40{,}000}{2(19.5 + 37)5.5} = 64.4 \text{ psi}$$

$$v_p(\text{E-W}) = \frac{40{,}000}{2(21.5 + 35)5.5} = 64.4 \text{ psi}$$

This is not the critical section defined as b_o.

(*c*) If the reinforcing steel is #4's at 12", the area of steel per foot of width is 0.20 in². For a $d = 5.5''$ the resisting moment is $Rbd^2 = 6.8$ ft-kips/ft, or

$$A_s f_s jd = 0.20(20{,}000)(0.872)(5.5)/12{,}000 = 1.60 \text{ ft-kips/ft}$$

There are not an exact number of reinforcing bars in the 35" and 37" wide sections. However, it will be assumed that the area of reinforcement is evenly distributed. Then

$$M_R(\text{N-S}) = 37(1.6)/12 = 4.93 \text{ ft-kips} \qquad M_R(\text{E-W}) = 35(1.6)/12 = 4.67 \text{ ft-kips}$$

The moment that must be transmitted by torsion will be 10.0 ft-kips less the above values of the resisting moments. The resisting moment is based on top bars only.

$$M_t(\text{N-S}) = 10.0 - 4.93 = 5.07 \text{ ft-kips} \qquad M_t(\text{E-W}) = 10.0 - 4.67 = 5.33 \text{ ft-kips}$$

(*d*) The shear stress due to torsion is $v_t = \dfrac{M_t}{J_c}\left(\dfrac{c}{2}\right)$, where $J_c = dc^3/6 + cd^3/6 + 2bd(c/2)^2$.

For the N-S direction, $c = 19.5''$, $b = 37''$, and $J_c = 46{,}000$ in⁴. For the E-W direction, $c = 21.5''$, $b = 35''$, and $J_c = 54{,}200$ in⁴.

The shear stress due to torsion in the N-S direction is

$$v_t(\text{N-S}) = \frac{5.07(12{,}000)(19.5)}{46{,}000(2)} = 12.9 \text{ psi}$$

The shear stress due to torsion in E-W direction is

$$v_t(\text{E-W}) = \frac{5.33(12{,}000)(21.5)}{54{,}200(2)} = 12.7 \text{ psi}$$

(*e*) Combining the shear stresses due to vertical shear and torsional shear in equation (*14.7*),

$$v = \frac{V}{A_c} \pm \frac{M_t}{J_c}\left(\frac{c}{2}\right)$$

$$v(\text{N-S}) = 64.4 \pm 12.9 = 77.3 \text{ and } 51.5 \text{ psi} \qquad v(\text{E-W}) = 64.4 \pm 12.7 = 77.1 \text{ and } 51.7 \text{ psi}$$

(*f*) If the torsional shear stresses are computed where the two critical sections cross, which is a $d/2$ from the column, and added to v_p determined in Problem 14.5, $v = 88.7 \pm 12.9 \pm 12.7 = 114.3$, 88.9, 88.5, and 63.1 psi.

(*g*) In Problem 14.5(*d*), the value for the maximum shear is approximately 10% less than the value in (*f*) above. In (*c*) above, it was assumed that the reinforcing steel was distributed uniformly across the panel. The reinforcement could be spaced closer together near the column and the value of M_R increased. If M_R is increased, M_t is decreased, the shear stress due to torsion would be reduced, and the results obtained from the two solutions could be made to agree more closely. In (*c*), the resistance of the concrete and bottom bars was neglected.

14.7. A 10-story apartment building is 61′-4″ wide by 201′-2″ long. It is to be constructed using flat plate floor and roof systems proportioned by working stress design techniques. The live load is 40 psf, the partition load is 20 psf of floor area, and the floor finish is 5 psf. Design a typical exterior and a typical interior panel for an intermediate floor. Assume $f'_c = 4000$ psi, $f_s = 20{,}000$ psi and $w = 150$ pcf. The story height is 9′-0″. The columns are 14″ by 16″ and the lateral forces will be resisted by shear walls.

(*a*) The slab will be designed by the empirical method if all the necessary requirements are met. The 1963 ACI Building Code specifies that the empirical method may be used if all the structure conforms to all of the following:

(1) There must be at least 3 continuous panels in each direction.
(2) The length to width ratio of the panels shall not be greater than 1.33.
(3) Successive span lengths shall not vary more than 20 percent.
(4) Columns shall not be offset more than 10 percent of the span.
(5) The building shall not have a height exceeding 125 feet.
(6) The maximum story height shall not exceed 12.5 feet.

If the panels are 20′-0″ by 20′-0″, all the above conditions can be met. See Fig. 14-19.

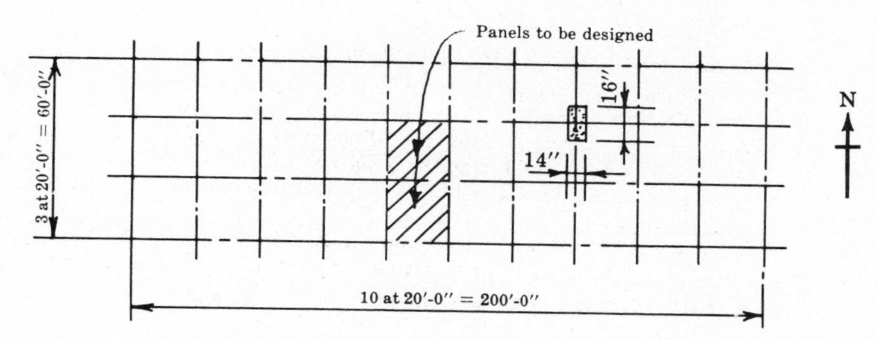

Fig. 14-19

(*b*) The minimum slab thickness without drop panels is $L/36$ or 5″.

Assume $t = 8'' > 20(12)/36 > 5''$. The loads would be

D.L. — Slab		=	100 psf
	Floor Finish	=	5 psf
	Partitions	=	20 psf
Total D.L.		=	125 psf
L.L.		=	40 psf
Total Load		=	165 psf

(*c*) The 1963 ACI Building Code specifies a minimum moment of inertia for the columns if there is no modification of the moments and slab thicknesses. If $W_D = (20.0)^2(125) = 50{,}000$ lb., $W_L = (20.0)^2(40) = 16{,}000$ lb, $t = 8.0''$, and $H = 9.0''$, then $I_c = \dfrac{t^3 H}{0.5 + W_D/W_L} = 1270\text{ in}^4$.

The minimum moment of inertia of the 14″ by 16″ column is $I = bd^3/12 = 16(14)^3/12 = 3660\text{ in}^4 > 1270$. It would be possible to furnish a column of lesser stiffness if the positive and negative moments in the slab were increased by the factors R_p and R_n, respectively.

(*d*) If the slab has no drop panels and is designed by the empirical method, an additional requirement of minimum thickness must be checked:

$$t \geqq 0.028L\left(1 - \frac{2c}{3L}\right)\sqrt{\frac{w'}{f'_c/2000}} + 1\tfrac{1}{2}$$

For check of thickness, the minimum value of c should be used.

$$t = 0.028(20.0)\left[1 - \frac{2(1.17)}{3(20.0)}\right]\sqrt{\frac{165}{4000/2000}} + 1\tfrac{1}{2} = 6.39'' < 8$$

(*e*) If it is assumed that $d = 6.5''$, the critical sections for shear around the columns are as shown in Fig. 14-20.

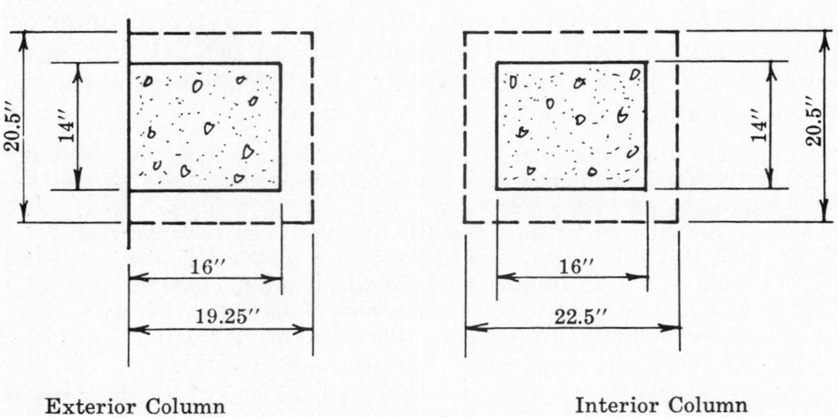

Fig. 14-20

For an interior column, the net shear is approximately $V = [(20.0)^2 - 20.5(22.5)/144]165 =$ 65,600 lb. If $b_o = 2(22.5 + 20.5) = 86''$, the unit shear stress is $v = 65{,}600/(86)(6.5) = 117$ psi. The allowable shear stress is $v = 2\sqrt{f'_c} = 126 \text{ psi} > 117$.

For an exterior column, the net shear is approximately

$$V = [(20.0)^2/2 - 19.25(20.5)/144]165 = 32{,}600 \text{ lb}$$

If $b_o = 2(19.25) + 20.5 = 59.0''$, the unit shear stress is $v = 32{,}600/(59)(6.5) = 85.0 \text{ psi} < 126$.

(*f*) If the slab is considered a wide beam, the shear acting on one edge at a distance d from the face of the column is approximately

$$V = [20.0 - 13/12 - 16/12][20.0 - 13/12 - 14/12](165/4) = 12{,}900 \text{ lb}$$

The unit shear stress is $v = 12{,}900/(17.58)(12)(6.5) = 9.4$ psi. The allowable shear stress is $v = 1.1\sqrt{f'_c} = 70 \text{ psi} > 9.4$. See Problem 14.1.

(*g*) The sum of the average of the negative moments and positive moment is $M_o = 0.09WLF(1 - 2c/3L)^2$, where $c = (14 + 16)/2 = 15''$ and $F = 1.15 - c/L = 1.15 - 15/(12)(20) = 1.09$. Then $M_o = 119{,}000$ ft-lb $= 119$ ft-kips.

When determining the distribution of M_o to the various portions of the slab, the calculations are made simpler by using a tabular form as shown in Table 14.4. In the table the E-W marginal column strip is the half column strip along the south exterior column line. Since there are no marginal beams, the exterior panel has support type A.

It should be noted that both the positive and negative moments for both panels are the same in the E-W direction.

The maximum spacing of bars in flat slabs shall not exceed two times the slab thickness. Then, maximum spacing $= 2(8) = 16''$, and $10'\text{-}0''/1'\text{-}4'' = 8$ spaces, or 9 bars.

The minimum area of reinforcing for flat slabs with $f_y = 40{,}000$ psi, is minimum $A_s = 0.0020bt = 0.0020(12)(8) = 0.19 \text{ in}^2/\text{ft}$, or $10(0.19) = 1.90 \text{ in}^2/\text{strip}$.

The maximum coefficient in the table is 0.50. Therefore the maximum moment that must be resisted by a strip of the slab is $M = 0.50(119) = 59.5$ ft-kips. The flexural capacity of the slab must be checked to determine if it can resist 59.5 ft-kips.

From Chapter 4, the capacity of a 6.5″ slab with $f_c = 1800$ psi and $f_s = 20{,}000$ psi is 13.7 ft-kips/ft. Because of the concentration of flexural stresses at the supports, the width of a strip that may be considered effective in resisting bending is limited to 75% of a strip width. Hence the resisting moment is $M_R = 0.75(10.0)(13.7) = 103 \text{ ft-kips} > 59.5$.

(*h*) In the table, the average depth to the reinforcing steel was used for d. The depth to the centroid of the top layer is $6\frac{1}{8}''$, and to the centroid of the bottom layer is $6\frac{7}{8}''$. The average of

TABLE 14.4

Direction of Span	Strip		Exterior Panel			Interior Panel	
			$M_o = 119$			$M_o = 119$	
			$-M$	$+M$	$-M$	$+M$	$-M$
North-South	Column Strip	$\% M_o$	40	28	50	22	46
		M	47.6	33.3	59.5	26.2	54.7
		d	6.5	6.5	6.5	6.5	6.5
		A_s	5.09	3.56	6.36	2.80	5.85
		Reinf.	12-#6	9-#6	15-#6	9-#6	14-#6
	Middle Strip	$\% M_o$	10	20	18	16	16
		M	11.9	23.8	21.4	19.0	19.0
		d	6.5	6.5	6.5	6.5	6.5
		A_s	1.90	2.54	2.29	2.03	2.03
		Reinf.	10-#4	13-#4	12-#4	10-#4	10-#4
			$M_o = 119$			$M_o = 119$	
East-West	Column Strip	$\% M_o$	—	22	46	22	46
		M	—	26.2	54.7	26.2	54.7
		d	—	6.5	6.5	6.5	6.5
		A_s	—	2.80	5.85	2.80	5.85
		Reinf.	—	9-#6	14-#6	9-#6	14-#6
	Middle Strip	$\% M_o$	—	16	16	16	16
		M	—	19.0	19.0	19.0	19.0
		d	—	6.5	6.5	6.5	6.5
		A_s	—	2.03	2.03	2.03	2.03
		Reinf.	—	10-#4	10-#4	10-#4	10-#4
	Marginal Column Strip	$\% M_o$	—	11	23	—	—
		M	—	13.1	27.4	—	—
		d	—	6.5	6.5	—	—
		A_s	—	1.40	2.93	—	—
		Reinf.	—	5-#6	7-#6	—	—

$6\frac{1}{2}''$ varies approximately 6% from the "exact" values. These values are based on a $\frac{3}{4}''$ cover and a maximum bar size of #6. If smaller bars are used, the variation would be even less. Some designers prefer to use the minimum depth in order to be conservative. Other designers prefer to use the two different depths and require specific placement sequences of the reinforcing steel.

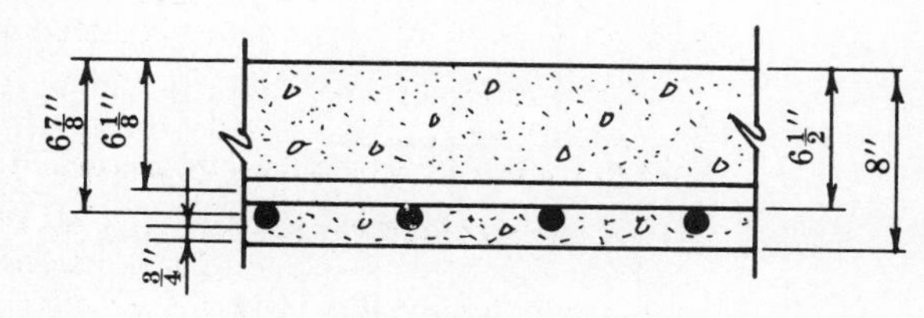

Fig. 14-21

Due to the uncertainty of stress variations within the slab and because of the many assumptions made in the derivations and design, the use of an average value of d in this problem should be reasonable and in keeping with sound structural design.

(*i*) It is not always possible to select the number and size of reinforcing bars to exactly meet the flexural requirements.

If trussed bars are used, it is sometimes convenient to select the same bar size in the table for both positive and negative moment regions. This practice coupled with the maximum spacing requirement, many times necessitates the furnishing of an excess of reinforcement. In this example, the 9-#6's in the positive moment region of the interior column strips are somewhat more than required. The use of mixed bar sizes may be considered in order to reduce this excess.

(*j*) The 1963 ACI Building Code requires that the bending moments for which columns must be designed shall not be less than

$$M = (WL_1 - W_D L_2)/f$$

where L_1 and L_2 are lengths of adjacent panels and f is 30 for exterior columns and 40 for interior columns. The moment shall be divided between the upper and lower columns in proportion to their stiffnesses. $W = (20.0)^2(165) = 66{,}000$ lb and $W_D = 50{,}000$ lb. Then

$$M = [66.0(20.0) - 50.0(20.0)]/40 = 8.0 \text{ ft-kips for interior columns}$$

$$M = [66.0(20.0) - 50.0(0)]/30 = 44.0 \text{ ft-kips for exterior columns}$$

The values 8.0 and 44.0 ft-kips compare to 5.0 and 49.6 ft-kips by differencing the column strip negative moments in the table.

(*k*) The transfer of bending moments to the columns will be checked based on the 20-80 distribution. The critical sections are shown in (*e*) above.

For an interior column for a moment in the N-S direction, $c = 22.5''$ and $b = 20.5''$. And from equation (*14.8*), $J_c = 47{,}130$ in^4. The shear stress due to torsion is

$$v_t = \frac{kM}{J_c}\left(\frac{c}{2}\right) = \frac{0.2(8.0)(12{,}000)(22.5)}{47{,}130(2)} = 4.58 \text{ psi}$$

For an interior column for a moment in the E-W direction, $c = 20.5''$ and $b = 22.5''$. And from equation (*14.8*), $J_c = 41{,}050$ in^4. The shear stress due to torsion is

$$v_t = \frac{kM}{J_c}\left(\frac{c}{2}\right) = \frac{0.2(8.0)(12{,}000)(20.5)}{41{,}050(2)} = 4.79 \text{ psi}$$

The maximum shear at the critical section around an interior column is the sum of the vertical shear stress and the shear due to torsion about the N-S and E-W axes. From (*e*), $v_p = 117$ psi. Hence the total shear stress is $v = 117 + 4.58 + 4.79 = 126.4$ psi. This is approximately equal to the allowable of 126 psi.

(*l*) At the exterior column, the vertical shear has an eccentricity in the N-S direction. See Fig. 14-22.

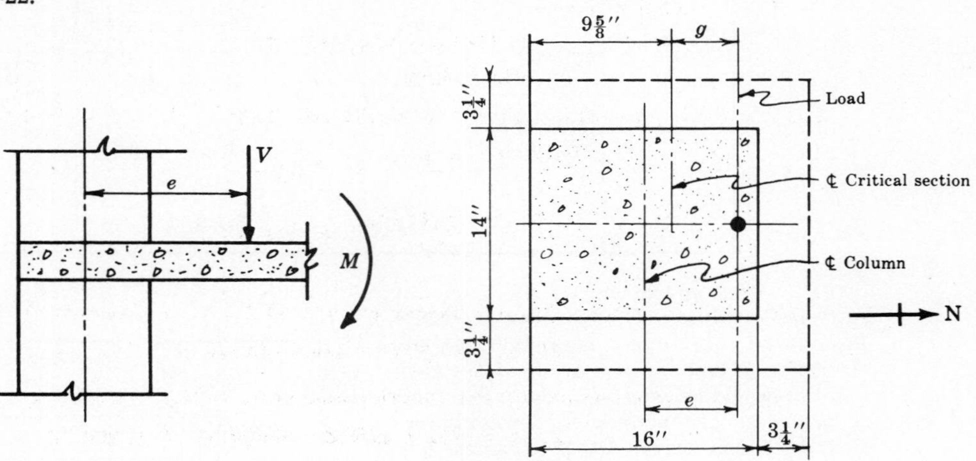

Fig. 14-22 **Fig. 14-23**

In plan, the critical section is as shown in Fig. 14-23. From the figure,

$$g = \frac{bc}{2(2c+b)} = \frac{20.5(19.25)}{2[2(19.25) + 20.5]} = 3.35'' \qquad \text{and} \qquad e = 9.62 + 3.35 - 8.0 = 4.97''$$

If the total moment in the column is M, the torsional moment in the slab is $T = 0.20(M - Ve) = 0.20[44.0 - 33.0(4.97)/12] = 6.06$ ft-kips.

At the exterior column in the N-S direction, $c = 19.25''$ and $b = 20.5''$. Then

$$J_c = dc^3/6 + cd^3/6 + 2cdg^2 + bd(c/2 - g)^2 = 16{,}650 \text{ in}^4$$

The maximum shear stress due to torsion in the N-S direction only is

$$v_t = Ty/J_c = 6.06(12{,}000)(9.62 + 3.35)/16{,}650 = 56.6 \text{ psi}$$

The above shear stress combined with the shear due to the vertical reaction exceeds the allowable. Or, $v = 56.6 + 85.0 = 141.6 \text{ psi} > 126$. Hence the design must be modified.

The unit shear stress could be reduced by increasing the slab thickness, increasing the column size, extending the slab to form a cantilever to the exterior, or a combination of the above. The concrete strength could be increased; this would permit an increase in the allowable shear stress.

It is assumed here that the column spacing and size, the overall building dimensions, and the concrete strength are fixed. Therefore the slab thickness must be increased.

(*m*) The slab thickness will be increased to 9″, $d = 7.5''$. The loads are

D.L. –	Slab	=	112 psf
	Floor Finish	=	5 psf
	Partitions	=	20 psf
Total D.L.		=	137 psf
L.L.		=	40 psf
Total Load		=	177 psf

(*n*) The column stiffness must be checked: $I_c = \dfrac{t^3H}{0.5 + W_D/W_L} = 1670 \text{ in}^4 < 3660.$

(*o*) The moment in the exterior columns is

$$M = (WL_1 - W_DL_2)/f = 70.8(20.0)/30 = 47.2 \text{ ft-kips}$$

The critical section for shear is as shown in Fig. 14-24.

$$g = \frac{bc}{2(2c+b)} = \frac{21.5(19.75)}{2[2(19.75) + 21.5]} = 3.48''$$

and

$$e = 9.88 + 3.48 - 8.0 = 5.36''$$

The torsional moment is

$$T = 0.20(M - Ve) = 0.20[47.2 - 35.4(5.36)/12] = 6.28 \text{ ft-kips}$$

And $c = 19.75''$, $b = 21.5''$; then $J_c = 21{,}190 \text{ in}^4$. Then

$$v_t = \frac{Ty}{J_c} = \frac{6.28(12{,}000)(9.88 + 3.48)}{21{,}190} = 47.6 \text{ psi}$$

$$v_p = \frac{V}{bd} = \frac{35{,}400 - (19.75)(21.5)(177)/144}{[2(19.75) + 21.5](7.5)} = 76.3 \text{ psi}$$

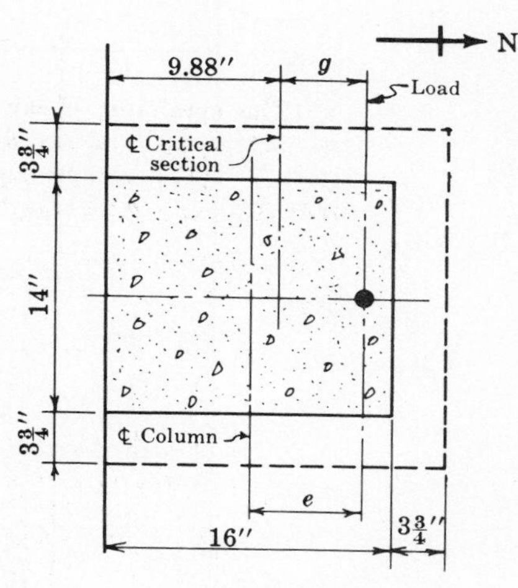

Fig. 14-24

(*p*) For the exterior column the unbalanced moment in the E-W direction is

$$M = [35.4(20.0) - 27.4(20.0)]/40 = 4.0 \text{ ft-kips}$$

In the E-W direction for the exterior column, $c = 21.5''$, $b = 19.75''$, and

$$J_c = dc^3/12 + cd^3/12 + 2bd(c/2)^2 = 41{,}200 \text{ in}^4$$

Then

$$v_t = \frac{kM}{J_c}\left(\frac{c}{2}\right) = \frac{0.2(4.0)(12{,}000)(21.5)}{41{,}200(2)} = 2.50 \text{ psi}$$

(*q*) Summing shear stresses due to the vertical reaction and the unbalanced torsion in both directions, $v = 76.3 + 47.6 + 2.5 = 126.4$ psi. This is approximately equal to the allowable of 126 psi.

(*r*) After the new slab thickness of 9″ was assumed in (*m*), parts (*a*), (*b*), (*c*), (*e*), (*j*) and (*l*) were revised. By inspection, parts (*d*), (*f*) and (*k*) were checked. The portion of the calculations to be revised is the determination of the bending moments and reinforcement requirements in (*g*). From equation (*14.2*),

$$M_o = 0.09(20.0)^2(177)(20.0)(1.09)\left[1 - \frac{2(15)}{3(12)(20)}\right]^2 = 128{,}000 \text{ ft-lb} = 128 \text{ ft-kips}$$

TABLE 14.5

Direction of Span	Strip		Exterior Panel			Interior Panel	
			$M_o = 128$			$M_o = 128$	
			$-M$	$+M$	$-M$	$+M$	$-M$
North-South	Column Strip	% M_o	40	28	50	22	46
		M	51.2	35.8	64.0	28.1	58.8
		d	7.5	7.5	7.5	7.5	7.5
		A_s	4.74	3.31	5.92	2.60	5.44
		Reinf.	11-#6	8-#6	14-#6	8-#6	13-#6
	Middle Strip	% M_o	10	20	18	16	16
		M	12.8	25.6	23.0	20.5	20.5
		d	7.5	7.5	7.5	7.5	7.5
		A_s	2.16	2.37	2.16	2.16	2.16
		Reinf.	11-#4	12-#4	11-#4	11-#4	11-#4
			$M_o = 128$			$M_o = 128$	
East-West	Column Strip	% M_o	—	22	46	22	46
		M	—	28.1	58.8	28.1	58.8
		d	—	7.5	7.5	7.5	7.5
		A_s	—	2.60	5.44	2.60	5.44
		Reinf.	—	8-#6	13-#6	8-#6	13-#6
	Middle Strip	% M_o	—	16	16	16	16
		M	—	20.5	20.5	20.5	20.5
		d	—	7.5	7.5	7.5	7.5
		A_s	—	2.16	2.16	2.16	2.16
		Reinf.	—	11-#4	11-#4	11-#4	11-#4
	Marginal Column Strip	% M_o	—	11	23	—	—
		M	—	14.1	29.4	—	—
		d	—	7.5	7.5	—	—
		A_s	—	1.31	2.72	—	—
		Reinf.	—	4-#6	7-#6	—	—

Maximum spacing $= 2(9) = 18''$ or $10'\text{-}0''/1'\text{-}6'' = 7$ spaces, or 8 bars.

Minimum $A_s = 0.0020(12)(9) = 0.216$ in²/ft; $10.0(0.216) = 2.16$ in²/strip.

(*s*) Due to the complexity of flat slabs, it is very difficult if not impossible to compute bond stresses with great precision. In fact, many designers do not check bond stresses in flat slabs.

One method proposed is to compute the bond stress at the support considering the slab as a wide beam. In this problem the shear at the support is approximately $V = 10.0(20.0)(177) = 35{,}400$ lb. The reinforcing at this section at the exterior wall is 11-#6's in the column strip and 11-#4's in the middle strip. Then $\Sigma o = 4A_s/D = 4(7.04)/0.75 = 37.5$ in. The bond stress is $u = V/\Sigma o\, jd = 35{,}400/(37.5)(0.86)(7.5) = 146$ psi. The allowable bond stress is $u_{All} = 4.8\sqrt{f'_c}/D = 405$ psi > 146.

Because of the uncertain nature of bond stresses in flat slabs, it is advisable to use the smallest diameter bars practical.

(*t*) The 1963 ACI Building Code contains specific requirements for bar lengths and bend points. They are as shown in Fig. 14-25 and 14-26.

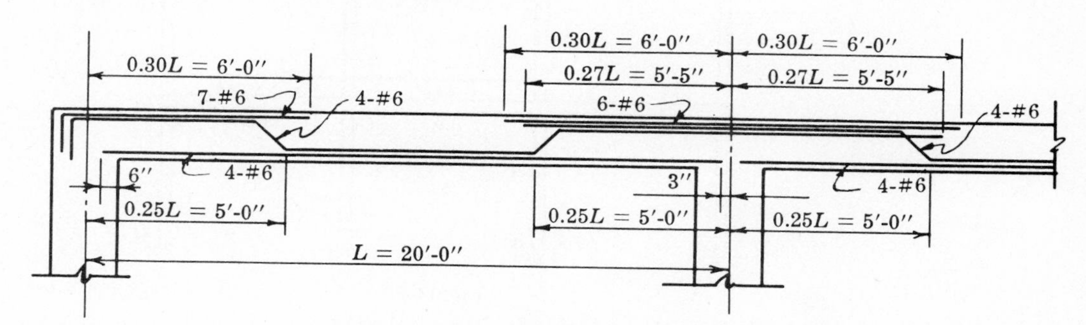

Fig. 14-25. North-South Column Strip

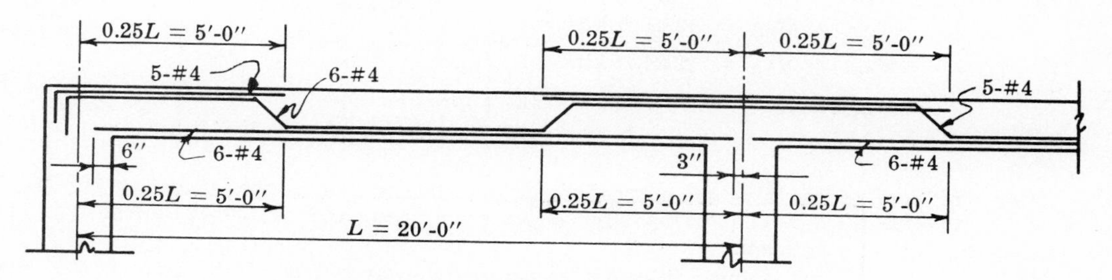

Fig. 14-26. North-South Middle Strip

(*u*) The reinforcing bars should be spaced approximately uniformly across each strip. However, in flat plates, at least 25 percent of the column strip negative reinforcement must cross the periphery at a distance d from the column.

(*v*) In this design, minimum thickness requirements have been met and working stress design techniques were used. Consequently a deflection check is not mandatory.

14.8. Using the member sizes determined in Problem 14.7, compute the design moments for the slab using the elastic analysis.

(*a*) If the elastic analysis is used, the structure is divided into strips which consist of a row of columns and slabs bounded laterally by the centerline of the panel on either side of the columns. The column-slab connection is considered rigid; and if each floor is analyzed individually, the columns are considered fixed at the remote ends.

The gross cross section of the members are used in computing the moments of inertia.

For a 9" slab, 20'-0" wide, $I = 20.0(9/12)^3/12 = 0.702$ ft^4.

For a 14" by 16" column in the N-S direction, $I = (14/12)(16/12)^3/12 = 0.231$ ft^4.

For a 14" by 16" column in the E-W direction, $I = (16/12)(14/12)^3/12 = 0.176$ ft^4.

For a N-S bent, the stiffness factors are $K_{slab} = I/L = 0.702/20 = 0.0350$ and $K_{col.} = I/L = 0.231/9 = 0.0257$.

For an E-W bent, the stiffness factors are $K_{slab} = 0.0350$ and $K_{col.} = I/L = 0.176/9 = 0.0196$.

(*b*) The elastic analysis may be based on the loading of alternate panels with 75% of the live load; or if the live load is less than 75% of the dead load, it may be based on the loading of all panels with full live load. The latter will be used here.

The unit loads are D.L. = 137(20.0) = 2740 lb/lin ft and L.L. = 40(20.0) = 800 lb/lin ft. The N-S bent with loads and relative stiffness factors are as shown in Fig. 14-27.

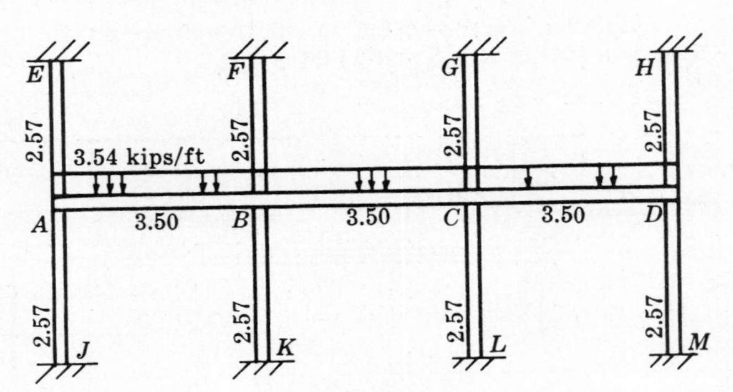

Fig. 14-27

A moment distribution yields

$M_{AB} = M_{DC} = 72.2$ ft-kips
$M_{BA} = M_{CD} = 133.4$ ft-kips
$M_{BC} = M_{CB} = 121.4$ ft-kips

$M_{EA} = M_{JA} = M_{HD} = M_{MD} = 18.5$ ft-kips
$M_{FB} = M_{KB} = M_{GC} = M_{LC} = 2.9$ ft-kips

The maximum positive moments are approximately

$+M = 74.2$ ft-kips in spans *AB* and *CD* $\qquad +M = 55.6$ ft-kips in span *BC*

(*c*) The sum of the average of the negative moments and positive moment is $M_o = (72.2 + 133.4)/2 + 74.2 = 177.0$ ft-kips for an exterior span, and $M_o = 121.4 + 55.6 = 177.0$ ft-kips for an interior span.

This elastic analysis does not recognize two-way action of the slab in that all the load was placed on a N-S bent. However, the 1963 ACI Building Code permits a modification of the elastic analysis to make it conform more closely with the empirical method.

(*d*) If a flat slab that is analyzed by the elastic analysis satisfies all the requirements of the empirical method, then the value of M_o determined in the former may be reduced to that determined in the latter. In Problem 14.7, it was determined that $M_o = 128$ ft-kips. Therefore the moments determined in (*b*) and (*c*) may be reduced by the ratio $128/177 = 0.723$. Then

$M_{AB} = M_{DC} = 0.723(72.2) = 52.2$ ft-kips
$M_{BA} = M_{CD} = 0.723(133.4) = 96.5$ ft-kips
$M_{BC} = M_{CB} = 0.723(121.4) = 88.0$ ft-kips

$+M = 0.723(74.2) = 53.7$ ft-kips for exterior span
$+M = 0.723(55.6) = 40.2$ ft-kips for interior span

(*e*) The percentage of the total moment distributed to the column and middle strips are obtained from Table 14.2.

For the exterior span column strip,
$-M = 0.80(52.2) = 41.8$ ft-kips
$+M = 0.60(53.7) = 32.2$ ft-kips
$-M = 0.76(96.5) = 73.4$ ft-kips

For the exterior span middle strip,
$-M = 0.20(52.2) = 10.4$ ft-kips
$+M = 0.40(53.7) = 21.5$ ft-kips
$-M = 0.24(96.5) = 23.2$ ft-kips

For the interior span column strip,
$-M = 0.76(88.0) = 66.9$ ft-kips
$+M = 0.60(40.2) = 24.1$ ft-kips

For the interior span middle strip,
$-M = 0.24(88.0) = 21.1$ ft-kips
$+M = 0.40(40.2) = 16.1$ ft-kips

(*f*) A comparison of methods is shown in the following table.

Panel		Column Strip		Middle Strip	
		Elastic	Empirical	Elastic	Empirical
Exterior Panel	$-M$ Exterior Support	−41.8	−51.2	−10.4	−12.8
	$+M$	+32.2	+35.8	+21.5	+25.6
	$-M$ Interior Support	−73.4	−64.0	−23.2	−23.0
Interior Panel	$-M$	−66.9	−58.8	−21.1	−20.5
	$+M$	+24.1	+28.1	+16.1	+20.5

Both methods yield comparable results. However, the greatest variations between the two methods occur at the supports in the column strips.

14.9. Proportion by working stress a typical 20′-0″ by 20′-0″ interior flat slab roof panel with drop panels and column capitals that must support a vertical live load of 60 psf. The story height is 9′-0″ and the columns are 12″ round. Assume $f'_c = 3000$ psi and $f_s = 20{,}000$ psi.

(*a*) The empirical method will be used. The minimum slab thickness with drop panels is $L/40$ or 4″. Assume $t = 6.5'' > 20(12)/40 > 4''$.

The minimum side or diameter of a drop panel is $0.33L$. Assume the drop panel $= 0.35(L) = 7'\text{-}0''$ square.

A "rule of thumb" for the diameter of the column capital is $0.20L$ to $0.25L$. Assume the column capital $= 0.20(20) = 4'\text{-}0''$ diameter.

The loads are

D.L. –	Slab	=	81 psf
	Drop Panel (Est.)	=	5 psf
	Roofing and Insulation	=	10 psf
	Total D.L.	=	96 psf
L.L.		=	60 psf
Total Load		=	156 psf

(*b*) Checking the minimum stiffness requirements for the columns:

$$I_c = \frac{t^3H}{0.5 + W_D/W_L} \geqq 1000 \text{ in}^4. \quad W_D = (20.0)^2(96) = 38{,}400 \text{ lb}, \quad W_L = (20.0)^2(60) = 24{,}000 \text{ lb}$$

$$I_c = \frac{(6.5)^3(9.0)}{0.5 + 38.4/24.0} = 1180 \text{ in}^4$$

The minimum moment of inertia of a 12″ round column is $I = \pi(12)^4/64 = 1020 \text{ in}^4$.

If the slab has no column above, I_c must be modified by the expression $(2 - 2.3h/H)$, where h is the distance from the top of the slab to the bottom of the capital as shown in Fig. 14-28 below. h is approximately 2.25 feet. Then $I_c = (2 - 2.3h/H)(1180) = 1680 \text{ in}^4 > 1020$, and $K = 1020/1680 = 0.607$.

(*c*) Because $K < 1.0$, the value for w' must be modified by R_N when determining the minimum thickness. The 1963 ACI Building Code also requires that the negative and positive slab moments must be multiplied by R_N and R_P respectively.

$$R_N = 1 + \frac{(1-K)^2}{2.2(1 + 1.4W_D/W_L)} \qquad R_P = 1 + \frac{(1-K)^2}{1.2(1 + 0.10W_D/W_L)}$$

from which $R_N = 1.0216$, $R_P = 1.111$, and $w' = 1.022(156) = 159$ psf.

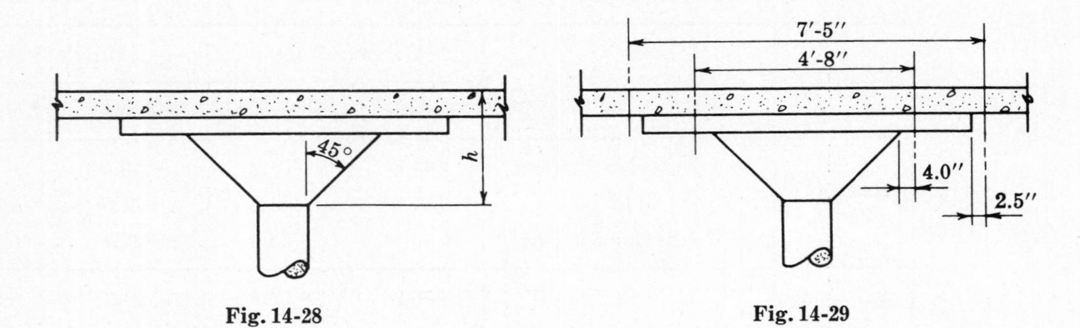

Fig. 14-28

Fig. 14-29

The minimum slab thickness through the drop panel is

$$t_1 \;=\; 0.028L\left(1-\frac{2c}{3L}\right)\sqrt{\frac{w'}{f_c'/2000}}+1.5 \;=\; 6.5'' \qquad \text{where} \qquad c \;=\; 4'\text{-}0''$$

The minimum slab thickness beyond the drop panel is

$$t_2 \;=\; 0.024L\left(1-\frac{2c}{3L}\right)\sqrt{\frac{w'}{f_c'/2000}}+1.0 \;=\; 5.3'' \;<\; 6.5$$

The thickness t_1 should be at least 1.25 times t_2. However, in flexural computations, the maximum effective value for t_1 shall be $1.5t_2$. Or, $1.25(t_2) = 1.25(6.5) = 8.12''$ and $1.50(t_2) = 1.50(6.5) = 9.75''$. Therefore assume $t_1 = 9.5''$, $d_1 = 8.0''$ and $t_2 = 6.5''$, $d_2 = 5.0''$.

(*d*) The critical sections for shear are at $d/2$ from the column capital and drop panel as shown in Fig. 14-29.

The net shear acting around the drop panel is approximately $V = [(20.0)^2 - (7.42)^2]151 = 52{,}100$ lb. And $b_o = 4(7.42)(12) = 366''$, $d = 5.0''$. Then $v = V/bd = 28.5$ psi.

The net shear acting around the column capital is approximately $V = [(20.0)^2 - \pi(4.67)^2/4]156 = 59{,}700$ lb. And $b_o = \pi(4.67)(12) = 176''$, $d = 8.0''$. Then $v = V/b_od = 42.4$ psi.

The allowable shear stress is $v = 2\sqrt{f_c'} = 110$ psi > 42.4.

(*e*) By inspection, the shear stress computed assuming the slab as a wide beam is not critical.

(*f*) If $W = 62.4$ kips, $W_D = 38.4$ kips and $f = 40$, then $M = (WL_1 - W_DL_2)/f = 12.0$ ft-kips.

Assuming a 20-80 distribution of the moment, the torsional moment is $M_t = 0.20(12.0) = 2.4$ ft-kips.

(*g*) It is assumed that the critical section for shear is at $d/2$ from the capital. Hence $c = 56''$, $b = 56''$, $d = 8.0''$. Substituting in equation (*14.8*),

$$J_c \;=\; 8.0(56)^3/6 + 56(8.0)^3/6 + 2(56)(8.0)(56/2)^2 \;=\; 940{,}800 \text{ in}^4$$

Then $$v_t \;=\; M_ty/J_c \;=\; 2.4(12{,}000)(28)/940{,}800 \;=\; 0.86 \text{ psi}$$

and $v = v_p + v_t = 28.5 + 0.9 + 0.9 = 30.3$ psi < 110.

(*h*) The sum of the average of the negative moments and positive moment is $M_o = 0.09WLF(1-2c/3L)^2$ where $F = 1.15 - c/L = 1.15 - 4.0/20.0 < 1.0$. Use $F = 1.0$. Then $M_o = 84.4$ ft-kips.

The positive and negative moments in each strip must be multiplied by R_P and R_N determined in (*c*). The maximum spacing of bars $= 2t = 13''$. The minimum $A_s = 0.002bt = 0.156$ in²/ft.

The coefficients in the following tabulation are obtained by multiplying the values in Table 14.3 by R_N or R_P. The average value of the effective depth at each section is used.

The maximum moment occurring in the 6.5" slab is 18.7 ft-kips. The maximum moment occurring in the 9.5" slab is 43.2 ft-kips. The resisting moments are $M_R = 0.75(10.0)(5.6) = 42.0$ ft-kips > 18.7 and $M_R = 0.75(7.0)(14.4) = 77.5$ ft-kips > 43.2.

Direction of Span	Strip		Exterior Panel			Interior Panel	
						$M_o = 84.4$	
			$-M$	$+M$	$-M$	$+M$	$-M$
Both	Column Strip	% M_o	—	—	—	22.2	51.1
		M	—	—	—	18.7	43.2
		d	—	—	—	5.0	8.0
		A_s	—	—	—	2.60	3.75
		Reinf.	—	—	—	10-#5	12-#5
Both	Middle Strip	% M_o	—	—	—	16.7	15.3
		M	—	—	—	14.1	12.9
		d	—	—	—	5.0	5.0
		A_s	—	—	—	1.96	1.79
		Reinf.	—	—	—	10-#4	9-#4

(*j*) The shear acting on one edge of the panel is $V = 10.0(20.0)(156) = 31{,}200$ lb. The number of bars crossing the edge is 12-#5's in the column strip and 9-#4 in the middle strip. Hence $\Sigma o = 4A_s/D = 4(5.52)/0.625 = 35.3$ in. Assuming the minimum depth,

$$u = \frac{V}{\Sigma o\, jd} = \frac{31{,}200}{35.3(0.86)(5.0)} = 206 \text{ psi}$$

The allowable bond stress is $u_{\text{All}} = 4.8\sqrt{f'_c}/D = 421 \text{ psi} > 206$.

The 1963 ACI Building Code specifies that flexural bond stresses need not be considered if anchorage bond stresses are less than 0.8 of that permitted. The required anchorage length is $l = f_s A_s/(0.8\ \Sigma o u)$. For the #4 and #5 bars this length is 6.4″ and 8.1″ respectively if $f_s = 20{,}000$ psi. Because of the complexity of the nature of shear and bond stresses in flat slabs, the anchorage bond check may assure a better design than a flexural bond check.

(*k*) The bend points and cutoff points are as shown in Fig. 14-30 and 14-31.

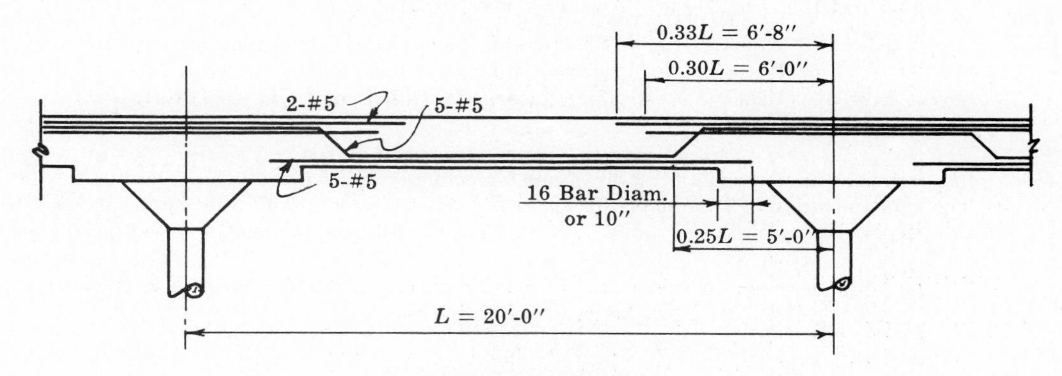

Fig. 14-30. Column Strip

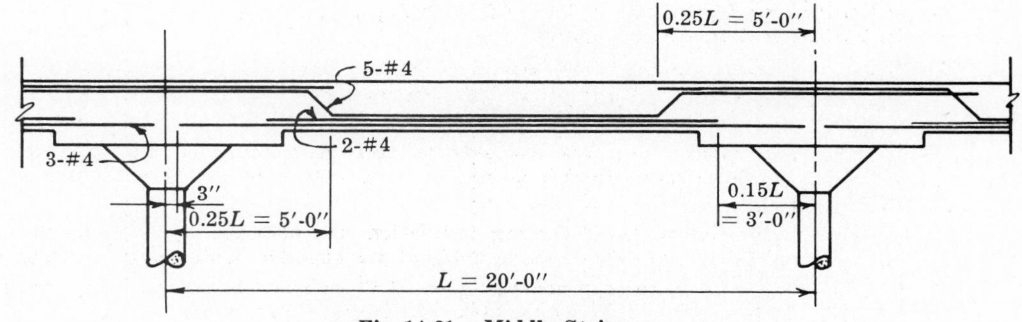

Fig. 14-31. Middle Strip

(*l*) The bars should be spaced so that at least 50% of the column strip negative reinforcement will cross the drop panel.

(*m*) Minimum thickness requirements have been met and working stress design techniques were used. Hence a deflection check is not mandatory.

Supplementary Problems

14.10. Repeat Problem 14.1 but assume USD. *Partial Ans.* $M_o = 115$ or 110 ft-kips

14.11. A 20′-0″ by 20′-0″ interior flat plate is supported on 18″ square columns and the total design load reaction is 50 kips. Determine the unit shear stress at the column due to vertical loads if the slab has an effective depth of 8.5″. *Ans.* $v = 55.5$ psi

14.12. Repeat Problem 14.11 assuming USD and that the dead and live loads are equal. *Ans.* $v = 91.8$ psi

14.13. A flat plate is supported on 14″ by 16″ columns that are subjected to a vertical reaction of 45 kips and a differential moment of 5 ft-kips. If $d = 6.0''$, determine the maximum shear stress at the column assuming the 20-80 moment distribution. *Ans.* $v = 96.1$ psi

14.14. A 17′-6″ by 17′-6″ typical interior flat slab is supported on 16″ round columns. Proportion the slab if $f'_c = 3000$ psi, $f_s = 24{,}000$ psi, L.L. = 40 psf, superimposed D.L. = 25 psf, and the story height = 9′-0″. Use WSD. *Partial Ans.* $t = 9''$

14.15. Repeat Problem 14.14 using USD and $f_y = 60{,}000$ psi. *Partial Ans.* $t = 8''$

14.16. A 17′-6″ by 17′-6″ flat plate roof with an effective depth of 7″ and a total thickness of $8\frac{1}{2}''$ is supported on 12″ round columns. Determine the reinforcement requirements for a typical interior span if $f'_c = 3000$ psi, $f_s = 24{,}000$ psi, L.L. = 30 psf, superimposed D.L. = 5 psf, and the story height is 9′-0″. Use WSD. *Partial Ans.* $K = 0.43$ and $M_o = 75$ ft-kips

14.17. Determine the shear stresses due to vertical reactions at the critical sections in a 20′-0″ by 20′-0″ flat slab that is supported on 3′-6″ diameter column capitals and with a 6′-0″ by 6′-0″ drop panel. The slab thickness is 7″ and the drop panel thickness is 9″. The total superimposed design load is 100 psf. Assume WSD. *Ans.* $v = 40.6$ psi at drop panel and $v = 64.0$ psi at column capital

Chapter 15

Retaining Walls

NOTATION

c = unit cohesion for clay soils, psf
e = eccentricity of resultant of vertical loads, feet
f = friction coefficient
h = height of wall, feet
h_s = surcharge height in equivalent soil, feet
H = horizontal force, kips
H_a = active pressure horizontal force, kips
H_p = passive pressure horizontal force, kips
H_s = total pressure force due to surcharge
k_a = coefficient of active soil pressure
k_p = coefficient of passive soil pressure
L = length of base of a footing, feet
M = bending moment, ft-kips
M_o = overturning moment, ft-kips
M_r = resisting moment, ft-kips
N = normal force, kips
p = unit pressure, psf
p_a = active soil pressure, psf
p_p = passive soil pressure, psf
p_s = surcharge pressure, psf
P_a = sloping active pressure force, kips (May also be horizontal)
P_p = sloping passive pressure force, kips (May also be horizontal)
R = resultant vertical force, kips
SF = safety factor
T = total footing thickness, inches; also a surcharge load on a footing, lb/ft
V = shear force, kips
w = weight of soil, pounds/ft^3
W = vertical load due to soil or concrete, kips
w_s = weight of surcharge, psf
α = natural angle of repose of soil
ϕ = angle of internal friction of soil; also, capacity reduction factor
Σo = sum of perimeters for reinforcing bars

USES OF RETAINING WALLS

Very often in the construction of buildings or bridges it is necessary to retain earth in a relatively vertical position. Whenever embankments are involved in the construction, retaining walls are usually necessary. In the construction of buildings having basements, retaining walls are mandatory. Fig. 15-1 indicates some of the various types of retaining walls which are commonly utilized.

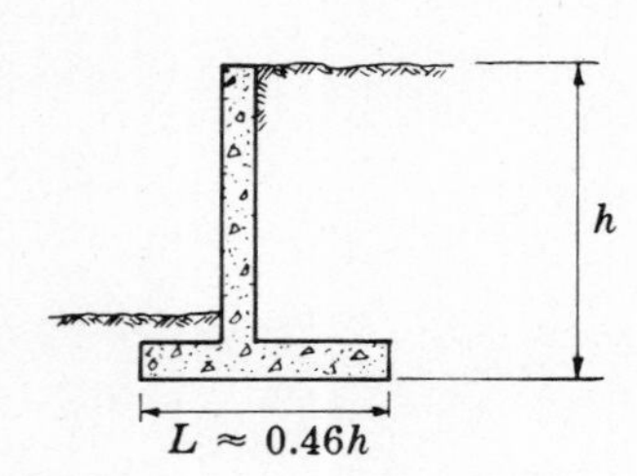

(*a*) T-shaped cantilever wall, horizontal fill

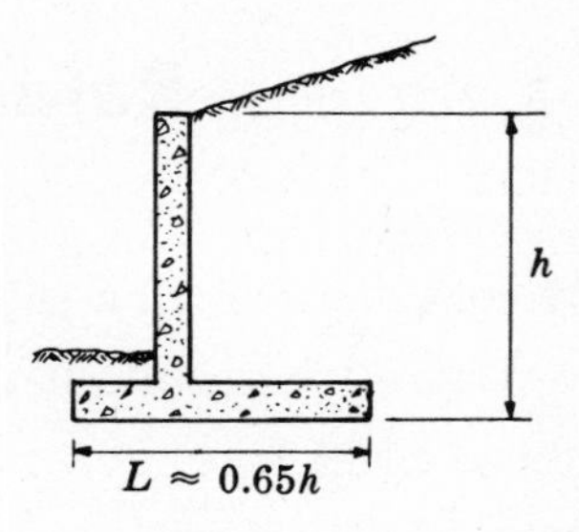

(*b*) T-shaped cantilever wall, sloping fill

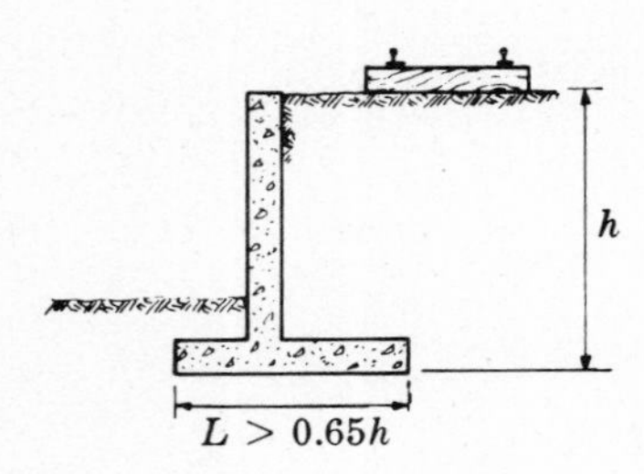

(*c*) T-shaped cantilever wall, with surcharge

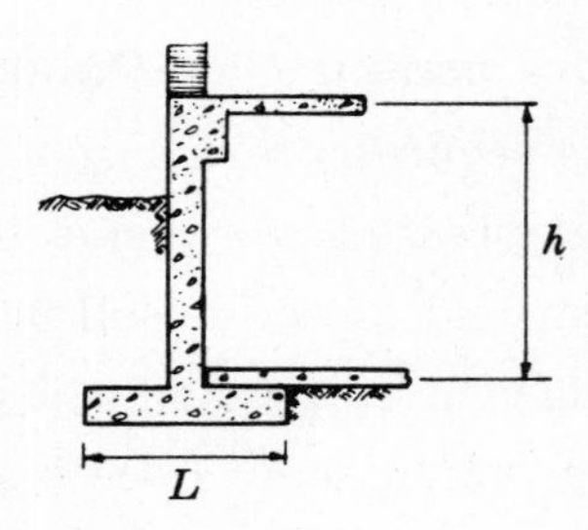

(*d*) Basement wall

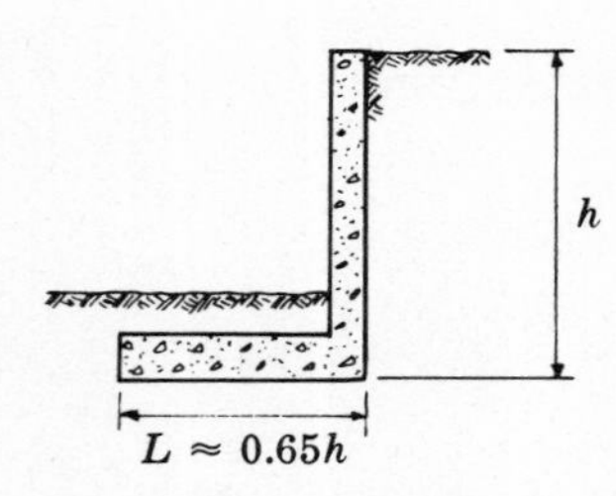

(*e*) L-shaped retaining wall, horizontal fill

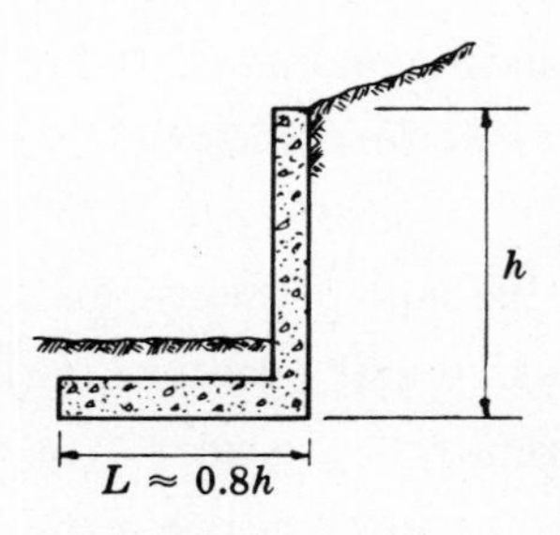

(*f*) L-shaped retaining wall, sloping fill

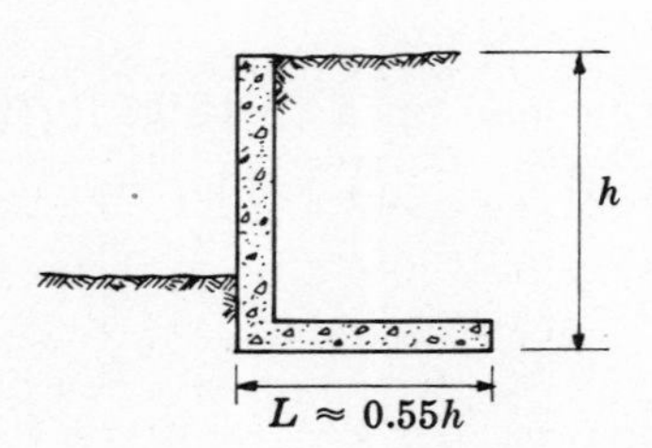

(*g*) Reversed L-shaped retaining wall, horizontal fill

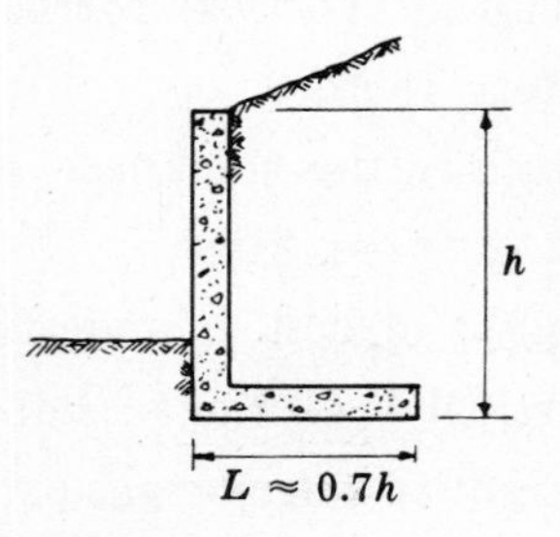

(*h*) Reversed L-shaped retaining wall, sloping fill

Fig. 15-1

FORCES ON RETAINING WALLS

The usual gravity loads due to the weights of the materials do not present great problems with respect to retaining walls. Actually, the *horizontal pressures due to the retained soil* will present the greatest problems.

If we construct a box having a sliding wall, as shown in Fig. 15-2, fill the box with sand, and then suddenly lift the wall, the sand will *slide along a shear plane* and form a slope as shown in Fig. 15-3.

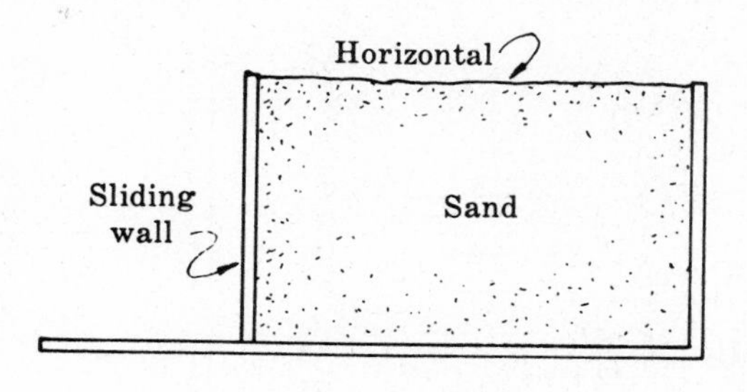

Fig. 15-2

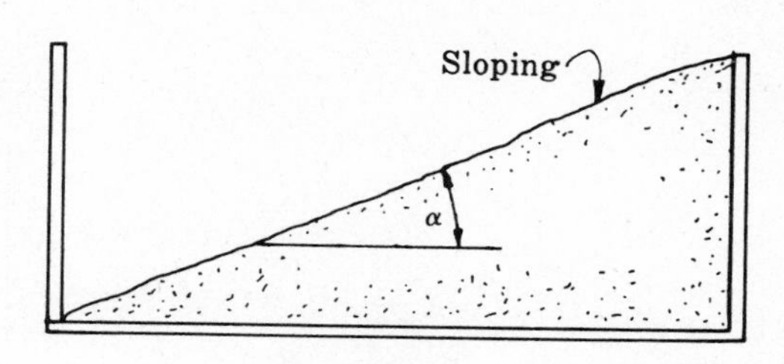

Fig. 15-3

The angle α formed by the sand is called the *natural angle of repose* of the material. Different materials exhibit widely varying slopes of repose. Further, the *moisture content* of the material is an important factor with respect to the slope of repose.

If the moisture content of the sand in Fig. 15-2 would be at an *optimum value,* the sand might stand vertically for a short time.

Granular materials such as sand or gravel behave differently than *cohesive materials* such as clay, when retained in some manner. Materials which contain combinations of the two types of soil will act similar to the predominant material. Since the percentages of cohesive and non-cohesive materials vary extensively in nature, it is necessary to conduct tests in order to determine the properties of a natural deposit of soil.

Although a soil will assume its natural angle of repose when not confined, it is improper to use the angle of repose in design for confined material. Under confinement, the soil has a tendency to slide in a manner similar to that discussed with respect to Fig. 15-2 and 15-3, but somewhat modified. The sliding surface will be more like those shown in Fig. 15-4 and 15-5.

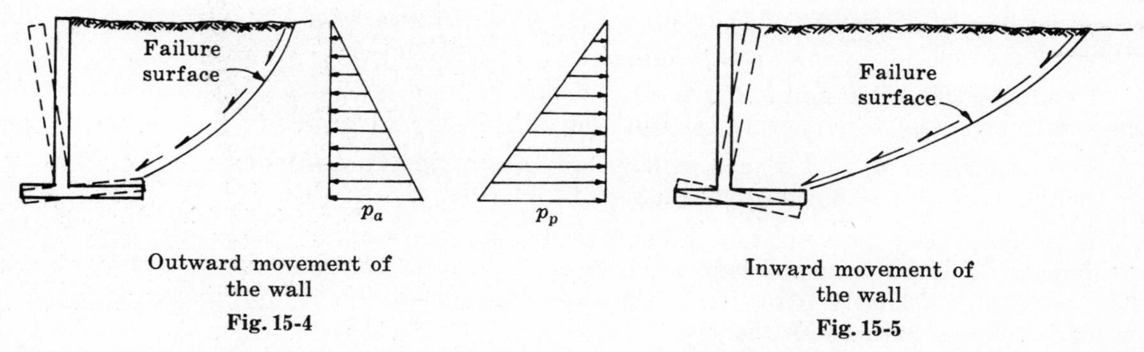

Outward movement of the wall

Fig. 15-4

Inward movement of the wall

Fig. 15-5

If the wall is absolutely rigid, *earth pressure at rest* will develop. If the wall should deflect or move a very small amount away from the earth, *active earth pressure* will develop, as shown in Fig. 15-4. If the wall moves toward the earth, *passive earth pressure* will develop, as shown in Fig. 15-5. The magnitude of earth pressure at rest lies at some value between active and passive earth pressure.

Under normal conditions earth pressure at rest is so intense that the wall deflects, relieving itself of this type of pressure, and active pressure results. For this reason, most retaining walls are designed for active pressure due to the retained soil.

Although the actual pressure intensity diagram is very complex, it is usual to assume a *linear pressure distribution* due to active or passive pressure. The intensity is assumed to increase with depth as a function of the weight of the soil, so that the horizontal pressure of the earth against the wall is often called *equivalent fluid pressure.*

Fig. 15-6 shows the usual types of pressure diagrams assumed in designing retaining walls.

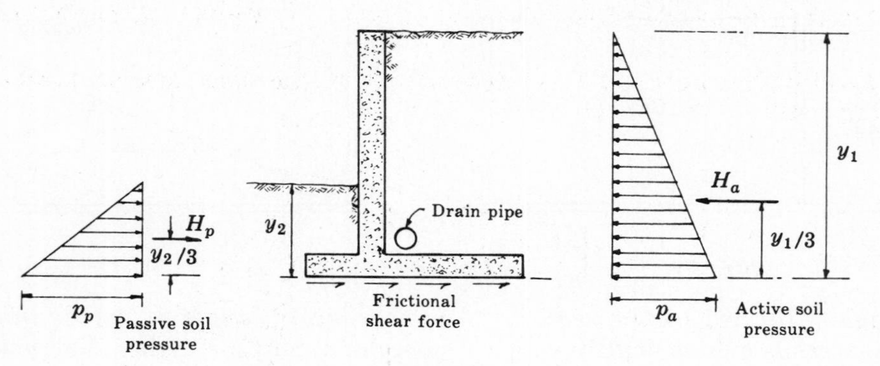

Fig. 15-6

For granular, non-cohesive materials which are dry, the assumed linear pressure diagram is nearly correct. Cohesive soils do not behave in this manner, and it is not good practice to *backfill* a wall with cohesive soils. Granular materials are used for backfill whenever possible to provide a linear pressure distribution, and also to provide for *drainage* to release water pressure from behind the wall.

When the materials used behind a wall are such that a linear pressure distribution may be assumed, the pressure intensity at any depth y may be stated as

$$p_a = k_a wy \tag{15.1}$$

for active pressure, and as

$$p_p = k_p wy \tag{15.2}$$

for passive pressure, where k_a = coefficient of active pressure, k_p = coefficient of passive pressure, w = unit weight of the soil in pcf, y = distance from the surface to the plane in question.

Fundamentally, $k_a w$ and $k_p w$ are *equivalent liquid pressures* to be used in determining the total horizontal force acting against the wall at any given depth.

The coefficients k_a and k_p are usually only approximate, and should never be treated as though they are scientifically exact.

Experience has proven that walls can be safely designed using the approximate coefficients obtained from mathematical theories such as those of *Coulomb* and *Rankine*. *The angle of internal friction* ϕ is obtained experimentally or approximated from experience for use with those theories.

Fig. 15-7 and 15-8 illustrate results of test data for cohesionless soils and cohesive soils, respectively. It is important to note from the figures that the friction angle line passes through the origin for cohesionless soils, and for cohesive soils the line springs from a non-zero value. The latter value is called the *cohesion* c and represents the greatest portion of the shear strength of the cohesive soil.

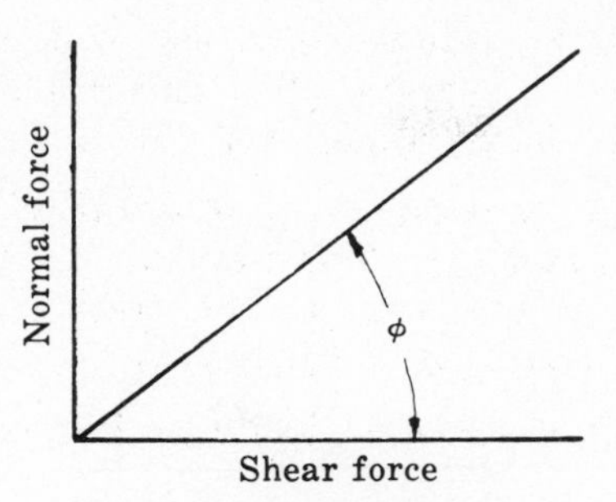

Fig. 15-7. Cohesionless Soils

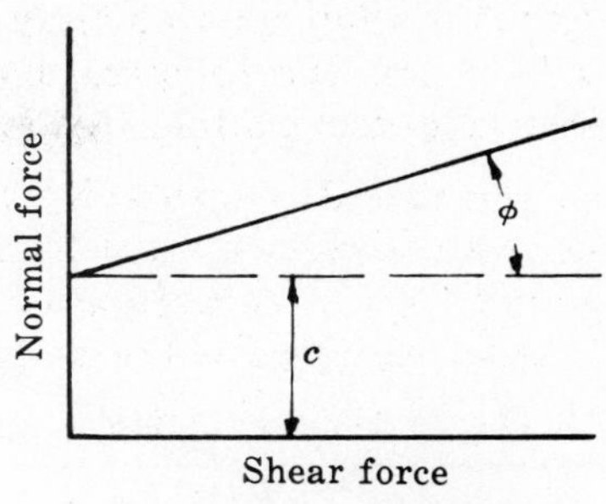

Fig. 15-8. Cohesive Soils

The general case concerning the forces acting on the back face of a retaining wall is shown in Fig. 15-9.

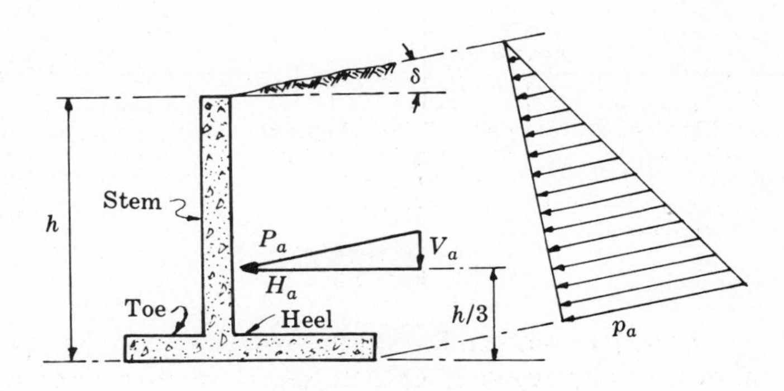

Fig. 15-9

It is significant to note that Coulomb's theory accounts for the frictional force between the wall and the soil while Rankine's theory neglects this force. Since it is rather difficult to accurately determine the frictional force on the back face of the wall, Coulomb's theory is not widely used in retaining wall design.

For a cohesionless soil a triangular pressure diagram is assumed to develop. This assumption is applied to both passive and active pressure. The nature of the triangular pressure diagram shown in Fig. 15-6 is such that the total force which develops at a depth y from the surface will be

$$H_a = (k_a w)y^2/2 \tag{15.3}$$

for active pressure and

$$H_p = (k_p w)y^2/2 \tag{15.4}$$

for passive pressure.

Although it is an approximation, for sloping backfill as shown in Fig. 15-9 it is usually assumed that the force P_a is parallel to the surface of the fill so the components of the force are easy to determine. In such cases equations (*15.3*) and (*15.4*) are usually applied in practice.

A more detailed study of the Rankine theory indicates that the force is not only dependent on the slope of the surface of the fill, but is also dependent on the angle between the face of the wall and the fill surface. For active pressure the angle between the back face of the wall and the horizontal is used.

The more accurate equations are shown in Fig. 15-10 for the Rankine theory. When the wall is vertical and the soil surface is horizontal the detailed equations are identical to the equations previously stated for active pressure. Thus

$$C = k_a = (1 - \sin\phi)/(1 + \sin\phi) \tag{15.5}$$

which is often replaced by the identity

$$k_a = \tan^2(45° - \phi/2) \tag{15.6}$$

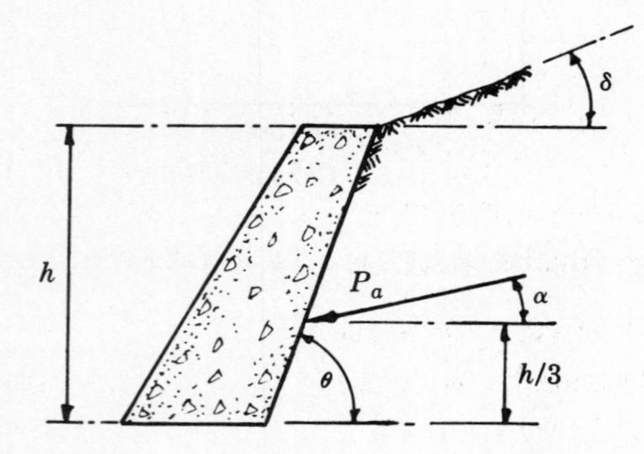

Case I. Wall leaning toward fill ($\theta < 90°$)

$\beta = \phi =$ Angle of internal friction for the soil

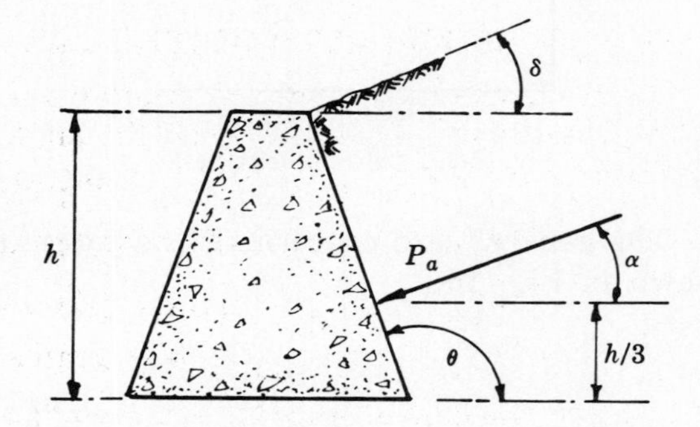

Case II. Wall leaning away from fill ($\theta > 90°$)

$\beta = \delta =$ Slope of fill surface

$$\tan\alpha = \frac{C\sin\delta - \cot\theta}{C\cos\delta}$$

$$C = \cos\delta\left\{\frac{\cos\delta - \sqrt{\cos^2\delta - \cos^2\phi}}{\cos\delta + \sqrt{\cos^2\delta - \cos^2\phi}}\right\}[1 - (\cot\theta)(\tan\beta)]$$

$$P_a = \frac{Cwh^2}{2}\left\{\frac{\cos\delta}{\cos\alpha}[1 - (\cot\theta)(\tan\beta)]\right\}$$

For $\theta = 90°$ and $\delta = 0$,

$$\alpha = 0 \quad \text{and} \quad P_a = H_a = \frac{wh^2}{2}\left[\frac{1 - \sin\phi}{1 + \sin\phi}\right]$$

Fig. 15-10. Rankine Soil Pressure and Force for Cohesionless Soils

Passive pressure occurs at the front of a retaining wall and on *shear keys,* as shown in Fig. 15-11. The passive pressure on the front of the wall is usually neglected, since it is possible that this soil will erode away or that shearing will occur, thus eliminating this passive pressure. It is always safe to neglect this passive pressure.

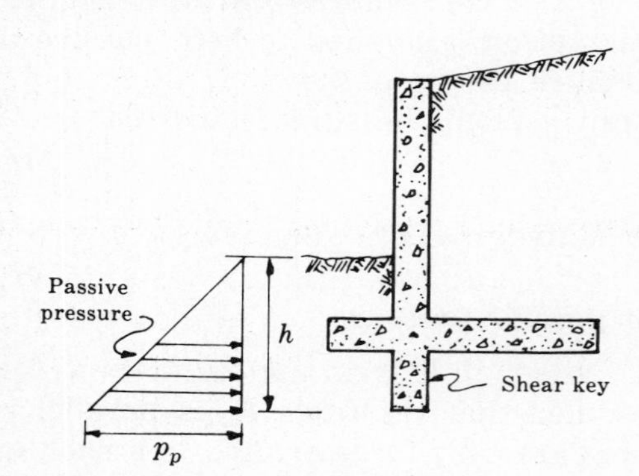

Fig. 15-11

The active horizonal soil pressure must be resisted by opposite forces so that the wall will not slide. Those forces are provided by *friction of the footing base on the soil* and by passive forces on the shear key when such is provided.

The unit passive pressure at any depth h is

$$p_p = k_p wh \tag{15.7}$$

where

$$k_p = (1 + \sin\phi)/(1 - \sin\phi) \tag{15.8}$$

which may also be written as

$$k_p = \tan^2(45° + \phi/2) \tag{15.9}$$

Although it is common practice to ignore the erodable soil above the toe of the wall, it is satisfactory to use the least magnitude of the pressure on the toe in front of the key as a *surcharge* in calculating the passive pressure on the shear key. This will be discussed in detail in connection with shear key design.

ANGLE OF INTERNAL FRICTION, ϕ

True values of the angle of internal friction can only be obtained by tests of the soil. In the absence of laboratory tests, the angle ϕ may be approximated as follows. For

Dry, loose sand with round grains, uniform gradation, $\phi = 28.5°$

Dry, dense sand with round grains, uniform gradation, $\phi = 35.0°$

Dry, loose sand, angular grains, well graded, $\phi = 34.0°$

Dry, dense sand, angular grains, well graded, $\phi = 46.0°$

Dry, loose silt, $\phi = 27°$ to $30°$

Dry, dense silt, $\phi = 30°$ to $35°$

When the soil is saturated, the angles of internal friction may be used as stated, considering the *buoyed unit weight* of the soil particles and, in addition, the *hydrostatic pressure of the water.*

FRICTION ON THE FOOTING BASE

The true friction factor which should be used in obtaining the force which resists sliding is that of the shear strength of the soil. In practice, the coefficient used is that of soil on concrete for coarse granular soils, and the *shear strength* or cohesion of the soil for cohesive soils. In the absence of tests, the following values may be used. For

Coarse grained soil without silt, $f = 0.55$

Coarse grained soil with silt, $f = 0.45$

Silt, $f = 0.35$

The total frictional force on the base of a footing is

$$F = fN \tag{15.10}$$

where f is the friction coefficient and N the normal force of the footing against the soil.

When the foundation rests on clay the cohesion should be used to determine the force resisting sliding. This cohesion force may be calculated using 1/2 of the *unconfined compressive strength* of the soil as the unit cohesion c.

When *pile footings* are used the outer rows should be *battered* to resist the horizontal forces.

For any type of retaining wall, the safety factor against sliding should never be less than 1.5, and should preferably be 2.0 or more.

OVERTURNING

The most hazardous mode of failure of retaining walls is due to overturning because of unbalanced moments. Considering Fig. 15-12, it is apparent that the horizontal force H will tend to overturn the footing about point A. The *overturning moment* will be

$$M_o = Hy_1 \qquad (15.11)$$

The weight of the stem, the soil and the base will resist the overturning by providing a moment about A equal to

$$M_r = W_1 x_1 + W_2 x_2 + W_3 x_3 \qquad (15.12)$$

which is the *righting moment.* See Fig. 15-12. The safety factor against overturning is

$$\text{SF} = M_r/M_o \qquad (15.13)$$

and should always exceed 1.5, with 2.0 being a desirable value.

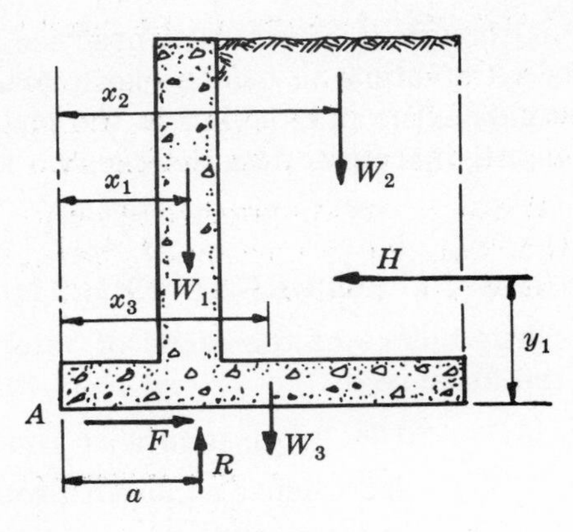

Fig. 15-12

The resultant vertical load R should be so located that the stability of the wall is insured. This is usually accomplished by dimensioning the wall so that the resultant force R falls within the *middle third of the base.* However, it is not absolutely necessary that the resultant fall within the middle third of the base, even though this provides the most economical solution. Under certain conditions such as *hydrostatic uplift,* it may not be feasible to have the resultant in the middle third. Also, when several loading conditions must be checked, it may be economical to have the resultant force outside of the middle third for one or more conditions.

When ultimate strength design procedures are used for proportioning the concrete and steel in the wall and footing, the safety factors are usually of the order 1.75. It is proper that safety factors for sliding and overturning should also be of this order.

When working stress design methods are used for proportioning the concrete and steel, the safety factors are about 2.0 to 2.5. It is proper that the safety factors against sliding and overturning should also be in this range.

SURCHARGE PRESSURES

Loads are often imposed on the soil surface behind a retaining wall. Such forces may be due to loading dock slabs, railroad tracks, roadways, etc. It is common practice to consider such loads as a *surcharge,* and to transform the load into an *equivalent height of soil.*

Fig. 15-13 illustrates a concrete slab supported on the soil. The height of the surcharge is

$$h_s = w_s/w \qquad (15.14)$$

where h_s = equivalent height of soil, feet

w_s = pressure of the surcharge, psf

w = weight of the soil, pcf.

It is seen that the horizontal pressure due to the surcharge is constant throughout the depth of the soil to the base of the footing. The intensity of p_s is obtained from the equation

$$p_s = wh_s k_a \qquad (15.15)$$

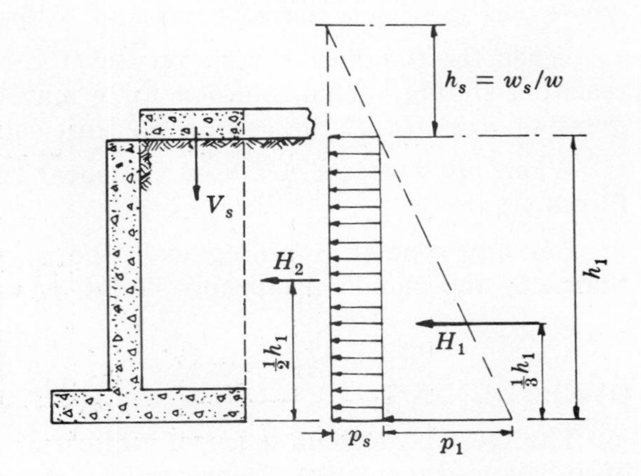

Fig. 15-13

The vertical pressure of the surcharge is equal to w_s.

In the case of a *wheel load,* the weight of the wheel is assumed to be distributed over a stated area. For example, train wheels are usually assumed to act over a width of 14′ and a length of 5′, considering two wheels (two rails) to act over this area.

When a concentrated load or partial uniform load acts at some distance from the rear of the wall, only a portion of the total load affects the wall. A reasonable approach devised by Ketchum is shown in Fig. 15-14.

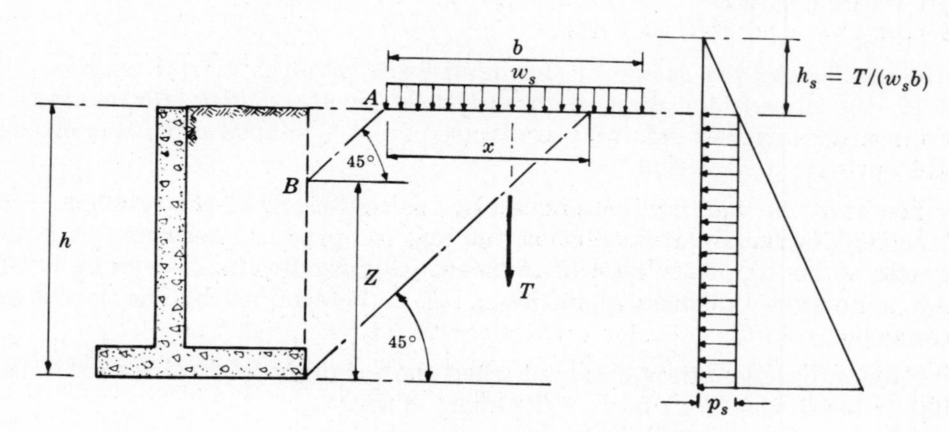

Fig. 15-14

The surcharge pressure p_s may be obtained from

$$p_s = wh_s'k_a \tag{15.16}$$

and h_s' may be determined from

$$h_s' = h_s(x/h) \tag{15.17}$$

using the equivalent height of surcharge in feet of soil,

$$h_s = w_s/w \tag{15.18}$$

where

$$w_s = T/b \tag{15.19}$$

Many engineers assume that the effective height of pressure is that shown as dimension Z in Fig. 15-14 and 15-15. Terzaghi and Peck recommend this method, considering that line AB forms an angle of 45° with the horizontal. This procedure is also recommended by the authors.

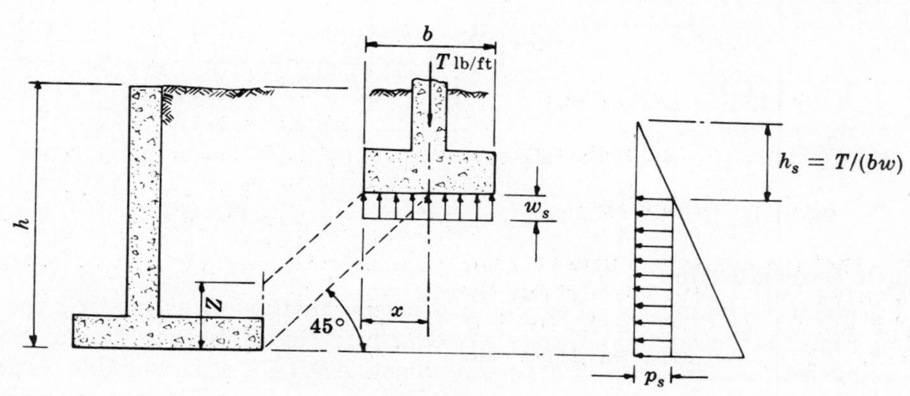

Fig. 15-15

BASE PRESSURE

The base pressure which will develop on the footing of a retaining wall may be calculated in the same manner as that described for eccentrically loaded footings in Chapter 12. Generally, the pressure diagram will be trapezoidal, as shown in Fig. 15-16. The pressure diagram may be triangular, but will rarely be rectangular due to uniform pressure.

DESIGN PROCEDURE

Dimensioning

In the usual case, the design of retaining walls involves a trial process. Referring to Fig. 15-16, the dimensions may be discussed in general. (See, for example, Fig. 15-1.) Since the vast majority of retaining walls are of the T shaped type, this discussion will be concentrated in that direction.

The height of the wall will be dictated by the conditions of the problem. The thickness GH of the *stem* is usually at least 8″ at the top, for practical reasons. A larger thickness may be used at the top if the base of the stem is unusually thick or many reinforcing bars are used. The stem thickness at its base, DE, is determined by the forces and moments to be resisted.

A minimum base length is obtained when the resultant vertical load strikes the ground immediately below the front face of the stem, or at D'.

If the *toe* is extended such that CD is exactly 1/2 of DF, a triangular pressure diagram results if the vertical resultant acts at D'.

If CD is more than 1/2 of DF, a trapezoidal pressure diagram results when the vertical resultant strikes at D'.

It is important to note that the conditions stated above usually provide a safety factor of about 2 against overturning.

The base thickness T will average from about 7 to 10 percent of the overall height of the wall. A minimum thickness of 10″ to 12″ is considered to conform to good practice.

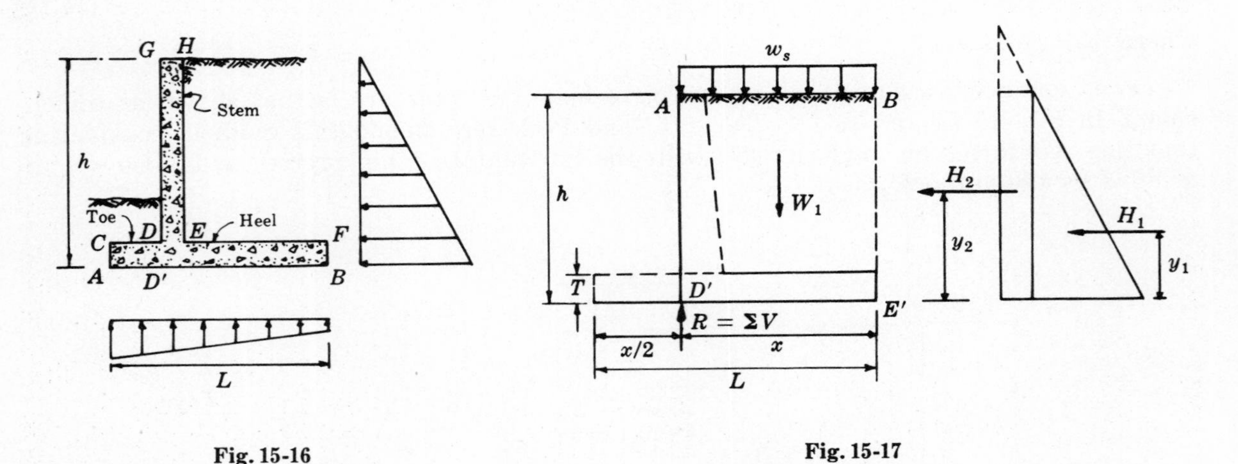

Fig. 15-16

Fig. 15-17

Base Length Determination

Prof. Ferguson recommends an excellent practical method of estimating the base length. Considering Fig. 15-17 and temporarily disregarding the difference in weight of concrete and soil, using W_1 as the weight of a soil block $ABD'E'$, a reasonable estimate of the length X can be obtained. (When surcharge exists, it is considered as added soil.) If

$W_1 = whX$ lb/ft of wall, the resultant will pass through D' if $\Sigma M_{D'} = 0$. Then $whX^2/2 = H_1y_1 + H_2y_2$, from which

$$X = \sqrt{2(H_1y_1 + H_2y_2)/wh} \tag{15.20}$$

The footing base length L may then be *closely estimated* by

$$L = 1.5\sqrt{2(H_1y_1 + H_2y_2)/wh} \tag{15.21}$$

Since the concrete weight is about 1.5 times that of the earth, a slight additional safety factor is introduced against overturning by disregarding the differences in the weight of soil and concrete.

For the simple special case in which there is no surcharge, that is, in which $H_2 = 0$, $H_1 = k_a wh^2/2$ and $y_1 = h/3$, the equation simplifies to

$$L = 1.5\sqrt{k_a h^2/3} \tag{15.22}$$

Stem Design

After the initial computations have been completed, an *estimate of the length of the base is known.* The location of the point of intersection of the front face of the wall and the base have also been established.

In order to design the stem, an *estimate of the depth of the base is made.* Considering the base thickness T to be about 7 to 10 percent of the total height, the preliminary wall height h_w can be established. The stem is considered to be a vertical cantilever, springing from the base as shown in Fig. 15-18. The bending moment and shear are obtained per foot of width of wall, and the usual equations for obtaining d, A_s and Σ_o are applied. Working stress design or ultimate strength design may be employed for this purpose.

The basic soil pressure, hydrostatic pressure and surcharge pressure due to *permanent* load are considered as *dead load.* Intermittently applied surcharge loads (such as wheel loads or storage loads) are considered to be live load.

In the case of walls of the type shown in Fig. 15-18, the stem reinforcement is required on the *earth side* of the wall. Adequate anchorage of the reinforcement into the base is absolutely necessary.

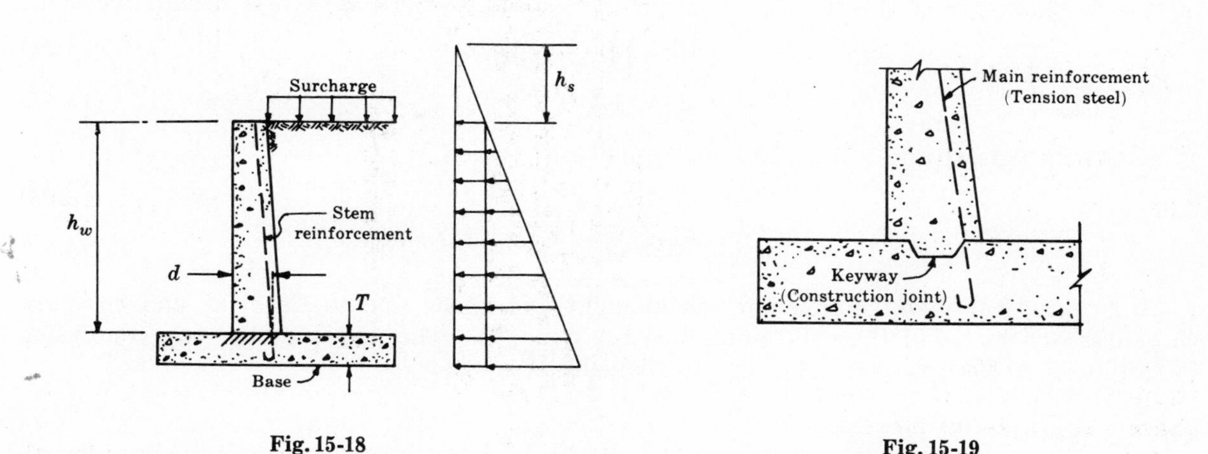

Fig. 15-18

Fig. 15-19

During construction, a *construction joint* is used at the base of the stem so that the footing and the wall may be constructed separately and not as a unit. The construction joint often takes the form of a *keyway* as shown in Fig. 15-19, but may simply be a roughened surface. The latter is used rather frequently in modern practice because of simplicity. Further, some tests indicate the roughened joint to be stronger than the keyway. This construction joint must be capable of transmitting the stem shear into the footing. For this reason the shear force is calculated *at the support*, rather than at a distance d from the support as in other types of beams. It may be necessary to provide dowels for both faces of the wall in order to resist excess shear at the construction joint.

Heel Design

The heel is considered to be a cantilever beam beyond the stem, fixed at the location of the tension steel in the stem. This location is chosen rather than the face of the stem since the tension bars in the stem provide for the transmittal of the reaction forces into the stem. This condition can be approximated fairly accurately by adding 3″ to the projection of the base slab beyond the rear face of the stem. Since the thickness of the stem has been established on the basis of an assumed base thickness, several trial solutions may be necessary in order to make the stem and base calculations compatible.

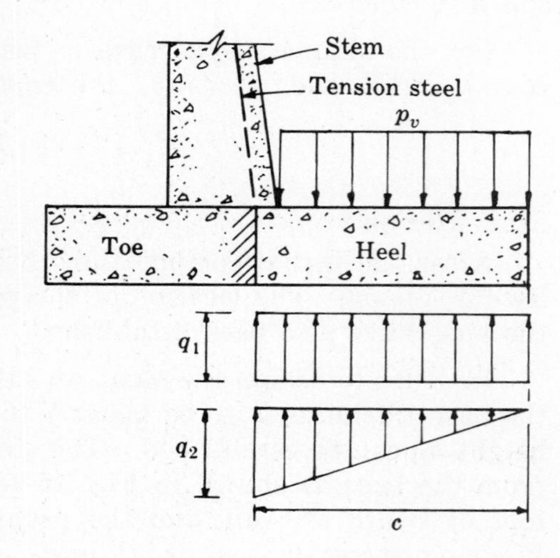

Fig. 15-20

The moment and shear in the heel are taken at the intersection of the stem tension steel and the assumed top of the heel. Two conditions are considered (see Fig. 15-20):

(1) Moment and shear due to the downward load p_v plus the weight of the footing acting downward, and

(2) Moment and shear due to the net effect of the downward load p_v and the upward loads q_1 and q_2. In this case, the weight of the footing causes downward load and upward pressure which counteract one another.

In common engineering practice, Case (1) is permitted to cause 150 percent of the *allowable stresses*. If this is considered with respect to working stress design methods,

$$M_a/S = 1.5f_c = 3f_c/2 \tag{15.23}$$

$$V_a/bd = 1.5v = 3v/2 \tag{15.24}$$

If equations (*15.23*) and (*15.24*) are multiplied by $\frac{2}{3}$, then

$$\tfrac{2}{3}M_a/S = f_c \tag{15.25}$$

$$\tfrac{2}{3}V/bd = v \tag{15.26}$$

It is practical then, to consider the moment and shear due to Case (2) and compare those quantities to $\frac{2}{3}$ of the values obtained for Case (1). The larger of the two conditions is employed in design using the actual allowable stresses.

Toe Design

The toe of the base is considered to be cantilevered from the *edge* of the front face of the stem. The loads employed are those of the base weight acting downward and the base pressures acting upward. The weight of the backfill soil *above the toe* is usually neglected since erosion could cause this counterbalancing load to be diminished.

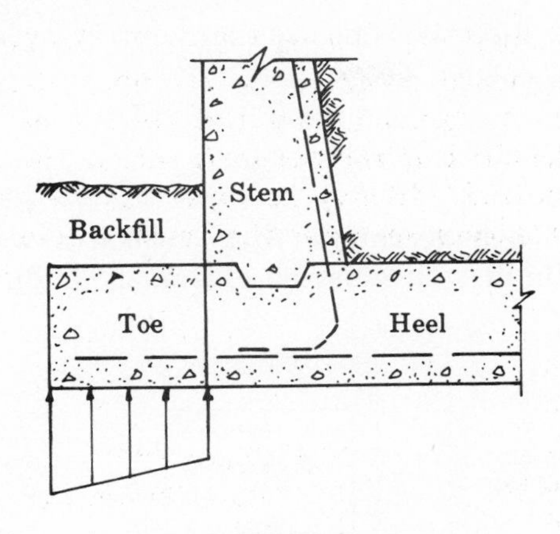

Fig. 15-21

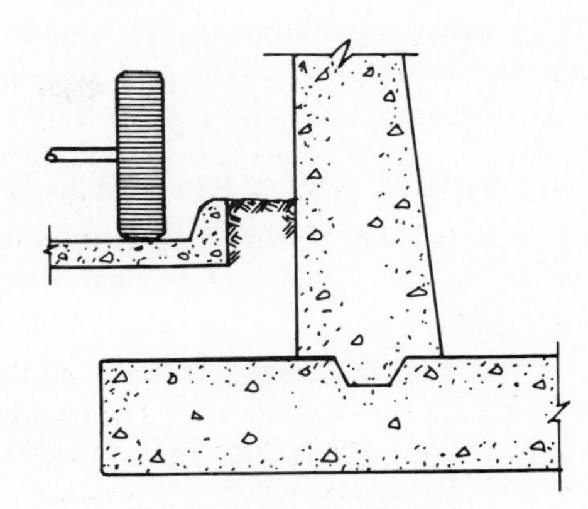

Fig. 15-22

In cases in which the backfill is *confined* under a slab, these loads should be considered as a second condition in order to determine which case governs the design. Fig. 15-22 shows another condition which could exist. Such construction usually produces complex design calculations which are merely approximate even under the most simple conditions. This situation should be avoided whenever possible.

Preliminary Dimensions

For reasonable preliminary estimates, the base length is usually about 40 to 65 percent of the wall height, and the base thickness may be obtained as 7 to 10 percent of the total height.

Shear Keys

When frictional resistance of the footing against the soil is insufficient to provide a satisfactory safety factor against sliding, *shear keys* are employed. The shear key must develop a sufficient passive pressure force to resist the excess lateral force. In order that the passive pressure might develop, the depth of the key and its location along the base must be such that the distance C in Fig. 15-23 will be at least

$$C = a \tan(45^\circ + \phi/2) \qquad (15.27)$$

since $(45 + \phi/2)$ is the *shearing angle for passive pressure*. The *passive pressure*, p_p, may be computed as

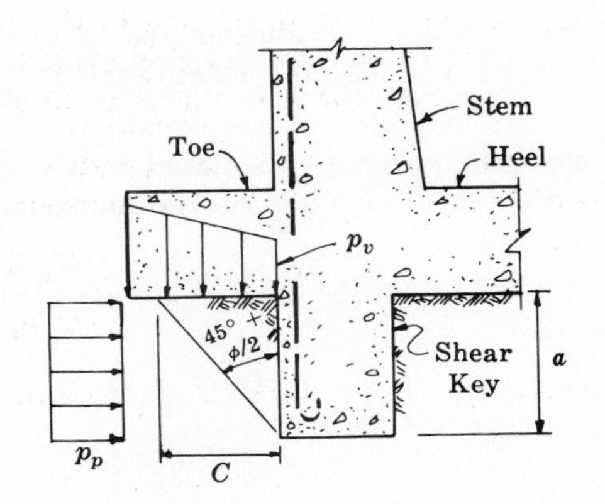

Fig. 15-23

$$p_p = p_v \tan^2(45^\circ + \phi/2) \qquad (15.28)$$

or

$$p_p = p_v(1 + \sin\phi)/(1 - \sin\phi) \qquad (15.29)$$

which are identical. The multiplier of p_v is the passive pressure coefficient k_p.

Inasmuch as there will be a pressure gradient under the footing in front of the key, the smaller value p_v should be used.

It should be noted that p_v is the pressure beneath the footing, which is pressing downward on the soil as well as upward on the base of the footing.

When possible, the shear key should be located directly beneath the stem in order to provide additional anchorage for the stem reinforcement.

COUNTERFORT RETAINING WALLS

Often it is desirable to use retaining walls which have *counterforts*. Such a wall is illustrated in Fig. 15-24 and is used when the retaining wall must be extremely high.

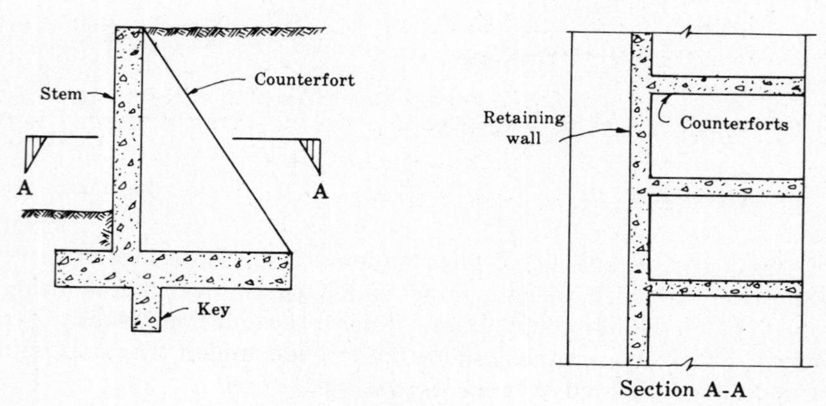

Fig. 15-24

The structural analysis and design of counterfort walls is quite different than for cantilever walls. The counterforts serve as supports for the wall and footing, both of which are designed as continuous slabs supported at the counterforts. The main reinforcement is placed horizontally in this type of wall.

The spacing of the counterforts depends on a number of factors, but usually ranges from 1/3 to 1/2 of the height of the wall.

Overturning and sliding are handled in the same manner as for cantilever walls, as is the design of the toe when such is used.

The stem and heel are considered to be one-way slabs, and the ACI Code coefficients for continuous beams are used in the design. The uniform load on any 12″ wide strip of vertical wall is taken as the average horizontal soil pressure over that strip, as shown in Fig. 15-25.

The counterforts are designed as triangular shaped cantilever beams, fixed at the base slab.

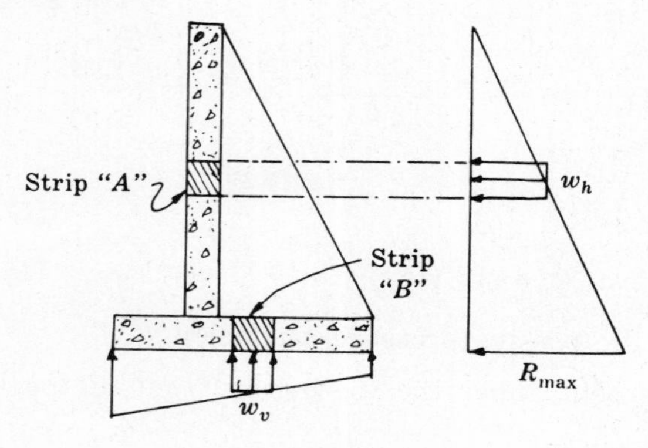

Fig. 15-25

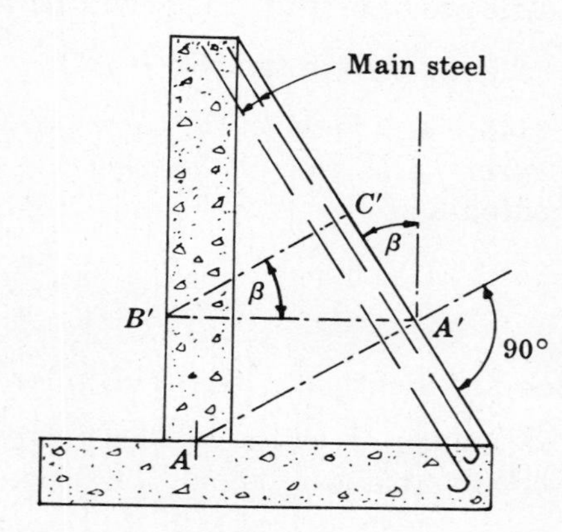

Fig. 15-26

The stem may be used as the flange of a T beam for which the counterfort acts as a web. The maximum depth of the T beam is shown as section A-A′ in Fig. 15-26. The design loads on the counterfort are the accumulated reactions from the stem slab which actually vary linearly. The Code does not specify a critical section for shear for counterforts, but in practice a section A′B′ is often used. The shear force should be calculated using the equation

$$V'' = V + (M/d)\tan\beta \qquad (15.30)$$

The effective depth d is obtained using section B′-C′, considering d to be the distance from B′ to the centroid of the main steel.

Horizontal and vertical bars are provided in both faces of the counterfort to provide for shrinkage. The horizontal bars may be utilized as stirrups and should be formed in the shape of a U in order to hold the main bars in place. It should be noted here that when the counterforts are placed in front of the wall, the term *buttress* is used. The wall is then called a *buttressed wall.*

BASEMENT WALLS

The method used for designing basement walls will differ in accord with the method of construction. Four-way support may be obtained from columns or pilasters and the basement and first floor slabs, as shown in Fig. 15-27(a). Two-way support will occur if the slabs alone support the wall, as shown in Fig. 15-27(b).

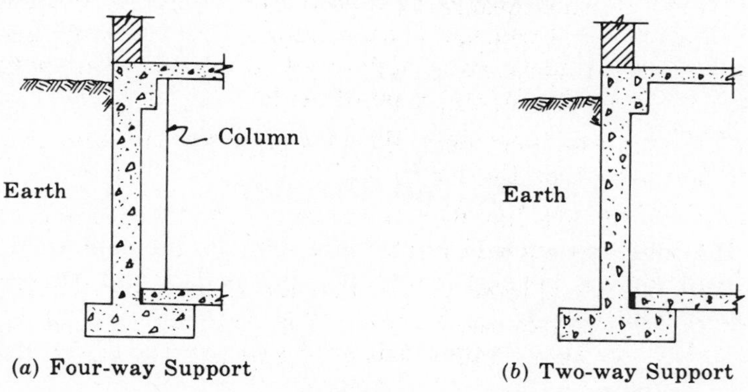

(a) Four-way Support (b) Two-way Support

Fig. 15-27

Axial loads on the wall may be large, in which case *beam-column* action should be considered, as shown in Fig. 15-28.

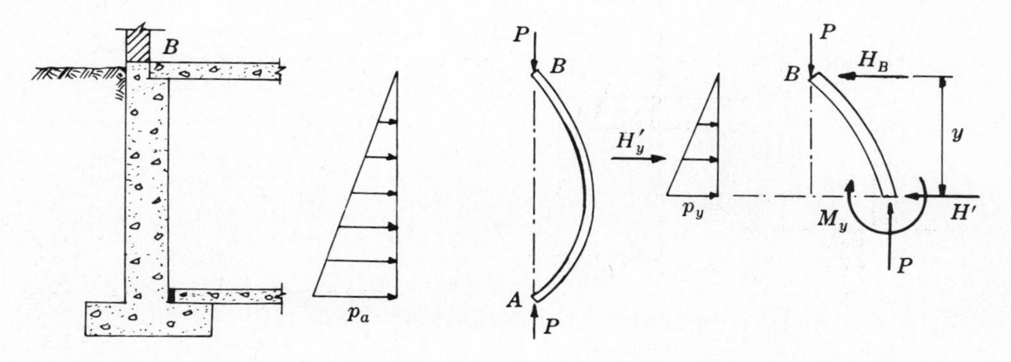

Fig. 15-28

The ACI Code provides empirical methods for wall design, but those methods are generally not applicable to basement wall design. The Code does provide requirements which apply to bearing walls, which includes most basement walls. Those requirements may be listed as follows:

(1) The wall thickness must be at least 1/25 of the unsupported height or width, whichever is the larger.

(2) The uppermost 15′ of wall must be at least 6″ thick, and the minimum thickness must be increased 1″ for every additional 25′ or fraction thereof.

(3) The area of horizontal steel must be at least $0.0025bt$ and the area of vertical reinforcement not less than $0.0015bt$ when bars are used, and not less than 3/4 of that area when wire mesh is used. Mesh may not be less than No. 10 AW & S gage.

(4) Walls shall be anchored to the floors, pilasters and intersecting beams using steel area at least equal to No. 3 bars at 12″ oc, for each layer of wall steel.

(5) Exterior walls and basement walls shall not be less than 8″ thick.

Many other provisions are stated for empirically designed walls in the ACI Code. Inasmuch as those provisions do not apply generally to retaining walls, those provisions will not be listed in detail here. The reader is referred to Chapter 22 of the ACI Code.

Further, many aspects of retaining walls have not been covered in this chapter, since only the most common factors could be presented in a text of this type. The reader should refer to standard texts on Foundation Design, and texts on Walls, Bins and Grain Elevators.

Solved Problems

15.1. Determine the total horizontal active force and the moment tending to overturn the cantilever wall shown in Fig. 15-29 if $\sin\phi = 0.5$ and the soil weighs 100 pcf. (Use 1 foot of wall width.)

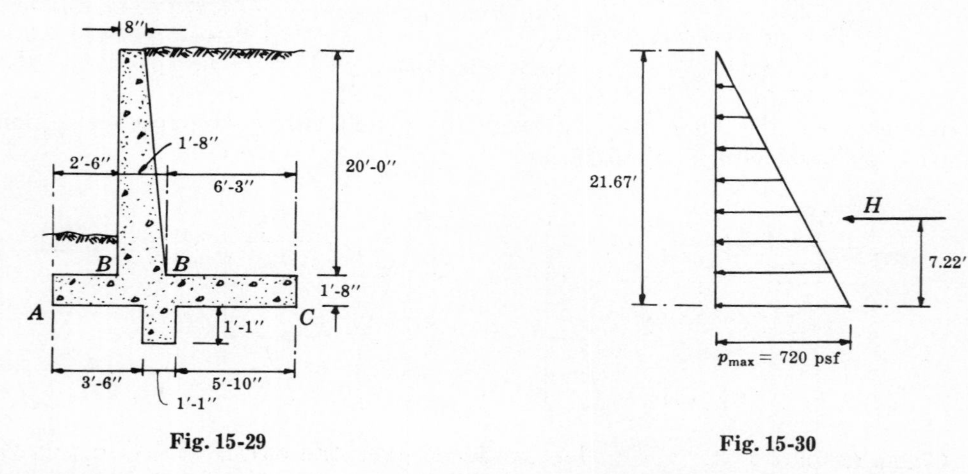

Fig. 15-29 Fig. 15-30

$k_a = (1 - \sin\phi)/(1 + \sin\phi) = (1 - 0.5)/(1 + 0.5) = 0.33.$

$p_{max} = (0.33)(100)(21.67) = 720$ psf, $H = (720)(21.67)/2 = 7800$ lb/ft.

Thus $M_o = (7800)(7.22) = 56{,}300$ ft-lb/ft about the toe (point A).

15.2. Determine the individual vertical forces and the total vertical force acting on the structure shown in Fig. 15-29.

Refer to Fig. 15-31.

Earth:	$W_1 = (6.25)(20)(100)$	$=$	12,500 lb/ft
	$W_2 = (1)(20/2)(100)$	$=$	1,000
Concrete:	$W_3 = (1)(20/2)(150)$	$=$	1,500
	$W_4 = (0.67)(20)(150)$	$=$	2,000
	$W_5 = (1.67)(10.43)(150)$	$=$	2,600
	$W_6 = (1.08)(1.08)(150)$	$=$	175
	Total Load	$=$	19,775 lb/ft

15.3. Determine the righting moment (about A) caused by the vertical loads for the structure of Problems 15.1 and 15.2. The lever arm for each force is shown in Fig. 15-31.

The appropriate values are tabulated as follows:

Force, lb	Arm, ft	Moment, ft-lb/ft
12,500	7.29	91,125
1,000	3.83	3,830
1,500	3.50	5,250
2,000	2.83	5,660
2,600	5.21	13,550
175	4.04	710
$\Sigma F = 19{,}775$ lb/ft		$M_r = 120{,}125$ ft-lb/ft

The total vertical load is 19,775 lb/ft and the righting moment is 120,125 ft-lb/ft.

15.4. Determine the safety factor against overturning for Problem 15.3.

From Problem 15.1 and Fig. 15-30, $M_o = 56{,}300$ ft-lb/ft.

From Problem 15.3 and Fig. 15-31, $M_r = 120{,}125$ ft-lb/ft.

Thus $\text{SF} = M_r/M_o = 120{,}125/56{,}300 = 2.13$.

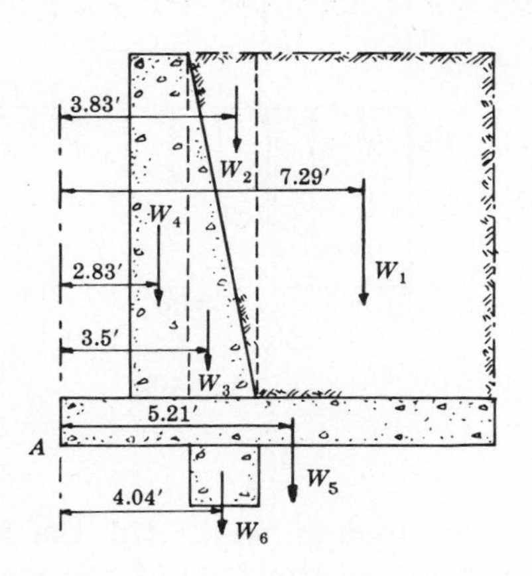

Fig. 15-31

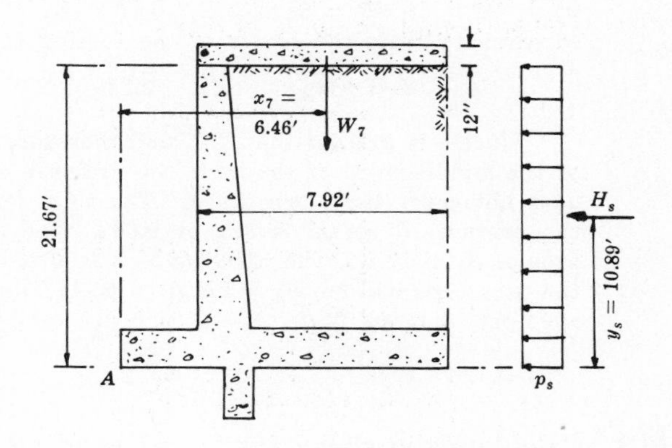

Fig. 15-32

15.5. Consider that a 12″ thick slab is placed over the surface of the ground behind the retaining wall in Fig. 15-29. Determine the added horizontal pressure against the wall, the changes in M_o and M_r, and find the new safety factor against overturning. Use results of Problem 15.4 and $k_a = 0.33$ with $w = 100$ psf.

Refer to Fig. 15-32.

The slab weight produces a vertical pressure of 150 psf. The equivalent height of soil is $h_s = 150/100 = 1.5'$, and the resulting change in horizontal pressure is $p_s = (100)(1.5)(0.33) = 50$ psf.

The change in the vertical force is $W_7 = (150)(1)(7.92) = 1188$ lb/ft, and the change in the horizontal force is $H_s = p_s h = (50)(21.67) = 1083$ lb/ft. The changes in the moments are

$$\Delta M_o = H_s y_s = (1083)(10.89) = 11{,}800 \text{ ft-lb/ft}$$

$$\Delta M_r = W_7 x_7 = (1188)(6.46) = 7{,}680 \text{ ft-lb/ft}$$

and total moments are

$$M_o = 56{,}300 + 11{,}800 = 68{,}100 \text{ ft-lb/ft}$$

$$M_r = 120{,}125 + 7{,}680 = 127{,}805 \text{ ft-lb/ft}$$

The safety factor against overturning is $\text{SF} = M_r/M_o = 127{,}805/68{,}100 = 1.88$.

15.6. **Determine the moment at the base of the stem for the wall shown in Fig. 15-29. Use the data given in Problem 15.1 and the results of that problem.**

Refer to Fig. 15-29. The pressure is proportional to the distance from the top; thus $p = (0.33)(100)(20) = 660$ psf.

The horizontal force on the stem is $H = (660)(20/2) = 6600$ lb/ft, and the moment of that force about the base of the stem is $M = Hy = Hh/3 = (6600)(20/3) = 44{,}000$ ft-lb/ft. This moment would be used to design the stem.

15.7. **Locate the resultant force R for Problem 15.1 and find the soil pressure at the edges of the toe and heel.**

Refer to Fig. 15-33.

The net moment about point A is $M_r - M_o = 120{,}125 - 56{,}300 = 63{,}825$ ft-lb/ft.

The total vertical force R is 19,775 lb, so the location of the resultant with respect to point A is $x_A = (M_r - M_o)/R = 63{,}825/19{,}775 = 3.22'$. Thus the eccentricity from the center of the footing is

$$e = L/2 - x_A = 5.22 - 3.22 = 2.0'$$

Since e is greater than $L/6$ and therefore outside of the middle third of the base, the pressure diagram does not cover the entire base. The total length of the pressure diagram is $3x_A$ or 9.66′. The effective area of the base will therefore be $A = 9.66$ ft²/ft, and the pressures will be $p_A = 2R/A = (2)(19{,}775)/9.66 = 4094$ psf and $p_{C'} = 0$.

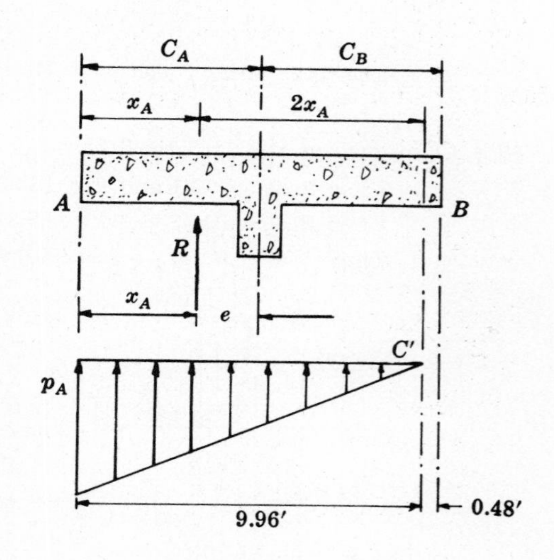

Fig. 15-33

15.8. **Calculate the shear force and bending moment per foot of width for the retaining wall toe shown in Fig. 15-34 below. Point E represents the face of the stem. The pressures given are *net* values. The difference in weight between the concrete and the earth displaced by the base has been deducted from the gross values. (Many designers ignore this quantity, while others ignore the displaced soil and consider the weight of the footing as a downward load.)**

The pressure at A is 3980 psf, so the pressure at E by proportion will be

$$p_E = (3980)(7.46)/9.96 = 2980 \text{ psf}$$

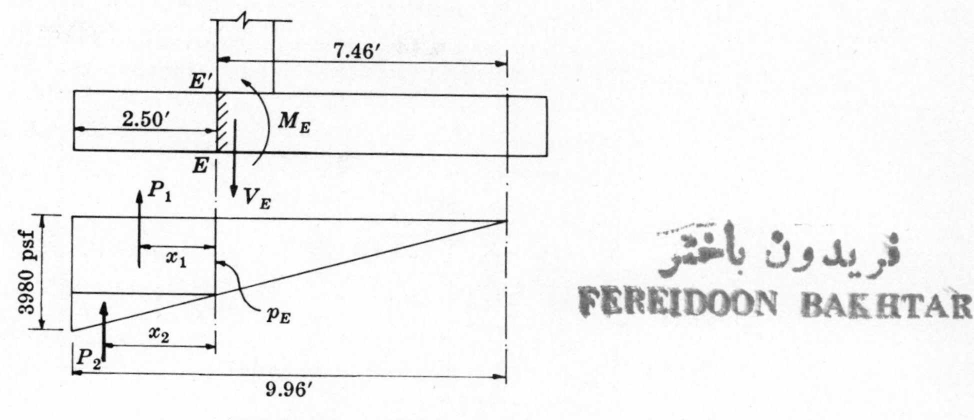

Fig. 15-34

The resulting forces are calculated as the volume of the pressure prisms per foot of width; thus

$$P_1 = (2980)(2.5)(1) = 7450 \text{ lb/ft}, \qquad P_2 = (3980 - 2980)(2.5/2)(1) = 1250 \text{ lb/ft}$$

The shear force at E is $V_E = P_1 + P_2 = 7450 + 1250 = 8700$ lb/ft.

The lever arms for forces P_1 and P_2 respectively are $x_1 = \frac{1}{2}(2.5) = 1.25'$, $x_2 = \frac{2}{3}(2.5) = 1.67'$.

The moment at E is $M_E = P_1 x_1 + P_2 x_2 = 11{,}400$ ft-lb/ft.

The shear force and moment would be used to design the toe structurally using methods explained in previous chapters.

15.9. Determine the preliminary design dimensions for the retaining wall shown in Fig. 15-35(a). The granular backfill weighs 100 pcf and has an angle of internal friction $\phi = 30$ degrees. Use $f'_c = 3.0$ ksi and $f_y = 40.0$ ksi. Use USD.

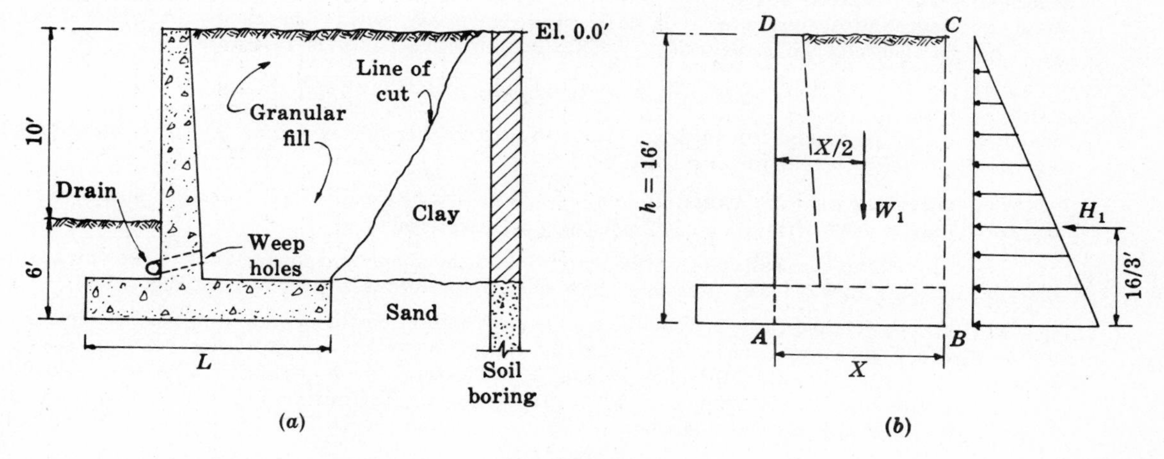

Fig. 15-35

Refer to Fig. 15-35(b) and temporarily assume no difference in the weight of soil and concrete. The total weight of block $ABCD$ is $W_1 = hwX = 1600X$, and the lever arm for this force about A is $X/2$.

The active pressure coefficient is $k_a = (1 - \sin\phi)/(1 + \sin\phi) = 0.5/1.5 = 1/3$.

The total horizontal force is $H_1 = k_a w h^2/2 = (1/3)(100)(16)^2/2 = 4260$ lb/ft.

The resultant may be established at point A by using equation (*15.22*) to determine the base length L. Thus $L = 1.5\sqrt{k_a h^2/3} = 8.0'$.

Try $L = 8.0'$ with a toe overhang of $(8/3)'$ or 2'-8".

Assume the thickness of the footing to be approximately 8% of the total height, or about 1.3'. Thus the cantilever height of the stem will be approximately 14.7', as shown in Fig. 15-36.

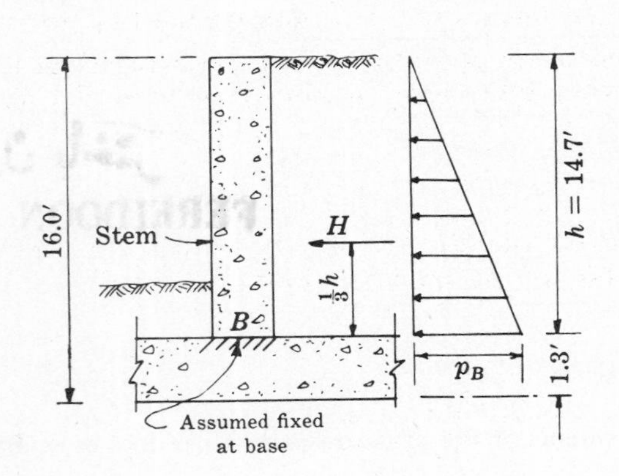

Fig. 15-36

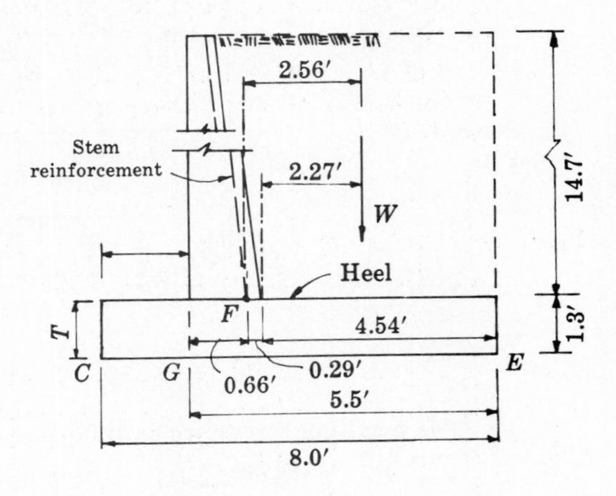

Fig. 15-37

The pressure at the base of the stem is $p_B = k_a hw = (1/3)(14.7)(100) = 490$ psf.

The horizontal force applied to the stem is $H = p_B h/2 = (490)(14.7/2) = 3600$ lb/ft. Thus the working load shear force at B is 3600 lb/ft. The moment about the base of the cantilever for working loads is $M_B = Hh/3 = (3600)(4.9) = 17{,}650$ ft-lb/ft.

Since the forces are related to dead loads, the ultimate values for use in design are

$$V_u = (1.5)V_B = (1.5)(3600) = 5400 \text{ lb/ft}$$

$$M_u = (1.5)M_B = (1.5)(17{,}650) = 26{,}500 \text{ ft-lb/ft}$$

For $f'_c = 3.0$ ksi and $f_y = 40.0$ ksi, Table 5.2, Page 103, indicates that deflections must be checked if ϕR_u exceeds 434.4. For retaining walls the deflections are quite difficult to calculate with any degree of accuracy, so it is desirable to keep ϕR_u within the range for which deflections need not be checked. Thus, use $\phi R_u = 434.4$. The effective depth is therefore

$$d = \sqrt{M_u/\phi R_u b} = \sqrt{(26{,}500)(12)/[(434.4)(12)]} = 7.8''$$

which will require a total thickness of 11.5" (to next higher $\frac{1}{2}$") including 3" cover. If No. 8 bars are used, the effective depth furnished will be 8".

The allowable shear stress is $v_{cu} = 2\phi\sqrt{f'_c} = 93.1$ psi. The actual shear stress is $v_u = V_u/(bd) = 5400/[(12)(8)] = 56.2$ psi, which is satisfactory.

The heel behind the wall is designed using 2/3 of the shear and moment due to downward loads, disregarding the upward soil pressure. The design conditions are shown in Fig. 15-37.

The total downward force is that due to the earth weight on the overhanging heel beyond the face of the stem. The moment, however, is usually taken about the center of the vertical (or slightly inclined) bars in the stem on the back side, at F. The force is $W = (4.54)(14.7)(100) = 6670$ lb/ft, so the ultimate load is $W_u = (1.5)(6670) = 10{,}000$ lb/ft. The ultimate shear is therefore $V_u = 10{,}000$ lb/ft, and the ultimate moment about point F is $M_u = (10{,}000)(2.56) = 25{,}600$ ft-lb/ft.

Since 2/3 of the shear and moment are used in the design, the effective design values are

$$V_u = (2/3)(10{,}000) = 6670 \text{ lb/ft}$$

$$M_u = (2/3)(25{,}600) = 17{,}100 \text{ ft-lb/ft}$$

Using $\phi R_u = 434.4$ as before,

$$d = \sqrt{(17{,}100)(12)/[(434.4)(12)]} = 6.3''$$

For shear, a rearrangement of the equation permits calculation of the value of d required, or

$$d = V_u/[(v_{cu})(b)] = 6670/[(93.1)(12)] = 5.98''$$

Moment governs, and the total thickness must be at least $T = 6.3 + 0.5 + 3.0 = 9.8''$, considering that No. 8 bars will probably be used. This thickness is based on the measurements including 1/2 bar diameter and 3″ clear cover. For practical purposes use $T = 10''$, so the d furnished is 6.5″.

The initial assumption for the thickness of the footing was 8% of the base length or 1.3′, so the height of stem used was $(1.30 - 0.83) = 0.47'$ smaller than the value provided by the 10″ depth.

The slight error involved would be eliminated in a *final review* of the designed section. The review is also made considering the total downward force and the total upward force acting together.

15.10. Determine the safety factor against overturning for the retaining wall shown in Fig. 15-38. Use $k_a = 1/3$ and $w = 100$ pcf. Concrete weighs 150 pcf.

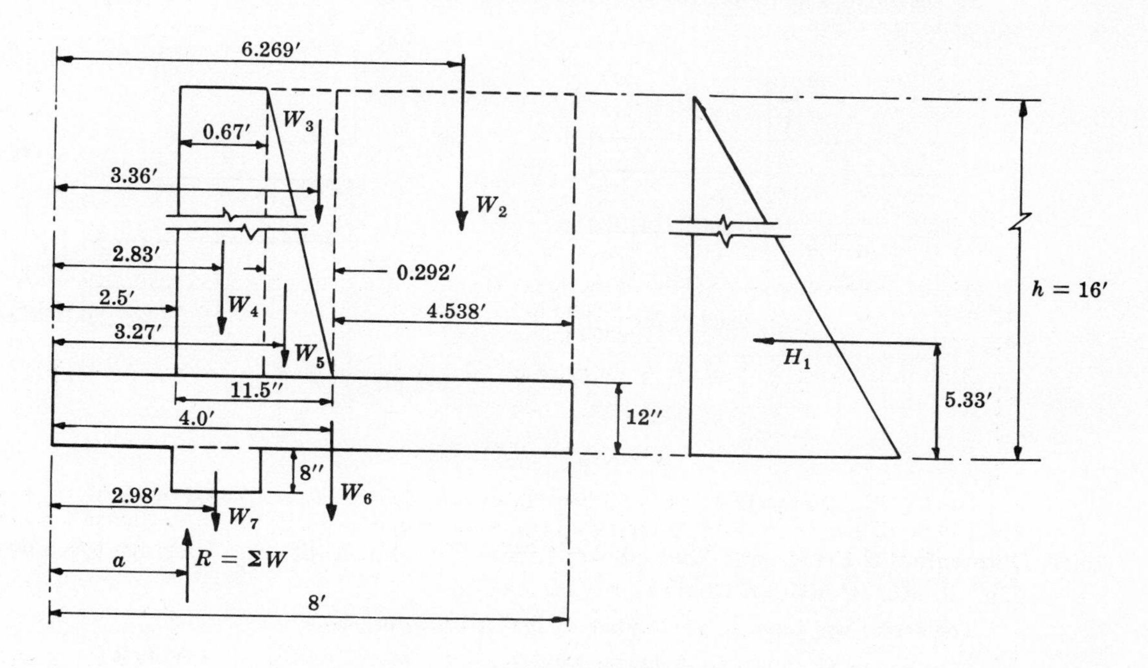

Fig. 15-38

Refer to Fig. 15-38 for dimensions and forces.

Vertical forces and righting moments per foot of wall are tabulated as follows:

	Calculations	Force, lb/ft	Arm, ft	M_r, ft-lb/ft
W_2 =	(4.538)(100)(15)	= 6,807	6.269	42,700
W_3 =	(0.292)(100)(15/2)	= 219	3.360	740
W_4 =	(0.67)(150)(15)	= 1,508	2.830	4,250
W_5 =	(0.292)(150)(15/2)	= 329	3.270	1,080
W_6 =	(1)(150)(8)	= 1,200	4.000	4,800
W_7 =	(0.67)(150)(11.5/12)	= 97	2.980	290
	Totals:	R = 10,160 lb/ft		M_r = 53,860 ft-lb/ft

The horizontal soil force $H_1 = (1/3)(100)(16)^2/2 = 4270$ lb/ft.

The overturning moment $M_o = H_1h/3 = (4270)(5.33) = 22,760$ ft-lb/ft.

For overturning, $\text{SF} = M_r/M_o = 53,860/22,760 = 2.37$, which is satisfactory.

15.11. For the retaining wall footing shown in Fig. 15-39, the resisting moment is 52,610 ft-lb/ft and the overturning moment is 23,600 ft-lb/ft. The total vertical load is 9867 lb/ft. Calculate the soil pressures at points C and E.

The location of the resultant force with respect to point C is

$$a = (M_o - M_r)/R = (52{,}610 - 23{,}600)/9867 = 2.94'$$

so the eccentricity from the center of the footing is $e = 4.0 - 2.94 = 1.06'$.

The moment of inertia of the footing base per foot width is $I = (1)(8)^3/12 = 42.6 \text{ ft}^4/\text{ft}$, and the area of the base is $A_F = (8)(1) = 8.0 \text{ ft}^2/\text{ft}$.

At any point the pressure is $p = R/A_F \pm Rec/I$ where c is the distance of the point from the center of the footing. Hence

$$p_C = (9867/8) + (9867)(1.06)(4)/42.6 = 2218 \text{ psf}, \qquad p_E = 1235 - 983 = 252 \text{ psf}$$

As shown in Fig. 15-39, the soil pressure is assumed to vary linearly from one end of the footing to the other.

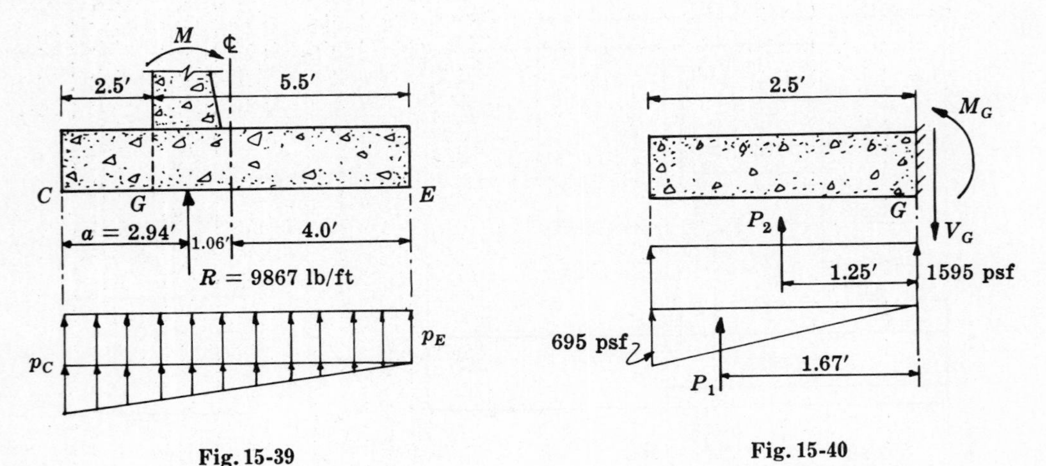

Fig. 15-39 Fig. 15-40

15.12. Determine the moment and shear force for design of the footing toe shown in Fig. 15-40. Consider working stress design.

The forces are equal to the volume of the pressure prism per foot; thus

$$P_1 = (695)(2.5/2) = 870 \text{ lb/ft}, \qquad P_2 = (1595)(2.5) = 4000 \text{ lb/ft}$$

The working load shear force is $V_G = P_1 + P_2 = 4870$ lb/ft.

The working load moment is $M_G = (870)(1.67) + (4000)(1.25) = 6455$ ft-lb/ft.

15.13. The basement wall shown in Fig. 15-41 is backfilled with a granular material for which $\phi = 30$ degrees. Determine the design shear and bending moment for the wall if the soil weighs 90 pcf.

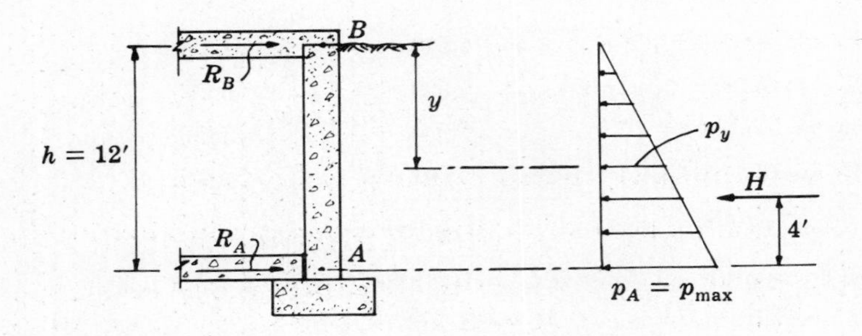

Fig. 15-41

$p_{max} = (90)(12)(1-1/2)/(1+1/2) = 360$ psf. $R_A = (2/3)(12)(360) = 2880$ lb/ft.

Maximum moment occurs at $y = (0.5774)(h) = 6.93'$ from B.

$M_{max} = (0.1283)Hh = (0.1283)(0.5)(360)(12)(12) = 3326$ ft-lb/ft.

Note. The coefficients for reactions and moments were obtained from texts on Strength of Materials.

15.14. The counterfort retaining wall shown in Fig. 15-42 is backfilled with a material for which $w = 100$ pcf and $\phi = 30$ degrees. Determine the shear and bending moments for designing the wall at its base A at points B and F, considering the wall continuously supported over the counterforts. Assume that restraint develops at A and E.

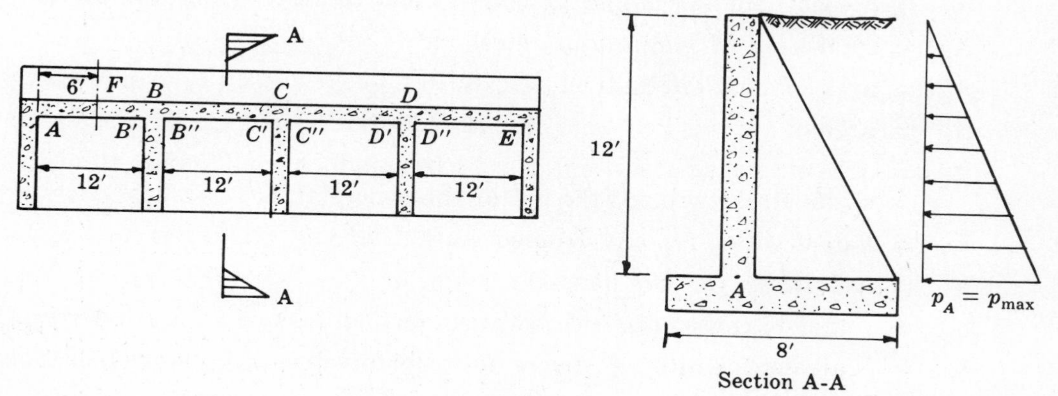

Fig. 15-42

$$k_a = (1 - \sin\phi)/(1 + \sin\phi) = (1 - 0.5)/(1 + 0.5) = 0.33$$

$$p_{max} = k_a wh = (0.33)(100)(12) = 400 \text{ psf}$$

Consider the load in the lateral direction to be equal to the maximum pressure at the lower end of the wall and constant for the first foot of height; then $w_H = 400$ lb/ft.

Using the ACI Code coefficients (Chapter 2), the moments are

$$M_{B'} = -(1/10)w_H(L')^2 = -(0.1)(400)(12)^2 = -5750 \text{ ft-lb/ft}$$

$$M_F = (1/14)w_H(L')^2 = (1/14)(400)(12)^2 = 4110 \text{ ft-lb/ft}$$

Since the pressure varies linearly from the top of the wall, the uniform load in the transverse direction will also vary in the same way. Thus the moments vary linearly from the top of the wall to the base. Since the pressure is zero at the top, the moments are also zero at the top of the wall.

Supplementary Problems

15.15. Solve Problem 15.1 if $\phi = 28$ degrees and the soil weighs 90 pcf. *Ans.* $M_o = 55{,}000$ ft-lb/ft

15.16. Solve Problem 15.3 if the soil weighs 90 pcf and the stem has a constant thickness of 15″. The length of toe is unchanged. *Ans.* $M_r = 111{,}300$ ft-lb/ft

15.17. Determine the safety factor against overturning for Problem 15.15. *Ans.* SF = 2.02

15.18. Solve Problem 15.5 if a live load of 250 lb/ft^2 is placed over the slab. *Ans.* SF = 1.60

15.19. Determine the design moment for the base of the stem of the wall in Fig. 15-29 for $\phi = 35$ degrees and $w = 110$ pcf. *Ans.* $M = 39{,}700$ ft-lb/ft

15.20. Solve Problem 15.8 if the pressure at A is 2600 psf. *Ans.* $M_E = 7460$ ft-lb/ft

Chapter 16

Torsion

NOTATION

a_T = a coefficient for design of longitudinal reinforcement for torsion
A_L = total area of longitudinal steel, in^2
A'_L = area of one longitudinal bar, in^2
A_v = area of two vertical legs of a closed stirrup, in^2
A'_v = area of one leg of an *equivalent stirrup* which would resist all of the torsional shear stress without the aid of the concrete, in^2
b = flange width for any flanged section, and in particular for an L-beam, in.
b' = web width for any flanged section, in.
b_c = center-to-center distance between vertical legs of a closed stirrup, in.
b_E = equivalent width of flange in transforming any flanged section into an equivalent L-beam, in.
C_T = coefficient for obtaining the equivalent polar moment of inertia for flanged or rectangular sections
d_c = center-to-center distance between horizontal legs of a closed stirrup, in.
d'' = distance from the center of a stirrup leg to the nearest face of the concrete, in.
f_s = working stress allowed in longitudinal bars, psi
f_v = working stress allowed in stirrup legs, psi
f'_v = stress in the equivalent stirrup area, A'_v, psi
f_y = yield stress for steel, psi
F_p = force due to principal stresses caused by torsion
h = total depth of web for any section, in.
J_E = equivalent polar moment of inertia for any section, in^4
K_T = a torsional coefficient in Nylander's equation
M_t = torsional moment at service loads, in-lb
M_{tu} = torsional moment at ultimate strength conditions, in-lb
s = longitudinal spacing of stirrups, in.
t = flange thickness, in.
t' = average length of stirrup per longitudinal bar, or $(b_c + d_c)/2$, in.
V = volume contained under the analogous sand-heap for torsion, in-lb
V = shear force, lb
V_{DL} = dead load shear force, lb
V_{LL} = live load shear force, lb

v_c = allowable working load shear stress in concrete, psi

v_{cu} = allowable ultimate shear stress in concrete, psi

v_t = torsional shear stress, working stress level, psi

v_{tu} = torsional shear stress, ultimate strength design, psi

v' = shear stress to be taken by stirrups, psi (WSD)

v'_u = shear stress to be taken by stirrups, psi (USD)

v_u = total shear stress, USD, psi

X = distance over which stirrups are theoretically required, in.

GENERAL CONSIDERATIONS

The 1963 ACI Code requires that torsional stresses must be added to stresses due to flexure, shear and direct load. However, the Code does not specify a method of approach to this problem. The engineer is therefore obligated to choose a method for designing structural members to withstand the effects of torsion.

The Code merely states that stirrups in edge beams or spandrel beams must be of the closed type, but does not demand that stirrups be provided. The Code does, however, require that at least one longitudinal bar must be placed in each corner of spandrel beams. The size of such bars may not be less than the size of the stirrups if such are provided, and may not be less than 1/2″ in diameter. Since longitudinal bars are required, the use of stirrups is implied inasmuch as stirrups are needed to hold the bars in place. Further, the purpose of the bars is to develop the longitudinal component of the forces due to torsion. If there is a longitudinal component, there is of necessity a transverse component which must be either resisted by the concrete or by stirrups.

Torsion has been discussed previously with respect to flat slabs and flat plates, so this chapter will deal only with twisting of beams.

OTHER BUILDING CODES

Many building codes specify provisions for torsion. The codes of Mexico, Egypt, Poland, Sweden, Russia and Australia are examples. One may review the various provisions of different building codes by referring to a paper by Fisher and Zia, Journal of the American Concrete Institute, January 1964.

Although the equations for torsion vary in the codes of different countries, most of the expressions are related to the elastic membrane analogy.

RECENT DEVELOPMENTS

Research has indicated that the elastic theory is not entirely satisfactory for accurately predicting the stresses in concrete due to torsion. However, Nylander's experiments in Sweden have indicated that the *theory of plasticity* provides equations which are in good agreement with test data with respect to *pure torsion.* A recent paper (prepared for presentation in 1966) by Everard and Cohen provides equations for torsional shear stresses based on the theory of plasticity. The plastic *sand-heap analogy* was utilized to develop a basic equation for the torsional stresses in L beams. It was then shown that the torsional shear stress equation for any open-end, flanged section can be transformed into that of the basic equation for the L beam. Using this process, it is only necessary to prepare design tables for one type of section. Such design tables appear herein as Tables 16.1 through 16.9. Fig. 16-1 below illustrates the *equivalent flange width* b_E for use in converting various cross-sections into an equivalent L beam.

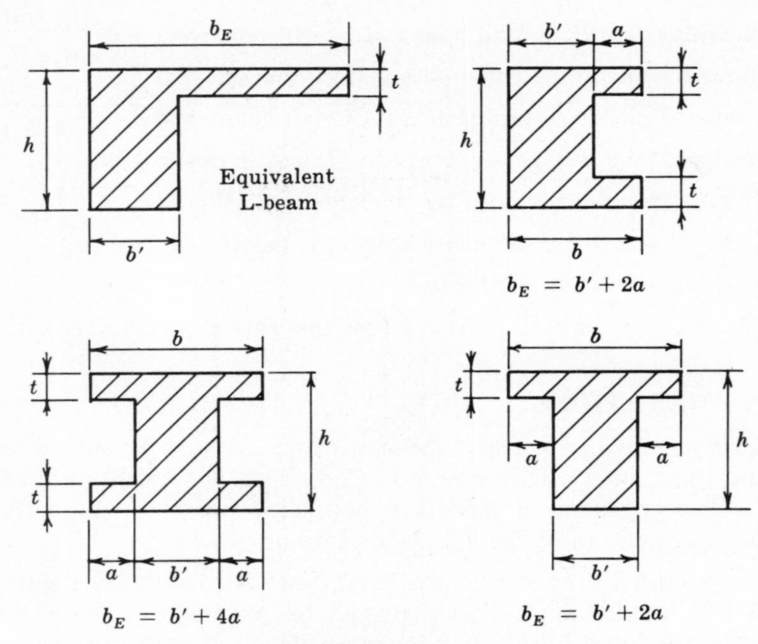

Fig. 16-1

THE SAND-HEAP ANALOGY AND THE MEMBRANE ANALOGY

The plastic theory sand-heap analogy, like the elastic theory membrane or soap-film analogy, provides a solution for Laplace's equation which relates torsional shear stress and the applied torsional moment. In both methods the total torsional moment M_{tu} or M_t is proportional to twice the volume contained under the sand-heap or the membrane. Further, the shear stress at any point is proportional to the slope of the sand-heap or the slope of the membrane. For the theory of elasticity the membrane has a constantly changing slope since the shear stress varies over the cross-section. However, inasmuch as the plastic shear stress is constant in magnitude throughout the cross-section, the slope of the sand-heap is constant everywhere. This simplifies the torsion problem materially.

For irregular sections such as the L beam, the sand-heap may also be visualized as a hipped-roof over a building for which the floor plan is identical to that of the shape of the cross-section. For a rectangular section the heap assumes the form of a pyramid. Similarly, for a circular cross-section the sand-heap takes the form of a cone.

THE EQUIVALENT L BEAM EQUATION

As previously stated, the sand-heap or hipped roof for the L beam or other flanged sections can be easily visualized as having ridges, hips and valleys such as occur on a sloping roof for a building. The slopes are constant everywhere. There are, however, *junction effects* at the intersections of the various elements, but those effects are quite small and can be disregarded for practical purposes.

Fig. 16-2 illustrates the plan of the *roof* over an L beam with isometric views of the various component parts. The segments can be manipulated in such a way as to represent the total volume as the sum of those for a *pyramid*, a *large tent* and a *small tent*. The volumes are easily computed as follows:

For the pyramid $\quad V_1 = Y_2(b')^2/3 \qquad (16.1)$

For the large tent $\quad V_2 = Y_2(b')(h - b')/2 \qquad (16.2)$

For the small tent $\quad V_3 = Y_1(t)(b - b')/2 \qquad (16.3)$

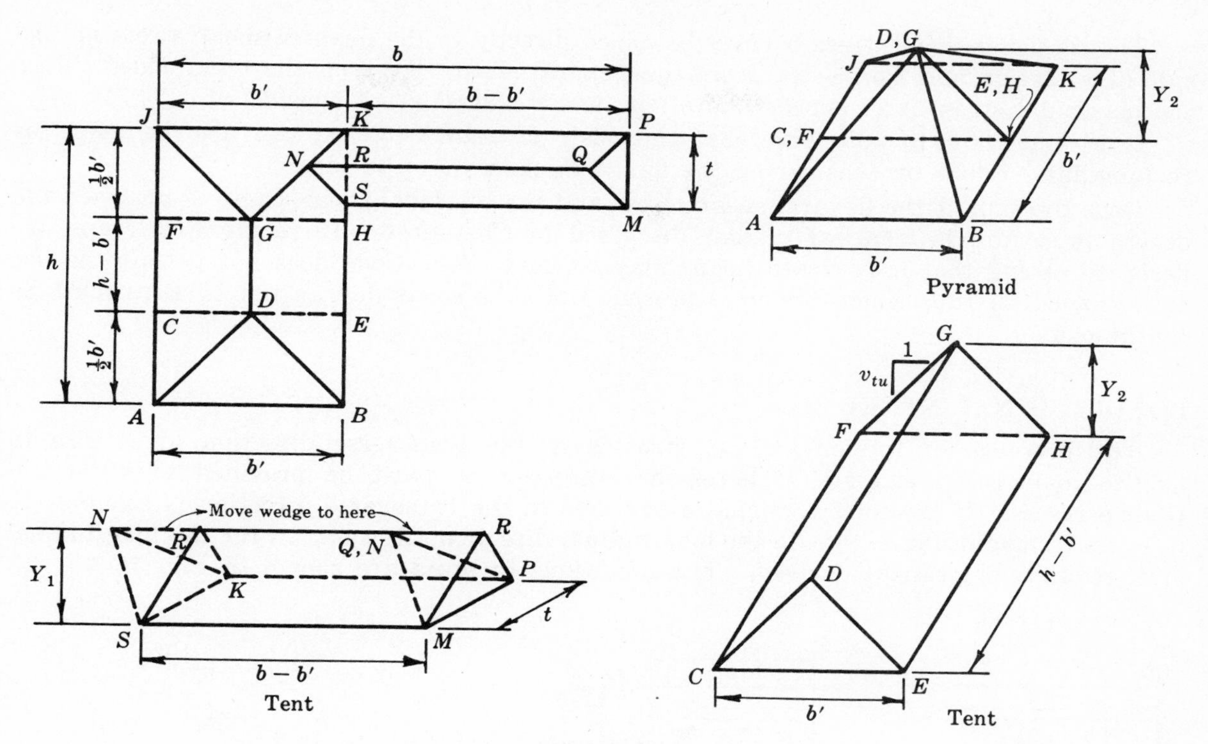

Fig. 16-2

The constant slopes of the sides of the roof can be expressed as $2Y_1/t$ and $2Y_2/b'$ which are identities. The volumes and the slope are dependent on the same constant of proportionality, so the constant can be assumed to be unity. Thus since the shear stress is proportional to the product of the constant and the slope,

$$Y_1 = v_{tu}(t/2) \tag{16.4}$$

and

$$Y_2 = v_{tu}(b'/2) \tag{16.5}$$

By applying equations (*16.4*) and (*16.5*) to *twice* the sum of the volumes from equations (*16.1*), (*16.2*) and (*16.3*), a relationship between the ultimate torsional shear stress v_{tu} and the ultimate torsional moment M_{tu} can be established as

$$v_{tu} = \frac{M_{tu}}{[(b')^2/6][3h - b'] + [(t^2/2)(b - b')]} \tag{16.6}$$

If both the numerator and the denominator on the right side of equation (*16.6*) are divided by $(b')^2h^2$ and appropriate algebraic rearrangements are applied, the equation can be expressed as

$$v_{tu} = \frac{M_{tu}h/[(b')^2h^2]}{[(1/6)(3 - b'/h)] + \{(1/2)[(b/h) - (b'/h)](t/b')^2\}} \tag{16.7}$$

or

$$v_{tu} = \frac{M_{tu}h}{J_E} \tag{16.8}$$

where

$$C_t = [(1/6)(3 - b'/h)] + \{(1/2)(t/b')^2[(b/h) - (b'/h)]\} \tag{16.9}$$

and

$$J_E = C_t(b')^2h^2 \tag{16.10}$$

Note that the *torsional coefficient* C_t depends only on the ratios of the various dimensions while the *equivalent polar moment of inertia* J_E depends on C_t and the actual dimensions.

The torsional shear stress v_{tu} may be added directly to the flexural shear stress on one side of a section and subtracted from the flexural shear stress on the other side. Since the maximum stress is of concern here, the lesser stress is not considered.

It is important to note here that the basic L beam equation can also be used for rectangular sections by considering the flange thickness t to be zero.

Once the sum of the flexural shear stress and the torsional shear stress is obtained the design procedure is identical to that discussed in Chapter 6. If required, stirrups are designed noting that only closed hoops may be used. The Code does not permit the use of U shaped stirrups when torsion is present since the shear flow in the stirrups must be continuous.

LONGITUDINAL STEEL

The presence of torsional shear stresses in the transverse direction gives rise to longitudinal tensile stresses. Therefore reinforcement must be provided to withstand those stresses. If the total torsional *shear flow* in the transverse direction is assumed to be equal in magnitude to that in the longitudinal direction, an equation for the longitudinal steel requirement can be derived. The conditions involved are shown in Fig. 16-3.

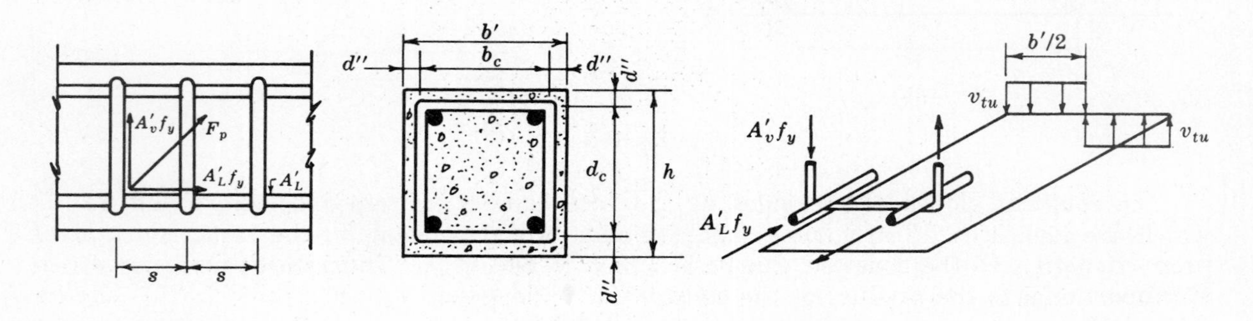

Fig. 16-3

Since the shear flow is identical in the orthogonal directions in Fig. 16-3,

$$A'_v f_y/s = A'_L f_y/t' \qquad (16.11)$$

or

$$A'_v = A'_L s/t' \qquad (16.12)$$

where t' is the average length of stirrup per longitudinal bar, or $t' = (b_c + d_c)/2$, and $A'_L = A_L/4$ for four longitudinal bars.

It should be noted here, however, that any number of bars may be used as long as the steel is symmetrically placed. Once A'_L has been determined the total steel required is $4A'_L$ and this quantity may be placed in any symmetrical pattern using an equal number of bars per face of the beam.

Using the definitions of A'_L and t' in equation (*16.12*) we obtain

$$A'_v = A_L s/[2(b_c + d_c)] \qquad (16.13)$$

The total force that develops in the concrete and in the stirrups over one-half the width of the beam on the most highly stressed side is

$$F_s = v_{tu} b's/2 = A'_v f_y \qquad (16.14)$$

Thus

$$F_s = A_L s f_y/[2(b_c + d_c)] \qquad (16.15)$$

and

$$A_L = v_{tu} b'(b_c + d_c)/f_y \qquad (16.16)$$

For a flanged section the component rectangles may be considered separately and a proportional part of the required longitudinal steel may be placed in each. Thus considering $t = 0$ in equation (*16.7*), we obtain the appropriate expression for the torsional shear stress in a rectangular section as

$$v_{tu} = \frac{6M_{tu}}{(b')^2(3h - b')} \tag{16.17}$$

Then, if the value of v_{tu} from equation (*16.17*) is substituted into equation (*16.16*) we obtain

$$A_L = \frac{6M_{tu}b'(b_c + d_c)}{f_y(b')^2(3h - b')} \tag{16.18}$$

Using $b_c = b' - 2d''$, $d_c = h - 2d''$, and multiplying both the numerator and denominator in equation (*16.18*) by h^2, we obtain

$$A_L = \frac{6M_{tu}[1 + (b'/h) - 4(d''/h)]}{(b'/h)[3 - (b'/h)]f_y h} \tag{16.19}$$

We may recall now that in Chapters 4 and 5 the expressions for tension steel were derived as

$$A_s = M/ad \tag{16.20}$$

for working stress design and

$$A_s = M_u/\phi a_u d \tag{16.21}$$

for ultimate strength design, where M and M_u are in ft-kips and d is in inches, so $a = f_s j/12{,}000$ and $\phi a_u = \phi f_y j_u/12{,}000$.

In equation (*16.19*) the ultimate torsional moment M_{tu} is in inch-pounds. If M_{tu} is stated as ft-kips × 12,000, a form similar to those of equations (*16.20*) and (*16.21*) can be developed. In this way equation (*16.19*) can be rearranged and finally stated as

$$A_L = M_{tu}/a_t h \tag{16.22}$$

where

$$a_t = \frac{(b'/h)[3 - (b'/h)]f_y}{72{,}000[1 + (b'/h) - 4(d''/h)]} \tag{16.23}$$

Thus the longitudinal steel requirement for torsion may be obtained using the same general process as is used for calculating the tension reinforcement for beams. Note that the factor ϕ has not been included in equation (*16.23*).

TABLES FOR USE IN TORSIONAL DESIGN

In order to simplify torsional design, tables have been prepared for C_t and a_t. Values of C_t for rectangular beams appear in Table 16.1, and for equivalent L beams in Tables 16.2 through 16.5. Tables 16.6 through 16.9 contain values for a_t. Since a_t is a function of f_y, separate tables are required for each type of steel. Note that $f_y = 63{,}750$ psi is used in Table 16.9. This corresponds to steel having a yield point stress of 75,000 psi and has been reduced to conform with the ACI Code requirement that the greater of 60,000 or $0.85f_y$ should be used in tension bars when f_y exceeds 60,000 psi unless proof tests of the reinforcement are made. Such proof tests must show that the strain does not exceed 0.003 in/in at the specified yield point stress. Further, cracking is definitely a problem when torsion exists, so the Code provisions relative to *full* scale tests apply here. Such full scale tests must show that the average crack widths at the extreme tension edge do not exceed 0.015″ for interior members and 0.010″ for exterior members. Since full scale tests will seldom if ever be made, the designer should not use high strength steel (f_y greater than 60,000 psi) without specifying the required full scale tests. Since a_t is linearly related to A_L, any strength steel may be used and the appropriate a_t obtained by selecting a value of a_t from one of the tables and multiplying that value by the ratio of actual f_y to the tabular value of f_y.

It is important to, note that anchorage bond may be critical when torsion exists, so stirrups should be composed of small size deformed bars. One may check anchorage bond using methods provided in Chapter 7, with the bar force equal to $v_{tu}b's/2$. The anchorage length involved then becomes $b_c + d_c$.

Lacking Code provisions, ϕ has not been used at all in the equations for torsion. The reader should, however, consider using $\phi = 0.85$ for consistency in design procedures. Longitudinal steel would be selected using $A_L/0.85$.

TABLE 16.1
RECTANGULAR SECTIONS
COEFFICIENTS C_t FOR TORSION DESIGN, $t/b' = 0.0$

(All values of b/h, since $b = b'$, $t = 0$)										
b'/h	0.1	0.2	0.3	0.4	0.5	0.6	0.7	0.8	0.9	1.0
C_t	.483	.466	.450	.433	.416	.400	.383	.366	.350	.333

TABLE 16.2
COEFFICIENTS* C_t FOR L-BEAMS, $t/b' = 0.20$

b/h	b'/h									
	0.1	0.2	0.3	0.4	0.5	0.6	0.7	0.8	0.9	1.0
.30	.487	.468	.450							
.50	.491	.472	.454	.435	.416					
.70	.495	.476	.458	.439	.420	.402	.383			
.90	.499	.480	.462	.443	.424	.406	.387	.368	.350	
1.10	.503	.484	.466	.447	.428	.410	.391	.372	.354	.335
1.30	.507	.488	.470	.451	.432	.414	.395	.376	.358	.339
1.50	.511	.492	.474	.455	.436	.418	.399	.380	.362	.343
1.70	.515	.496	.478	.459	.440	.422	.403	.384	.366	.347
1.90	.519	.500	.482	.463	.444	.426	.407	.388	.370	.351
2.10	.523	.504	.486	.467	.448	.430	.411	.392	.374	.355
2.30	.527	.508	.490	.471	.452	.434	.415	.396	.378	.359
2.50	.531	.512	.494	.475	.456	.438	.419	.400	.382	.363
2.70	.535	.516	.498	.479	.460	.442	.423	.404	.386	.367
2.90	.539	.520	.502	.483	.464	.446	.427	.408	.390	.371
3.10	.543	.524	.506	.487	.468	.450	.431	.412	.394	.375
3.30	.547	.528	.510	.491	.472	.454	.435	.416	.398	.379
3.50	.551	.532	.514	.495	.476	.458	.439	.420	.402	.383
3.70	.555	.536	.518	.499	.480	.462	.443	.424	.406	.387
3.90	.559	.540	.522	.503	.484	.466	.447	.428	.410	.391
4.10	.563	.544	.526	.507	.488	.470	.451	.432	.414	.395
4.30	.567	.548	.530	.511	.492	.474	.455	.436	.418	.399
4.50	.571	.552	.534	.515	.496	.478	.459	.440	.422	.403
4.70	.575	.556	.538	.519	.500	.482	.463	.444	.426	.407
4.90	.579	.560	.542	.523	.504	.486	.467	.448	.430	.411
5.10	.583	.564	.546	.527	.508	.490	.471	.452	.434	.415
5.30	.587	.568	.550	.531	.512	.494	.475	.456	.438	.419
5.50	.591	.572	.554	.535	.516	.498	.479	.460	.442	.423
5.70	.595	.576	.558	.539	.520	.502	.483	.464	.446	.427
5.90	.599	.580	.562	.543	.524	.506	.487	.468	.450	.431

*For values above stepped line, use Table 16.1.

TABLE 16.3
COEFFICIENTS* C_t FOR L-BEAMS, $t/b' = 0.40$

b/h	b'/h									
	0.1	0.2	0.3	0.4	0.5	0.6	0.7	0.8	0.9	1.0
.30	.499	.474	.450							
.50	.515	.490	.466	.441	.416					
.70	.531	.506	.482	.457	.432	.408	.383			
.90	.547	.522	.498	.473	.448	.424	.399	.374	.350	
1.10	.563	.538	.514	.489	.464	.440	.415	.390	.366	.341
1.30	.579	.554	.530	.505	.480	.456	.431	.406	.382	.357
1.50	.595	.570	.546	.521	.496	.472	.447	.422	.398	.373
1.70	.611	.586	.562	.537	.512	.488	.463	.438	.414	.389
1.90	.627	.602	.578	.553	.528	.504	.479	.454	.430	.405
2.10	.643	.618	.594	.569	.544	.520	.495	.470	.446	.421
2.30	.659	.634	.610	.585	.560	.536	.511	.486	.462	.437
2.50	.675	.650	.626	.601	.576	.552	.527	.502	.478	.453
2.70	.691	.666	.642	.617	.592	.568	.543	.518	.494	.469
2.90	.707	.682	.658	.633	.608	.584	.559	.534	.510	.485
3.10	.723	.698	.674	.649	.624	.600	.575	.550	.526	.501
3.30	.739	.714	.690	.665	.640	.616	.591	.566	.542	.517
3.50	.755	.730	.706	.681	.656	.632	.607	.582	.558	.533
3.70	.771	.746	.722	.697	.672	.648	.623	.598	.574	.549
3.90	.787	.762	.738	.713	.688	.664	.639	.614	.590	.565
4.10	.803	.778	.754	.729	.704	.680	.655	.630	.606	.581
4.30	.819	.794	.770	.745	.720	.696	.671	.646	.622	.597
4.50	.835	.810	.786	.761	.736	.712	.687	.662	.638	.613
4.70	.851	.826	.802	.777	.752	.728	.703	.678	.654	.629
4.90	.867	.842	.818	.793	.768	.744	.719	.694	.670	.645
5.10	.883	.858	.834	.809	.784	.760	.735	.710	.686	.661
5.30	.899	.874	.850	.825	.800	.776	.751	.726	.702	.677
5.50	.915	.890	.866	.841	.816	.792	.767	.742	.718	.693
5.70	.931	.906	.882	.857	.832	.808	.783	.758	.734	.709
5.90	.947	.922	.898	.873	.848	.824	.799	.774	.750	.725

*For values above stepped line, use Table 16.1.

TABLE 16.4
COEFFICIENTS* C_t FOR L-BEAMS, $t/b' = 0.60$

b/h	b'/h									
	0.1	0.2	0.3	0.4	0.5	0.6	0.7	0.8	0.9	1.0
.30	.519	.484	.450							
.50	.555	.520	.486	.451	.416					
.70	.591	.556	.522	.487	.452	.418	.383			
.90	.627	.592	.558	.523	.488	.454	.419	.384	.350	
1.10	.663	.628	.594	.559	.524	.490	.455	.420	.386	.351
1.30	.699	.664	.630	.595	.560	.526	.491	.456	.422	.387
1.50	.735	.700	.666	.631	.596	.562	.527	.492	.458	.423
1.70	.771	.736	.702	.667	.632	.598	.563	.528	.494	.459
1.90	.807	.772	.738	.703	.668	.634	.599	.564	.530	.495
2.10	.843	.808	.774	.739	.704	.670	.635	.600	.566	.531
2.30	.879	.844	.810	.775	.740	.706	.671	.636	.602	.567
2.50	.915	.880	.846	.811	.776	.742	.707	.672	.638	.603
2.70	.951	.916	.882	.847	.812	.778	.743	.708	.674	.639
2.90	.987	.952	.918	.883	.848	.814	.779	.744	.710	.675
3.10	1.023	.988	.954	.919	.884	.850	.815	.780	.746	.711
3.30	1.059	1.024	.990	.955	.920	.886	.851	.816	.782	.747
3.50	1.095	1.060	1.026	.991	.956	.922	.887	.852	.818	.783
3.70	1.131	1.096	1.062	1.027	.992	.958	.923	.888	.854	.819
3.90	1.167	1.132	1.098	1.063	1.028	.994	.959	.924	.890	.855
4.10	1.203	1.168	1.134	1.099	1.064	1.030	.995	.960	.926	.891
4.30	1.239	1.204	1.170	1.135	1.100	1.066	1.031	.996	.962	.927
4.50	1.275	1.240	1.206	1.171	1.136	1.102	1.067	1.032	.998	.963
4.70	1.311	1.276	1.242	1.207	1.172	1.138	1.103	1.068	1.034	.999
4.90	1.347	1.312	1.278	1.243	1.208	1.174	1.139	1.104	1.070	1.035
5.10	1.383	1.348	1.314	1.279	1.244	1.210	1.175	1.140	1.106	1.071
5.30	1.419	1.384	1.350	1.315	1.280	1.246	1.211	1.176	1.142	1.107
5.50	1.455	1.420	1.386	1.351	1.316	1.282	1.247	1.212	1.178	1.143
5.70	1.491	1.456	1.422	1.387	1.352	1.318	1.283	1.248	1.214	1.179
5.90	1.527	1.492	1.458	1.423	1.388	1.354	1.319	1.284	1.250	1.215

*For values above stepped line, use Table 16.1.

TABLE 16.5

COEFFICIENTS* C_t FOR L-BEAMS, $t/b' = 0.80$

b/h	b'/h									
	0.1	0.2	0.3	0.4	0.5	0.6	0.7	0.8	0.9	1.0
.30	.547	.498	.450							
.50	.611	.562	.514	.465	.416					
.70	.675	.626	.578	.529	.480	.432	.383			
.90	.739	.690	.642	.593	.544	.496	.447	.398	.350	
1.10	.803	.754	.706	.657	.608	.560	.511	.462	.414	.365
1.30	.867	.818	.770	.721	.672	.624	.575	.526	.478	.429
1.50	.931	.882	.834	.785	.736	.688	.639	.590	.542	.493
1.70	.995	.946	.898	.849	.800	.752	.703	.654	.606	.557
1.90	1.059	1.010	.962	.913	.864	.816	.767	.718	.670	.621
2.10	1.123	1.074	1.026	.977	.928	.880	.831	.782	.734	.685
2.30	1.187	1.138	1.090	1.041	.992	.944	.895	.846	.798	.749
2.50	1.251	1.202	1.154	1.105	1.056	1.008	.959	.910	.862	.813
2.70	1.315	1.266	1.218	1.169	1.120	1.072	1.023	.974	.926	.877
2.90	1.379	1.330	1.282	1.233	1.184	1.136	1.087	1.038	.990	.941
3.10	1.443	1.394	1.346	1.297	1.248	1.200	1.151	1.102	1.054	1.005
3.30	1.507	1.458	1.410	1.361	1.312	1.264	1.215	1.166	1.118	1.069
3.50	1.571	1.522	1.474	1.425	1.376	1.328	1.279	1.230	1.182	1.133
3.70	1.635	1.586	1.538	1.489	1.440	1.392	1.343	1.294	1.246	1.197
3.90	1.699	1.650	1.602	1.553	1.504	1.456	1.407	1.358	1.310	1.261
4.10	1.763	1.714	1.666	1.617	1.568	1.520	1.471	1.422	1.374	1.325
4.30	1.827	1.778	1.730	1.681	1.632	1.584	1.535	1.486	1.438	1.389
4.50	1.891	1.842	1.794	1.745	1.696	1.648	1.599	1.550	1.502	1.453
4.70	1.955	1.906	1.858	1.809	1.760	1.712	1.663	1.614	1.566	1.517
4.90	2.019	1.970	1.922	1.873	1.824	1.776	1.727	1.678	1.630	1.581
5.10	2.083	2.034	1.986	1.937	1.888	1.840	1.791	1.742	1.694	1.645
5.30	2.147	2.098	2.050	2.001	1.952	1.904	1.855	1.806	1.758	1.709
5.50	2.211	2.162	2.114	2.065	2.016	1.968	1.919	1.870	1.822	1.773
5.70	2.275	2.226	2.178	2.129	2.080	2.032	1.983	1.934	1.886	1.837
5.90	2.339	2.290	2.242	2.193	2.144	2.096	2.047	1.998	1.950	1.901

*For values above stepped line, use Table 16.1.

TABLE 16.6

VALUES OF a_t FOR LONGITUDINAL STEEL FOR TORSION. $f_y = 40.0$ ksi

b/h	d''/h								
	0.04	0.06	0.08	0.10	0.12	0.14	0.16	0.18	0.20
.10	.171	.187	.206	.230	.259	.298	.350	.423	.537
.20	.299	.324	.353	.388	.432	.486	.555	.648	.777
.30	.394	.424	.459	.500	.548	.608	.681	.775	.900
.40	.465	.498	.534	.577	.628	.687	.760	.849	.962
.50	.518	.551	.588	.631	.680	.738	.807	.890	.992
.60	.555	.588	.625	.666	.714	.769	.833	.909	1.000
.70	.580	.612	.648	.688	.733	.784	.843	.912	.993
.80	.596	.626	.660	.698	.740	.788	.842	.905	.977
.90	.603	.632	.664	.700	.739	.783	.833	.889	.954
1.00	.603	.631	.661	.694	.730	.771	.816	.868	.925
1.10	.598	.624	.652	.683	.716	.753	.795	.841	.893
1.20	.588	.612	.638	.666	.697	.731	.769	.810	.857
1.30	.573	.596	.620	.646	.674	.705	.739	.777	.818
1.40	.555	.576	.598	.622	.648	.676	.707	.740	.777
1.50	.534	.553	.573	.595	.618	.644	.672	.702	.735
1.60	.510	.527	.545	.565	.587	.610	.634	.661	.691
1.70	.483	.499	.515	.533	.553	.573	.596	.620	.646
1.80	.454	.468	.483	.500	.517	.535	.555	.576	.600
1.90	.423	.436	.450	.464	.479	.496	.513	.532	.552
2.00	.391	.402	.414	.427	.440	.455	.470	.487	.505

TABLE 16.7

VALUES OF a_t FOR LONGITUDINAL STEEL FOR TORSION. $f_y = 50.0$ ksi

b/h	d''/h								
	0.04	0.06	0.08	0.10	0.12	0.14	0.16	0.18	0.20
.10	.214	.234	.258	.287	.324	.372	.437	.529	.671
.20	.373	.405	.441	.486	.540	.607	.694	.810	.972
.30	.493	.530	.573	.625	.685	.760	.852	.969	1.125
.40	.582	.622	.668	.722	.785	.859	.950	1.062	1.203
.50	.647	.688	.735	.789	.851	.923	1.009	1.112	1.240
.60	.694	.735	.781	.833	.892	.961	1.041	1.136	1.250
.70	.726	.765	.810	.860	.916	.980	1.054	1.140	1.242
.80	.745	.783	.825	.873	.925	.985	1.053	1.131	1.222
.90	.754	.790	.830	.875	.924	.979	1.041	1.112	1.193
1.00	.754	.789	.826	.868	.913	.964	1.021	1.085	1.157
1.10	.748	.780	.815	.853	.895	.942	.994	1.051	1.116
1.20	.735	.765	.797	.833	.872	.914	.961	1.013	1.071
1.30	.717	.745	.775	.807	.843	.882	.924	.971	1.023
1.40	.694	.720	.747	.777	.810	.845	.883	.925	.972
1.50	.667	.691	.716	.744	.773	.805	.840	.877	.919
1.60	.637	.659	.682	.707	.733	.762	.793	.827	.864
1.70	.604	.623	.644	.667	.691	.717	.745	.775	.807
1.80	.568	.585	.604	.625	.646	.669	.694	.721	.750
1.90	.529	.545	.562	.580	.599	.620	.642	.665	.691
2.00	.489	.503	.518	.534	.551	.569	.588	.609	.631

TABLE 16.8

VALUES OF a_t FOR LONGITUDINAL STEEL FOR TORSION. $f_y = 60.0$ ksi

b/h	d''/h								
	0.04	0.06	0.08	0.10	0.12	0.14	0.16	0.18	0.20
.10	.257	.281	.309	.345	.389	.447	.525	.635	.805
.20	.448	.486	.530	.583	.648	.729	.833	.972	1.166
.30	.592	.636	.688	.750	.823	.912	1.022	1.163	1.350
.40	.698	.747	.802	.866	.942	1.031	1.140	1.274	1.444
.50	.777	.826	.882	.946	1.021	1.108	1.211	1.335	1.488
.60	.833	.882	.937	1.000	1.071	1.153	1.250	1.363	1.500
.70	.871	.918	.972	1.032	1.099	1.176	1.265	1.369	1.490
.80	.894	.940	.990	1.047	1.111	1.182	1.264	1.358	1.466
.90	.905	.948	.996	1.050	1.109	1.175	1.250	1.334	1.431
1.00	.905	.946	.992	1.041	1.096	1.157	1.225	1.302	1.388
1.10	.897	.936	.978	1.024	1.075	1.130	1.192	1.262	1.339
1.20	.882	.918	.957	1.000	1.046	1.097	1.153	1.216	1.285
1.30	.860	.894	.930	.969	1.011	1.058	1.109	1.165	1.227
1.40	.833	.864	.897	.933	.972	1.014	1.060	1.111	1.166
1.50	.801	.829	.860	.892	.928	.966	1.008	1.053	1.102
1.60	.765	.790	.818	.848	.880	.915	.952	.992	1.037
1.70	.725	.748	.773	.800	.829	.860	.894	.930	.969
1.80	.681	.703	.725	.750	.775	.803	.833	.865	.900
1.90	.635	.654	.675	.696	.719	.744	.770	.798	.829
2.00	.586	.603	.621	.641	.661	.683	.706	.730	.757

TABLE 16.9

VALUES OF a_t FOR LONGITUDINAL STEEL FOR TORSION. $f_y = 63.75$ ksi*

b/h	d''/h								
	0.04	0.06	0.08	0.10	0.12	0.14	0.16	0.18	0.20
.10	.273	.298	.329	.366	.414	.475	.558	.675	.855
.20	.476	.516	.563	.619	.688	.774	.885	1.032	1.239
.30	.629	.676	.731	.796	.874	.969	1.086	1.236	1.434
.40	.742	.793	.852	.920	1.000	1.096	1.211	1.354	1.534
.50	.825	.878	.937	1.006	1.085	1.177	1.286	1.418	1.581
.60	.885	.937	.996	1.062	1.138	1.225	1.328	1.448	1.593
.70	.925	.976	1.032	1.096	1.168	1.250	1.344	1.454	1.583
.80	.950	.998	1.052	1.113	1.180	1.256	1.343	1.442	1.558
.90	.961	1.008	1.059	1.115	1.178	1.248	1.328	1.418	1.521
1.00	.962	1.006	1.054	1.106	1.165	1.229	1.302	1.383	1.475
1.10	.953	.994	1.039	1.088	1.142	1.201	1.267	1.340	1.423
1.20	.937	.975	1.017	1.062	1.111	1.166	1.225	1.292	1.366
1.30	.914	.949	.988	1.029	1.075	1.124	1.178	1.238	1.304
1.40	.885	.918	.953	.991	1.032	1.077	1.126	1.180	1.239
1.50	.851	.881	.913	.948	.986	1.026	1.071	1.119	1.171
1.60	.812	.840	.869	.901	.935	.972	1.011	1.054	1.101
1.70	.770	.795	.822	.850	.881	.914	.949	.988	1.029
1.80	.724	.747	.771	.796	.824	.853	.885	.919	.956
1.90	.675	.695	.717	.740	.764	.790	.818	.848	.881
2.00	.623	.641	.660	.681	.702	.725	.750	.776	.804

*0.85 × 75.0 ksi

Solved Problems

16.1. Use the sand-heap analogy of the plastic theory to derive an equation for the torsional shear stress v_{tu} in a square section subjected to a torsional moment M_{tu}.

Refer to Fig. 16-4.

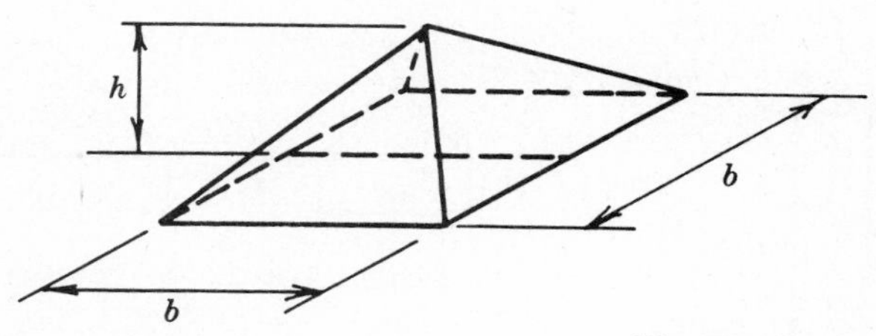

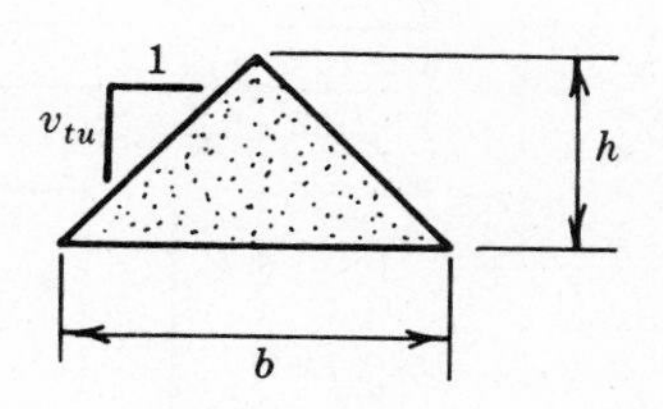

Fig. 16-4

For the pyramidal sand heap the double volume is $2b^2h/3$, where $h = v_{tu}b/2$. Thus

$$v_{tu} = 3M_{tu}/b^3$$

16.2. Solve Problem 16.1 using statics and a constant shear stress v_{tu}.

Refer to Fig. 16-5.

Since the individual stressed areas are triangular and v_{tu} is constant, $F = v_{tu}b^2/4$. The lever arm for each force F is $b/3$ from the centroid A. Thus summing moments about A, obtain

$$M_{tu} = 4Fb/6 = (4v_{tu}b^2/4)(b/3)$$

or

$$v_{tu} = 3M_{tu}/b^3$$

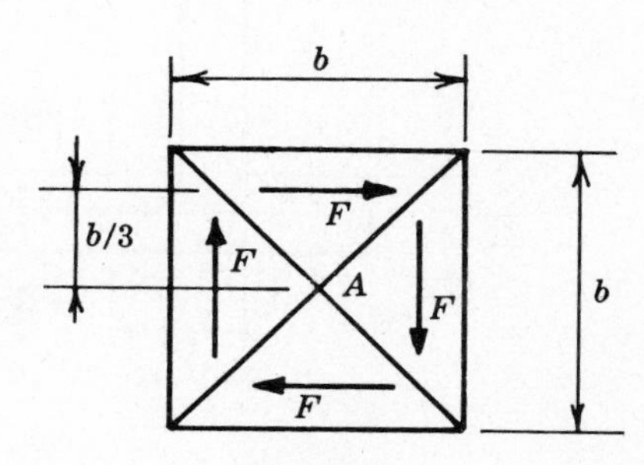

Fig. 16-5

16.3. Determine the relationship between M_{tu} and v_{tu} for the L beam shown in Fig. 16-6. Use the basic equation (*16.6*) and also the Tables.

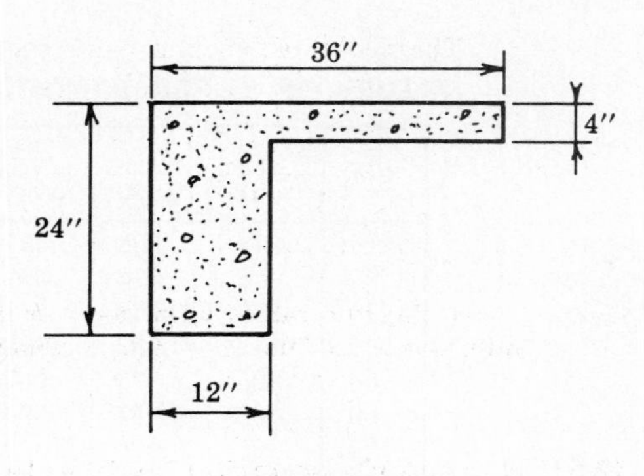

Fig. 16-6

Using equation (*16.6*), $b = 36''$, $h = 24''$, $b' = 12''$ and $t = 4''$. Thus

$$v_{tu} = \frac{M_{tu}}{(144/6)(72-12) + (16/2)(24)} = M_{tu}/1632$$

Using the Tables and interpolation, $t/b' = 0.33$, $b/h = 1.5$ and $b'/h = 0.5$, obtain $C_t = 0.473$. Thus

$$v_{tu} = \frac{M_{tu}}{C_t(b')^2h^2} = M_{tu}/1634$$

16.4. Find the relationship between M_{tu} and v_{tu} for the T beam shown in Fig. 16-7 using the basic equation (*16.6*) and the Tables. Consider $M_{tu} = 200{,}000$ in-lb.

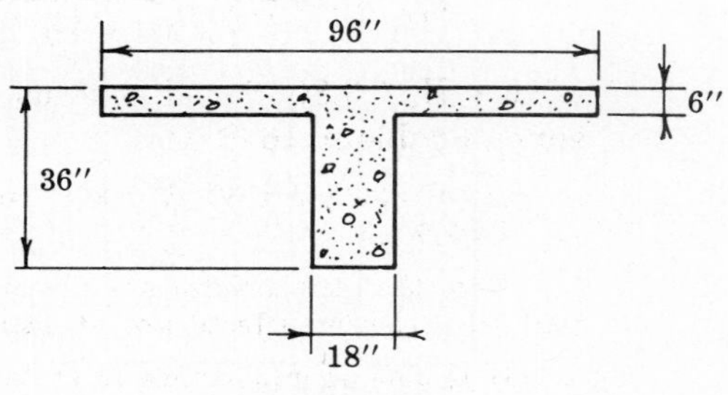

Fig. 16-7

The equivalent flange width is 96″, $t = 6''$, $b' = 18''$ and $h = 36''$. Hence using the equation,

$$v_{tu} = \frac{200{,}000}{[(18)^2/6][(3)(36)-18] + [(6)^2/2][(96-18)]} = 31.9 \text{ psi}$$

Using the Tables with $t/b' = 0.333$, $b/h = 2.667$ and $b'/h = 0.5$, find by interpolation that $C_t = 0.538$. Then

$$J_E = 0.538(18)^2(36)^2 = 6275 \text{ in}^4 \quad \text{so} \quad v_{tu} = M_{tu}h/J_E = 31.9 \text{ psi}$$

16.5. Determine the relationship between M_{tu} and v_{tu} for the I beam shown in Fig. 16-8 using the double volume under the sand heap and also using the Tables.

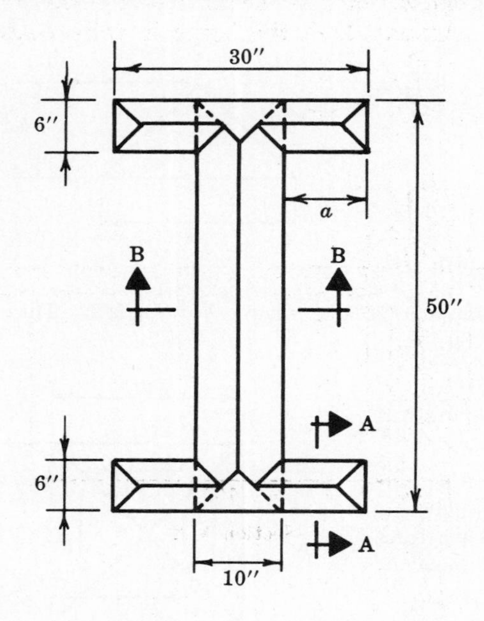

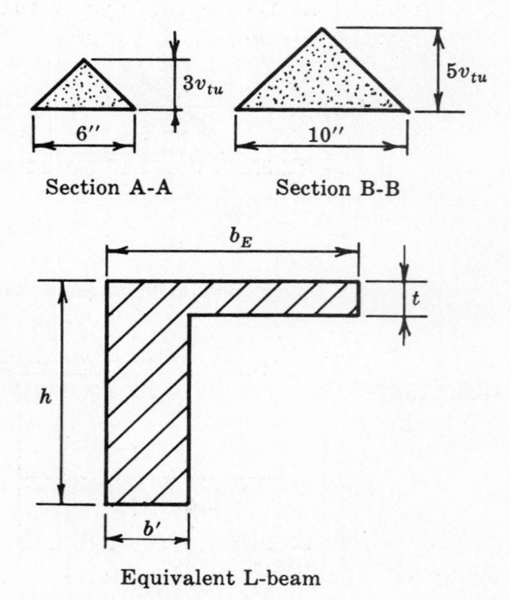

Fig. 16-8

The double volumes are calculated as follows:

(1) Pyramid from web ends:	$2V = (10)(10)(2)(1/3)(5v_{tu})$	$=$	$333\ v_{tu}$
(2) Long tent from web:	$2V = (2)(40)(10)(1/2)(5v_{tu})$	$=$	$2000\ v_{tu}$
(3) Four short tents from flanges:	$2V = (2)(4)(10)(1/2)(6)(3v_{tu})$	$=$	$720\ v_{tu}$
	$\Sigma 2V$	$=$	$3053\ v_{tu}$

Thus, $M_{tu} = 3053\ v_{tu}$.

Using the tables with $b_E = b' + 4a = 10 + 40 = 50''$, $t/b' = 0.6$, $b'/h = 0.2$ and $b_E/h = 1.0$, interpolate to find $C_t = 0.61$. Thus, $J_E = (b')^2h^2 = 152{,}500\ \text{in}^4$ and $M_{tu} = v_{tu}J_E/h = 3050v_{tu}$.

16.6. Nylander's equation, which takes into account junction effects at the intersection of rectangular elements is stated as

$$v_{tu} = \frac{M_{tu}}{[(t^2/2)(b - t/3)] + (1/2)(b')^2\{h - t - (b'/3) + K_t[(b')^2t/[(b')^2 + t^2]]\}}$$

where $K_t = 0.86$ for T beams and $K_t = 0.79$ for L beams. Use this equation to solve Problem 16.4.

If $b = 96''$, $t = 6''$, $h = 36''$ and $b' = 18''$, then by substitution with $K_t = 0.86$ find that $v_{tu} = 31.9$ psi.

Note that the results are identical to those obtained in Problem 16.4. We conclude therefore that the junction effects are of little significance.

It is also of importance to state that this problem was solved by Henry J. Cowan in his paper "Design of Beams Subject to Torsion Related to The New Australian Code", Journal of the American Concrete Institute, January 1960.

16.7. The cantilever slab shown in Fig. 16-9 supports a roof dead load of 7.5 psf and a live load of 20 psf. The wall over the beam is 10′ high and weighs 50 psf of wall surface. Use ultimate strength design to proportion the section for shear due to direct load and torsion. Use an effective flange overhang of $6t$, $f_c' = 5.0$ ksi and $f_y = 60.0$ ksi. Assume that the torsion center of the beam lies on the vertical centroidal axis of the web of the L beam. Assume that the beam is fully restrained against torsion at A and B.

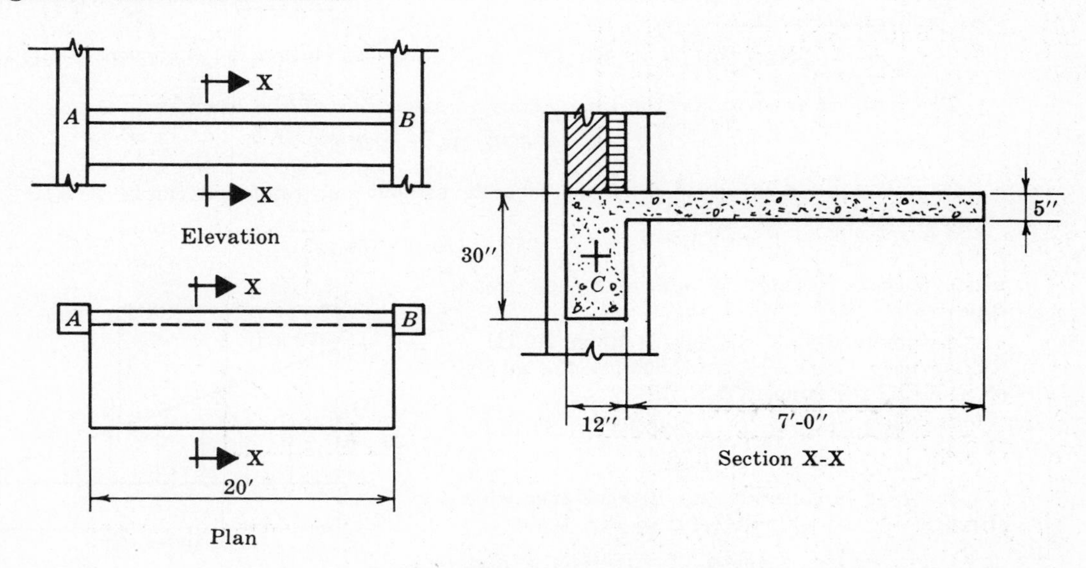

Fig. 16-9

The loads on the *slab* are as follows:

$$\text{Roof dead load} = 7.5 \text{ psf}$$
$$\text{Slab dead load} = 62.5 \text{ psf}$$
$$\text{Total } w_{DL} = 70.0 \text{ psf} \qquad w_{LL} = 20.0 \text{ psf}$$

The torsional moment about C due to *service loads* is

$$\text{for dead load,} \quad M_t = 7(70)(3.5 + 0.5) = 1960 \text{ ft-lb/ft}$$
$$\text{for live load,} \quad M_t = 7(20)(3.5 + 0.5) = 560 \text{ ft-lb/ft}$$

Hence for *ultimate conditions,* $M_{tu} = 1.5(1960) + 1.8(560) = 3950$ ft-lb/ft, so at the face of eithe column

$$M_{tu} = 10(3950) = 39{,}500 \text{ ft-lb}$$

The direct service loads delivered to the beam (including its weight) will be:

$$\text{for dead load,} \quad \text{slab and roofing } w_{DL} = 7(70) = 490 \text{ lb/ft}$$
$$\text{beam } w_{DL} = 2.5(1)(150) = 375 \text{ lb/ft}$$
$$\text{wall } w_{DL} = 10(50) = 500 \text{ lb/ft}$$
$$\text{Total } w_{DL} = 1365 \text{ lb/ft}$$
$$\text{and for live load,} \quad w_{LL} = 7(20) = 140 \text{ lb/ft}$$

Now, the critical section for shear according to the ACI Code occurs at a distance d from th face of the support for direct shear. However, when torsion is present the face of the suppor may be a more realistic location for the critical section. Considering the face of the support, th reactions are

$$R_{DL} = 10(1365) = 13{,}650 \text{ lb} \qquad \text{and} \qquad R_{LL} = 10(140) = 1400 \text{ lb}$$

Hence, the ultimate shear force is

$$V_u = 1.5(13{,}650) + 1.8(1400) = 23{,}000 \text{ lb}$$

The shear stress due to the vertical loads is

$$v_u = V_u/(bd) = 23{,}000/(12)(28) = 68 \text{ psi}$$

Considering an effective flange overhang for torsional resistance to be 6 times the flange thickness, then $b_E = (b' + 6t) = 12 + 30 = 42''$.

Using equation (*16.9*), $C_t = 0.52$ for $b = 42''$, $t = 5''$, $b' = 12''$ and $h = 30''$.

Using the Tables with the same values, find $C_t = 0.52$ which checks the formula solution. (See Problem 16.8.) Thus

$$J_E = (0.52)(12)^2(30)^2 = 67{,}390 \text{ in}^4 \qquad \text{and} \qquad v_{tu} = 39{,}500(12)(30)/67{,}390 = 211 \text{ psi}$$

The total shear stress on the most highly stressed side of the beam is

$$v_u = 68 + 211 = 279 \text{ psi}$$

The permissible shear stress on the concrete without web reinforcement is $2\phi\sqrt{f'_c} = 120$ psi, so the excess stress is

$$v'_u = 279 - 120 = 159 \text{ psi}$$

which must be resisted by *stirrups of the closed type.*

Assuming No. 3 stirrups, $A_v = 2(0.11) = 0.22$ in²/hoop and for $f_y = 60{,}000$ psi the spacing of hoops at the support is

$$s = A_v f_y/[(v_u)(b)] = 0.22(60{,}000)/[159(12)] = 6.92''$$

In order to calculate the distance over which stirrups will be required, refer to Fig. 16-10.

Using similar triangles, $X = 159(10)(12)/279 = 68.4''$ *theoretically.*

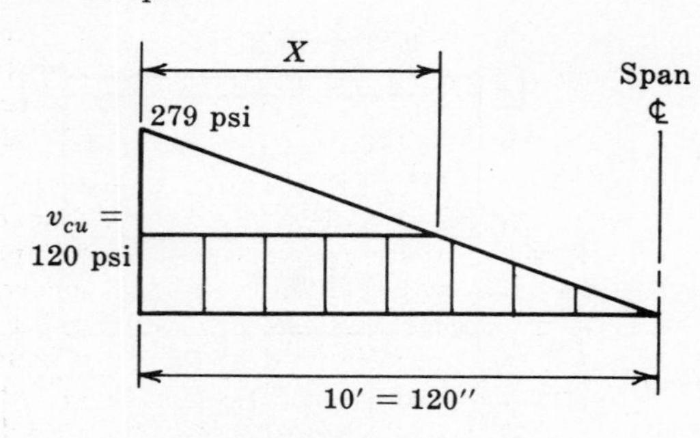

Fig. 16-10

However, the Code requires that stirrups be continued for a distance d beyond the theoretical stopping point, or $X' = 68.4 + 28 = 96.4''$. Use 8'-1" for practical consideration.

The stirrup spacing increases as the shear stress v_u' decreases, so that a variable spacing is used. The following table shows the distance X'' from the face of the support, v_u' and the theoretical spacing s corresponding thereto. The table is used to plot a curve of distance vs. theoretical stirrup spacing. The curve is used to select practical spacing values. The stepped curve of practical spacing must remain below the theoretical curve at all points. The practical spacing may not exceed $d/2$ or 14", according to the ACI Code.

X'', inches	v_u', psi	Theoretical s, inches
0.0	159.0	6.92
17.1	119.25	9.21
34.2	79.5	13.83
51.3	59.6	18.46
59.4	39.75	27.67
68.4	0.0	Infinite

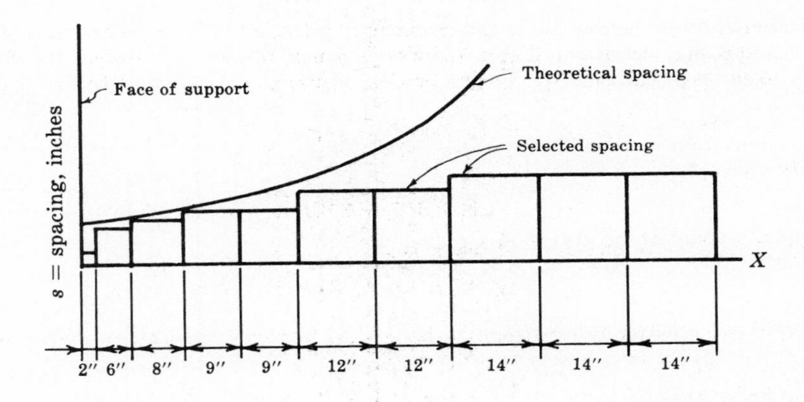

Fig. 16-11

The practical spacing, of course, is one of preference, and many patterns are possible. Many designers prefer to use a constant spacing when torsion is present, but in this case a variable spacing is satisfactory since the torsional moment increases from the center of the beam toward the support.

If the total shear stress at the support would have exceeded $6\phi\sqrt{f_c'} = 360$ psi, the maximum spacing at the support would be $d/4$ or 7.5" in that vicinity.

The longitudinal steel requirement will be determined using only the web of the beam, and disregarding the flange. This is conservative as long as *some* steel is placed in the slab in the longitudinal direction. Temperature and shrinkage steel will serve this purpose.

Referring to Fig. 16-12, note that $b_c = 9''$, $d_c = 27''$ and $d'' = 1.5''$. Using equation (*16.19*) find that $A_L = 1.815$ in^2 for $M_{tu} = 39{,}500$ in-lb at the support. Considering 8 bars, if the corner bars are No. 5 and the bars along the side and end faces are No. 4, the total $A_L = 2.04$ in^2. The corner bars must be continued for the full length of the beam but the side and end bars may be discontinued at a distance d beyond which they are no longer needed to resist torsional stress.

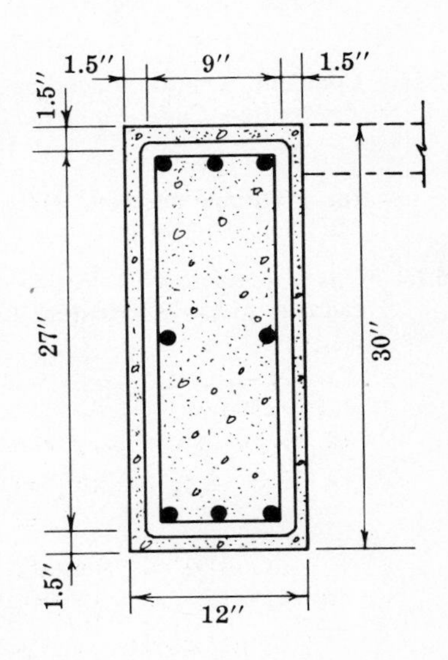

Fig. 16-12

Note that if the present ACI Code provisions had been used, only 4 No. 4 bars would have been required, one at each corner of the beam. Further, if $\phi = 0.85$ had been used for finding s, the theoretical spacing would be $0.85s$ since the appropriate equation would be $s = \phi A_v f_y/(v'b)$. Then, A_L would be increased to $A_L/0.85$.

In the slab the required steel for temperature and shrinkage is $A_s = 0.002bt = 0.12$ in²/ft. It is assumed that this steel will satisfy the needs for torsion in the slab.

16.8. Find C_t and A_L for Problem 10.7 using the Tables.

Using the data stated in Problem 10.7, calculate $t/b' = 0.4167$, $b'/h = 0.4$ and $b/h = 1.4$; find $C_t = 0.52$ by interpolation.

Then for $d'/h = 0.04$ find $a_t = 0.698$; and for $d'/t = 0.06$ find $a_t = 0.698$.

Thus by interpolation, for $d'/h = 0.05$ find $a_t = 0.7235$. Hence $A_L = M_{tu}/(a_t h) = 1.815$ in².

Supplementary Problems

16.9. For the beam shown in Fig. 16-13, find the relationship between v_{tu} and M_{tu} using the Tables and equation (*16.6*). *Ans.* $v_{tu} = M_{tu}/5510$

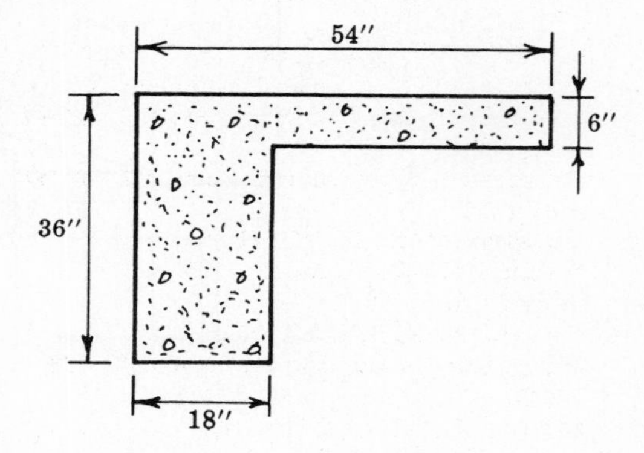

Fig. 16-13

16.10. Use the double volume method of the sand-heap analogy to solve Problem 16.9.

16.11. Use Nylander's equation as stated in Problem 16.6 to solve Problem 16.9. *Ans.* $v_{tu} = M_{tu}/5514$

16.12. Use the sand-heap analogy to prove that

$$v_{tu} = 3M_{tu}/(2\pi R^2)$$

for a circular beam of radius R.

16.13. Use statics to solve Problem 16.12. (See Problem 16.2.)

16.14. Consider a square section $(b = b' = h$ and $t = 0)$ to show that equation (*16.6*) reduces to the *Australian Code Equation* for working stress design

$$v_t = 3M_t/(b')^3$$

for a square section. Note that v_t and M_t are used as working stress values.

16.15. Use the results of Problem 16.14, $f_y = 2f_s$ and $b'h = 1.2b_c d_c$ to show that equation (*16.19*) reduces to the *Australian Code Equation* for working stress design

$$A_L = \frac{M_t(b_c + d_c)}{0.8 b_c d_c f_s}$$

INDEX